PHYSICAL CHEMISTRY
A MODERN INTRODUCTION
SECOND EDITION

Second Edition Updated and Revised by
William M. Davis

Original Work Authored by
Clifford E. Dykstra

CRC Press
Taylor & Francis Group
Boca Raton London New York

CRC Press is an imprint of the
Taylor & Francis Group, an **informa** business

CRC Press
Taylor & Francis Group
6000 Broken Sound Parkway NW, Suite 300
Boca Raton, FL 33487-2742

First issued in paperback 2018

ISBN-13: 978-1-4398-1077-4 (hbk)
ISBN-13: 978-1-138-11399-2 (pbk)

Library of Congress Cataloging-in-Publication Data

Davis, William M. (William Michael), 1969-
　　Physical chemistry : a modern introduction / William M. Davis and Clifford E. Dykstra.
　　　　p. cm.
　　Previous ed.: Physical chemistry / Clifford E. Dykstra. Upper Saddle River, NJ : Prentice Hall, c1997.
　　Includes bibliographical references and index.
　　ISBN 978-1-4398-1077-4
　　1. Chemistry, Physical and theoretical. I. Dykstra, Clifford E. II. Title.

QD453.3.D38 2012
541--dc23

2011033736

Visit the Taylor & Francis Web site at
http://www.taylorandfrancis.com

and the CRC Press Web site at
http://www.crcpress.com

Contents

Preface...vii
Acknowledgments xi
Authors xiii
Guide for Students xv
List of Special Examples xvii

1. World of Atoms and Molecules 1
1.1 Introduction to Physical Chemistry 1
1.2 Theory and Experiment in Physical Chemistry 2
1.3 Atomic and Molecular Energies 3
1.4 Configurations, Entropy, and Volume 7
1.5 Energy, Entropy, and Temperature...10
1.6 Distribution Law Derivation 13
1.7 Conclusions...18
Point of Interest: James Clerk Maxwell 19
Exercises 20
Bibliography...22

2. Ideal and Real Gases...23
2.1 The Ideal Gas Laws..23
2.2 Collisions and Pressure 27
2.3 Nonideal Behavior..33
2.4 Thermodynamic State Functions..35
2.5 Energy and Thermodynamic Relations 37
2.6 Conclusions..45
Point of Interest: Intermolecular Interactions...46
Exercises 48
Bibliography...50

3. Changes of State 51
3.1 Pressure–Volume Work..51
3.2 Reversibility, Heat, and Work 55
3.3 Entropy...62
3.4 The Laws of Thermodynamics 65
3.5 Heat Capacities..68
3.6 Joule–Thomson Expansion 73
3.7 Conclusions..75
Point of Interest: Heat Capacities of Solids 76
Exercises 78
Bibliography...80

4. Phases and Multicomponent Systems...81
4.1 Phases and Phase Diagrams..81
4.2 The Chemical Potential 86
4.3 Clapeyron Equation..89

4.4 First- and Second-Order Phase Transitions...93
4.5 Conclusions...95
Point of Interest: Josiah Willard Gibbs...96
Exercises 97
Bibliography..99

5. Activity and Equilibrium of Gases and Solutions 101
5.1 Activities and Fugacities of Gases 101
5.2 Activities of Solutions..106
5.3 Vapor Pressure Behavior of Solutions..108
5.4 Equilibrium Constants 111
5.5 Phase Equilibria Involving Solutions...114
5.6 Conclusions..118
Point of Interest: Gilbert Newton Lewis..119
Exercises 121
Bibliography..123

6. Chemical Reactions: Kinetics, Dynamics, and Equilibrium...........................125
6.1 Reaction of Atoms and Molecules...125
6.2 Collisions and Transport...129
6.3 Rate Equations..135
6.4 Rate Laws for Complex Reactions...138
6.5 Temperature Dependence and Solvent Effects...142
6.6 Reaction Thermodynamics...144
6.7 Electrochemical Reactions 151
6.8 Conclusions..157
Point of Interest: Galactic Reaction Chemistry..158
Exercises 160
Bibliography..163

7. Vibrational Mechanics of Particle Systems 165
7.1 Classical Particle Mechanics and Vibration 165
7.2 Vibration in Several Degrees of Freedom..170
7.3 Quantum Phenomena and Wave Character 176
7.4 Quantum Mechanical Harmonic Oscillator...180
7.5 Harmonic Vibration of Many Particles 185
7.6 Conclusions..187
Point of Interest: Molecular Force Fields...188
Exercises 189
Bibliography..191

8. Molecular Quantum Mechanics..193
8.1 Quantum Mechanical Operators 193
8.2 Information from Wavefunctions 197
8.3 Multidimensional Problems and Separability...203
8.4 Particles with Box and Step Potentials...206
8.5 Rigid Rotator and Angular Momentum 216
8.6 Coupling of Angular Momenta 224
8.7 Variation Theory 228
8.8 Perturbation Theory 232

8.9 Conclusions..238
Point of Interest: The Quantum Revolution..239
The Solvay Conference...239
Exercises 241
Bibliography...245

9. **Vibrational–Rotational Spectroscopy** 247
9.1 Molecular Spectroscopy and Transitions.....................................247
9.2 Vibration and Rotation of a Diatomic Molecule.........................254
9.3 Vibrational Anharmonicity and Spectra.......................................260
9.4 Rotational Spectroscopy..272
9.5 Harmonic Picture of Polyatomic Vibrations 276
9.6 Polyatomic Vibrational Spectroscopy 281
9.7 Conclusions..285
Point of Interest: Laser Spectroscopy 286
Exercises 287
Bibliography...289

10. **Electronic Structure** 291
10.1 Hydrogen and One-Electron Atoms 291
10.2 Orbital and Spin Angular Momentum 297
10.3 Atomic Orbitals and Atomic States 301
10.4 Molecules and the Born–Oppenheimer Approximation..........310
10.5 Antisymmetrization of Electronic Wavefunctions 313
10.6 Molecular Electronic Structure...317
10.7 Visible–Ultraviolet Spectra of Molecules....................................324
10.8 Properties and Electronic Structure...330
10.9 Conclusions...336
Point of Interest: John Clarke Slater 337
Exercises 338
Bibliography...340
Advanced Texts and Monographs...341

11. **Statistical Mechanics** 343
11.1 Probability...343
    11.1.1 Classical Behavior 345
11.2 Ensembles and Arrangements...346
11.3 Distributions and the Chemical Potential 347
    11.3.1 High-Temperature Behavior...352
    11.3.2 Low-Temperature Behavior..352
    11.3.3 Dilute Behavior 353
11.4 Molecular Partition Functions...353
11.5 Thermodynamic Functions...358
11.6 Heat Capacities...362
11.7 Conclusions...365
Point of Interest: Lars Onsager...366
Exercises 367
Bibliography...369

12. Magnetic Resonance Spectroscopy 371
    12.1 Nuclear Spin States 371
    12.2 Nuclear Spin–Spin Coupling 378
    12.3 Electron Spin Resonance Spectra 386
    12.4 Extensions of Magnetic Resonance 391
    12.5 Conclusions 393
    Point of Interest: The NMR Revolution 394
    Exercises 396
    Bibliography 397

13. Introduction to Surface Chemistry 399
    13.1 Interfacial Layer and Surface Tension 399
    13.2 Adsorption and Desorption 402
    13.3 Langmuir Theory of Adsorption 407
    13.4 Temperature and Pressure Effects on Surfaces 408
    13.5 Surface Characterization Techniques 409
    13.6 Conclusions 411
    Point of Interest: Irving Langmuir 412
    Exercises 413
    Bibliography 414

Appendix A: Mathematical Background 415

Appendix B: Molecular Symmetry 437

Appendix C: Special Quantum Mechanical Approaches 453

Appendix D: Table of Integrals 465

Appendix E: Table of Atomic Masses and Nuclear Spins 469

Appendix F: Fundamental Constants and Conversion of Units 473

Appendix G: List of Tables 479

Appendix H: Points of Interest 481

Appendix I: Atomic Masses and Percent Natural Abundance of Light Elements 483

Appendix J: Values of Constants 485

Appendix K: The Greek Alphabet 487

Answers to Selected Exercises 489

Index 493

Preface

This text has been designed and written especially for use in a one-year, two-course sequence in introductory physical chemistry. For semester-based courses, Chapters 1 through 6 can be covered in the first semester and Chapters 7 through 12 in the second. For quarter-based courses, Chapters 1 through 5 can be covered in the first quarter, Chapters 6 through 8 in the second quarter, and Chapters 9 through 12 in the third quarter. Chapter 13 has been written to enhance this edition and can be added to either semester or quarter as time permits. This text may also be used in one-semester surveys of physical chemistry if you select among the sections in each chapter.

The text is organized in a way that minimizes extraneous material unnecessary to understand the fundamental concepts while focusing on a strong molecular approach to the subject.

As you will see, this text has a novel approach. The ideas, organization, emphasis, and examples herein have evolved from the experience of teaching several different physical chemistry courses at several North American universities.

Distinguishing Features of This Text

A unifying molecular approach: The foremost goal of this text is to provide a unifying molecular view of the core elements of contemporary physical chemistry. This is done with a topically connected and focused development—in effect, a story line about molecules that leads one through the major areas of modern physical chemistry. At some places, this means a somewhat nontraditional organization of subtopics. The advantage is much improved retention of working knowledge of essential material. After finishing with this text, your students should have a good grasp of the concepts of physical chemistry and should be able to analyze problems and deal with new developments that occur during their careers. Seeing physical chemistry as a continuous story about molecular behavior helps accomplish that.

Focus on core concepts: Throughout this text, fundamental issues are stressed and basic examples are selected rather than the myriad of applications often presented in other, more encyclopedic books. Physical chemistry need not appear as a large assortment of different, disconnected, and sometimes intimidating topics. Instead, students should see that physical chemistry provides a coherent framework for chemical knowledge, from the molecular level to the macroscopic level.

That this text offers a streamlined introduction to the subject is apparent in the presentation of thermodynamics at the start of the text. As gas laws are first considered, a thorough yet concise development of real gases and equations of state is given. State functions are introduced in a global fashion. This organization offers students the strongest sophistication in the least amount of time. It prepares them for tackling more challenging topics.

Novel organization to foster student understanding: The first three chapters provide the foundation material for thermodynamics, always tying it to a molecular point of view. The approach in these chapters is to understand the behavior of thermodynamic systems and to express that in mathematical terms. This means, for instance, that gas kinetics is used as needed to understand pressure, reaction rates, and so on, rather than being collected as an isolated segment. A more usual organization considers the first law, then the second law, and then the third law of thermodynamics; however, that structure does not always bring out what thermodynamics is meant to explain, and it is not always an effective organization for remembering the material. It is the understanding of molecular behavior, such as that leading to chemical reactions, that is the focus of the development here. Most instructors will not find this organization much of a departure from a traditional approach, and yet the differences that do exist should benefit students.

To streamline the presentation of quantum mechanics, notions that are more of historical interest than pedagogical value are removed. For example, the Bohr atom, important as it was in the development of quantum theory, was not correct. The photoelectric effect was part of the quantum story, but a detailed discussion is not essential to introducing the material. Also, a primary example, the one-dimensional oscillator, is introduced at the outset in order to have it serve as a continuing example as we build sophistication. The usual first problem, the particle in a box, is set aside for later because it simply is not as applicable as a model of chemical systems as the harmonic oscillator. This is another way to connect the new concepts to molecular behavior. It is easier to understand that molecules vibrate than to contemplate a potential becoming infinite at some point.

Point of interest essays: Each chapter ends with a short Point of Interest. These essays discuss a selected set of the historical aspects of physical chemistry (to give students an appreciation of certain of the people whose clever and creative thinking have moved this discipline forward) as well as insights into modern applications and a few of the current areas of active research. These essays are set off from the technical story line; they are the roadside stops with interesting glimpses of individuals, revolutionary developments, and a few special areas for future study.

Strong problem-solving emphasis: Working on exercises is a key to mastering all topics in physical chemistry. The end of each chapter provides numerous practice Exercises (mostly the "plug-and-chug" variety) as well as numerous Additional Exercises (which are more challenging and test your students' understanding of concepts and ability to apply the material covered in the chapter). In addition, over two dozen special worked examples are interspersed with the topical development in the text. These special examples are boxed and usually on individual pages. They augment the in-line examples in the discussion but do not interrupt the flow of the material. They offer detail at the level of a chalkboard work-up of an exercise in a classroom. Finally, a number of exercises are included that are best handled using a spreadsheet of your choosing. These exercises are clearly identified.

Mathematics review for students who have forgotten their calculus: Appendix A provides some quick review of (or even initial training in) the mathematics needed to follow the material in the text. In writing this book, it is assumed that students have completed two or three semesters of calculus and that they can differentiate simple functions easily and can form differentials. The material in Appendix A is meant to supplement mathematical preparation since, in our experience, that is often the biggest difficulty for students beginning their study of physical chemistry. It is strongly recommended that students review Appendix A as they begin to use this text.

Powerful streamlined development of group theory and advanced topics in quantum mechanics: Appendix B (Molecular Symmetry) and Appendix C (Special Quantum Mechanical Approaches) cover topics that many physical chemistry courses include and that could each in fact be their own chapter. However, they are not essential to the flow of the remaining material, and so these appear at the end for inclusion in a course at the instructor's discretion.

Acknowledgments

WMD first and foremost thanks Clifford E. Dykstra for allowing him the privilege of revising an already excellent text into a second edition. Special thanks go to Barbara Glunn and Pat Roberson at Taylor & Francis for their help in bringing this textbook to fruition.

My teaching style and ultimate interest in teaching physical chemistry is first and foremost influenced by my quantum chemistry professor at the University of Western Ontario, Dr. William J. Meath. I was astounded at how he could make such a complicated mathematical subject so fascinating. I can only strive to match his level of teaching ability. I would be remiss if I did not mention my graduate advisor at the University of Guelph, Dr. John D. Goddard. His patient guidance during my graduate days was always much appreciated. I must also thank my postdoc advisor at York University, Dr. Huw O. Pritchard, for allowing me the opportunity to teach my first lecture class and help me realize my passion for undergraduate teaching.

Finally, I wish to thank my students, both at the University of Texas at Brownsville and my new home at Texas Lutheran University (TLU). Their comments, questions, and complaints have always helped me to refine my lectures and examples. A very special thanks goes out to the Fall 2009/Spring 2010 class of Chemistry 344/345 at TLU who used a draft of this book and helped to find errors and omissions in the text. You are all a continuing inspiration to teach.

Comments from readers are most heartily encouraged and can be sent to me at wdavis@tlu.edu.

William M. Davis

Authors

William M. Davis received his BSc (honors) in chemistry from the University of Western Ontario, London, Ontario, Canada, and his MSc and PhD from the University of Guelph, Guelph, Ontario, Canada. He taught lecture and laboratory sections of general, physical, and inorganic chemistry at several Canadian universities before moving to Texas to take up a tenure-track position at The University of Texas at Brownsville, Texas, where he taught general, physical, inorganic, analytical, organic, and environmental chemistry for 10 years. In 2008, he moved to Texas Lutheran University, where he is currently associate professor and chair of chemistry and holds the George Kieffer Fellowship in Science. Dr. Davis's research interests include application of computational and analytical chemistry techniques to systems of environmental and biochemical interest.

Clifford E. Dykstra received a BS in chemistry and a BS in physics from the University of Illinois at Urbana–Champaign in 1973, and he received his PhD from the University of California, Berkeley in 1976. He joined the faculty at the University of Illinois at Urbana–Champaign in 1977. His research has focused on computational electronic structure theory with particular attention to molecular properties and weak intermolecular interaction. In 1990, he moved to Indiana University–Purdue University Indianapolis where he served as Associate Dean of Science (1992–1996) and was named Chancellor's Professor (2001). He served as chair of the Department of Chemistry at Illinois State University from 2006 to 2009, and then returned to the University of Illinois at Urbana–Champaign. He serves as editor of the journal *Computational and Theoretical Chemistry*.

Guide for Students

Physical chemistry is a required course in most undergraduate chemistry curricula and in most chemical engineering curricula. Many students in other areas, such as biochemistry or materials science, find it useful preparation. Students take physical chemistry with different aims, different objectives, and different backgrounds. In all cases, there are some ways to optimize the learning process. Here are some ideas for those using this text for undergraduate physical chemistry.

Math background and proficiency are immensely important. Algebra, analytical geometry, and calculus are unavoidable in this subject. Before beginning your course, read the first four sections of Appendix A. This should provide a thorough review of the mathematics needed with the material in this text. If any of it seems to be more than a review, it may be worthwhile to study a mathematics text in that area.

Next, familiarize yourself with the information available in the text. Each appendix is a source of information on which you may wish to draw throughout your course. In particular, look at Appendix D on units, since unit conversion can sometimes obscure the scientific concepts.

As you get started going through the chapters of the text, try to anticipate upcoming subjects in your lecture sessions. Read chapter material on those subjects shortly before the lecture. In other words, read ahead. On a first reading, it is not necessary to strive for 100% comprehension, and you may choose to ignore a lot of the mathematical detail. Simply try to get the direction of the presentation and a general feel for the subtopics. Then, follow your instructor's lecture closely, especially to see the areas that your instructor has selected to emphasize.

After a lecture presentation on some topic or subtopic, do a thorough reading of the corresponding text sections. Attempt to follow all mathematical steps and go through the special boxed examples. At this stage, follow closely the examples where numerical values are used to obtain numerical results (i.e., make sure you can "plug-and-chug" the key formulas right away).

The next recommended step in the learning process is one of the most important ones in physical chemistry—working exercises. A set of exercises are given at the end of each chapter. The first set includes the more straightforward problems, typically those that require applying a particular formula to achieve a numerical answer. These problems help solidify understanding of newly introduced quantities and functions. Answers for the majority of these problems can be found at the end of the text, and you can check your work right away. Realize that it is less advisable to simply look at and mull over a problem so as to convince yourself "I can do it" than it is to carry out the work in full. Actually working an exercise rather than merely "thinking a problem through" consistently builds better understanding and strengthens retention—retention you may value during that exam 3 weeks from now.

Approach additional exercises in each chapter to develop a solid, working knowledge of the material. These problems may involve derivations, complicated calculations, analysis of new problems, and challenges that will call for a mastery of the subject material.

As you work exercises, refer back to text material. Reread paragraphs if something is unclear. Consult other books, such as those listed in the bibliography at the end of each chapter. The concepts in physical chemistry may seem formidable. The first presentation you see on some particular topic may not do the job, even if it should happen to be the best presentation around. Different individuals will generate different questions about any given topic. Checking alternate sources might offer a different perspective, a different approach, or a different derivation, and that may be what it takes for you to achieve understanding.

Want to enhance your sophistication, understanding, and physical chemistry abilities even further? Then work with other students. Try to devise your own exercises with your own solutions, and then try to explain them to others. Nothing challenges the solidness of your knowledge as much as trying to explain it to others.

Prepare for exams by quick rereads of chapters. Go over new terms (in bold when first used or defined) and the conclusions to each chapter. Rework problems you have already done.

Learning and studying in a subject that is highly mathematical is different than in a subject that is wholly conceptual and qualitative. The sequence of seeing material, hearing it, looking at it in detail, and working with it through exercises seems to be the best learning process available. Obviously, this is not a passive learning approach. Few can grasp the richness and complexity of physical chemistry by simply listening to lectures. The guts of the material is not a stream of facts and some definitions. It is a physical-mathematical description of the world around, a description with many connected concepts, theories, formulas, and abstractions. You have to work with it to understand it all.

What if you cannot do everything advised here? The best use of a limited amount of time is to focus on reading the chapters and studying the special examples. With any remaining time, work exercises, work exercises, and work exercises.

Good luck! Physical chemistry tends to grab the interests of a good fraction of students who are required to take it as a course. The reason seems to be that physical chemistry is fundamental to chemistry. Nature has not made overly simple the structure of physical chemistry, and thus, there will be difficulties and frustrations for many in studying it. Many overcome initial frustrations and find clear sailing thereafter. From the effort to study the physics of molecules—physical chemistry—should come a sense of wonder, perhaps fascination, that mankind has obtained such incredible insight into the workings of things (atoms and molecules) that no human has ever directly seen.

List of Special Examples

1.1 Energy Level Populations

2.1 Most Probable Speed of Gas Particles

2.2 Isothermal Compressibility

2.3 Thermodynamic Relations

2.4 Thermodynamic Compass

3.1 Work in a Stepwise Gas Expansion

3.2 ΔU for an Adiabatic Expansion

3.3 ΔS for an Adiabatic Expansion

3.4 ΔS of an Engine Cycle

4.1 The Solid–Liquid Phase Boundary

5.1 Fugacity and Activity of a Real Gas

6.1 Integrated Rate Expression for an A + B Reaction

6.2 Temperature Dependence of a Reaction Enthalpy

6.3 ΔS and ΔG of a Reaction

6.4 ΔG of an Electrochemical Cell

6.5 Effect of Temperature on Cell Voltage

8.1 Position Uncertainty for the Harmonic Oscillator

8.2 Degenerate Energy Levels

8.3 Particle in a Three-Dimensional Box

8.4 Variational Treatment of a Quartic Oscillator

9.1 Diatomic Molecule Vibrational Spectrum

10.1 Diatomic Molecule Electronic Absorption Bands

11.1 Products of Partition Functions of Independent Systems

11.2 Internal Energy of an Ideal Diatomic Gas

12.1 NMR Energy Levels of Methane

1

World of Atoms and Molecules

Physical chemistry is the study of the physical basis of phenomena related to the chemical composition and structure of substances. It has been pursued from two levels: the macroscopic and the molecular. The theories and laws of physical chemistry provide a rich, comprehensive view of the world of atoms and molecules that connects their nature with the macroscopic properties and phenomena of materials and substances. A starting point for an introduction to physical chemistry is the concept of energy levels in atoms and molecules, distributions among these energy levels, and something of familiar use in everyday life, temperature.

1.1 Introduction to Physical Chemistry

Physical chemistry, or chemical physics, is an area of molecular science with boundaries that are still being enlarged. In many ways, it is at the core of chemical science because it is concerned, in part, with achieving the most detailed, quantitative view of molecules and of chemical phenomena. This means it covers the structure of molecules, starting from a description of electrons and nuclei and the nature of chemical bonds. It covers dynamics (the changes in a molecular system with time), and this includes chemical reactions. It also covers properties of assemblies of atoms and molecules. Beyond that, the subject deals with the properties and phenomena of gases, liquids, and solids. Surely, this is a subject with applications in every area of molecular science, and to study physical chemistry is to pursue a very fundamental understanding of chemistry.

Because it is developed from basic physical laws, physical chemistry deals with most issues quantitatively and mathematically. Even the qualitative notions that emerge usually rely on mathematical arguments. Often the theories used in physical chemistry are presented most concisely as mathematical expressions, making mathematical sophistication advantageous. The mathematical basis of physical chemistry allows the derived theories and laws to be powerfully predictive tools in science.

The modern atomic theory of matter is almost two centuries old. It was in the early nineteenth century that Dalton's work (John Dalton, England, 1766–1844) advanced the proposal that matter is not continuously divisible and that there is some fundamental type of particle, the atom. The line of thought that began with the atomic theory of matter took its next major step in the early twentieth century when experiments pointed to the existence of subatomic particles. In a few more decades, it became clear that there are even smaller particles. Even today, the search for exotic subatomic particles continues. As matter is viewed using more and more powerful techniques, we can see that all matter is composed of discrete building blocks (particles) rather than continuous materials.

In 1905, Einstein's (Albert Einstein, 1879–1955) special theory of relativity connected the property of mass with energy with his now infamous equation $E = mc^2$. This mass–energy connection makes it less surprising that scientists in the early twentieth century found that many strange observations could be explained if energy came in discrete packages or **quanta**. In other words, it is not only matter but also energy that comes in discrete building blocks in the tiny world of atoms and molecules. The problems that led to the hypothesis of quantization of energy involved the spectrum of atomic hydrogen, the photoelectric effect, the temperature dependence of the heat capacities of solids, and others. One after another, unexpected phenomena were explained by a quantum hypothesis, and this hypothesis eventually grew into what we now refer to as **quantum mechanics**. After one becomes familiar with quantum mechanics and its chemical implications, it is fascinating to look back at the early developments—they mark the start of a major scientific revolution.

We know today that the constituents of atoms and molecules are electrons, neutrons, and protons. These constituents are particles, very small entities that have mass. They are so small and so light that they are beyond the limits of our own senses and experience; we cannot hold a single atom in our hands and look at it. Likewise, the mechanics of systems of such particles are outside everyday experience. Even so, there is a correspondence between our macroscopic world and the subatomic world, and we will analyze systems of particles in both worlds. The picture in the macroscopic world is generally referred to as a *classical* picture. It has been established by human perception and observation. The picture of small, light particles is termed a *quantum* picture because quantization of energy—the partitioning of energy into discrete blocks—is the distinction between this world and the macroscopic world. Because energy quanta tend to have such tiny amounts of energy, a macroscopic system involving numerous quanta appears to behave as if energy is continuous. Nonetheless, there are manifestations of quantum features that are detectable by macroscopic instruments and are quite recognizable.

Many properties and qualities of substances, such as the temperature dependence of the pressure of gases, were well understood before the development of quantum theory. With the detailed molecular view obtained with quantum mechanical analysis, an even more fundamental basis for macroscopic chemical phenomena is at hand.

1.2 Theory and Experiment in Physical Chemistry

The pursuit of understanding in most branches of science is a process of observation and analysis. In physical chemistry, laboratory experiments are the means for observation, which is to say that experiment is the means for probing and measuring. The analysis of the data may be carried out for different reasons. For one, it might use the data with some generally accepted theory so as to deduce some useful quantity that is not directly measurable. We will find, for instance, that bond lengths are not measured the way an object's length is measured in our macroscopic world; instead they are often determined on the basis of measurements of energy changes in molecules. In this way, the established physical understanding provides the means for utilizing experimental information in examining molecular systems.

A possible reason for carrying out an experiment and analyzing the data is to test one notion, concept, model, hypothesis, or theory, or else possibly to select from among several competing theories. If the data do not conform to what is anticipated by some particular theory, then its validity is challenged. In such a circumstance, one may devise a new notion, concept, or theory to better fit the data, or possibly reject one concept in favor of another. Whatever new idea emerges is then tested in still further experiments.

Many problems in physical chemistry are analyzed with approximations or idealizations that make the mathematics of the analysis less complicated or that offer a more discernible physical picture. Experimental data and analysis offer a validation or a rejection of the approximation.

There is a vital interplay between observation and understanding, or between experiment and theory. Of course, this is true throughout science, but in physical chemistry the interplay very much affects the way developments are viewed. The goal is always a physical understanding of how systems behave, and that means understanding molecules and their reactions in terms of physical laws, especially those laws familiar to us through our everyday experience. Ultimately, the knowledge is embodied in theories that at some point have been well tested by experiments. In many respects, a textbook presentation of physical chemistry is a presentation of theories, and yet, experiments are very much the basis for the story. We cannot properly explain our best physical picture of chemical and molecular behavior without knowing the means of observation (experiment). Therefore, the direction for this text is to present understanding (theories) integrated with the means for observation and measurement, though usually without detailed discussion of experimental techniques. An ideal introduction to the subject combines this grounding in theory with hands-on laboratory experience.

1.3 Atomic and Molecular Energies

Our everyday experience tells us that energy can be stored continuously in mechanical systems. A child moving back and forth on a swing is a system with mechanical energy that can be set continuously. We can give a small push or a big push or anything in between. A baseball can be thrown at any desired velocity, subject only to the thrower's ability, and thus, it can have any amount of kinetic energy. Any moving particle in our everyday world can be given as little or as great a kinetic energy desired without restriction. In the world of very small particles, such as atoms and molecules, the situation is different. Systems that are **bound** (connected together over time) store energy continuously but in a stepwise manner. Their energies are said to be **quantized**.

Quantum mechanics, a subject for Chapters 8 through 10, deals with the mechanical behavior of systems whose size makes quantization of their mechanical energy a significant feature. Quantum mechanics may be regarded as, and may be shown to be, a more complete mechanical picture of our universe than the classical mechanics (e.g., Newton's laws) that we use to analyze systems in our everyday world. That is, classical mechanics may be developed as a specialized type of mechanics corresponding to a heavy-particle limit of a quantum mechanical description. As well, classical mechanics may be regarded as an approximation to quantum mechanics, an approximation that is highly accurate for massive particles and for macroscopic systems. Both types of mechanics turn out to be useful in different ways in the understanding of atomic and molecular behavior. For now, it is not the full mechanical description that is of interest but rather one special concept, energy storage.

Energy quantization means that a system can store energy only in certain fixed amounts. A **harmonic oscillator** is a standard, useful example. The harmonic oscillator system consists of a mass attached to a spring whose opposite end is connected to an infinitely heavy wall. Imagine a tennis ball attached to a lightweight spring hanging from a ceiling, and you have some idea of a harmonic oscillator system. Also imagine an atom attached by a chemical bond to a metal surface, and you are thinking about an analogous system in the small world of atoms and molecules. Pull the mass (tennis ball) away from its equilibrium position, its rest position, and release it. The system oscillates, which is to say that the mass moves back and forth. Such an oscillator system is harmonic if the spring has certain ideal properties, which are not yet of concern. As we pull the tennis ball and stretch the spring, we are adding energy (potential energy) to the mechanical system. We can stop stretching as desired, which means we can add any amount of energy. That does not hold for the atom attached to the metal surface. That system can accept (store) only certain specific increments of energy.

Quantum mechanics dictates that we cannot add energy continuously to the tennis ball–spring system. This is not in conflict with our observations, however. For the tennis ball system, the energy is quantized into such small pieces that we cannot distinguish energy storage via numerous small pieces from continuous energy storage. Classical mechanics serves very well in describing the tennis ball system, though not the atomic system. This can be appreciated by applying without derivation one result of quantum mechanical analysis. It is that energy may be stored in a harmonic oscillator system in quanta equal in size to the fundamental constant known as **Planck's constant**, h, times the frequency of oscillation, ν

$$E_{quanta} = h\nu \tag{1.1}$$

Planck's constant is a very small fundamental constant with a value of 6.626069×10^{-34} J s. Thus, in our macroscopic, everyday world, the energy quanta are very tiny. The tennis ball suspended by a spring might have a frequency of around $1\,\text{s}^{-1}$ ($1\,\text{Hz}$). This would make the energy quanta on the order of 10^{-33} J relative to a total mechanical energy approaching $1\,\text{J}$ given an initial displacement of several hundredths of a meter. The quanta are so small in relation to the behavior we can perceive that it is as if the stored energy varies continuously. In contrast, an atom vibrating against a metal surface to which it is bonded may have a frequency on the order of $10^{13}\,\text{s}^{-1}$. Then, the amount of energy that can be added or removed from the vibrational motion would be on the order of 10^{-20} J. This is much bigger than the quanta for the tennis ball system but still a very small amount of energy in our everyday world. It is, however, a large amount in the world of atoms and molecules. For instance, if the atom had a bond dissociation energy of $300\,\text{kJ}\,\text{mol}^{-1}$, which is a representative value for chemical bonds, then on dividing this by Avogadro's number ($N_A = 6.022142 \times 10^{23}$), we would find that the required energy to break one bond is on the order of 10^{-18} J. This means that the vibrational quanta of energy of the atom system are sizable relative to chemical energies of bond breaking, the quanta being about 1% of the bond energy in this hypothetical case.

Quantum mechanical analysis may be used for mechanical systems with numerous particles, including atomic and molecular systems. The analysis usually shows that there are many ways a system can exist, and associated with each way is a certain amount of stored energy. The distinct ways in which a quantum mechanical system can exist are referred to as **quantum states**. There is always a lowest energy state referred to as the **ground state** of the system. The ground state is not necessarily a state with zero energy; it simply corresponds to the lowest possible allowed energy for the system.

Quantum states other than the ground state of a system are called **excited states**. There may be an infinite number of excited states. **Quantum numbers** are values used to distinguish or label the different states. Mostly, though not always, these are whole numbers, and they often arise in the course of the mathematics (differential equation solving) that goes along with a quantum mechanical analysis.

It is possible for two or more states to have the same energy, in which case the states are said to be **degenerate**. The different energies that are possible are referred to as the **energy levels** of the system. Just as there can be an infinite number of states, there can be an infinite number of energy levels. It turns out that the energy levels of a number of model systems can be expressed in simple formulas involving the quantum numbers of the system. For example, the energies of a harmonic oscillator depend on the vibrational frequency, v, and the quantum number, n.

$$E_n = \left(n + \frac{1}{2} \right) hv \qquad (1.2)$$

To use this expression, we must know or must obtain the vibrational frequency of the specific system at hand, and we must know another result from the quantum mechanical analysis: The quantum number n can take on values of 0, 1, 2, 3, and so on to infinity. Notice that the lowest allowed energy is $hv/2$; this is the energy of the ground state. Thus, the ground state for a simple oscillator corresponds to the quantum number n being 0. The next lowest allowed energy for this system is $3hv/2$, and this corresponds to the quantum number choice of $n = 1$. This is an energy step of hv up from the ground state energy. Likewise, the next step to the $n = 2$ level requires another hv in energy, and therefore, the size of the quanta that the oscillator may store is hv, as in Equation 1.1.

Another common system in quantum mechanics is the so-called rigid rotator. The energy levels of a rotating linear molecule are given to good approximation by the following expression.

$$E_J = BJ(J+1) \qquad (1.3)$$

J is the quantum number associated with rotation, and the values it can take are the positive integers and zero. B is called the **rotational constant** and is specific to each molecule. It depends on the moment of inertia of the molecule and thereby on the bond lengths and atomic masses. Notice that this expression shows a different dependence on the quantum number than the dependence in Equation 1.2. The energy associated with molecular rotation increases quadratically with the quantum number J. Figure 1.1 illustrates the increasing energetic separations among rotational energy levels versus the uniform separation for a harmonic oscillator.

Quantum mechanical analysis reveals that a rotating linear molecule can exist in several states with the same energy; another quantum number, M, distinguishes among the rotational states of the same energy. M is associated with the orientation of the angular momentum vector and does not affect the energy since the energy of a freely rotating body does not depend on orientation of the angular momentum vector. The number of states that may have the same energy is related to the quantum number J; it is simply the value $2J + 1$. Thus, if $J = 1$, then there are three degenerate states for which the energy of the system is $B(1)(2) = 2B$. The total number of states of a given energy level is the **degeneracy** of the level.

A diatomic molecule both vibrates and rotates. Strictly speaking, the motions are coupled, but to a good approximation the energies of the diatomic molecule are simply the

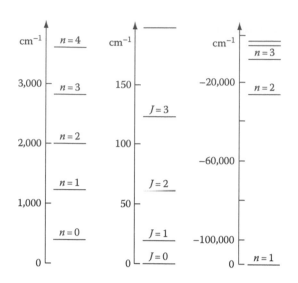

FIGURE 1.1

The energy levels of an idealized harmonic oscillator (left), a rotating diatomic molecule (center), and the hydrogen atom (right). Each horizontal line is drawn according to the energy scale given by the vertical lines and following Equations 1.2 through 1.4, respectively. Notice that the oscillator levels are evenly spaced, whereas the rotational levels become increasingly separated with increasing energy. Also, notice the different energy scales. Typically, rotational energy levels are more closely spaced than vibrational energy levels, and vibrational energy levels are more closely spaced than the levels that develop from the electrons' motions in an atom or molecule.

sum of the rotational contributions following Equation 1.3 and the vibrational contributions following Equation 1.2. Thus, the overall energy levels depend on two quantum numbers, n and J.

A final set of energy levels to consider are those of the hydrogen atom. The hydrogen atom is a two-particle system consisting of an electron and a much heavier particle, a proton. The electron "orbits" the proton, and the energies associated with that depend on one quantum number, n, in the following way.

$$E_n = -\frac{R}{n^2} \tag{1.4}$$

R is the **Rydberg constant** (1.09737×10^7 m^{-1}). The hydrogen atom's quantum number n can take on the values of all the positive integers; it obviously cannot be zero. The degeneracy of each energy level is n^2. If $n = 2$, the energy is $-R/4$, which is one-quarter of the energy of the ground state, and the degeneracy of the level is 4. As seen in Figure 1.1, the energy levels of the hydrogen atom get closer together as the quantum number increases, and this behavior is different from that of both the harmonic oscillator energy levels and the rotational energy levels.

Translation of a quantum mechanical particle in a container corresponds to a **bound** state with the energy quantized. Translation of a particle, such as an atom or molecule, in an unrestricted space (no walls and no potentials) is an **unbound** motion. The mechanics of such a free particle have certain similar elements in both the quantum and classical pictures. In both, the energy may vary continuously. In chemistry then, energy storage depends on the nature of the mechanical system and on whether or not there are bound states. We will revisit all of the quantum mechanical systems discussed earlier later in the book.

1.4 Configurations, Entropy, and Volume

Depositing energy in a single atom or molecule depends on the existing energy level structure of the atom or molecule. But then, what happens in a very large collection of atoms and molecules? The answer to how energy is stored is an important property of a macroscopic substance, and it is one that reflects somewhat the behavior at the molecular level. To understand this, we shall consider statistical arguments in order to deal with the large numbers of particles found in a macroscopic system. This subfield is called **statistical mechanics**, a subject that will be covered in more detail in Chapter 11.

The first step is to consider configurations or arrangements within systems. Ultimately, we will be concerned with arrangements of the particles in their individual energy levels, but we can start with a non-quantum mechanical example to illustrate the analysis. Consider three pairs of dice arranged on a table, all with a 6 showing. There is no energetic preference for this arrangement over any other, such as the arrangement with two 6s, a 5, a 4, and two 2s. The analogy with a molecular system is that the arrangements of dice are like the different degenerate quantum states. If the dice were collected from the table and dropped, we would be surprised if each had a 6 showing. Rather we would expect some "random" arrangement. If we picked up the dice a second time and threw them again, it would be just as much a surprise to see six 6s. Our experience tells us to expect some mix of 6s, 5s, 4s, 3s, 2s, and 1s on any throw. Thus, we find that there is a tendency toward certain arrangements even without any energy difference. Six 6s showing is very unlikely, but three 2s, a 4, a 5, and a 6, for instance, does not seem to be as rare. If we had first started with a throw yielding six 6s, subsequent throws would likely have taken us away from that arrangement and not returned us to it.

The reasons for the behavior of the dice are not energetic but lie in statistics. Let us use the term state, but not exactly in the quantum mechanical sense of the last section, to refer to any specific arrangement of the dice. A state of the dice may be represented by a list of six integers, n_1–n_6, which tells how many dice have 1–6, respectively, showing. Thus, the initial state in our example is $(n_1, n_2, n_3, n_4, n_5, n_6) = (0, 0, 0, 0, 0, 6)$. It is different from the state $(1, 0, 0, 0, 0, 5)$ which would be five 6s and a 1. For the state $(1, 0, 0, 0, 0, 5)$, the 1 could be showing on any of the six dice, and so there are six different ways for this state to exist. In contrast, there is only one way for the state $(0, 0, 0, 0, 0, 6)$ to exist. We shall refer to these ways of existing as **configurations** or as **microstates**. For each possible state there are a specific number of configurations. Here is a tabulation of a few of them.

State	Number of Configurations	Meaning
(6, 0, 0, 0, 0, 0)	1	Six 1s showing occurs once
(0, 6, 0, 0, 0, 0)	1	Six 2s showing occurs once
(1, 0, 0, 0, 0, 5)	6	A 1 and five 6s occurs 6 times
(5, 1, 0, 0, 0, 0)	6	Five 1s and a 2 occurs 6 times
(4, 2, 0, 0, 0, 0)	15	Four 1s and two 2s occurs 15 times
(4, 1, 1, 0, 0, 0)	30	Four 1s, a 2, and a 3 occurs 30 times
(1, 1, 1, 1, 1, 1)	720	One of each occurs 720 times

There is a simple formula that gives the number of configurations for all possible states. If N is the number of dice and C is the number of configurations, then

$$C(N, n_1, n_2, ...) = \frac{N!}{n_1! n_2! ...} \tag{1.5}$$

C has been written as a function of N and of the integers that specify the state. We expect that each individual configuration is equally likely in any throw of the dice. But this means that the state (1, 1, 1, 1, 1, 1) is 720 times as likely to turn up as the state (0, 0, 0, 0, 0, 6).

The dice throwing example is an illustration from common experience that systems tend to change from states with low C values to states with large values. They change from states with little probability to states with high probability. Ludwig Boltzmann (Austria, 1844–1906) associated this with the natural occurrence of spontaneous changes, arguing that all spontaneous processes go in a direction leading to a larger number of configurations. This is an important concept in chemistry because it associates a probability with a change. The melting of a cube of ice immersed in liquid water at room temperature and the lowering of the temperature of the liquid, then, comprise an occurrence with an overwhelmingly greater probability than the reverse spontaneous process of freezing some of the liquid water into a cube and raising the remaining liquid to room temperature. It is not that the latter cannot occur, but that it has a vanishingly small probability of occurring.

A collection of molecules exists in any one configuration for only an instant as collisions and interactions among the molecules lead to changes that are the equivalent of continuously throwing the dice in our example. Thus, a system fluctuates in time, changing from one instantaneous configuration to another. However, for a large collection of molecules, one configuration tends to dominate. For a set of N dice, this is the configuration for which C in Equation 1.5 is a maximum.

Entropy is a thermodynamic quantity that is related to this tendency of a system to be in the state of maximum probability. In the case of our hypothetical system of many dice, it must be a function of the C values, for they indicate the probability. For two noninteracting or separate systems, the overall tendency to be in the state of maximum probability (the total entropy) must be the same as the sum of their individual entropies. This requirement dictates the relationship between entropy, S, and the number of configurations, C, of a given state. First, though, we must realize that the number of configurations for combining two systems is not additive but is multiplicative. If system A has 3 possible configurations and system B has 4, then the number of configurations for the composite system AB is 12. This is because for every choice for the configuration of part A of the composite system (i.e., 1, 2, or 3), there are 4 choices for part B; in total, this gives 12 for the composite system. We may state the relationships between the composite and individual systems in the following way.

$$C_{AB} = C_A C_B$$

$$S_{AB} = S_A + S_B$$

Now, for the entropy to be a function of the configurations, that is, for $S = f(C)$, the following relations must hold.

$$S_A = f(C_A)$$

$$S_B = f(C_B)$$

$$S_{AB} = f(C_{AB})$$

These relationships are satisfied if and only if S is proportional to the logarithm of C. Notice the effect of taking the logarithm of the expression for C_{AB}.

$$\ln C_{AB} = \ln(C_A C_B) = \ln C_A + \ln C_B$$

With S taken to be proportional to $\ln C$, the entropy for the composite system, S_{AB}, is an additive result of contributions from A and B, as required.

Absolute entropy is related to the number of configurations of a system, with the specific constant of proportionality being Boltzmann's constant, k.

$$S = k \ln C \tag{1.6}$$

k is a fundamental physical constant with a value of $1.3806504 \times 10^{-23}$ J K^{-1}. Since $\ln C$ is dimensionless, then entropy has units of energy per unit of temperature. We shall examine why S has such units shortly, but we can appreciate that the tendency of a system to be in a state of maximum probability will be related to energy quantities and to temperature.

It is perhaps worth noting at this point that we have not defined entropy on the basis of "randomness" or "disorder" as is commonly done. In most chemical systems, the state of maximum probability is directly related to the state of maximum "disorder." However, entropy, in the strictest sense, should always be associated with probability.

The number of configurations, C, of a given state represents a probability of being in that state. The bigger the value of C for the three pairs of dice, the greater the probability that a throw of the dice will leave them in that state. In some cases, the states of a system may be somewhat abstract or unknown, and yet statistical arguments can still be quite revealing and useful. For instance, consider a single molecule that is free to move in a box with dimensions $l \times l \times l$. If the width, height, and length of the box were suddenly doubled, would the entropy change? We expect that it would increase because the space available has increased and somehow the number of configurations should increase, even if we do not know exactly what the configurations are. To analyze this situation, we can consider the original box to have been divided into small boxes, each $l/N \times l/N \times l/N$, where N is an integer. A configuration is taken to consist of the particle being found in one of the boxes. For instance, if the original box were 10 m on a side, a choice of $N = 10$ would yield boxes 1 m on a side. There would be 1000 such small boxes within the original box's volume. In the expanded box, where the sides have been doubled, there would have to be $(2l/N)^3$ small boxes of the same size. Clearly, the ratio of the number of small boxes varies as the ratio of the volumes.

$$\frac{(2l/N)^3}{(l/N)^3} = \frac{(2l)^3}{l^3} = \frac{V_{expanded}}{V_{original}}$$

If we consider the number of small boxes to be proportional to the number of configurations, then the change in absolute entropy from expanding the box must be related to the volume since V is directly proportional to the number of configurations C, that is, $V \propto C$.

$$\Delta S_{expansion} = S_{expanded} - S_{original}$$

$$= k \ln(C_{expanded}) - k \ln(C_{original})$$

$$= k \ln \frac{C_{expanded}}{C_{original}}$$

$$= k \ln \frac{V_{expanded}}{V_{original}} \tag{1.7}$$

This result has no dependence on N, the number of divisions we made along the sides of the box, and it is a general result.

If there are n moles of the molecules in the volume instead of just one molecule, then there will be nN_A times this entropy change, with N_A being Avogadro's number (i.e., the number of molecules in a mole).

$$\Delta S_{expansion} = nN_A k \ln \frac{V_{expanded}}{V_{original}}$$

$$= nR \ln \frac{V_{expanded}}{V_{original}} \tag{1.8}$$

This is a thermodynamic relation for ideal gases composed of non-interacting particles. R, the constant introduced to replace the product of Boltzmann's constant and Avogadro's number, is called the gas constant. Its value is 8.314472 J mol^{-1} K^{-1}. For a sample of gas particles free to move in some volume, the tendency to be in a state of maximum probability is related to the volume of the sample.

1.5 Energy, Entropy, and Temperature

Entropy is related to the probability that a collection of atoms or molecules exists in a given state of a system. This relationship is used to understand the distribution of molecules among accessible molecular quantum states. That distribution is important because it ultimately serves to define temperature fundamentally. Of course, we all have a phenomenological understanding of temperature. As temperature increases around us, the mercury in a thermometer expands, and using a linear scale printed on the thermometer, we measure that increase while feeling warmer. The thermometer provides reproducible measurements and serves as a device for observation and experiment. However, it does not establish a basis for temperature change at the molecular level, only a way to reference temperature to the volume of a certain mercury sample. A more detailed basis for temperature can be developed by considering the distribution of atoms and molecules among their available states.

If we know the probability that a system exists in a certain state i is the value P_i, this value can be used in place of the number of configurations in Equation 1.6. P and C must be proportional to each other, and so it is sufficient to know P even if we do not know the absolute numbers of configurations, C. Therefore,

$$\Delta S = k \ln \frac{P_2}{P_1} \tag{1.9a}$$

Dividing by k and taking the exponential of this equation gives

$$P_2 = P_1 e^{\Delta S/k} \tag{1.9b}$$

The probability of being in state 2 is related to the probability of being in state 1 via an exponential in the entropy difference, ΔS.

Let us now **postulate**, or assume without any proof, that temperature, T, relates energy changes, ΔE, and entropy changes, ΔS, of a system via the equation $\Delta E = T\Delta S$. Using a postulate is like trying to work a jigsaw puzzle by putting a puzzle piece in place with nothing connected and then fitting pieces around it. If the guess was right, we have solved the puzzle. If not, we have to remove the piece and use a different one in that spot. Postulates may be regarded as making a choice of a piece of a scientific puzzle. If everything else fits, that is, if there are no contradictions, then we can accept the postulate as valid. If something does not fit, we must reject the postulate. Here, we are postulating the existence of something we are calling temperature by presuming or guessing that it relates entropy and energy changes, two things we already understand at a fundamental level. In exploring the implications of this postulate, we look for contradictions with other established ideas or theories.

Now consider a hypothetical system with one molecule embedded in a continuous medium (a heat bath) that is maintained at a constant temperature, T. We shall think of the molecule as component A and the surrounding heat bath as component B. One specific change in the system to consider is for the quantum state of the molecule to go from state 1 to state 2. With this comes a change in the energy of component A which is $\Delta E(A) = E_2 - E_1$. For energy to be conserved, the energy change of the heat bath and the energy change of the molecule must sum to zero.

$$\Delta E(A) + \Delta E(B) = 0$$

The entropy of the molecule does not change, assuming that all the states are nondegenerate, because there is only one configuration for each state. That is, $\Delta S(A) = 0$. Therefore, the entropy change for the system is the same as $\Delta S(B)$, which is $\Delta E(B)/T$ according to our postulate.

$$\Delta S = \Delta S(B) = \frac{\Delta E(B)}{T} = -\frac{\Delta E(A)}{T} = \frac{E_1 - E_2}{T}$$

This can be related to the probabilities that the system exists in the two states being considered.

$$\Delta S = \frac{E_1 - E_2}{T} = k \ln \frac{P_2}{P_1}$$

Dividing by k and taking the exponential of this equation yields

$$\frac{P_2}{P_1} = \frac{e^{-E_2/kT}}{e^{-E_1/kT}} \tag{1.10}$$

Since states 1 and 2 of this hypothetical system differ only in the quantum state of the molecule in component A, this law states that the probability of existing in a given quantum state for a molecule in contact with a heat bath diminishes exponentially with the energy of the state. Increasing the temperature raises the probability of existing in higher energy states. This implication of our postulate is experimentally confirmed, and the equation is one form of the **Maxwell–Boltzmann distribution law**. Temperature is thereby connected to the distributions found among the states of a system.

For a sample of many molecules, under the assumption of no interaction between them, the probabilities are directly proportional to the numbers of molecules in each state, the **populations**, which we designate as n. For a large collection of A molecules at some temperature T, the number found in the ith energy level of an A molecule is $n_i = NP_i$, where N is the total number of A molecules. Is it possible for a large collection of molecules to have populations of molecular quantum states other than those dictated by the Maxwell–Boltzmann law? Yes. But in that event, the system is not at equilibrium, which means that it is not stable and is undergoing change. The distribution law holds for equilibrium conditions, and under those conditions, it can be used to determine the number of molecules in particular energy level states.

A subtlety that we shall not yet consider in detail occurs when the quantum energy levels of the molecule are degenerate. In that case, the distribution law becomes

$$\frac{P_2}{P_1} = \frac{g_2 e^{-E_2/kT}}{g_1 e^{-E_1/kT}} \tag{1.11}$$

where g_1 and g_2 are the degeneracies of the respective levels. Equation 1.11 is a more general form of Equation 1.10. Notice that for systems in which the degeneracy happens to increase on moving up from one energy level to the next higher level, the g_i factors in Equation 1.11 contribute to making the higher state more probable, thereby working against the exponentially diminishing energy factor.

Example 1.1: Energy Level Populations

Problem: Relative to the ground state population, find the populations of the $n = 1$ and $n = 2$ quantum states for an equilibrium system of identical harmonic oscillators at a temperature $T = 100\,K$. The oscillator vibrational frequency, ν, is such that $h\nu = 2.0 \times 10^{-21}$ J.

Solution: Since the populations are probability times the number of molecules, the ratio of the population of one state to the ground state population is the same as the corresponding ratio of the probabilities. To obtain that ratio, we use Equation 1.10 because the degeneracies of the harmonic oscillator energy levels all happen to be 1 (i.e., nondegenerate). All that is needed to complete the problem is the energy level expression for harmonic oscillators, and that is Equation 1.2.

$$\frac{P_1}{P_0} = \frac{e^{-3h\nu/2kT}}{e^{-h\nu/2kT}} = e^{-h\nu/kT}$$

$$\frac{P_2}{P_0} = \frac{e^{-5h\nu/2kT}}{e^{-h\nu/2kT}} = e^{-2h\nu/kT}$$

With $T = 100\,K$, $kT = 1.38 \times 10^{-21}$ J. With the given value of $h\nu$, the following ratios are obtained.

$$\frac{P_1}{P_0} = 0.23 \quad \text{and} \quad \frac{P_2}{P_0} = 0.055$$

1.6 Distribution Law Derivation

The distribution law of Equation 1.11 was obtained by postulating that $\Delta E = T\Delta S$. This section is aimed at achieving the same result but following a lengthier statistical argument that does not require making that postulate.

The most probable arrangement for a system is the one with maximum entropy, and that means the state where $\ln C$ is a maximum, with C being the number of configurations. Let us maximize $\ln C$ and determine the distribution that results for a system of N non-interacting particles. An energy, ε_i, is associated with the ith quantum state of a particle. As we did in considering arrangements of dice, let us use a set of integers such that n_0 is the number of particles in the lowest state, n_1 is the number in the first excited state, and so on. The total energy of the system, E, is a fixed value given that no energy is being added or removed from the system. E is necessarily the sum of the energies of the individual particles.

$$E = \sum_{i=0}^{\infty} n_i \varepsilon_i \tag{1.12}$$

Equation 1.12 is a constraint on the system. We are interested only in those arrangements $(n_0, n_1, n_2, \ldots)$ that have this energy. Some other arrangement $(n_0', n_1', n_2', \ldots)$ is of interest only if Equation 1.12 yields the same energy with that arrangement. This means that the n_i are not freely adjustable. In addition, there is the constraint that the number of particles is the unchanging value, N.

$$N = \sum_{i=0}^{\infty} n_i \tag{1.13}$$

Now, the mathematical task is to maximize $\ln C$, where C is obtained from Equation 1.5, under the constraints of Equations 1.12 and 1.13.

Equation 1.5 involves factorials in N and n_i, and this makes the logarithm of C a complicated function. However, there exists a very useful approximation for the logarithm of a factorial, **Stirling's approximation.**

$$\ln N! \cong N \ln N - N + \frac{1}{2}\ln(2\pi N) \tag{1.14}$$

And if N is very large, the smallest term in Equation 1.14 may be dropped as a further approximation.

$$\ln N! \cong N \ln N - N \tag{1.15}$$

Using Stirling's approximation for large N and large n_i, $\ln C$ is approximated in the following way.

$$\ln C \cong N \ln N - \sum_i n_i \ln n_i - \left(N - \sum_i n_i\right) \tag{1.16a}$$

The term in parentheses in Equation 1.16a is zero according to Equation 1.13, and so it may be deleted.

$$\ln C \cong N \ln N - \sum_i n_i \ln n_i \qquad (1.16b)$$

Now, in this form, we can proceed to maximize ln C. A good number of mathematical steps are required to accomplish this, and the result has general implications.

When a function, $f(x)$, is at a maximum or minimum value, the first derivative is zero, that is, $\partial f(x_{max})/\partial x = 0$. This is because the first derivative gives the slope of a line tangent to the function; at a maximum or minimum, the tangent should be a horizontal line, which is a line of zero slope (see Appendix A). To find the maximum of ln C, we need to find the choice of the set of n_i values where all the first derivatives of ln C are zero. The values in this special set will be designated n_i' to distinguish them from the variables n_i. (This is the same as saying that x is the variable in $f(x)$ but that x_{max} or x' means the specific value at which f is a maximum.) We can use differentials to find the point at which the first derivatives are zero. The mathematical complication lies in the fact that Equations 1.12 and 1.13 are constraints; the n_i values are not entirely independent.

Since N is a constant, taking the differential of Equation 1.16b yields

$$d(\ln C) = -\sum_i (1 + \ln n_i)dn_i = -\sum_i dn_i - \sum_i \ln n_i dn_i = -\sum_i \ln n_i dn_i \qquad (1.17)$$

since the sum of the dn_i is dN, which is zero. Likewise, taking the differentials of Equations 1.12 and 1.13 yields

$$dE = 0 = \sum_i \varepsilon_i dn_i \qquad (1.18)$$

$$dN = 0 = \sum_i dn_i \qquad (1.19)$$

The left-hand side is zero in both cases because N and E are constants. Something that is zero may be freely added or subtracted, and so it is possible to state that for any constant α and any constant β,

$$d(\ln C) = -\sum_i (\ln n_i)dn_i - \alpha \sum_i dn_i - \beta \sum_i \varepsilon_i dn_i \qquad (1.20)$$

Though this step may seem arbitrary, it is the standard mathematical technique of using **Lagrange multipliers** to build external constraints into differential expressions. At the maximum of ln C, $d(\ln C) = 0$. We will designate the n_i values at that point as n_i', and then Equation 1.20 at the maximum of ln C is

$$0 = -\sum_i (\ln n_i')dn_i - \alpha \sum_i dn_i - \beta \sum_i \varepsilon_i dn_i \qquad (1.21)$$

$$= -\sum_i (\alpha + \beta\varepsilon_i + \ln n_i')dn_i \qquad (1.22)$$

This equation will be satisfied if each summation term in parentheses is independently zero. That is,

$$\ln n_i' = -\alpha - \beta \varepsilon_i \tag{1.23}$$

Taking the exponential of both sides of Equation 1.23 gives an expression for n_i'.

$$n_i' = e^{-\alpha} e^{-\beta \varepsilon_i} \tag{1.24a}$$

α and β remain unknown at this point in the mathematical procedure.

We can replace α in Equations 1.24a and b by using Equation 1.13; that is, we replace the factor involving α by using the expression for N when $\ln C$ is a maximum.

$$N = \sum_i n_i' = \sum_i e^{-\alpha} e^{-\beta \varepsilon_i} = e^{-\alpha} \sum_i e^{-\beta \varepsilon_i} \tag{1.24b}$$

Rearranging this equation gives

$$e^{-\alpha} = \frac{N}{\sum_i e^{-\beta \varepsilon_i}} \tag{1.25}$$

The summation in the denominator is referred to as the **partition function** and will be designated by the symbol Q.

$$Q = \sum_i e^{-\beta \varepsilon_i} \tag{1.26}$$

Therefore,

$$n_i' = \frac{N}{Q} e^{-\beta \varepsilon_i} \tag{1.27}$$

In effect, α has now been expressed in terms of β though β remains unknown. The effect of the partition function amounts to applying a common factor to each n_i such that the sum of the n_i's will be N.

We can combine Equation 1.27 with the energy expression, Equation 1.12, to write

$$E = \sum_i \frac{N}{Q} \varepsilon_i e^{-\beta \varepsilon_i} \tag{1.28}$$

Noticing that

$$\frac{dQ}{d\beta} = -\sum_i \varepsilon_i e^{-\beta \varepsilon_i} \tag{1.29}$$

we have an alternate expression for the energy.

$$E = -\frac{N}{Q}\frac{dQ}{d\beta} \tag{1.30}$$

Hence, if we know Q or something proportional to Q, we can immediately find E.

The development to this point is quite general, and Equations 1.28 and 1.30 can be applied anywhere. If we apply them to one specific problem that we understand or can analyze fully, we shall achieve a relation between β and E that is more explicit and better helps us understand and use β.

An idealized problem that can be analyzed fully is that of a large number of non-quantum mechanical particles possessing only translational energy and free to move about without interaction between particles (a gas of non-interacting particles). For this system, we will be able to show that the energy is inversely proportional to β. We have already seen from Equation 1.28 that β determines the distribution, and so it seems to be a good choice for a thermodynamic variable. Is it the temperature? The answer is "no" because we expect energy to increase as temperature increases, not as temperature decreases. In other words, temperature could be the inverse of β to within some constant of proportionality, and that turns out to be the case.

The energies of non-interacting gas atoms (particles) are associated with their individual translational motions. From a classical point of view, these energies are continuous because the velocities of the particles can be any value. From a quantum mechanical view, we can say that the kinetic energy of motion of each particle is not discrete. Under these conditions, the summation in Equation 1.26 needs only to be converted to a sum of infinitesimals, that is, converted to an integral, an integral over all the mechanical degrees of freedom. However, the energy depends only on the momenta of the particles because the kinetic energy of a free gas atom of mass m is

$$\varepsilon_{atom} = \frac{1}{2m}\left(p_x^2 + p_y^2 + p_z^2\right) \tag{1.31}$$

Integrating over only p_x, p_y, and p_z, and not over the spatial degrees of freedom, yields in this case a value proportional to Q. We shall let C be the constant of proportionality. Thus, the integral expression for Q is

$$Q = C\int_{-\infty}^{\infty}\int_{-\infty}^{\infty}\int_{-\infty}^{\infty} e^{-(p_x^2+p_y^2+p_z^2)\beta/2m}\,dp_x dp_y dp_z \tag{1.32}$$

This can be expressed as a product of integrals over the momentum components.

$$Q = C\int_{-\infty}^{\infty} e^{-\beta p_x^2/2m}\,dp_x \int_{-\infty}^{\infty} e^{-\beta p_y^2/2m}\,dp_y \int_{-\infty}^{\infty} e^{-\beta p_z^2/2m}\,dp_z \tag{1.33}$$

Evaluating these integrals (see Appendix D) yields

$$Q = C\left(\frac{2m\pi}{\beta}\right)^{3/2} \tag{1.34}$$

Notice that the three integrals in Equation 1.33 have the same value, and this leads to the 3 in the exponent value of 3/2 in Equation 1.34.

The derivative of Equation 1.34 can be used in Equation 1.30 to finally establish the relationship between the energy of the system and β.

$$\frac{dQ}{d\beta} = -C(2m\pi)^{3/2}\frac{3}{2}\left(\frac{1}{\beta}\right)^{5/2} \tag{1.35}$$

$$E = \frac{3}{2}\frac{N}{\beta}$$

This indicates that the total internal energy for this particular case is inversely proportional to β. As already derived, β is more than this for it generally dictates populations among available energy levels. Any new quantity we define that is directly related to β will serve as well as β, and one such choice is to define a new quantity, T, to be $(k\beta)^{-1}$, where k is Boltzmann's constant. With that, T is what we usually refer to as temperature. If we replace β by its dependent quantity T in Equation 1.35, which is specific for a gas of non-interacting particles, then

$$E = \frac{3}{2}NkT = \frac{3}{2}nRT \tag{1.36}$$

Likewise, substituting for β in Equation 1.27 yields the Maxwell–Boltzmann distribution law, which is a general result.

$$n_i = \frac{N}{Q}e^{-\varepsilon_i/kT} \tag{1.37}$$

Again, this indicates that the population diminishes exponentially with energy.

Returning to the example of a gas of non-interacting particles, we note that the average energy per gas particle is the total energy divided by the number of particles. Using Equation 1.36, the average energy is

$$\bar{E} = \langle E \rangle = \frac{E}{N} = \frac{3}{2}kT \tag{1.38}$$

where two designations, an overbar and $\langle \rangle$, have been used to identify the average or the mean of the value E. From the energy expression in Equation 1.31, we relate this average to the average of the square of the momentum.

$$\frac{1}{2m}\left(p_x^2 + p_y^2 + p_z^2\right) = \frac{3}{2}kT \tag{1.39}$$

Since the space for a moving free particle is isotropic (the same in each direction), the mean-squared components of the momenta must be the same for each direction: $\langle p_x^2 \rangle = \langle p_y^2 \rangle = \langle p_z^2 \rangle$. Using this expression with Equation 1.39 yields

$$\langle p_x^2 \rangle = \langle p_y^2 \rangle = \langle p_z^2 \rangle = mkT \tag{1.40}$$

Or, for velocities, v (recalling that $p_x = mv_x$),

$$\langle v_x^2 \rangle = \langle v_y^2 \rangle = \langle v_z^2 \rangle = \frac{kT}{m} \tag{1.41}$$

Speed, s, is the magnitude of a velocity vector, and the square of the speed is the sum of the squares of the velocity components. Thus,

$$\langle s^2 \rangle = \frac{3kT}{m} \tag{1.42}$$

The square root of this value, $\langle s^2 \rangle^{1/2}$, is the root-mean-squared speed of the non-interacting gas particles. The root-mean-squared speed increases as the square root of temperature. A manifestation of this is that as air warms, the speed at which sound may be transmitted increases because of the greater average speed of the particles.

1.7 Conclusions

Physical chemistry is the exploration of the world of atoms and molecules in terms of fundamental physical laws; it is the physics of chemistry. It embraces properties and dynamics of isolated species and of bulk systems, and it seeks to provide a fundamental connection between both realms. Atoms and molecules are such small species that their mechanical behavior (i.e., vibrations, rotations, electron orbital motion, and so on) is governed by quantum mechanics. Atoms and molecules exist in different states with distinct energies. Statistical analysis of a large sample of quantum mechanical particles tells us that there are many ways in which they can be arranged among the available states. Equilibrium corresponds to existence largely in the most probable arrangement, and this condition is characterized by maximizing a quantity called entropy. At equilibrium, the distribution of particles among available states is dictated by the thermodynamic property of temperature following the Maxwell–Boltzmann distribution law. Temperature is not ad hoc but is fundamentally defined by the distribution of particles among energy levels in a condition of equilibrium.

Point of Interest: James Clerk Maxwell

James Clerk Maxwell

James Clerk Maxwell is an interesting figure both in the development of physical chemistry, a subject at the boundary of his chosen field of physics, and in his execution of scientific discovery. He was born in Edinburgh, Scotland, on June 13, 1831, and he died in Cambridge, England, on November 5, 1879. He held professorships first in Aberdeen, Scotland, then at King's College in London, and finally at Cambridge University. Despite his rather short life, he had a remarkable and lasting impact on physical science.

Maxwell is perhaps most known for his work on electricity, magnetism, and the electromagnetic theory of light. This work began following his graduation from Cambridge in 1854 and included the development of the electromagnetic field relations known today as Maxwell's equations.

The use of analogy was a powerful tool in Maxwell's hands, and it was something in which he strongly believed. His contributions to electromagnetic theory were aided by analogies Maxwell recognized between electric current, heat conduction, and fluid flow. These were analogies between mathematical relations more than similarities between the nature of different things.

An interesting problem area that occupied Maxwell's attention from 1855 to 1859 was the nature of the rings of Saturn. He showed that certain ideas of the time, such as that Saturn had solid or rigid rings, were not possible. Instead, the stability of the rings require that they consist of concentric circles of small objects, the orbital speed of each circle being dependent on its distance from the planet. In 1895, the differential rotation of the rings Maxwell predicted almost 40 years earlier was confirmed by observation.

The study of Saturn's rings led Maxwell to the problem of the motions of large numbers of colliding bodies, such as would be found in the rings. This in turn led him to the study of gas kinetics. Here he introduced the use of statistical methods, not for data analysis but for a description of the physical process. He recognized that there must be a distribution of velocities of gas particles, and by 1860 he had developed a statistical formula for that distribution (the first expression of what today we refer to as the Maxwell–Boltzmann distribution law). One idea of the time was that the pressure of a gas resulted from static repulsion between gas particles. Maxwell's kinetic picture, on the other hand, correctly associated pressure with collisions of molecules with the walls of the vessel in which the gas is held. His theory led to successful predictions about viscosity, diffusion, and other properties of gases; and it launched the study of statistical mechanics.

Many things were of interest to Maxwell, and he published papers on a wide range of topics (his first at the age of 14). He studied optics and optical properties of materials, and he developed the fish-eye lens. One of Maxwell's interests and areas of investigation was color vision. He projected the first color photograph, and he explained color blindness as a deficiency in one or two of the three types of color receptors in the eye. He was interested in the stability of the earth's atmosphere and its thermodynamics and in stress in building frameworks.

Maxwell's success, and the impact of his work on physical chemistry, came from his belief in the power of analogy and also from his interest in studying an assortment of problems.

Exercises

1.1 The rotational constant, B, for carbon monoxide is 3.836×10^{-23} J. Use Equation 1.3 to find the energy of the lowest three rotational state levels (i.e., $J = 0, 1,$ and 2) of a carbon monoxide molecule and find the degeneracies of each of these levels.

1.2 List the 15 configurations that are possible for three pair of dice in a state with two 6s and four 5s showing.

1.3 Apply Equation 1.5 to the arrangements of four dice to find the total probability that three and only three of the dice will have the same number showing.

1.4 Verify Equation 1.7 for an expansion where only one side of the original box is increased from l to $2l$.

1.5 For a gas sample of non-interacting particles at $T = 100\,\text{K}$, apply Equation 1.10 to find the ratio of probabilities for two states with energies $E_1 = 3.0 \times 10^{-21}$ J and $E_2 = 9.0 \times 10^{-21}$ J.

1.6 Obtain approximations to $\ln N!$ for $N = 10, 1{,}000,$ and $100{,}000$ by applying Equations 1.14 and 1.15. How significant are the differences?

1.7 Find the root-mean-squared speed of atoms of helium, neon, and argon at temperatures of 10, 300, and 500 K using Equation 1.42.

1.8 What is the mass of the particles of a hypothetical gas whose root-mean-squared speed at $T = 300\,\text{K}$ happens to match the speed of sound in air at 300 K (about 350 m s^{-1})?

1.9 Calculate the amount of energy required to ionize 1 mol of hydrogen atoms from their ground state.

1.10 What is the ratio of populations in the ground and second excited state of a harmonic oscillator with a vibrational frequency of 5.0×10^{12} Hz at a temperature of 300 K?

1.11 For a rigid rotator with a rotational constant of 8.5 cm^{-1}, what is the temperature at which the population of the $J = 2$ rotational state is 50% of the $J = 1$ rotational state?

1.12 Use Equation 1.5 to find the number of configurations for all the states of a system consisting of three four-sided pyramidal bodies numbered 1 through 4 (i.e., four-sided dice).

1.13 Make a plot of the number of configurations with all dice showing different numbers for a set of dice versus the number of dice in the set, with that number ranging from 1 to 6.

1.14 Given the following expression for the entropy change that accompanies a temperature change of a liquid, find the probability that a 1 kg sample of water initially at equilibrium at 300 K will be found to exist with half the molecules at 310 K and half at 290 K.

$$\Delta S = nR \ln\left(\frac{T_2}{T_1}\right)$$

1.15 Use the condition given in Exercise 1.11 and find a value for the temperature difference, ΔT, where half of the 1 kg sample of water is at a temperature of $300 + \Delta T$ K and the other half is at $300 - \Delta T$ K, such that the probability of this occurring is 1 in 100.

1.16 Test the two forms of Stirling's approximation, Equations 1.14 and 1.15, by directly computing ln $N!$ and comparing it with the results from the approximate expressions for $N = 5, 10, 15, 20$, and 25.

1.17 Consider a system of 10 identical harmonic oscillators. The state of the system is given by a list of integers $(n_0, n_1, n_2, \ldots)$ that indicate the number of oscillators in their 0, 1, 2, 3, ..., quantum states. Of course, the sum of these integers must be 10, the number of oscillators. If the system happens to be in the state $(2, 1, 1, 5, 0, 1, 0, 0, \ldots)$, make a list of at least five other states with the same energy; show that the constraints of Equations 1.12 and 1.13 are satisfied. Then, find a state where Equation 1.13 holds but Equation 1.12 does not.

1.18 Find the value of kT in joules at room temperature (about 300 K). Convert this value to cm^{-1} (see Appendix F). At this temperature, what must be the energy difference (in cm^{-1}) between two nondegenerate molecular energy levels such that the relative population of the higher state to the lower state is 0.5? Does this energy difference seem typical of the separations between rotational states or between vibrational states?

1.19 The energy difference associated with excitation of the electronic structure of atoms and molecules is typically 20,000 cm^{-1} and above. Assuming no degeneracy of energy levels, find the ratio of the population of an excited state at 20,000 cm^{-1} to the population of the ground state at 300 K. Next, find the temperature at which this ratio becomes 0.1.

1.20 Using Equation 1.11, make a plot of the population of quantum states versus the quantum number, J, of a sample of a gas at 300 K given that $J = 0, 1, 2, 3, \ldots$, the degeneracy is $g_J = 2J + 1$, and the energy in units of cm^{-1} (see Appendix F) is

$$E_J = 20\,\mathrm{cm}^{-1}J(J+1)$$

1.21 Consider a system of N non-interacting particles trapped so that they are free to move in only two dimensions. Find Q for this system and then find an expression for the energy of the system.

Bibliography

Hill, T. L., *An Introduction to Statistical Thermodynamics* (Dover Publications, New York, 1987). *This is a concise introductory level text.*
Ihde, A. J., *The Development of Modern Chemistry* (Dover Publications, New York, 1984).
Kuhn, T. S., *The Structure of Scientific Revolutions* (University of Chicago Press, Chicago, IL, 1970).
Servos, J. W., *Physical Chemistry from Ostwald to Pauling* (Princeton University Press, Princeton, NJ, 1990).
Smith, E. B., *Basic Chemical Thermodynamics*, 5th edn. (World Scientific Books, Singapore, 2004).

2

Ideal and Real Gases

Examination of the gaseous state of matter offers an excellent means for understanding certain basics of thermodynamics and for seeing connections with atomic and molecular level behavior. This chapter is concerned with explaining the differences between the behavior of real gases and the behavior of a hypothetical gas called an ideal gas because of its particularly simple behavior. The ideal gas is a model that under certain conditions can serve as a good approximation of real gas behavior, and we begin by considering the relationship of temperature to other properties of an ideal gas.

2.1 The Ideal Gas Laws

Experiments that are now centuries old have revealed relationships among three properties of gases, the volume, V, the pressure, P, and the temperature, T. Volume is simply the three-dimensional space occupied by the gas sample. It is fixed by the geometry of the container holding the gas. Pressure is a force per unit area, and it can be measured by balancing against an external force of known size. As shown in Figure 2.1, a piston assembly provides one means of measuring pressure. The gas pressure is exerted against the piston, and the piston is loaded until its position is unchanging, that is, until it reaches the point of balance. The force exerted by the piston is the gravitational force of the mass (m) loaded on it, and that force is the gravitational constant g times m. As discussed in the previous chapter, one fundamental definition of temperature is related to distributions among the available energy states. In practice, certain mechanical changes have been shown to vary linearly with temperature, at least over certain ranges. The volume of mercury over a fairly wide range around room temperature is one of these. Hence, measuring the volume of mercury in a tube (a thermometer) serves to measure the temperature.

As the temperature of a sample of fixed volume increases, the pressure increases. If the external pressure acting on a gas sample contained in a piston assembly is increased, the volume will diminish. These are statements of observed phenomena, and they are consistent with our everyday experience. For instance, an automobile tire that has reached a higher than ambient temperature gives a higher reading on a tire pressure gauge than it would at the ambient temperature. Or, a small balloon that can fit between our hands can be diminished in size by squeezing. Experiments can establish mathematical relationships that quantify these observations.

If we measured an automobile tire's temperature and pressure a number of times before and after a number of different trips, we would develop a collection of data points, one pressure value for every temperature value. Could temperature be expressed as a function of pressure from these data? Could pressure be expressed as a function of temperature? Both of these questions ask for a mathematical relationship between pressure and temperature. If a relationship is found, it should be predictive. Once a relationship has been

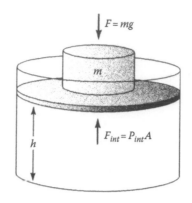

FIGURE 2.1
The pressure of a gas sample, P_{int}, can be measured by balancing the force due to the pressure, F_{int}, against an external force, F. The point of balance is when there is no contraction or expansion of the sample. The external force is applied to the system via a piston such that the force F is the gravitational force of the mass m loaded onto the piston. The internal force is the pressure times the area of the piston, A.

mathematically established, the pressure can be immediately derived if one knows the temperature, and vice versa.

Experiments done long ago on fixed-mass samples of air showed that the relationship between volume, V, and temperature, T, was a linear one.

$$V = cT + d \tag{2.1}$$

where
  c is a constant of proportionality
  d depends on the choice of the temperature scale

For instance, taking $T = 0$ to be the lowest possible temperature means that d is the smallest possible volume for the gas. A different scale could take $T = 0$ to be room temperature, and then d would be the volume of the sample at room temperature, not the smallest possible volume. d can be eliminated by using an absolute scale whereby $T = 0$ corresponds to the temperature at which $V = 0$. c can be eliminated (replaced) by invoking some reference temperature, T_0, at which the volume is known to be some value, V_0.

$$\frac{T}{T_0} = \frac{cV}{cV_0} = \frac{V}{V_0} \tag{2.2}$$

Equations 2.1 and 2.2 are statements of **Charles' law** (Jacques Alex Caesar Charles, France, 1746–1823). This temperature–volume relationship tends to hold reasonably well for a sample of air at temperatures and pressures not too different from those we experience every day.

At constant temperature, pressure is inversely proportional to volume, and so with increasing external pressure (balanced by the pressure of the gas) there is a diminishing volume. Pressure and volume are always expressed on an absolute scale, not on a relative scale, and the inverse proportionality means that their product is a constant at a given temperature.

$$PV = b \tag{2.3}$$

The proportionality constant, b, may be eliminated by referencing to a specific volume, V_0, and pressure, P_0.

$$\frac{P}{P_0} = \frac{b/V}{b/V_0} = \frac{V_0}{V} \tag{2.4}$$

Equations 2.3 and 2.4 are statements of **Boyle's law** (Robert Boyle, England, 1627–1691).

The temperature and pressure of an air sample at fixed volume are found to be related in direct proportion.

$$T = aP \tag{2.5}$$

Eliminating the proportionality constant by using a reference temperature and pressure yields

$$\frac{T}{T_0} = \frac{P}{P_0} \tag{2.6}$$

Thus, an increase in temperature leads to an increase in pressure at a fixed volume. Equations 2.5 and 2.6 are statements of **Gay-Lussac's Law** (Joseph Louis Gay-Lussac, France, 1778–1850).

The three expressions in Equations 2.1, 2.3, and 2.5 can be combined into one expression that has a single proportionality constant, R, the molar gas constant. The usual units for R are joules per Kelvin per mole (J K^{-1} mol^{-1}). For 1 mol of gas, the combined relation is $PV = RT$. If there are n moles of gas, the proportionality constant must be nR rather than R, and thereby

$$PV = nRT \tag{2.7}$$

Equation 2.7 is called the **ideal gas law**, and the reasons for this name will emerge as we consider the range of behavior of real gases.

The relationship in Equation 2.7 is a functional relationship among three variables, two of which are independent. When we choose the temperature and pressure for a system, the volume is dictated by the relationship. If we choose pressure and volume, then there is only one temperature at which a system obeying Equation 2.7 can be found to exist.

How can we verify Equation 2.7 by experiment? One possibility is to use the piston assembly shown in Figure 2.1. Let us assume that we can immerse the whole system in a heat bath that can be maintained at any desired temperature. We start out selecting one temperature, T_1, for the bath and the piston system. As we load mass onto the piston, we can measure the volume of the sample. That is, at T_1 we can measure several (P, V) data points. According to Equation 2.7, as long as the temperature is constant, the product of P and V is constant. Thus, a plot of the data at T_1 will be the same as an x–y plot of the curve $xy = c$, where c is a constant. If we repeat the experiment at $T_2 = T_1 + 10$, we will expect to find a similar curve for our (P, V) data points, though it is a curve that will be shifted from the first. A series of curves generated in this manner are pressure–volume isotherms. They can be presented on the same graph, as shown in Figure 2.2.

The isotherm plot in Figure 2.2 is equivalent to a contour plot for some three-dimensional function. If we had a relationship among variables in Cartesian geometrical space, or (x, y, z) space, such that $z = f(x, y)$, a graph of z versus x and y could be given with z as the height above the x–y plane. Contours are simply lines of constant z. Equation 2.7 is this type of relation, and we can think of T the way we think of z, with P and V being like x and y.

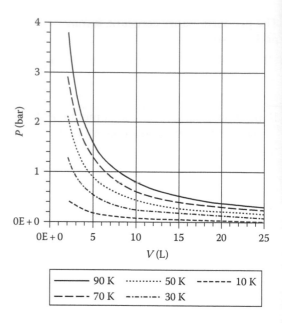

FIGURE 2.2

Pressure–volume isotherms for an ideal gas. Pressure is given by the vertical axis, and volume is given by the horizontal axis. Each curve shows the possible pressure and volume as they vary together at one fixed temperature, those temperatures being 10, 30, 50, 70, and 90 K.

The isotherm or contour data can also be viewed as a three-dimensional display, and this is shown in Figure 2.3. This representation of Equation 2.7 is that of one-quarter of the outside of a smooth, inverted bowl.

Does air or a sample of pure nitrogen or a sample of pure argon display the regular isotherms in Figure 2.2 (Equation 2.7) over all ranges of temperature, pressure, and volume? The strict answer according to careful measurement is "no"; however, Equation 2.7 is a very good approximation for real gases at high temperature and/or low pressure. The only gas for which Equation 2.7 holds for all temperatures and pressures is a hypothetical gas called an ideal gas. The hypothetical behavior of perfectly obeying Equation 2.7 is considered to be the ideal by virtue of the mathematical simplicity of the *PVT* relationship.

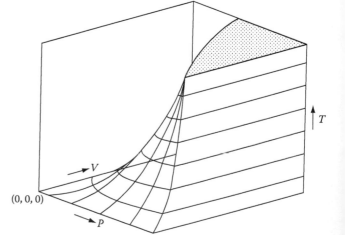

FIGURE 2.3

Three-dimensional representation of Equation 2.7. Pressure and volume are the *x* and *y* variables, and temperature is the vertical or *z* direction. The surface depicts how temperature varies with pressure and volume according to Equation 2.7.

2.2 Collisions and Pressure

Gas kinetics is the analysis of the mechanics of gas particle collisions, and it provides a fundamental, molecular level basis for the relationships involving pressure that we have just considered. To carry out the analysis at the molecular level, we need to consider the distribution law discussed in the last chapter, Equation 1.38:

$$n_i = \frac{N}{Q} e^{-\varepsilon_i/kT}$$

Q, the partition function, is specific to the system under consideration. The analysis in Chapter 1 determined Q to within a proportionality constant, C, for ideal monatomic gas particles: $Q = C(2\pi mkT)^{3/2}$ according to Equation 1.34. The distribution law can give the average number of particles that possess a given discrete energy, ε_i, but for the gas kinetic analysis needed here, the distribution needs to be expressed in terms of other values such as velocity components, momentum, or speed.

A classical monatomic particle's kinetic energy in terms of velocity components v_x, v_y, and v_z, is

$$\varepsilon = \frac{1}{2} m \left(v_x^2 + v_y^2 + v_z^2 \right) \tag{2.8}$$

This is used in Equation 1.37 along with the Q to give the number of molecules, n, in a sample of N molecules at temperature T that has the energy ε, which in turn is a function of the velocity components.

$$n\left(\varepsilon(v_x, v_y, v_z)\right) = \frac{N}{C(2\pi mkt)^{3/2}} e^{-m(v_x^2 + v_y^2 + v_z^2)/2kT} \tag{2.9}$$

The quantity $n(\varepsilon)/N$ is the fraction of molecules with the energy ε. It is the probability that a molecule in the sample will have that energy. It can be expressed as a product of functions.

$$\frac{n(\varepsilon)}{N} = f(v_x)f(v_y)f(v_z) \tag{2.10}$$

$$f(v_i) = \frac{C^{-1/3}}{\sqrt{2\pi mkT}} e^{-mv_i^2/2kT} \tag{2.11}$$

At this point, we shall define a function, ρ, to be a **probability density**. It gives the probability per unit velocity.

$$\rho(v_i) = \frac{f(v_i)}{\displaystyle\int_{-\infty}^{\infty} f(v_i)dv_i} \tag{2.12a}$$

$$= \sqrt{\frac{m}{2\pi kT}} e^{-mv_i^2/2kT} \tag{2.12b}$$

The differential probability, as opposed to the probability density, for the ith component of the velocity to be between the values v_i and $v_i + dv_i$ is $\rho(v_i)dv_i$. A net probability results from integrating the differential probability over some range in the coordinates. The definition of ρ in Equation 2.12a insures that

$$\int_{-\infty}^{\infty} \rho(v_i)dv_i = 1$$

that is, there is unit probability (no chance of anything else) that a particle's velocity component will be between $-\infty$ and ∞. Equation 2.12b follows from carrying out the integration in Equation 2.12a. Notice that the probability density is independent of the sign of the velocity component. There is an identical probability of moving in one direction as of moving in exactly the opposite direction.

The mean or average value of some variable (observable quantity) is obtained from a known distribution by multiplying that variable by the corresponding differential probability and integrating. For instance, the mean or average value of v_x, denoted $\langle v_x \rangle$, is obtained by multiplying v_x by $\rho(v_x)dv_x$ and integrating over the entire range of v_x:

$$\langle v_x \rangle = \int_{-\infty}^{\infty} v_x \rho(v_x)dx = \frac{1}{\sqrt{2\pi}}\left(\frac{m}{kT}\right)^{1/2}\int_{-\infty}^{\infty} v_x e^{-mv_x^2/2kT}dv_x = 0 \tag{2.13}$$

This mean value turns out to be zero, a result consistent with our expectation that at any instant there is as much of a chance for any one particle to be moving to the left as for it to be moving to the right. A nonzero result is obtained for $\langle v_x^2 \rangle$.

$$\langle v_x^2 \rangle = \sqrt{\frac{m}{2\pi kT}}\int_{-\infty}^{\infty} v_x^2 e^{-mv_x^2/2kT}dv_x = \sqrt{\frac{m}{2\pi kT}}\frac{1}{2}\sqrt{\frac{8\pi k^3 T^3}{m^3}} = \frac{kT}{m} \tag{2.14}$$

The result in Equation 2.14 is the same as Equation 1.41.

The distribution and mean values of **speed**, the magnitude of the velocity vector, are a useful characterization of gas behavior. Since speed, s, has no associated direction, a change of variables that separates the orientation (direction) of velocity from its magnitude is helpful. Hence, three coordinates, s, θ, and φ, are now defined in relation to velocity vector components, as shown in Figure 2.4.

$$v_x = s\sin\theta\cos\varphi$$

$$v_y = s\sin\theta\sin\varphi \tag{2.15}$$

$$v_z = s\cos\theta$$

This substitution for v_x, v_y, and v_z in the differential probability leads to

$$\rho(v_x)\rho(v_y)\rho(v_z)dv_x dv_y dv_z = \left(\frac{m}{2\pi kT}\right)^{3/2} e^{-ms^2/2kT} s^2\,ds\sin\theta\,d\theta\,d\varphi \tag{2.16}$$

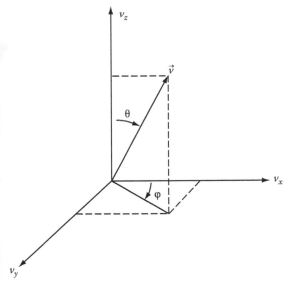

FIGURE 2.4
For three orthogonal axes, v_x, v_y, v_z, coordinates, s, q, f, are defined such that s is the magnitude of a vector in (v_x, v_y, v_z) space, q is the elevation angle relative to the $v_x - v_y$ plane ($q = 0$ corresponds to a vector pointing perpendicular and up from the $v_x - v_y$ plane), and f is the angle about the v_z-axis.

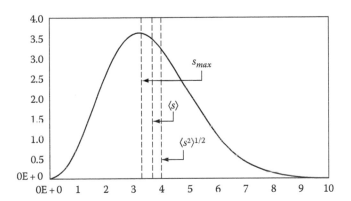

FIGURE 2.5
The functional form of the distribution of speeds, s (horizontal axis), of ideal gas particles. This is simply a plot of the function $f(s) = s^2 \exp(-\alpha s^2)$ (f on the vertical axis) versus s with an arbitrarily chosen value of $\alpha = m/2kT$. The three vertical lines mark, from left to right, the s-values of the most probable speed, the average speed, and the root-mean-squared speed. Notice that this function is not symmetrical about its maximum, and as a consequence, the average speed is somewhat greater than the most probable speed.

The s dependence in this expression is shown in Figure 2.5. Multiplying this differential probability by a particular value, such as s or s^2, and then integrating yields the mean of the particular value. So, the **mean speed**, $\langle s \rangle$, is

$$\langle s \rangle = \left(\frac{m}{2\pi kT}\right)^{3/2} \int_0^\infty s^3 e^{-ms^2/2kT}\, ds \int_0^\pi \sin\theta\, d\theta \int_0^{2\pi} d\varphi$$

$$= \left(\frac{m}{2\pi kT}\right)^{3/2} \frac{1}{2} \frac{4k^2 T^2}{m^2} (2)(2\pi) = \sqrt{\frac{8kT}{\pi m}} \tag{2.17}$$

The mean speed squared, $\langle s^2 \rangle$, obtained this way yields the same result as in Equation 1.42.

Example 2.1: Most Probable Speed of Gas Particles

Problem: Find an expression for the most probable speed of ideal gas atoms. Assuming ideal behavior, what is the most probable speed for argon atoms at 300 K?

Solution: The distribution of particle speeds is given by the s dependence in the differential probability expression, Equation 2.16. That is,

$$f(s) = s^2 e^{-ms^2/2kT}$$

The most probable speed is the speed for which $f(s)$ is at its maximum value. To find the maximum of f, we evaluate the first derivative of $f(s)$ and then find the specific value of s, designated s_{max}, at which the derivative is zero (see Appendix A) and at which f is maximized.

$$\frac{d}{ds} f(s) = 2s e^{-ms^2/2kT} - \frac{m}{kT} s^3 e^{-ms^2/2kT}$$

$$\frac{d}{ds} f(s_{max}) = 0 \Rightarrow \left(2s_{max} - \frac{m}{kT} s_{max}^3 \right) e^{-ms^2/2kT} = 0$$

$$2 - \frac{m}{kT} s_{max}^2 = 0 \Rightarrow s_{max} = \sqrt{\frac{2kT}{m}}$$

Argon atoms have a mass of 39.96 amu (Appendix E). A value in grams is obtained by dividing this by Avogadro's number. With $T = 300$ K,

$$\sqrt{\frac{2kT}{m}} = \sqrt{\frac{2(1.381 \times 10^{-23}\,\text{J K}^{-1})(300\,\text{K})}{39.96 \times 10^{-3}\,\text{kg}/6.022 \times 10^{23}}} = 353\,\text{m s}^{-1}$$

So, the most probable speed for argon atoms at 300 K is 353 m s^{-1}. If N$_2$ molecules were taken to be ideal gas particles, the most probable speed would be greater, 422 m s^{-1}, because of the smaller mass of N$_2$ compared to that of Ar. Notice that the most probable speed is less than the mean speed and the root-mean-squared speed (Figure 2.5).

The internal pressure of a collection of atoms or molecules is a macroscopic manifestation of numerous collisions, specifically collisions of the particles with the walls of a vessel. Because ideal gas particles exhibit no interaction potential with each other, their collisions are perfectly elastic. An **elastic collision** is one in which no energy is converted to internal energy of the atoms or molecules involved in the collision. Elasticity may be contrasted with **inelasticity** in mechanical systems by considering a hypothetical tennis ball with the property of being perfectly elastic. If this ball were dropped from a height of 2 m onto a perfectly rigid hard surface, it would bounce back up to its original height of 2 m. Its kinetic energy following the impact would be the same, though its direction of motion has been reversed. A real tennis ball shows different behavior because it bounces back to less than its original height. The ball flexes when it hits, and some of the energy is converted to internal energy, mostly heat, in the ball. (You might be able to demonstrate this conversion by rapidly bouncing a tennis ball off of a wall and noticing that it warms.) A collision between two particles, rather than between a particle and an unmovable wall, is elastic if the total kinetic energy of the two particles is unchanged. About the closest one comes to seeing truly elastic collisions between particles in our everyday world is on a billiard table

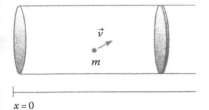

$\vec{v}$

m

$x = 0$

FIGURE 2.6
A hypothetical system consisting of a vessel with a wall that can move freely to any position along the x-axis. Inside the vessel is only one moving particle.

or an air hockey table. As objects on these tables collide with one another, the total kinetic energy is almost unchanged; that is, the sum of $mv^2/2$ (one-half mass times the square of velocity) for all the objects remains (almost) the same even though velocities of individual objects change.

A gas particle that collides with a wall may transfer momentum and kinetic energy to the wall if the wall can move. Let us consider a vessel with one wall of mass M that can move in the x direction as shown in Figure 2.6. Within the vessel is one gas particle of mass m moving with velocity components v_x, v_y, and v_z. At some instant, the gas particle strikes the wall, and assuming an elastic collision, we know that momentum and kinetic energy are conserved. Let V_x be the wall's velocity after the collision, and v'_x, v'_y, and v'_z be the particle's velocity components. Equating the components of momentum before and after the collision yields

$$mv_x = mv'_x + MV_x \tag{2.18a}$$

$$mv_y = mv'_y \tag{2.18b}$$

$$mv_z = mv'_z \tag{2.18c}$$

This shows that only the x-component of the particle's velocity changes. To determine that change, we equate the kinetic energy before and after the collision.

$$\frac{1}{2}m\left(v_x^2 + v_y^2 + v_z^2\right) = \frac{1}{2}MV_x^2 + \frac{1}{2}m\left(v_x'^2 + v_y'^2 + v_z'^2\right) \tag{2.19}$$

$$\therefore mv_x^2 = MV_x^2 + mv_x'^2 \tag{2.20}$$

Equation 2.18a with Equation 2.20 determines the unknown velocity components after collision in terms of v_x.

$$V_x = \frac{m}{M}\left(v_x - v'_x\right) \tag{2.21}$$

$$v'_x = \frac{m - M}{m + M}v_x \tag{2.22}$$

The velocity of the gas particle will be diminished, which is consistent with energy having been transferred.

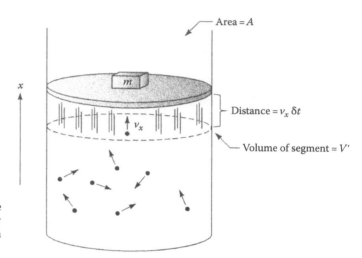

FIGURE 2.7
With a piston area of A, the volume that includes all particles with velocity component v_x that will hit the piston in time dt is $V' = Av_x\delta t$.

The hypothetical system in Figure 2.6 is not entirely relevant to the thermodynamics of a gas because there is only one particle. Also, the vessel wall moves to the right with a velocity according to Equation 2.22 and does not stop. To devise a more realistic picture, we need to allow for there being N particles in the vessel, and we need to change the vessel to be a piston type of system as in Figure 2.1. Then, for a large number of particles colliding with the wall, the average force depends on the average number of particles hitting the wall per unit time. Figure 2.7 helps illustrate how this can be determined. If a particle has a velocity v_x, then it will hit the wall in some small increment of time δt if $v_x\delta t$ is greater than or equal to its distance from the wall. Thus, any particle within a hypothetical section of the piston system, with a volume, V', equal to the area of the piston, A, times the distance from the piston, $v_x\delta t$, will collide with the piston provided the velocity is v_x or greater. How many particles are in this volume with that velocity? Equation 2.12 gives the probability density, $\rho(v_x)$, that any one particle will have a given x-component velocity. That becomes a probability when it is multiplied by the differential in velocity, dv_x, and integrated. Then, that probability times the number of particles in the volume V' is the answer to the question. And so how many particles are in the volume V'? That is the number of particles in the entire vessel, N, times the ratio of V' to the total volume V. Therefore, the following is the number of particles with velocity v_x in the volume V', now using an infinitesimal increment of time in place of δt.

$$\rho(v_x)dv_xN\frac{V'}{V} = \rho(v_x)\frac{N}{V}v_xAdtdv_x \tag{2.23}$$

The next step is to account for the momentum transfer due to all the collisions.
Each colliding particle experiences a change in momentum of $2mv_x$, where m is the mass of the gas particle. (The initial momentum in the x direction for a particle moving in the $+x$ direction is mv_x, and so on colliding with the wall, its momentum changes to $-mv_x$ for a net change of $2mv_x$.) Multiplying this by the quantity in Equation 2.23 gives dX, the differential momentum change for particles with velocity v_x in the time interval dt.

$$dX = 2mv_x\rho(v_x)\frac{N}{V}v_xAdtdv_x \tag{2.24}$$

The force is the momentum change per unit time, and thus, the differential force exerted by particles with velocity v_x is simply dX divided by dt. If this is integrated over all velocities, the result will be the total force exerted by the colliding particles on the wall.

$$F = 2m\frac{NA}{V}\int_0^\infty v_x^2 \rho(v_x)dv_x \qquad (2.25)$$

The pressure is the force divided by the area of the piston, A.

$$P = 2m\frac{N}{V}\int_0^\infty v_x^2 \rho(v_x)dv_x \qquad (2.26)$$

The integral can be evaluated using Equation 2.12.

The result of integration over the x component of velocity gives the following expression for the pressure.

$$P = 2m\frac{N}{V}\sqrt{\frac{m}{2\pi kT}}\int_0^\infty v_x^2 e^{-mv_x^2/2kT}dv_x = 2m\frac{N}{V}\frac{kT}{2m} = \frac{NkT}{V} \qquad (2.27)$$

Notice that from this detailed analysis of particles in a piston system obeying a velocity distribution law, we achieve the ideal gas equation of state, $PV = NkT = nRT$.

2.3 Nonideal Behavior

Deviations from ideal behavior are related to the atomic and molecular nature of a gas and to the existence of intermolecular forces. There are fundamentally two assumptions undertaken with the ideal gas equation of state. It is assumed that the gas particles have no volume and that there are no interactions between the gas particles. For instance, at $T = 0$, the ideal gas law implies $PV = 0$, which for any fixed pressure means $V = 0$. Atoms and molecules are not hypothetical points in space that can collapse into a zero-volume existence; their volume does not approach zero as the temperature approaches absolute zero. Thus, a proper description of a real gas requires a correction to the ideal gas law. As one example of such a correction, we could introduce a constant, V_0, that is the gas volume at $T = 0$. This means $P(V - V_0) = nRT$. Another difference is that ideal gas particles exhibit no attraction, whereas real atoms and molecules do. So, in many ways, the PVT behavior of a real gas tends to differ from that of an ideal gas, and this is manifested in different forms for $P-V$ isotherms.

The most complete knowledge about the behavior of some real gas offers the same predictive capability as for an ideal gas, that is, a mathematical relationship among P, V, and T. Finding such a formula for a particular real gas might be accomplished through measurement of numerous P, V, and T values for a given sample of the gas. Such a PVT data set might be used in a curve fitting procedure to extract a formula relating P, V, and T (see Appendix A). Here is one approach. For each different laboratory data point (P_i, V_i, T_i), let us state

$$X_i = P_iV_i - nRT_i \tag{2.28}$$

If the gas were ideal, all X_i values would be zero, and so X represents the nonideality of the real gas. We can regard each X_i as the value of some function $X(P, V, T)$ at the point (P_i, V_i, T_i), and then the objective in fitting is to find that function. There is a mathematical basis for claiming that this function in general can be written as an infinite power series of the three variables, though there may be simpler expressions that are more suitable.

$$X = c_{000} + c_{100}P + c_{200}P^2 + \cdots$$

$$+ c_{010}V + c_{020}V^2 + \cdots$$

$$+ c_{001}T + c_{002}T^2 + \cdots$$

$$+ c_{110}PV + c_{101}PT + c_{011}VT + \cdots$$

where
    the cs are adjustable coefficients obtained from fitting the PVT data
    c_{ijk} is the coefficient associated with a term that is to the ith power in P, the jth power in V, and the kth power in T

Appendix A includes a section on curve fitting that outlines a means to find the cs for a *truncated* power series expansion. We have, then, a general approach for finding an expression for X in terms of P, V, and T. When that is obtained, we return to Equation 2.28 and write $X = PV - nRT$, now using the fitting expression in P, V, and T in place of X. Such a result is a *real* gas law based on the PVT laboratory data that are available.

The general power series expansion often proves to be more elaborate than is necessary. For some gases, the deviation from ideality depends mostly on the pressure, and then it is appropriate to limit the general power series expansion to a simpler form:

$$X = c_1P + c_2P^2 + c_3P^3 + \cdots$$

If we truncate this series at the cubic (P^3) term, there are but three coefficients to obtain via techniques such as those explained in Appendix A. Once these are found from laboratory data and curve fitting, we will have an equation for the gas that follows from Equation 2.28.

$$PV = nRT + c_1P + c_2P^2 + c_3P^3 \tag{2.29}$$

The cs are often referred to as **virial coefficients**, and this is the type of predictive relationship one seeks. By specifying n, P, and T, Equation 2.29 yields V for the real gas being studied. The accuracy of the value predicted for V depends on the accuracy of the laboratory data used in the fitting (finding the virial coefficients) and on the degree to which the truncated expansion can represent the data. Usually, this means that the function is reliable in the temperature, volume, and pressure regions for which there were data.

There are a number of relationships that have been found to be concise and accurate functional forms for representing the PVT behavior of real gases over limited P and T ranges. Some of these are listed in Table 2.1. The lowercase letters in these equations (a, b, c, and so on) are coefficients that must be obtained from PVT data for a given gas. The same

TABLE 2.1

Equations of State for Gases

Ideal gas equation

$PV = nRT$ (2.7)

Berthelot gas equation

$PV = nRT + aP\left(1 - \dfrac{b}{T^2}\right)$ (2.30)

van der Waals gas equation

$\left(p + a\dfrac{n^2}{V^2}\right)(V - nb) = nRT$ (2.31)

Others

$PV = n(RT + bP)$ (2.32)

$PV = nRT + c_1P + c_2P^2 + c_3P^3$ (2.29)

$\dfrac{PV}{nRT} = 1 + \dfrac{b}{V} + \dfrac{c}{V^2} + \dfrac{d}{V^3}$ (2.33)

$\dfrac{PV}{nRT} = 1 + \dfrac{b}{V} + \dfrac{c}{V^2} - \dfrac{d}{VT} - \dfrac{e}{V^2T} - \dfrac{f}{VT^3}$ (2.34)

set of coefficient letters are used throughout, but this does not mean that the coefficients are transferable from one equation to another. These expressions are called **equations of state** for gases. Table 2.2 gives representative values for these coefficients for the widely used van der Waals equation of state.

2.4 Thermodynamic State Functions

A gas sample or other substance found to be at a certain temperature, pressure, and volume and not changing in time is in a particular **thermodynamic state**. If the pressure and volume are changed, then the state of the system will be changed. In general, existence in a thermodynamic state refers to a thermodynamic system (not just gases) with **thermodynamic state functions** that have definite values that do not change with time. A change of state is accompanied by changes in the thermodynamic state functions, with such changes being independent of the path by which the change is made to a system. As a simple way of understanding this last defining feature of a thermodynamic state function, consider several people hiking from one spot on a plateau to another designated spot on a nearby mountain. Each may find different paths to hike, and if we asked the distance each hiked, there would be a number of different values. On the other hand, if we asked what their net change in elevation was, the results would all be the same. The change in elevation is analogous to a state function, but the distance traveled is not.

Pressure, volume, and temperature are all state functions, and that is why the expressions given in Table 2.1 are termed equations of state. They are equations that relate the various state functions. We can also regard pressure, volume, and temperature as variables;

TABLE 2.2

Coefficients in the van der Waals Equation
of State for Selected Real Gases

Gas	a (bar L^2 mol^{-2})	b (L mol^{-1})
He	0.0346	0.0238
Ne	0.208	0.0167
Ar	1.355	0.0320
Kr	5.193	0.0106
Xe	4.192	0.0516
H_2	0.2453	0.0265
N_2	1.370	0.0387
O_2	1.382	0.0319
Cl_2	6.343	0.0542
O_3	3.570	0.0487
NH_3	4.225	0.0371
H_2O	5.537	0.0305
H_2S	4.544	0.0434
CH_4	2.303	0.0431
C_2H_6	5.580	0.0651
C_3H_8	9.39	0.0905
C_2H_2	4.516	0.0522
C_2H_4	4.612	0.0582
C_6H_6	18.82	0.1193
CH_3OH	9.476	0.0659
CO	1.472	0.0395
CO_2	3.658	0.0429
CS_2	11.25	0.0726
CCl_4	20.01	0.1281
NO	1.46	0.0289
NO_2	5.36	0.0443
N_2O	3.852	0.0444
SO_2	6.865	0.0568
SiH_4	4.38	0.0579

Source: Values from Lide, D.R., *The Handbook of Chemistry and Physics*, 86th edn., CRC Press, Boca Raton, FL, 2005.

when pressure is expressed as a function of volume and temperature, for example, $P(V, T)$, then V and T are variables of the function P. There are two kinds of variables. **Intensive state variables**, such as temperature and pressure, are those that would be the same if the system were replicated, whereas **extensive state variables**, such as volume, are those that would increase in proportion to the number of times the system was replicated. When we have a gas sample at a certain temperature, pressure, and volume, replication means forming an exact copy of the system and combining it with the original system. So the volume would be doubled, but the temperature and pressure would be the same. Another way in which intensive and extensive variables differ is that intensive variables are not related to the number of particles, whereas extensive variables are.

In general, there are a certain number of state variables that may be adjusted independently. For pure gases, we already know that that number is 2. We may be able to pick a pressure and temperature for a gas sample at hand, but the sample will then exist at one unique volume, that given by the appropriate equation of state. Any function we construct entirely from thermodynamic state variables is necessarily a thermodynamic state function. For instance, P and V are independent of the path taken to reach a given state, and so is the function PV, the function P^2/V, the function $P + V$, and so on.

Mathematical relationships among thermodynamic state functions—how one or more state functions vary with another—are the very workings and utility of thermodynamics. In some cases, it is of interest only to know how one function varies with a small change in another, and then it is derivatives of the function that receive attention. Note that in analytical geometry, the standard expression for a straight line on an x–y graph is $y = mx + b$. The slope, m, is the value that tells how y changes with a change in x. The slope is also the first derivative with respect to x of the function $mx + b$. For more complicated functions of x, the first derivative gives the slope of a straight line tangent to the function. For sufficiently small changes in x from some point, the tangent line at that point is a good approximation to the true curve. The first derivative at the point gives the linear dependence of the function (y) on the variable (x) in the vicinity of the point. Likewise, the second derivative gives the quadratic dependence of the function on the variable (see Appendix A). Partial differentiation of an equation of state of some thermodynamic system gives analogous information. The first derivative of P with respect to T is the "slope" of pressure with respect to temperature, a value that gives the linear dependence of pressure on temperature.

Throughout our development of chemical thermodynamics, we will be considering various partial derivatives of thermodynamic state functions. Two of these can be introduced now as examples. The first is often designated α, and it is the **constant-pressure coefficient of thermal expansion.**

$$\alpha = \frac{1}{V}\left(\frac{\partial V}{\partial T}\right)_P \tag{2.35}$$

The value of α for a particular substance tells how volume depends on temperature at a fixed pressure. Usually, α is a positive value meaning that volume increases with increasing temperature. How volume changes with pressure at a fixed temperature is the isothermal compressibility, β.

$$\beta = -\frac{1}{V}\left(\frac{\partial V}{\partial P}\right)_T \tag{2.36}$$

With this definition, the isothermal compressibility has a positive value corresponding to volume decreasing with increasing pressure.

2.5 Energy and Thermodynamic Relations

Conservation of total energy is a law of mechanical systems, and it serves as the first law of thermodynamics. From the conservation of energy of a closed, isolated system, we can identify certain energy quantities that are state functions; their values do not depend on the path the system followed to reach its current state.

There are many sources of energy in a chemical system and many modes of energy change. There are kinetic energy of the moving atoms and molecules, potential energy from their interaction with each other, energy of interaction with electromagnetic radiation, and energy stored in an atom or molecule because of its quantum mechanical state. We will identify the total of all the energy stored within a closed system as U, the **internal energy**. This excludes energy of incident electromagnetic radiation before it enters from outside the system, and it excludes kinetic energy of motion of the mass center of the entire system since that is energy not within the system.

The internal energy, U, is a state function and can be expressed as a function of two other thermodynamic state variables for a gas sample of uniform composition. We might have $U(P, V)$, or $U(T, V)$, and so on. The gas equation of state prescribes the entire behavior of a gas, and so U is implicitly determined by a PVT equation of state. The analysis of velocity distributions in Chapter 1 gave the internal energy of an ideal monatomic gas as $U = 3nRT/2$ (Equation 1.36). For real gases, U is almost always a somewhat different expression. As we have already considered, the ideal gas law implies that this hypothetical substance is infinitely compressible: As P is increased, V diminishes. We interpret this to mean that the particles that make up the ideal gas have no intrinsic volume. In contrast, consider a collection of glass marbles floating weightlessly inside a vessel in some interstellar environment. If the marbles have some kinetic energy and if they collide elastically with each other (i.e., without loss of kinetic energy), then via collisions with the box, a pressure is exerted that balances the external pressure (squeezing on the box). If the external pressure is increased, the volume of the box may be diminished; however, this will continue only so far. Once the marbles are tightly packed so that they are in contact with each other, the pressure required for a further volume reduction will be much more substantial. The intrinsic volume of the marbles, which simply refers to their strongly repulsive short-range potential, prevents the system from being infinitely compressible. The ideal gas has no short-range interaction potential. Furthermore, there is no long-range interaction, even though this is found for real particles.

Example 2.2: Isothermal Compressibility

Problem: Find an expression for the isothermal compressibility of a gas that obeys Equation 2.30, a Berthelot gas.

Solution: The isothermal compressibility, β in Equation 2.36, requires finding the derivative of V with respect to P at constant temperature. We start with the equation of state for the Berthelot gas, Equation 2.30.

$$PV = nRT + aP\left(1 - \frac{b}{T^2}\right) \quad \text{or} \quad V = \frac{nRT}{P} + a - \frac{ab}{T^2}$$

Partial differentiation yields

$$\left(\frac{\partial V}{\partial P}\right)_T = -\frac{nRT}{P^2}$$

Substituting this into Equation 2.36 yields β for a Berthelot gas.

$$\beta = \frac{nRT}{VP^2}$$

This answer may be sufficient; however, we could go back to the original equation of state and use it to express β in terms of only two variables, not three. From taking the inverse of the second form of the equation of state, $1/V$ is expressed in terms of P and T.

$$\frac{1}{V} = \frac{PT^2}{aPT^2 - abP + nRT^3}$$

Factoring $1/V$ from $nRT/(VP^2)$ and substituting via the previous equation yields β. Therefore,

$$\beta = \frac{nRT^3}{aP^2(T^2 - b) + nRT^3 P}$$

The compressibility of a gas can be measured, and by doing so at different pressures and temperatures, the resulting set of (β, P, T) data can be used to find the coefficients a and b. In other words, measurement of the isothermal compressibility can be used to determine the equation of state, in this case, for a gas that is well represented by Equation 2.30.

It is not always easy or even possible to measure U directly. Also, different experiments hold different variables constant in measuring energy changes, and thus, in thermodynamics there are several "energies" or energy state functions that are used along with U. We shall start by simply defining these energies, and then later we will see how they are used and how they are useful. At this stage, we are simply making up thermodynamic state functions by the combination of thermodynamic state functions already defined. The **enthalpy** function is H.

$$H = U + PV \tag{2.37}$$

The **Helmholtz energy** function (Hermann Ludwig Ferdinand von Helmholtz, Germany, 1821–1894) is A.

$$A = U - TS \tag{2.38}$$

The **Gibbs energy** function (Josiah Willard Gibbs, United States, 1839–1903) is G.

$$G = A + PV = H - TS \tag{2.39}$$

The Helmholtz and Gibbs energies are often referred to as **free energies**. Both are defined using the entropy, S. In Chapter 1, we developed a molecular level definition of the absolute entropy. From that definition, it should be clear that entropy is a thermodynamic state function, too. Therefore, A and G are state functions.

For pure gas samples of fixed mass, the number of independent variables is 2. Of P, V, and T, the equations of state in Table 2.1 can be used to obtain the third from the other two. The introduction of various energy state functions does not change this. There are still two independent state variables. This fact will help us establish relations among the derivatives of the state functions, including U, H, A, and G. We will start with a general, solely mathematical treatment of the relations and then consider certain physical implications.

The differential of some function $f(x,y)$ in terms of the differentials in x and y is the following (see Appendix A).

$$df = \left(\frac{\partial f}{\partial x}\right)_y dx + \left(\frac{\partial f}{\partial y}\right)_x dy \qquad (2.40)$$

We can use this to express the differential of any thermodynamic state function in terms of any two other variables. For instance, by stating $P = P(T, V)$, we have

$$dP = \left(\frac{\partial P}{\partial T}\right)_V dT + \left(\frac{\partial P}{\partial V}\right)_T dV \qquad (2.41)$$

Or, taking H to be a function of P and V gives

$$dH = \left(\frac{\partial H}{\partial P}\right)_V dP + \left(\frac{\partial H}{\partial V}\right)_P dV \qquad (2.42)$$

Furthermore, S, H, G, A, and U are all state functions (variables), and we can select from them, as well as from P, V, and T, two that will be taken as the independent variables of some other function. So, for example, we are free to state that $A = A(T,S)$ just as much as we can state that $A = A(U,P)$. Obviously then, quite a number of differential expressions, such as those in Equations 2.41 and 2.42, can be written.

Starting from differential relations, we can carry out various mathematical manipulations (see Appendix A) to obtain any partial derivative, $(\partial X/\partial Y)_Z$, where X, Y, and Z are any three different elements of the set $\{P, V, T, S, H, U, G, A\}$. However, not all of these correspond to the way behavior is usually observed. For instance, we do not usually expect to measure how pressure varies with free energy under the condition of constant entropy. The relations of greatest interest are those that involve certain natural choices of variables. These happen to be $H = H(S,P)$, $U = U(S,V)$, $A = A(T,V)$, and $G = G(T,P)$; other choices will be of interest as well.

Pressure and volume are natural choices for dependent variables of energy functions because two types of calorimetry experiments measure energy change, one with fixed pressure and one with fixed volume. The fixed-pressure system is usually an open system so that the ambient air pressure is the fixed pressure, whereas fixed-volume experiments are performed with a closed calorimeter. With volume or pressure as one of the dependent variables, and with that variable fixed by experimental conditions, measurements can be made subject to only one variable. They can be repeated for different choices of the fixed variable.

There is a certain symmetry in the so-called natural choices of dependent variables, and a simple diagram can be used to generate all the derivative relations very concisely. That is, we can use a diagram to look up a derivative relationship rather than carrying out the calculus steps ourselves. We can arrange all the variables on a circle, as shown in Figure 2.8, placing the energy state functions outside the circle and arranging the others inside such that those on either side of an energy are the natural choices. Thus, the natural

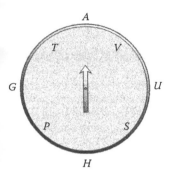

FIGURE 2.8
A graphical device, the "thermodynamic compass," for identifying the so-called natural choices of dependent variables and for generating thermodynamic relations without derivation. With the rules given in the discussion, this graph encodes all the Maxwell relations of thermodynamics.

choice of dependent variables for A is V and T. It is often useful to express the differential of A in terms of differentials in T and V.

$$dA = \left(\frac{\partial A}{\partial T}\right)_V dT + \left(\frac{\partial A}{\partial V}\right)_T dV \qquad (2.43)$$

Using this expression to determine precisely how A varies with an incremental change in volume or an incremental change in temperature requires that we know $(\partial A/\partial T)_V$ and $(\partial A/\partial V)_T$. In the absence of an explicit expression for $A(V,T)$, we must find some equivalent derivatives that can be evaluated via the equation of state for the gas. It turns out that the circle in Figure 2.8 can yield simple expressions for these partial derivatives and for all others we generate, but first let us examine how these expressions are derived.

A crucial relation to establish involves the differential of the energy, dU. In Chapter 1, entropy was related to energy for a system of one molecule in a heat bath. This is actually a system of constant volume, and the relationship $\Delta U = T\Delta S$ (E was used in Chapter 1 for what we now call U) is equivalent to the differential relationship $dU_V = TdS_V$. This tells us that

$$\left(\frac{\partial U}{\partial S}\right)_V = T \qquad (2.44)$$

$$\left(\frac{\partial S}{\partial U}\right)_V = \frac{1}{T} \qquad (2.45)$$

We also know from the analysis of configurations, entropy, and volume in Chapter 1 that the change in entropy (for an isothermal or constant-temperature expansion of an ideal gas) is $\Delta S = nR\Delta(\ln V)$ as in Equation 1.7. The corresponding differential expression is $dS_T = nRd \ln V_T$. Since $d \ln x = dx/x$, this yields

$$\left(\frac{\partial S}{\partial V}\right)_T = \frac{nR}{V} = \frac{P}{T} \qquad (2.46)$$

where the rightmost expression results from invoking the ideal gas law in the form $nR/V = P/T$. From the distribution law in Equation 1.28, we can establish that the constraint $dU = 0$ is the same as $dT = 0$. Thus, Equation 2.46 provides the derivative for constant U

as well as constant T. Now, we take S to be a function of U and V, a choice required to obtain the derivative of U with respect to V at constant S. With $S = S(U, V)$ and then using Equations 2.45 and 2.46,

$$dS = \left(\frac{\partial S}{\partial U}\right)_V dU + \left(\frac{\partial S}{\partial V}\right)_U dV = \frac{1}{T}dU + \frac{P}{T}dV \tag{2.47}$$

with the last step achieved by substitution using Equations 2.45 and 2.46. Equation 2.47 insures that

$$\left(\frac{\partial U}{\partial V}\right)_S = -P \tag{2.48}$$

This is achieved by setting dS equal to zero (constant S) and multiplying through by T.

The natural choice of dependent variables for U are S and V. Writing the differential of U in terms of these two variables and then substituting according to Equations 2.44 and 2.48 yields an important result that is general, even though ideal gas relations were used to simplify the preceding analysis.

$$dU = \left(\frac{\partial U}{\partial S}\right)_V dS + \left(\frac{\partial U}{\partial V}\right)_S dV \tag{2.49}$$

$$dU = TdS - PdV \tag{2.50}$$

Like expressions can be generated for other energy state functions. For instance, with the enthalpy in terms of U, we can obtain an expression for dH:

$$H = U + PV$$

$$dH = dU + PdV + VdP$$

$$= (TdS - PdV) + PdV + VdP$$

$$= TdS + VdP \tag{2.51}$$

Notice that Equation 2.51 offers a very direct means of expressing dH. Using $H = H(S, P)$ and invoking Equation 2.40 yields

$$dH = \left(\frac{\partial H}{\partial S}\right)_P dS + \left(\frac{\partial H}{\partial P}\right)_S dP \tag{2.52}$$

Example 2.3: Thermodynamic Relations

Problem: Find $(\partial G/\partial T)_P$.

Solution: First, express G in terms of U via Equations 2.38 and 2.39.

$$G = A + PV$$

$$G = U - TS + PV$$

Find the differential of this expression.

$$dG = dU - TdS - SdT + PdV + vdP$$

Equation 2.50, an expression for dU, makes it possible to write the right-hand side in terms of only dT and dP.

$$dG = (TdS - PdV) - TdS - SdT + PdV + VdP$$

$$= -SdT + VdP$$

From that result, we can express partial derivatives for $dT = 0$ or $dP = 0$.

$$\left(\frac{\partial G}{\partial T}\right)_P = -S \quad \text{and} \quad \left(\frac{\partial G}{\partial P}\right)_T = V$$

Comparing Equation 2.51 with Equation 2.52 shows that $(\partial H/\partial S)_P = T$ and $(\partial H/\partial P)_S = V$.

The differential expressions in Equations 2.50 and 2.51, and all other such relations, are encoded in the thermodynamic compass in Figure 2.8. The rule for extracting these relations, or using the compass, is that for a particular partial derivative with respect to a variable inside the circle, we use the variable in the opposite compass direction (e.g., northeast is opposite to southwest), taking the sign to be positive if the direction is partly north (the direction of the arrow in the figure) or using a negative sign if the direction is partly south (opposite the arrow's direction: southeast, south, or southwest). Thus,

$$\left(\frac{\partial A}{\partial T}\right)_V = -S \quad \text{and} \quad \left(\frac{\partial A}{\partial V}\right)_T = -P \tag{2.53}$$

This is exactly what we would obtain by direct derivation from the definitions of the thermodynamic state functions. Thus, instead of Equation 2.43, we write $dA = -SdT - PdV$.

A set of relations identified as **Maxwell relations** can also be obtained from the thermodynamic compass. These are relations among the partial derivatives of the variables within the circle. We can relate any two partial derivatives in the following way.

1. Going clockwise around the compass, select three variables in sequence, such as V, S, and P. This selects the partial derivative of the first (V), with respect to the second (S), with the third (P) held constant, for example, $(\partial V/\partial S)_P$. If the middle variable is on the lower half of the compass (south), make the derivative the negative of itself.

2. From the same starting point from which the first three variables were selected, select three variables in sequence going counterclockwise. This corresponds to the partial derivative of the first variable with respect to the second, while the third is held constant. If the middle variable is south, apply a negative sign.

3. Equate the results of the first two steps to yield a Maxwell relation.

The power and utility of these derivative and differential relationships among thermodynamic state functions can be appreciated by going back to the equations of state.

A particular equation of state from Table 2.1, or some other equation of state, is sufficient to generate all the derivatives among P, V, and T. For instance, let us consider a system described by the following equation of state.

$$P = \frac{nRT}{V - nb} - \frac{n^2 a}{V^2} \tag{2.54}$$

Partial differentiation yields

$$\left(\frac{\partial P}{\partial T} \right)_V = \frac{nR}{V - nb} \tag{2.55}$$

Example 2.4: Thermodynamic Compass

Problem: Find $(\partial S/\partial P)_T$ for a gas that obeys the equation of state, Equation 2.33. Simplify the expression to involve as few thermodynamic variables as possible.

Solution: First, note that the equation of state, Equation 2.33, does not involve S. So direct differentiation of the equation of state does not yield the desired first derivative. Instead, we need an equivalent derivative, and the thermodynamic compass is an aid in finding it. S, P, and T are arranged clockwise on the compass, and counterclockwise from the same starting point are V, T, and P. Thus, the partial derivative of V with respect to T when P is constant is −1 times the derivative we seek. To find this partial derivative, divide the equation of state by V:

$$\frac{P}{nRT} = \frac{1}{V} + \frac{b}{V^2} + \frac{c}{V^3} + \frac{d}{V^4}$$

Next, differentiate with respect to T.

$$-\frac{P}{nRT^2} = -\left(\frac{1}{V^2} + \frac{2b}{V^3} + \frac{3c}{V^4} + \frac{4d}{V^5} \right) \left(\frac{\partial V}{\partial T} \right)_P$$

This can be rearranged to give $(\partial V/\partial T)_P$.

$$-\left(\frac{\partial S}{\partial P} \right)_T = \left(\frac{\partial V}{\partial T} \right)_P = \frac{V^5 \left(P/nRT^2 \right)}{V^3 + 2bV^2 + 3cV + 4d}$$

This equation can be simplified by substituting for P/nRT using the original equation of state.

$$\left(\frac{\partial S}{\partial P} \right)_T = -\frac{1}{T} \frac{V^4 + bV^3 + cV^2 + dV}{V^3 + 2bV^2 + 3cV + 4d}$$

Then, the Maxwell relationship among partial derivatives,

$$\left(\frac{\partial S}{\partial V}\right)_T = \left(\frac{\partial P}{\partial T}\right)_V \tag{2.56}$$

tells us that

$$\left(\frac{\partial S}{\partial V}\right)_T = \frac{nR}{V - nb} \tag{2.57}$$

From the Maxwell relationships we can generate, or extract from the thermodynamic compass, we have complete knowledge about all the other thermodynamic state functions of the system. Obviously, the important ingredient is the equation of state.

The equation of state is generally determined phenomenologically, which means it is extracted from laboratory data as opposed to being developed from fundamental principles. Hence, the forms of equations of state are different given the differences in behavior of real gases. Furthermore, real gases can change state—condense and solidify—whereas the ideal gas is always a gas. The *PVT* behavior of a real gas in the vicinity of a change from gas to liquid or from gas to solid tends to be much more complicated than the ideal gas behavior, and this will be considered in later chapters.

2.6 Conclusions

The temperature of a gas is associated with a distribution of gas particles among available energies. Gas particles collide with one another and exchange kinetic energy. Their collisions with walls of a vessel give rise to the pressure that a gas sample may exert.

The ideal gas of thermodynamics is a fictitious substance that has idealized gas behavior corresponding to the relation $PV = nRT$. A relation between thermodynamic state variables such as the ideal gas law is termed an equation of state. The ideal gas law can serve as a good approximation for real gases at very low pressures and high temperatures, but otherwise equations of state for real gases are more complicated. Such equations of state are established phenomenologically, that is, by measurement of P, V, and T values, for instance, and usually the result holds for a limited range of pressures and temperatures.

The ideal gas law corresponds to particles that have no interaction potential with each other. So, unlike the situation with real gases, there is no long-range attraction and there is no short-range repulsion; the volume of the ideal gas goes to zero as the temperature goes to zero.

Thermodynamic state functions are those that change in a way that does not depend on the path of the change. Pressure, volume, and temperature are state functions, and so is internal energy, U. State functions can also be used as variables of other functions. Functions made up solely of state variables are automatically state functions. Thus, other energy quantities, H, G, and A, which are defined in terms of U, T, P, and V, are also state functions. Via the

Maxwell relations, we can in principle connect all the state functions to the equation of state. The equation of state is the crucial thermodynamic information about a gas.

Conservation of energy is the first law of thermodynamics.

Point of Interest: Intermolecular Interactions

The difference between the ideal gas and any real gas is that a real gas is composed of particles (atoms or molecules) that interact both attractively and repulsively with one another, whereas the ideal gas is composed of hypothetical non-interacting point masses. Not only is intermolecular interaction important for gases, it is responsible for crystal structure, it plays a role in biological processes that take place in water, it gives rise to properties of certain materials, and it has an influence on chemical reactions in liquids. The study of intermolecular interaction began with the study of real gases.

Johannes Diderik van der Waals

It was Johannes Diderik van der Waals of the Netherlands whose work on equations of state captured the essence of interatomic or intermolecular interaction in real gases. van der Waals, who was born in Leiden on November 23, 1837, was a physicist with an early interest in the molecular kinetic theories of his time. His doctoral thesis (1873), which was strongly praised by Maxwell, gave the first satisfactory molecular level explanation of critical phenomena, which is the existence of vapor and liquid of a pure substance as a homogeneous phase. (We will consider critical phenomena in Chapter 4.) The heart of his thinking was the equation of state now identified as the van der Waals equation of state (Equation 2.31). He came to this equation by considering a gas to be composed of spheres (atoms or molecules) with a local density equal to the overall density. An attraction between spheres leads to the a/V^2 term in the van der Waals equation of state, while the parameter b in the equation comes from excluding overlap of the spheres. For this and related contributions, van der Waals received the 1910 Nobel Prize in physics. He died in Amsterdam on March 8, 1923. van der Waals' thinking provided the breakthrough needed for the liquefaction of gases that had previously been thought to be "permanent" gases. That his theories were so effective in application was a strong argument in favor of atomic theory. Though the modern idea of matter being composed of atoms goes back before van der Waals' time, it is fascinating to realize that challenges to atomic theory arose, or defects of atomism were being argued, well up to the end of the nineteenth century. One argument, made during the 1870s, was that the irreversibility of certain thermodynamic processes was incompatible with a molecular kinetic view involving individual collision events since such events are reversible under Newtonian mechanics. Ludwig Boltzmann is credited with resolving this paradox through a statistical treatment of molecular collisions. Another challenge made in the 1890s had to do with a theorem that a mechanical system in a finite volume eventually passes through its initial configuration. Thermodynamic systems did not exhibit this recurrence, but Boltzmann explained that the recurrence was actually possible as a fluctuation in the system. The probability of such a fluctuation, though, could be extremely small. Boltzmann defended atomic and molecular kinetic theories (the "theory of gases") against a number of leading figures, such as Ernst Mach, who denied it was necessary to regard atoms and molecules as real entities. Boltzmann eventually despaired of the argument, but in 1898 he took note of van der Waals' success with gas theory as important support.

Ludwig Eduard Boltzmann

van der Waals did not offer a fundamental idea about the attractive forces between molecules. That came with the work of Fritz London who was born in Breslau, Germany (Wroclaw, Poland today), on March 7, 1900. About 1925, London became interested in the new area of quantum mechanics, and as discussed in Chapter 10, one of his first contributions provided the fundamental physical basis for valence bonds. In 1930, London's attention was focused on intermolecular interaction as something apart from chemical bonding. From quantum mechanical analysis, he was able to show that attractive forces exist between atoms because of the time-averaged instantaneous interaction of the fluctuating charge fields arising from orbiting electrons in the atoms. He showed that the distance dependence of this interaction varies as the inverse sixth power (i.e., $1/r^6$). The forces associated with this attraction are often referred to as London forces. Fritz London left Germany in 1933. He became a professor of theoretical chemistry at Duke University in 1939 and died in Durham, North Carolina, on March 30, 1954.

   London forces did much to complete the picture of real gases that began with van der Waals. They could be used to explain why two argon atoms had an attraction, and that in turn correlated with the attractive term in the van der Waals equation of state. The weak attraction that atoms and molecules exhibit means that in a low-density gas sample, two or more atoms or molecules can be held together by a "weak bond" that does not come about through any change in chemical bonding. These weakly bound complexes are often referred to as van der Waals clusters or complexes. Weak attraction among molecules arises from London forces and other sources. By 1960, A. David Buckingham of Cambridge University had shown the importance and intricate role of the interaction of permanent, along with fluctuating, molecular charge fields in intermolecular interaction and related phenomena. About a decade later, William Klemperer of Harvard University started generating van der Waals clusters, such as Ar–HCl, and for the first time determined their structure and properties in the laboratory. The study of van der Waals clusters increased interest in intermolecular interaction at its most fundamental level, thereby furthering the links between molecular quantum behavior and macroscopic thermodynamics.

Exercises

2.1 Find the pressure in pascals (see Appendix F for information on units) in a closed 1 L container of air at 300 K if it contains 1 mol of gas particles and if the air is presumed to obey the ideal gas law.

2.2 Assuming Equation 2.7 holds, calculate the pressure of a sample of 3 mol of argon contained in a 100 L vessel at a temperature of 100 K.

2.3 For 1 mol of an ideal gas, plot the temperature of the gas against the volume for pressures of 1.0, 1.5, 2.0, 2.5, and 3.0 bar.

2.4 Consider a bicycle tire inflated to a pressure measured to be 40 lb in.$^{-2}$ with the ambient temperature being 70°F. If the bicycle is used for some time and the temperature of the tire rises to 105°F, what is the tire pressure assuming ideal gas behavior and assuming the volume of the tire was constant?

2.5 What are the mean, most probable, and root-mean-squared speeds of oxygen molecules at 300 and 400 K treating them as if they behave as ideal point-mass gas particles?

2.6 Consider 1 mol of a pure gas at a pressure of 10^5 Pa (see Appendix F for information on units) and a volume of 20.0 L. What is the temperature of the gas if it obeys the ideal gas law? What is the temperature if the gas were the following and obeyed the van der Waals equation of state (Equation 2.31): argon, methane, and acetylene. Use the data in Table 2.2.

2.7 Invoking the van der Waals equation of state and using the data in Table 2.2, evaluate the pressure of a 4.0 mol gas sample of propane at a temperature of 400 K contained in a vessel with a volume of 100 L.

2.8 Calculate the pressure of 1.67 mol of ozone occupying a volume of 8.9 L at 100.0°C assuming it behaves as (a) an ideal gas and (b) a van der Waals gas.

2.9 Calculate the constant-pressure coefficient of thermal expansion of 1.32 mol of xenon gas with a volume of 10.9 L at 15.6°C, assuming it is an ideal gas.

2.10 Calculate the coefficient of isothermal compressibility for the gas described in Exercise 2.9.

2.11 Develop an expression that gives the fraction of molecules in an ideal gas sample at temperature T that have speeds greater than the most probable speed.

2.12 Assuming ideality for a sample of nitrogen molecules at 300 K, what fraction has a kinetic energy greater than twice the mean kinetic energy?

2.13 Using trigonometric relations with Equation 2.15, express $v_x^2 + v_y^2 + v_z^2$ in terms of s, θ, and φ.

2.14 Derive an expression for $\langle s^4 \rangle$ for an ideal gas.

2.15 Calculate the average speed and the average energy of a collection of helium atoms at 2430 K.

2.16 Consider a hypothetical nonideal gas that obeys the equation of state $PV = nRT + aP$, where $a = 0.05$ L. Make a plot of the pressure–volume isotherms for this gas and compare them with those for an ideal gas.

2.17 Make a plot of the $T = 380$ K pressure–volume isotherm of water assuming it obeys the van der Waals equation of state and using coefficients given in Table 2.2.

2.18 Assuming the atmosphere has an average temperature of 0°C and the average molecular weight of air is 29.0 amu, calculate the atmospheric pressure at the cruising height of most jets (35,000 ft).

2.19 The compressibility factor, Z, of a gas is PV/nRT and thus $Z = 1$ for an ideal gas. For 1 mol of CO_2 at $T = 273$ K, the coefficients for the equation of state given as Equation 2.33 are approximately $b = -100$ cm³, $c = 7200$ cm⁶, and $d = 0$. Plot Z as a function of $1/V$ for this gas using the equation of state. Explain the significance of the coefficient b being less than zero.

2.20 Derive an expression for the constant-pressure coefficient of thermal expansion for a Berthelot gas, Equation 2.30.

2.21 Compare the isothermal compressibility of an ideal gas with that of a gas that obeys Equation 2.29.

2.22 Derive an expression for the ratio α/β for a van der Waals gas.

2.23 Calculate α and β for a gas that obeys the equation of state

$$P(V - nb) = nRT$$

2.24 An equation of state sometimes called the Dieterici equation is the following.

$$P(V - nb) = nRTe^{-an/VRT}$$

Replace the exponential with a power series expansion. Truncate the series to one term and then to two terms and compare the resulting equations to those in Table 2.1.

2.25 As a way of realizing the generality of the differential expressions for thermodynamic state functions, such as those in Equations 2.41 and 2.42, determine exactly how many such relations can be written. There are eight functions to work with, T, P, V, S, U, H, A, and G. A unique differential expression can be written by choosing one of these to be a function of any other two. How many such choices are there?

2.26 Express the partial derivatives $(\partial H/\partial P)_S$ and $(\partial G/\partial T)_P$ as functions of thermodynamic state variables.

2.27 Second partial derivatives of thermodynamic state functions may be evaluated order by order. For instance, $\partial(\partial A/\partial T)_V \partial S = -1$ because $(\partial A/\partial T)_V = -S$. Find $\partial(\partial U/\partial V)_S/\partial P$.

2.28 For a van der Waals gas, find $(\partial S/\partial V)_T$ in terms of the van der Waals coefficients.

Bibliography

Burshtein, A. I., *Introduction to Thermodynamics and Kinetic Theory of Matter* (Wiley-Interscience, New York, 1995).

Klotz, I. M. and R. M. Rosenberg, *Chemical Thermodynamics*, 5th edn. (Wiley, New York, 1994).

Lewis, G. N. and M. Randall, revised by Pitzer, K. S. and L. Brewer, *Thermodynamics* (McGraw-Hill, New York, 1961).

Pauli, W., *Thermodynamics and the Kinetic Theory of Gases* (Dover Publications, New York, 2000).

3

Changes of State

The prescription for the existence of a thermodynamic system at equilibrium is the thermodynamic equation of state. Systems may undergo a change in state, and in some cases the change may involve a transfer of energy. Equations of state are used to understand these changes. In this chapter, we consider work, heat, and entropy and their roles in the process of changing states. Following that, we examine the three laws of thermodynamics.

3.1 Pressure–Volume Work

The expansion of a gas against an external pressure does work on the surrounding medium. Also, the expansion of the gas is a change of state: The volume and at least one other thermodynamic state function are changed in the course of the process. The amount of work that is done can be analyzed with a piston setup, as illustrated in Figure 3.1. We can visualize a piston that is cylindrical, and that means that the volume, V, of the enclosed gas is the cross-sectional area of the cylinder, A, times the height, h. If the height of the top of the piston changes, the volume changes proportionately; that is, $\Delta V = A \Delta h$. Pressure, P, is force per unit area. Thus, the force of the gas on the piston is $F = PA$. When the gravitational force arising from the mass on top of the piston is in balance with the force of the gas, the piston does not move. This balancing of forces on the piston is achieved before and after an expansion process. So the gas is initially at a fixed volume and pressure dictated by the mass that has been placed on top of the piston. If some part of the mass is instantaneously removed, the downward force is changed, and the gas expands until the upward force it exerts (pressure) is again in balance with the downward force. To quantify how much work is involved in this process, we need to analyze the mechanics more closely.

Let us take m_1 as the initial mass on top of the piston. Δm is the amount of mass suddenly removed, and so the final mass is $m_2 = m_1 - \Delta m$. The initial and final forces due to the mass are

$$F_1 = m_1 g$$

$$F_2 = m_2 g$$

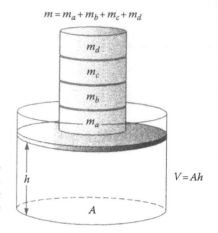

$$m = m_a + m_b + m_c + m_d$$

$$V = Ah$$

FIGURE 3.1
A cylindrical piston system. The volume of the gas inside the cylinder is the cross-sectional area of the cylinder, A, times the height, h.
The mass, m, is placed on top of the piston in increments, giving rise
to a downward force due to gravity, mg. (The piston is attached to the
bottom mass and its mass is included in the value m_a.)

where g is the gravitational constant. Let us take h_1 as the initial height of the piston and
h_2 as the final height. The work* that has been expended to move the mass on top of the
piston is

$$w = -m_2 g (h_2 - h_1)$$

This is simply the change in the gravitational potential energy of the mass, m_2, that is on
top of the piston as the gas expands. Since $m_2 g$ is the external force, F^{ext}, and since $F = PA$,

$$w = -F^{ext}(h_2 - h_1) = -P^{ext} A(h_2 - h_1) = -P^{ext}(V_2 - V_1) = -P^{ext} \Delta V$$

This says that the work expended in the course of the gas expansion is the product of
the change in volume and the external pressure. The external pressure is constant in this
expansion because the mass was removed instantaneously. If the change in mass were to
decrease, the change in volume would decrease as well. Thus, we can say that in the limit
of very small changes, the differential in the work is the external pressure times the differential in the volume.

$$\delta w_{gas} = -P^{ext} dV \tag{3.1}$$

Equation 3.1 uses the symbol δw instead of dw for an important mathematical reason that
we next consider.

Differentials are infinitesimally small increments. Integration over some variable that
has an **exact differential** yields a value that depends on the end points of the integration but not on the path that connects the end points. Integration over a quantity with an
inexact differential is path dependent. A differential quantity (exact or inexact) takes the
following form.

---

* A very precise analysis of this hypothetical system might call on us to take it as frictionless so that heat is not
  unintentionally generated. As well, the instantaneous removal of the mass could cause the piston to bounce
  and then oscillate, adding another complication. Perhaps a low-gravity environment would lead to very slow
  movement of the piston so as to avoid oscillation. We could restate the example in some limiting situation like
  that. However, rather than get into such details right now, let us simply think of this system as idealized, yet
  not too removed from our everyday experience. If a mass were removed from a piston in our laboratory, the
  piston would in fact move up to some fixed point, and that's what we seek to analyze.

$$\delta\Gamma = g(x,y,z,\dots)dx + h(x,y,z,\dots)dy + \cdots$$

where

  g and h are functions of the variables of the problem
  dx, dy, etc., are their differentials

Clearly, we can make up an arbitrary differential, $\delta\Gamma$, by choosing g, h, and so on, to be any functions we like. However, the differential so constructed will be exact *if and only if* there exists some function f such that

$$\delta\Gamma = df = \frac{\partial f}{\partial x}dx + \frac{\partial f}{\partial y}dy + \frac{\partial f}{\partial z}dz + \cdots$$

This requires that $g(x,y,z,\dots) = \partial f/\partial x$, that $h(x,y,z,\dots) = \partial f/\partial y$, and so on. In other words, a differential is exact if it corresponds to the infinitesimal change in a function of the variables of the problem. (For further discussion, see Appendix A.)

All thermodynamic state functions have exact differentials; they are functions of other thermodynamic state variables. The differential of work, though, is not exact. Work is not necessarily a function of the thermodynamic state variables, and this is shown by demonstrating its dependence on path.

In the example we have considered for the system in Figure 3.1, P^{ext} is constant during the course of the gas expansion. In this case, integration of the differential expression equation (3.1) goes as follows.

$$w = -\int_{V_1}^{V_2} P^{ext}dV = -\int_{V_1}^{V_2}\frac{m_2 g}{A}dV = -\frac{m_2 g}{A}(V_2 - V_1)$$

Now, consider the expansion of the gas in two steps. In the first step, the mass is changed by $\Delta m/2$, and after that expansion, there is another mass change of $\Delta m/2$. The final state of the piston system will be exactly as before. However, the total work will be different. Let us designate V_{1i} to be the intermediate volume achieved after the first step, and P^{ext-i} to be the external pressure during the first expansion. This intermediate pressure is simply the mass for the first-step expansion, $m_2 + \Delta m/2$, times the gravitational constant, divided by the area of the cylinder.

$$w_{1\to 1i\to 2} = -\int_{V_1}^{V_{1i}} P^{ext-i}dV - \int_{V_{1i}}^{V_2} P^{ext}dV = -\int_{V_1}^{V_{1i}}\frac{(m_2 + \Delta m/2)g}{A}dV - \int_{V_{1i}}^{V_2}\frac{m_2 g}{A}dV$$

$$= -\frac{(m_2 + \Delta m/2)g}{A}(V_{1i} - V_1) - \frac{m_2 g}{A}(V_2 - V_{1i}) = \frac{-m_2 g}{A}(V_2 - V_1) - \frac{\Delta m g}{2A}(V_{1i} - V_1)$$

$$= w_{1\to 2} - \frac{\Delta m g}{2A}(V_{1i} - V_1)$$

The result is that the work with this two-step path is different from the work with the single-step expansion. Work is not a state function; its differential is not necessarily exact. Equation 3.1 is the key relation for PV mechanical work done via expansion or compression of a gas. It is the expression for the incremental change in work.

The molecular level picture of pressure that was developed in Chapter 2 was based on an analysis of collisions of individual gas particles. Doing work to move a piston corresponds to a net exchange of kinetic energy of the numerous gas particles which collide with the piston mass. The piston is also composed of atoms, but the difference between collisions among gas particles and collisions with the particles in the piston is that the atoms of the piston are rigidly held together. Their restricted mechanical freedom is different from that of the particles making up the gas, and that difference is the key to doing work. The kinetic energy of particles of an ideal monatomic gas corresponds to the heat that can be transferred from the gas. When those particles collide with the piston's atoms, the kinetic energy (heat) of the gas particles can be transferred to the atomic framework of the piston, and thereby, molecular kinetic energy is converted to mechanical energy of the piston. At the molecular level, work and heat involve the same energy transfer process (collisions), and the sum of available work and heat is the internal energy, U, a thermodynamic state function. On the other hand, the division into heat dissipated and work expended depends on *macroscopic* features of the process; that is, heat and work depend on the path and are not thermodynamic state functions.

Example 3.1: Work in a Stepwise Gas Expansion

Problem: Find an expression for the amount of work done by an ideal gas expanding in a piston arrangement (Figure 3.1) if the total mass removed from the piston, Δm, is removed in n equal steps, $\Delta m/n$ mass being taken away at each step. m_1 is the initial mass on the piston.

Solution: Express the fact that the total work for the n steps is the sum of the work done at each step, using an index i to designate steps.

$$w_{1 \to n} = \sum_{i=1}^{n-1} w_{i \to i+1}$$

Since $\Delta m/n$ is removed at each step, at the ith step the mass in place on the piston is

$$m_i = m_1 - i\frac{\Delta m}{n}$$

At the end of the process, the mass is

$$m_n = m_1 - n\left(\frac{\Delta m}{n}\right) = m_1 - \Delta m$$

The gas expands during each step against a constant external pressure which is derived from the gravitational force acting on the mass remaining on the piston.

$$P_1 = \frac{g}{A}\left(m_1 - i\frac{\Delta m}{n}\right)$$

The work done in one step is

$$w_{i \to i+1} = \int_{V_i}^{V_{i+1}} P_i dV = -\frac{g}{A}\left(m_1 - i\frac{\Delta m}{n}\right)(V_{i+1} - V_i)$$

Sum this work for all the steps and combine terms in the various V_i.

$$w_{1 \to n} = -\frac{g}{A}\left[\left(m_1 - \frac{\Delta m}{n}\right)(V_2 - V_1) + \left(m_1 - \frac{2\Delta m}{n}\right)(V_3 - V_2) + \cdots\right]$$

$$= -\frac{g m_n}{A}(V_n - V_1) + \frac{g\Delta m}{A}V_1 - \frac{\Delta m}{n}\frac{g}{A}(V_1 + V_2 + \cdots + V_n)$$

The first term is that resulting from a single-step removal of Δm. The second term is a difference from carrying out multiple steps, and it diminishes the amount of work done. If we approximate the sum of the n volumes in the last term as n times the average volume, then

$$-\frac{g}{A}\left[m_n(V_n - V_1) - \Delta m V_1 + \Delta m\left(\frac{V_n + V_1}{2}\right)\right] = -\frac{g}{A}\left(m_n + \frac{\Delta m}{2}\right)(V_n - V_1)$$

So, there is more work from a stepwise process, and deductively this means work increases with increasing n (toward a limiting value).

3.2 Reversibility, Heat, and Work

In Chapter 1, equilibrium was seen as the most probable way in which a system exists. An equation of state, as introduced in Chapter 2, gives the thermodynamic relationship for a gas system at equilibrium. Our molecular level view, based on statistical arguments, allows that a system does not have to be at equilibrium at any given instant. There can be small fluctuations from equilibrium, which over time average to the equilibrium behavior. There may also be sudden changes, external perturbations, that disrupt equilibrium.

A process that leads to a change of state is said to be **reversible** if equilibrium is maintained throughout the course of the process. If not, the process is **irreversible**. The sudden expansion of a gas against a fixed external pressure is an example of an irreversible process. Let us consider a gas sample contained in a cylinder fitted with a piston such that the pressure is 2 bar. Suddenly, the external force on the piston is reduced from 2 to 1 bar. The gas expands rapidly to twice (or roughly twice, depending on the particular equation of state) the original volume so that its pressure is in balance with the external pressure. At any instant in the course of that expansion, do we know the pressure of the gas? Do we know the temperature? The answer to these questions is no, and that is because the system is not at equilibrium until the end. We must realize that we cannot directly measure a pressure for the gas because that would require balancing the gas pressure against an external pressure–achieving equilibrium. The piston is moving at every instant during the expansion process. A balance cannot be achieved in practice or in theory.

Let us consider a reversible isothermal expansion of the same gas. If for the moment we take the gas to be ideal, then reversibility means that $PV = nRT$ must hold at every instant of the expansion process. With temperature being held constant, this means that P will vary inversely and continuously with volume. For each infinitesimal change in volume, there will be an infinitesimal change in pressure. It is difficult to envision a means of accomplishing this type of expansion in the laboratory. We would need to remove mass continuously from the top of the piston, and the rate at which mass was removed would have to be slow enough for the system to remain at equilibrium at every instant. The only rate that could guarantee this is one that is infinitesimally close to zero. Though we may not be able to achieve this in practice—it would take forever—it is of interest to consider the work done in such a reversible process.

We obtain the work done in the reversible expansion by integration of the differential of work, which is $-PdV$. However, we must realize that the pressure is not constant in this type of expansion but is instead a function of the volume. It is determined by the equation of state since equilibrium is being maintained throughout the expansion.

$$w_{1 \to 2} = -\int_{V_1}^{V_2} P^{ext} dV$$

$$= -\int_{V_1}^{V_2} \frac{nRT}{V} dV \quad \text{from taking } P = P^{ext} \text{ and } PV = nRT$$

$$= -nRT \ln\left(\frac{V_2}{V_1}\right) \text{ assuming } T \text{ is constant} \tag{3.2}$$

This amount of work is the maximum amount that can be done by the gas isothermally. As illustrated in Figure 3.2, this work is the area under a segment of a particular pressure–volume isotherm. The only other pathways for expansion of the gas are pathways that follow segments below the isotherm. (Pathways above the isotherm must correspond at some point to expansion against a greater external pressure, which is unphysical.) The area under such segments is smaller than the area under the isotherm, showing that the maximum work is by a reversible expansion. Of course, since a strictly reversible process might take an infinite amount of time, in practice, work is maximized only by approaching the reversible limit of a process.

As already mentioned, doing work or transferring heat is accomplished by a gas via the same molecular level events, collisions. Work and heat are measures of energy that is transferred. Heating results in a new distribution of kinetic energies of the particles in the substance that is being heated, whereas work means that kinetic energy is converted to or from some macroscopic mechanical process. Incremental changes in heat and work give the incremental change in the state function U. The mathematical expression of this involves the differentials of U, of work (w), and of heat (q).

$$dU = \delta q + \delta w \tag{3.3}$$

For a given system, δw is negative if the system does work on the surroundings and thereby gives up energy. δq is negative if the system transfers energy away as heat. It is important

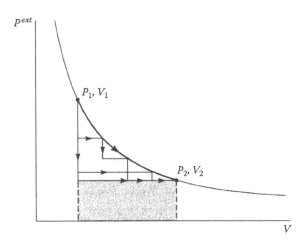

FIGURE 3.2
Expansion of an ideal gas at a specific temperature. The curve is the reversible path for the given temperature. Expansion from V_1 to V_2 follows this path only if equilibrium is maintained throughout. Then, the area under the segment from V_1 to V_2 is $\int P dV$, which is the work associated with the reversible process. Several irreversible expansions are indicated. A one-step irreversible process is for the external pressure to go instantaneously from P_1 to P_2, and then the gas expands from V_1 to V_2 against a pressure of P_2. The amount of work expended by the gas in this irreversible process is the area of the shaded rectangle. It is clear that the area under any of the irreversible paths is less than the area under the path that follows the isotherm where the gas pressure is always equal to the external pressure. The amount of reversible work is approached through a sequence of irreversible steps as the steps are made smaller and smaller.

to note that Equation 3.3 does not say that work and heat give us a value for U but rather that they indicate the change in U.

The work and heat that develop in the course of various gas expansion processes provide good examples of the thermodynamics of systems undergoing change. We shall first consider two special conditions. In one, the atoms and molecules in a system cannot directly transfer momentum to atoms and molecules outside the system. This means no heat can be transferred in the process, a condition referred to as **adiabatic**. For these, $\delta q = 0$. The other special condition is referred to as **isothermal**, meaning that the temperature is held constant in a process. To achieve an isothermal arrangement, the system is immersed in a much larger system that can instantaneously absorb all transferred energy without it changing temperature or can transfer heat back to the system so that the system remains at the same temperature. Both adiabatic and isothermal conditions are limiting cases that may be very nearly approached in practice even if they are not readily achievable.

Earlier in this chapter, we found that the maximum amount (absolute value) of work that could be done by a system undergoing isothermal expansion occurs when the expansion is carried out reversibly. Consider an ideal gas. Equation 3.2 can be used to calculate the work done. For isothermal expansion of the gas, the internal energy, U, does not change at any point in the expansion because the temperature does not change. This means that energy must be added to the system from the surroundings in the course of an isothermal expansion. That energy transfer is heat. Heat has been transferred from the surroundings to the system. From Equation 3.3 with $dU = 0$, we must have $\delta w = -\delta q$; equilibrium is maintained continuously in the course of the reversible expansion, and this relation between heat and work is maintained continuously as well.

An irreversible expansion carried out adiabatically (not isothermally) against a nonvanishing external pressure leads to a decrease in temperature of the system. Consider a gas in a cylinder fitted with a piston at some initial conditions (T_1, P_1, V_1). An example of an irreversible expansion is to instantaneously reduce the external pressure on the piston by half. The gas expands in volume until it reaches a new equilibrium (T_2, P_2, V_2) where $P_2 = P_1/2$. The adiabatic condition $(\delta q = 0)$ imposed on the system means that $dU = \delta w$. The work is the negative of the external pressure times the change in volume, $\delta w = -P_2 dV = dU$. Integrating dU yields $\Delta U = P_2(V_1 - V_2)$. Notice that ΔU is a negative value; the gas has done work and has given up energy. If it is an ideal monatomic gas, then this change in energy must equal $3nR(T_2 - T_1)/2$ because $U = 3nRT/2$. We now have three equations to solve for the unknown final temperature and final volume.

$$P_2(V_1 - V_2) = \frac{3}{2}nR(T_2 - T_1) \tag{3.4}$$

$$P_1V_1 = nRT_1 \quad \text{and} \quad P_2V_2 = nRT_2 \tag{3.5}$$

We begin solving these equations by substituting $P_2 = P_1/2$ in the left side of Equation 3.4 and then using the equations of state to substitute for PV products.

$$P_2(V_1 - V_2) = P_2V_1 - P_2V_2 = \frac{P_1V_1}{2} - P_2V_2 = \frac{nRT_1}{2} - nRT_2$$

Equating this with the right side of Equation 3.4 and dividing by nR yields

$$\frac{T_1}{2} - T_2 = \frac{3}{2}(T_2 - T_1)$$

$$T_2 = \frac{4}{5}T_1 \tag{3.6}$$

Next, we use the equation of state to solve for V_2.

$$V_2 = \frac{nRT_2}{P_2} = \frac{nR(4/5T_1)}{P_1/2} = \frac{8}{5}\frac{nRT_1}{P_1} = \frac{8}{5}V_1 \tag{3.7}$$

Example 3.2: ΔU for an Adiabatic Expansion

Problem: In energy units of L-atm, calculate ΔU for 1 mol of a monatomic ideal gas expanding adiabatically and irreversibly against a pressure of 1 atm starting from an initial volume of 20 L and an initial temperature of 300 K.

Solution: An adiabatic expansion is one for which $\delta q = 0$. Thus, $dU = \delta w$. Since the expansion is against a constant external pressure, $\delta w = -PdV$ and $\Delta U = P(V_{initial} - V_{final})$. Since the gas is a monatomic ideal gas, we also have $\Delta U = 3R(T_{final} - T_{initial})/2$. Equating these two expressions yields

$$P(V_{initial} - V_{final}) = 3R(T_{final} - T_{initial})/2$$

We know the pressure, the initial volume, and the initial temperature. We can express the product of the final pressure and volume as RT_{final} because $PV = RT$ (with $n = 1$).

That eliminates one unknown, V_{final}, from the problem. Then, we can solve for the remaining unknown, T_{final}. That value can be used to compute ΔU.

$$PV_{initial} - RT_{final} = \frac{3RT_{final}}{2} - \frac{3RT_{initial}}{2}$$

$$PV_{initial} + \frac{3RT_{initial}}{2} = \frac{5RT_{final}}{2}$$

$$(1 \text{ atm})(20 \text{ L}) + \frac{3(0.08206 \text{ L-atm K}^{-1})(300 \text{ K})}{2} = \frac{5(0.08206)T_{final}}{2}$$

$$T_{final} = 277.5 \text{ K} \quad \text{and so} \quad \Delta U_{gas} = 0.08206(277.5 - 300) \text{ L-atm} = -1.85 \text{ L-atm}$$

In this adiabatic irreversible process, the temperature of the gas decreases as the system does work. The volume increases but not as much as it would if the process were carried out isothermally. Halving the pressure under isothermal conditions would have doubled the volume; however, according to Equation 3.7, the volume turns out to be only 1.6 times the original volume, not 2.0 times.

We can relate the adiabatic and isothermal expansion behavior of a gas to the mechanics of gas particles; that is, we can connect the macroscopic view with a molecular level picture. Expansion of a gas against a piston is accomplished by many gas particles colliding with the piston and transferring momentum. After each such collision, a gas particle ends up with less momentum and hence less kinetic energy than it had before. It then continues its movement and before long collides with other gas particles and perhaps exchanges kinetic energy. But what happens as these "cold" particles coming from a collision with the piston collide with the on-average faster particles? There tends to be an energy exchange, mostly from the fast particles to the slower ones. So if the gas expansion is done adiabatically, then the collisions with the piston will result in a lower average velocity of all the particles. That means a lower energy, U, and a lower temperature, T. On the other hand, if the expansion is isothermal, then the slowing of particles from collisions with the piston will be offset by collisions with the walls that are maintaining the temperature at constant T. In this way, tiny amounts of energy are transferred from the surroundings (as heat) to the molecules to replenish the energy lost because of expansion. Again, at the level of particle collisions and momentum change, heat and work are similar.

A device or process that converts heat to work in a repeating cycle is an engine. Gas expansion can do work, and so one or more expansion steps are needed in an engine. To have a repeating cycle, the engine must return to its initial state. If the expansion were carried out reversibly and isothermally, the work done would be the area under the particular isotherm (Figure 3.2). Returning to the initial state by following the same isotherm would result in no net work, but returning by a reversible isothermal compression at a lower temperature would. This illustrates, but does not prove, that a temperature change must be part of an engine's cycle.

Sadi Carnot (France, 1796–1832), an engineer interested in engine efficiency, developed an engine cycle that is an important illustration of the application of thermodynamics. It is depicted in Figure 3.3 and consists of four steps. We can devise other types of cycles to convert heat to work, and these necessarily involve expansions and temperature changes. The efficiency of an engine is defined as the ratio of the net work done by an engine in a cycle divided by the heat transferred from the high-temperature reservoir.

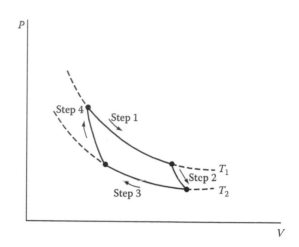

FIGURE 3.3

A hypothetical four-step process for an ideal gas known as a Carnot cycle. Step 1 is a reversible isothermal expansion that follows the T_1 isotherm from (P_1, V_1) to (P_2, V_2). Step 2 is a reversible adiabatic expansion from (P_2, V_2) to (P_3, V_3) which results in a drop in temperature. This is a move to a different isotherm. Step 3 is a reversible isothermal compression from (P_3, V_3) to (P_4, V_4), and step 4 is a reversible adiabatic compression from (P_4, V_4) to (P_1, V_1), leading to an increase in temperature. In this cycle, the gas does mechanical work on the surroundings using heat transferred from the surroundings. The differential of reversible work is $-PdV$, and so the area under the line corresponding to any given step is the work associated with that step. The entire process yields a net amount of work corresponding to the enclosed area.

Let us tabulate the thermodynamic quantities for the four steps of the Carnot cycle in Figure 3.3. We will assume that the engine operates with an ideal monatomic gas (i.e., that $U = 3nRT/2$). For steps 1 and 3, the work is obtained by integrating $-PdV$ with the ideal gas substitution, $P = nRT/V$, as in Equation 3.2. For steps 2 and 4, the change in U is used to find the work.

Step 1: Expansion

$$w_1 = -nRT_1 \ln\left(\frac{V_2}{V_1}\right)$$

$$\Delta U_1 = 0$$

$$\Rightarrow q_1 = -w_1$$

Step 2: Expansion

$$\Delta U_2 = \frac{3}{2}nR(T_2 - T_1)$$

$$q_2 = 0$$

$$\Rightarrow w_2 = \Delta U_2$$

Step 3: Compression

$$w_3 = -nRT_2 \ln\left(\frac{V_4}{V_3}\right)$$

$$\Delta U_3 = 0$$

$$\Rightarrow q_3 = -w_3$$

Step 4: Compression

$$\Delta U_4 = \frac{3}{2} nR(T_1 - T_2)$$

$$q_4 = 0$$

$$\Rightarrow w_4 = \Delta U_4$$

In step 1, since $V_2 > V_1$, $w_1 < 0$ and $q_1 > 0$; that is, work is expended by the gas and heat is absorbed. In step 2, since $T_2 < T_1$, $w_2 < 0$; work is expended by the gas. In step 3, since $V_4 < V_3$, $w_3 > 0$ and $q_3 < 0$; work is done on the gas and heat is released. In step 4, since $T_2 < T_1$, $w_4 > 0$; work is done on the gas.

Complete Cycle

$$w = -q = -nR\left(T_1 \ln \frac{V_2}{V_1} + T_2 \ln \frac{V_4}{V_3} \right)$$

$$\Delta U = 0$$

$$\text{Efficiency} = -\frac{w}{q_1} = \frac{-w_1 - w_3}{-w_1} = 1 + \frac{w_3}{w_1}$$

The heat exchanged with the surroundings in the steps of the Carnot cycle can be related to the temperature of the heat baths. Again assuming an ideal monatomic gas, the expression for U, $U = 3nRT/2$, leads to the differential relation

$$dU = \frac{3}{2} nR dT \tag{3.8}$$

At the same time, $dU = \delta q - PdV$. Combining this with Equation 3.8 yields

$$\frac{3}{2} nR dT = \delta q - PdV = \delta q - \frac{nRT}{V} dV \tag{3.9}$$

If we apply this equation to steps 2 and 4, for which $q = 0$, Equation 3.9 can be rearranged into the following expression.

$$\frac{3}{2} \frac{dT}{T} = -\frac{dV}{V} \tag{3.10}$$

Integrating this expression for step 2 and then for step 4 yields

$$\int_{T_1}^{T_2} \frac{3}{2} \frac{dT}{T} = -\int_{V_2}^{V_3} \frac{dV}{V} \tag{3.11a}$$

$$\int_{T_2}^{T_1} \frac{3}{2} \frac{dT}{T} = -\int_{V_4}^{V_1} \frac{dV}{V} \tag{3.11b}$$

The left-hand sides of these two equations differ only in sign because of the reversal in the integration limits. Thus, the right-hand sides differ from each other only in sign.

$$\int_{V_2}^{V_3} \frac{dV}{V} = -\int_{V_4}^{V_1} \frac{dV}{V} \tag{3.12}$$

Then, carrying out the integration leads to a relation among the four volumes.

$$\ln\frac{V_3}{V_2} = -\ln\frac{V_1}{V_4} \Rightarrow \frac{V_3}{V_2} = \frac{V_4}{V_1} \tag{3.13}$$

Hence, there is a common proportionality among the different volumes of the gas.

The Carnot cycle can be operated in the reverse direction, and then the signs are reversed for the work and heat in each step. This means that net work is done on the gas in one cycle, and heat is released into the higher-temperature bath (step 1) and absorbed from the lower-temperature bath (step 3). In this manner, the Carnot engine operates as a heat pump or refrigerator.

3.3 Entropy

The Carnot cycle returns a system to its initial thermodynamic state. In the course of this cycle, the value of any thermodynamic state function has a net change of zero, and thus, we can say not only that $\Delta P = 0$, $\Delta T = 0$, and $\Delta V = 0$ but also that $\Delta U = 0$ for one full cycle. The ratio of the heat exchanged with the surroundings in the two nonadiabatic steps reveals another function whose value does not change in the course of the full cycle.

$$\frac{q_{step1}}{q_{step3}} = \frac{nRT_1 \ln(V_2/V_1)}{nRT_2 \ln(V_4/V_3)} = -\frac{T_1}{T_2}\frac{\ln(V_2/V_1)}{\ln(V_2/V_1)} = -\frac{T_1}{T_2} \tag{3.14}$$

From this, we write

$$\frac{q_{step1}}{T_1} + \frac{q_{step3}}{T_2} = 0 \tag{3.15}$$

Since the other two steps are adiabatic ($q = 0$), we say that for a Carnot cycle,

$$\sum_{i}^{steps} \frac{q_{step\,i}}{T_{step\,i}} = 0 \tag{3.16}$$

It can be proved that this result is not subject to the gas behaving ideally. It holds for real gases. It also applies to any other reversible cyclical process because one can subdivide any reversible cyclical process of expansions and compressions into a sequence of individual Carnot cycles.

If we consider some reversible cyclical process to take place via infinitesimally small reversible steps, then the terms for each step in Equation 3.16 will be differentials, $\delta q_{rev}/T$, and the summation will become an integral. These differentials are differentials of a thermodynamic state function consistent with the fact that the value of that function does not change in the course of a reversible cyclical process. This thermodynamic state function is entropy, S, and thus,

$$dS = \frac{\delta q_{rev}}{T} \tag{3.17}$$

for a reversible process.

Chapter 1 introduced the thermodynamic state function called entropy, S. An argument was presented that for expansion of the volume occupied by one gas particle, the change in entropy is proportional to the natural log of the ratio of the initial and final volumes. If the sample consists of many molecules instead of one, we may also wish to know the entropy change. Let us restrict our attention to an ideal gas because then there are no interactions between particles and in a certain sense the particles are independent. With many particles instead of one, we must consider that there is a distribution of velocities. Since we wish to know the entropy change associated with only a change in volume, then we must consider a process where the velocity or energy distribution is fixed. Thus, this is an isothermal process. Given that, we simply sum the entropic contributions from each of the molecules (all the same as in the one-molecule case). We thereby obtain the entropy change for this hypothetical gas sample of many molecules. Thus, for an expansion from V_1 to V_2, the entropy change, according to Equation 1.8 with k replaced by nR, is

$$\Delta S = nR \ln \frac{V_2}{V_1} \tag{3.18}$$

Taking the expansion to be carried out reversibly, our analysis from the prior section gives the heat transfer and the work.

$$q = -w = \int_{V_1}^{V_2} P dV = \int_{V_1}^{V_2} \frac{nRT}{V} dV = nRT \ln \frac{V_2}{V_1} \tag{3.19}$$

Comparison of Equations 3.18 and 3.19 leads to the conclusion that for a reversible process

$$q_{rev} = T\Delta S \tag{3.20}$$

where the subscript "*rev*" is used to designate the path for a quantity associated with an inexact differential. The differential of this expression, $\delta q_{rev} = TdS$, is Equation 3.17.

If a process is reversible and adiabatic (i.e., $\delta q = 0$), then the change in entropy is zero. Also notice that whereas the product of two state functions is a state function, the product of a state function and a differential of a state function (i.e., TdS) do not have to be an exact differential. Integration of Equation 3.20 yields a value for ΔS in some overall process, but since S is a state function, ΔS is independent of the path. This means we will be able to find ΔS for *irreversible* processes by finding reversible paths with the same end points.

Let us now consider a specific irreversible process, an adiabatic ($\delta q = 0$) expansion of an ideal gas against zero external pressure until the final volume of the gas is 10 times its original volume. Zero external pressure is a limiting situation that in practice can only be approached. It is important to realize that if there is zero external pressure, no work can be done. To contemplate such a process, it may be helpful to consider a situation close to this limiting case. This involves a 10 L container that is empty except for a balloon filled with 1 L of gas. If the balloon were suddenly ruptured, there would be a free expansion (or very nearly free) and clearly no work would be done. If $w = 0$, and with $q = 0$, then $\Delta U = 0$ for the process. For an ideal gas, this means the temperature would not change. However, if the process were carried out reversibly instead, work would be done and there would be a change in heat, as given by Equation 3.19. The change in entropy for the reversible isothermal expansion is then $\Delta S = nR\ln(10)$ by Equation 3.18 and given that the ratio of the volumes is 10. The irreversible process ends up at exactly the same (P, V, T) point as the reversible process; S is a state function, and its value does not depend on the path; hence, this result for ΔS holds for the irreversible process, too.

Example 3.3: ΔS for an Adiabatic Expansion

Problem: Calculate ΔS_{gas} for 1 mol of ideal monatomic gas expanding adiabatically and irreversibly against a pressure of 1 bar starting from an initial volume of 20 L and an initial temperature of 300 K.

Solution: First, we should obtain all the initial and final conditions. From the ideal gas law, we have $P_1 = RT_1/V_1 = (8.314510 \text{ J K}^{-1})(300 \text{ K})/(0.020 \text{ m}^3) = 1.247 \times 10^5$ Pa = 1.247 bar. Following Example 3.2 with the slightly different conditions (1 bar instead of 1 atm), we find that $T_2 = 276.2$ K, and so $V_2 = RT_2/P_2 = 22.96$ L. Now, we need to find a reversible process that takes the gas from the initial to the final state and then obtain the heat transfer to or from the gas for that reversible process. Consider the following:

Step 1: Reversible isothermal expansion to V_2

Step 2: Reversible pressure decrease to 1 bar at fixed volume

For step 1, the general analysis of a reversible isothermal expansion can be applied, and so

$$\Delta S_{step1} = R\ln\frac{V_2}{V_1} = 1.149 \text{ J K}^{-1}$$

For step 2, we use a thermodynamic relation from Chapter 2:

$$dU = TdS - PdV$$

Of course, in this step, $dV = 0$, and so $dS = dU/T$. For an ideal monatomic gas, $U = 3nRT/2$. Therefore,

$$\Delta S_{step2} = \int_{300\text{K}}^{276.3\text{K}} \frac{dU}{T} = \int_{300\text{K}}^{276.2\text{K}} \frac{3}{2}\frac{RdT}{T} = \frac{3}{2}R\ln\frac{276.2\text{K}}{300\text{K}} = -1.031 \text{ J K}^{-1}$$

And for the entire process, done either irreversibly or as two reversible steps,

$$\Delta S_{gas} = \Delta S_{step1} + \Delta S_{step2} = 0.118 \text{ J K}^{-1}$$

3.4 The Laws of Thermodynamics

There are three major laws of thermodynamics. As already mentioned, the first law is the conservation of energy. When invoking this law, it is essential that all energy sources be considered. For any isolated system, the following statement of the first law always holds.

$$dU = 0 \qquad (3.21a)$$

For instance, a gas in a piston chamber can exchange energy with the piston, and in that case, the system for which Equation 3.21a holds must include both the gas and the piston. Isolated means that there is no energy or mass exchange with anything outside the system. However, we can also apply the first law to the gas in the piston chamber alone by accounting for energy exchange with the piston or any surroundings. We have seen that gas particles exchange energy via collisions, and depending on the collision partner(s), this exchange takes the form of heat transfer and/or work. Conservation of energy means that any change in the internal energy of the gas, ΔU, less the work done on the gas, w, and less heat transferred to the gas, q, by the piston or surroundings is zero:

$$\Delta U - q - w = 0 \qquad (3.21b)$$

If heat is transferred to the gas ($q > 0$) and work is done on the gas ($w > 0$), the internal energy increases such that $\Delta U = q + w$. This is an alternate expression of the first law.

The second law of thermodynamics involves entropy and deals with the tendency of systems to change spontaneously. In Chapter 1, we took a statistical view that led to a definition of absolute entropy. Entropy increases with the number of configurations or arrangements available to a system. Furthermore, as we recognized by considering arrangements of dice, a system tends to change in the direction of more possible arrangements. If there is a positive change in entropy, that is, if entropy increases via some process, the system must be moving to one of greater likelihood. A **spontaneous thermodynamic process** is one that occurs without outside influence or intervention, and such processes we associate with moving to a state of greater likelihood. Therefore, in some way, we can expect entropy to indicate the tendency of systems to change.

Is entropy a conserved quantity? Consider n moles of ideal gas trapped in a chamber of some volume V. Connected to this chamber by a valve is another chamber of the same volume, but it is evacuated. The gas chambers are an isolated system. If the valve were opened, thereby connecting the two chambers, we would expect gas particles to move into the empty chamber. Since the available volume has doubled, then by Equation 1.7 the change in entropy would be $nR\ln(2) \neq 0$. Though energy would be conserved in this process, that is, $\Delta U = 0$, entropy would increase. And if we added another chamber and allowed the gas to expand, entropy would continue to increase. It would not be conserved, and the fact that it would increase for this type of process is significant.

The flow of gas from occupying a volume V to occupying a volume $2V$ in the two chamber setup corresponds to our everyday experience. We do not walk into a room and from time to time find all the air is in one half of the room while the other half is evacuated; rather, the air has a uniform density throughout the room. It occupies the entire volume, not part of it, because that maximizes the number of configurations or arrangements of gas particles. In the hypothetical example, the expansion of the gas from occupying one chamber to occupying two is a spontaneous process. Once the valve is opened, it happens without external influence. As well, it is an irreversible process.

Therefore, the tendency for a spontaneous change is connected with the tendency of a system to maximize entropy.

The second law of thermodynamics is a statement about the tendency of a system to change spontaneously because of entropy. It says that for any change in the thermodynamic state of an isolated system, the entropy change is greater than or equal to zero.

$$\Delta S \geq 0 \quad \text{and} \quad dS \geq 0 \tag{3.22}$$

The entropy change is zero only for a reversible process. This makes sense because reversibility means that equilibrium is maintained throughout a process. To do that continuously seems possible only if the number of configurations available to the system is not changing. In a reversible isothermal expansion of a gas, there is a nonzero entropy change for the gas because it gains heat from the reservoir that maintains the temperature; however, the heat loss is a diminishment of the entropy of the reservoir equal and opposite to the entropy increase of the gas. On the other hand, an irreversible process, such as adiabatic free expansion, occurs spontaneously *in the direction of increasing total entropy*. Given this, one may take as a definition of equilibrium that it is the state of a system when it can no longer undergo any spontaneous processes.

For an engine that turns heat into mechanical work, the second law of thermodynamics is an important concern. Heat must be exchanged irreversibly. This requires a temperature differential, and it means that entropy is increased as the engine cycles. Consequently, entropy is sometimes associated with a capacity to do work. A system that has changed spontaneously to maximize entropy loses the capacity to do work. Consider again the two gas chambers connected by a valve. If the valve were replaced by a piston, there would be a capacity to do work under the initial conditions of the system. Gas in one chamber could move the piston in the direction of the evacuated chamber. However, by allowing for free expansion of the gas (opening the valve), there is no longer the same capacity to do work. This is true even though the energy of the system has not changed.

For processes that involve a constant-temperature heat bath, the entropy change associated with the heat bath or reservoir is simply $\delta q_{res}/T$, where T is the temperature of the reservoir. If the entire system consists of a gas and the reservoir, then the following holds.

Example 3.4: ΔS of an Engine Cycle

Problem: Show that $\Delta S_{gas} = 0$ for a Carnot engine cycle in which 1 mol of an ideal monatomic gas (1) expands isothermally and reversibly, (2) expands reversibly and adiabatically, (3) is compressed isothermally and reversibly, and (4) is compressed adiabatically and reversibly to the initial conditions.

Solution: Steps 2 and 4 are adiabatic and reversible. This means that for both the gas and for the surroundings, $\Delta S = 0$.

Steps 1 and 3 are reversible and isothermal. So

$$\Delta S_{gas(step1)} = \frac{q_{step1}}{T_{step1}}$$

$$\Delta S_{surr(step1)} = -\Delta S_{gas(step1)}$$

The same is done for step 3. From the analysis that was presented for the Carnot cycle

$$q_{step1} = RT_{step1} \ln \frac{V_2}{V_1}$$

and

$$\Delta S_{gas(step1)} = R \ln \frac{V_2}{V_1}$$

Likewise

$$\Delta S_{gas(step3)} = R \ln \frac{V_4}{V_3}$$

From the result given in Equation 3.13, we can write

$$\ln \frac{V_2}{V_1} = -\ln \frac{V_4}{V_3}$$

Therefore, $\Delta S_{gas(step1)} = -\Delta S_{gas(step3)}$, and for the complete cycle, $\Delta S_{gas} = 0$.

$$dS = dS_{gas} + dS_{res} = dS_{gas} + \frac{\delta q_{res}}{T} \geq 0 \qquad (3.23)$$

The relation between the heat change of the reservoir and of the gas yields a useful result.

$$\delta q_{res} + \delta q_{gas} = 0 \quad \text{or} \quad \delta q_{res} = -\delta q_{gas}$$

$$dS_{gas} + \frac{\delta q_{res}}{T} = dS_{gas} - \frac{\delta q_{gas}}{T} \geq 0 \Rightarrow dS_{gas} \geq \frac{\delta q_{gas}}{T} \qquad (3.24)$$

Equation 3.24, the **Clausius inequality** (Rudolf Julius Clausius, Germany, 1822–1888), serves as another way of expressing the second law of thermodynamics allowing "gas" to be anything that exchanges heat with surroundings.

The condition of equality in the second law, Equation 3.24, holds for systems undergoing a reversible change, whereas the inequality refers to irreversible changes. $dS > \delta q/T$ is a requirement for spontaneous change. Let us now consider irreversible changes under two sets of conditions. The first is that of constant temperature and constant volume. Constant volume means no pressure–volume work is done, that is, $\delta w = 0$, and thus, $dU = \delta q$. We can substitute for δq in Equation 3.24.

$$dS \geq \frac{dU}{T} \Rightarrow dU - TdS \leq 0 \qquad (3.25)$$

With $dT = 0$ because of the condition of constant temperature, the following results.

$$dU - TdS - SdT \leq 0$$

$$d(U - TS) \leq 0 \qquad (3.26)$$

$$dA \leq 0$$

This means that under the conditions of constant temperature and constant volume, a negative value for the differential in the Helmholtz free energy corresponds to a spontaneous process or indicates that a process will be spontaneous.

The second set of conditions is constant temperature and constant pressure. In this case, we use $\delta q = dU + P dV$ in Equation 3.24.

$$dS \geq \frac{dU + P dV}{T} \Rightarrow dU + P dV - T dS \leq 0 \tag{3.27}$$

Freely adding terms that are zero because $dT = 0$ and $dP = 0$ yields

$$dU + P dV + V dP - T dS - S dT \leq 0$$

$$d(U + PV - TS) \leq 0 \tag{3.28}$$

$$dG \leq 0$$

Therefore, under the conditions of constant temperature and constant pressure, the differential in the Gibbs energy is less than zero for a spontaneous change and equal to zero for a reversible change.

Given Equation 3.24, we may wonder what happens to entropy as the absolute zero of temperature is approached. The population distribution of a system of quantum mechanical species has a clear limiting form at $T = 0$. Assuming no degeneracies, the population of a particular excited state is related to the ground state population by the factor $e^{-\Delta\varepsilon/kT}$, where $\Delta\varepsilon$ is the energy difference between the ground state and that excited state. Then, $T = 0$ implies zero population for all excited states. In this case, our molecular view of entropy says there is only one configuration available to the system, and $S = nk \ln 1 = 0$. Indeed, this is one way of expressing the third law of thermodynamics. For any pure crystalline substance,

$$\lim_{T \to 0} S = 0 \tag{3.29}$$

Entropy approaches zero at $T = 0$. For systems maintained at $T = 0$, any change in entropy, ΔS, must be zero as well.

3.5 Heat Capacities

Heat capacities, designated C, are important properties of materials. They are the partial first derivatives of heat with respect to temperature, following some particular path, two convenient choices being paths where either volume or pressure is constant.

$$C_V = \left(\frac{\partial q}{\partial T} \right)_V \tag{3.30}$$

$$C_P = \left(\frac{\partial q}{\partial T} \right)_P \tag{3.31}$$

More useful expressions for these values can be derived. For a gas sample, it is always the case that $\delta w = -P^{ext}\,dV$. Then, from $dU = \delta q + \delta w$, we have

$$\delta q = dU + P^{ext}dV \tag{3.32}$$

Partial differentiation with respect to temperature, holding volume constant (i.e., $dV = 0$), yields a simple expression for C_V.

$$C_V = \left(\frac{\partial q}{\partial T}\right)_V = \left(\frac{\partial U}{\partial T}\right)_V \tag{3.33}$$

From the identity, $d(PV) = PdV + VdP$, we rewrite Equation 3.32.

$$\delta q = dU + PdV + [d(PV) - PdV - VdP]$$
$$= dU + d(PV) - VdP \tag{3.34}$$

From this, we have the partial derivative when pressure is constant ($dP = 0$).

$$C_P = \left(\frac{\partial q}{\partial T}\right)_P = \left(\frac{\partial[U + PV]}{\partial T}\right)_P = \left(\frac{\partial H}{\partial T}\right)_P \tag{3.35}$$

where we have recalled the definition of enthalpy, $H = U + PV$, to simplify the expression.

Equation 3.35 is another indication of why energy quantities other than U are useful and important in thermodynamics. If we were to measure the heat capacity of a copper rod, for instance, we would be more likely to carry out the experiment at atmospheric pressure (the constant pressure in the laboratory) than to try to insure that the rod's volume stays constant.

Beginning with Equation 3.35, we develop an important relation between C_P and C_V.

$$C_P = \left(\frac{\partial H}{\partial T}\right)_P = \left(\frac{\partial[U + PV]}{\partial T}\right)_P = \left(\frac{\partial U}{\partial T}\right)_P + P\left(\frac{\partial V}{\partial T}\right)_P \tag{3.36}$$

The partial derivative of U with respect to temperature at constant pressure may be obtained from differentiating $U = U(V, T)$. We "divide" the differential expression

$$dU = \left(\frac{\partial U}{\partial V}\right)_T dV + \left(\frac{\partial U}{\partial T}\right)_V dT$$

by dT_P (see Appendix A) to obtain

$$\left(\frac{\partial U}{\partial T}\right)_P = \left(\frac{\partial U}{\partial V}\right)_T\left(\frac{\partial V}{\partial T}\right)_P + \left(\frac{\partial U}{\partial T}\right)_V$$

$$= \left(\frac{\partial U}{\partial V}\right)_T\left(\frac{\partial V}{\partial T}\right)_P + C_V$$

Substituting this result into Equation 3.36 and bringing C_V to the left-hand side yields an expression for the difference between the heat capacities.

$$C_P - C_V = \left(\frac{\partial U}{\partial V}\right)_T \left(\frac{\partial V}{\partial T}\right)_P + P\left(\frac{\partial V}{\partial T}\right)_P \tag{3.37}$$

A similar derivation leads to

$$C_V - C_P = \left(\frac{\partial P}{\partial T}\right)_V \left(\frac{\partial H}{\partial P}\right)_T - V\left(\frac{\partial P}{\partial T}\right)_V \tag{3.38}$$

Still other such relations between C_V and C_P can be derived.

From an expression such as Equation 3.37, we can evaluate the difference for gases obeying different equations of state. For an ideal monatomic gas, the following results for the difference between C_P and C_V.

$$\left(\frac{\partial U}{\partial V}\right)_T = 0 \quad \text{because} \quad U = \frac{3nRT}{2}$$

$$\left(\frac{\partial V}{\partial T}\right)_P = \frac{nR}{P} \quad \text{because} \quad V = \frac{nRT}{P}$$

$$\therefore C_P - C_V = nR \tag{3.39}$$

For a nonideal gas, the difference may be a more complicated expression. To apply Equation 3.37, a van der Waals gas, which is one that obeys the equation of state given in Equation 2.31, two derivatives, $(\partial U/\partial V)_T$ and $(\partial V/\partial T)_P$, are needed. Using $dU = TdS - PdV$ (Equation 2.50), we write

$$\left(\frac{\partial U}{\partial V}\right)_T = T\left(\frac{\partial S}{\partial V}\right)_T - P$$

$(\partial U/\partial V)_T$ is often referred to as the **internal pressure** of a gas. Using a Maxwell relation which we can obtain from the thermodynamic compass yields

$$\left(\frac{\partial U}{\partial V}\right)_T = T\left(\frac{\partial P}{\partial T}\right)_V - P \tag{3.40}$$

Next, we rewrite Equation 2.31 and differentiate with respect to T at constant V:

$$PV + \frac{an^2}{V} - nbP - \frac{abn^3}{V^2} = nRT$$

$$\left(\frac{\partial P}{\partial T}\right)_V (V - nb) = nR \tag{3.41}$$

On dividing this equation by $(V - nb)$, an expression is obtained for $(\partial P/\partial T)_V$ which can then be used in Equation 3.40. The equation of state can be rearranged to give P in terms of T and V, and this expression can be substituted for P in Equation 3.40. Thus, we obtain

$$\left(\frac{\partial U}{\partial V}\right)_T = T\left(\frac{nR}{V-nb}\right) - \left(\frac{nRT}{V-nb} - \frac{an^2}{V^2}\right) = \frac{an^2}{V^2} \tag{3.42}$$

Next, we differentiate the equation of state, Equation 3.41, with respect to T at constant P and rearrange the result to obtain $(\partial V/\partial T)_P$.

$$P\left(\frac{\partial V}{\partial T}\right)_P = \frac{an^2}{V^2}\left(\frac{\partial V}{\partial T}\right)_P + \frac{2abn^3}{V^3}\left(\frac{\partial V}{\partial T}\right)_P = nR$$

$$\left(\frac{\partial V}{\partial T}\right)_P = \frac{nR}{P-(an^2/V^2)+2abn^3/V^3} \tag{3.43}$$

Using Equations 3.42 and 3.43 with Equation 3.37 yields the heat capacity difference for a van der Waals gas:

$$C_P - C_V = \frac{nR\left(P+(an^2/V^2)\right)}{P-(an^2/V^2)+2abn^3/V^3} \tag{3.44}$$

From its appearance, this expression may suggest behavior very different from that of an ideal gas. However, at low pressure, this expression approaches that of the ideal gas. As pressure is diminished, volume increases. Taking the limit of the expression as the volume goes to infinity shows that $C_P - C_V$ approaches nR. The difference between the real gas and the ideal gas at other than low pressures is due to the interaction between real atoms and molecules.

The ratio of the constant pressure heat capacity to the constant volume heat capacity is usually designated by the Greek letter γ.

$$\gamma = \frac{C_P}{C_V} \tag{3.45}$$

The heat capacity at constant pressure is always greater than the heat capacity at constant volume, and thus, $\gamma > 1$ holds for all gases. The internal energy of an ideal monatomic gas is $U = 3nRT/2$. Then, from Equation 3.33, the constant volume heat capacity of an ideal monatomic gas is $3nR/2$. With Equation 3.39, we obtain a value for γ of an ideal monatomic gas.

$$\gamma_{ideal, monatomic} = \frac{5}{3}$$

A difference between molecules and atoms is that only molecules can store energy in vibrational and rotational motion (vibrational–rotational quantum states). Thus, a molecular gas necessarily has heat capacities different from those of a monatomic gas. An ideal molecular gas, sometimes referred to as a **perfect gas**, is a hypothetical gas for which $dU = C_V dT$ and $PV = nRT$. For an adiabatic expansion of any ideal gas, $dU = -PdV$. Therefore,

$$C_V dT = -PdV$$

Substituting for P from the equation of state and rearranging yields

$$C_V \frac{dT}{T} = -P \frac{dV}{V}$$

Upon integration from (T_1, V_1) to (T_2, V_2), we obtain

$$C_V \ln \frac{T_2}{T_1} = -nR \ln \frac{V_2}{V_1} \tag{3.46}$$

From the equation of state, we can substitute for the ratio of temperatures.

$$\frac{T_2}{T_1} = \frac{P_2 V_2}{P_1 V_1} \Rightarrow \ln \frac{T_2}{T_1} = \ln \frac{P_2}{P_1} + \ln \frac{V_2}{V_1}$$

Next, substituting this expression in Equation 3.46 yields

$$C_V \ln \frac{P_2}{P_1} = -(nR + C_V) \ln \frac{V_2}{V_1} = -C_P \ln \frac{V_2}{V_1}$$

Taking the exponential of this equation and rearranging yields

$$\frac{P_2}{P_1} = \left(\frac{V_1}{V_2} \right)^{C_P/C_V} \tag{3.47}$$

$$P_2 V_2^\gamma = P_1 V_1^\gamma$$

Therefore, for any ideal gas, PV^γ is constant.

From statistical mechanics, discussed later in this text, one can directly obtain heat capacities of certain substances from information about the quantum mechanical energy levels of the gas particles. However, for temperature ranges from about 300 K to at least 1000 K higher, direct calorimetric measurements have shown very slight variation in the heat capacities of monatomic gases with temperature. Molecular gases tend to show a dependence on temperature that can often be well represented by a truncated power series expansion (see Appendix A) of the heat capacity per mole:

$$C_P(T) = a + bT + cT^2 \tag{3.48}$$

Some representative values of the coefficients in this expression for different gases are given in Table 3.1. As an example of the variation of the heat capacity with temperature, the data for Cl_2 show that the heat capacity per mole is 34.4 J K^{-1} at 300 K and 37.8 J K^{-1} at 1000 K.

TABLE 3.1

Heat Capacity Coefficients of Equation 3.48
for 1 mol of Selected Gases

	a (J K^{-1} mol^{-1})	b (J K^{-2} mol^{-1})	c (J K^{-3} mol^{-1})
H_2	29.066	−0.0008364	2.012×10^{-6}
O_2	25.72	0.01298	-3.86×10^{-6}
N_2	27.30	0.00523	-0.4×10^{-8}
Cl_2	31.696	0.010144	-4.0376×10^{-6}
CO	26.86	0.006966	-8.20×10^{-7}
CO_2	26.00	0.043497	-1.483×10^{-5}
HCl	28.1663	0.001810	1.547×10^{-6}

Sources: Values are for the temperature range of 300–1500 K
and have been converted to J from values reported
in cal by Spencer, H.M. and Justice, J.L., *J. Am.
Chem. Soc.*, 56, 2311, 1934 and Spencer, H.M., *J. Am.
Chem. Soc.*, 67, 1859, 1945.

3.6 Joule–Thomson Expansion

An experiment derived from the work of James P. Joule (England, 1818–1889) and William Thomson (England, 1824–1907; also known as Lord Kelvin) characterizes how temperature changes with pressure during an expansion process. An idealization of the experimental setup is the following. A gas of volume V_1 is placed in a cylinder with a piston on one side and a porous plug on the other. Another piston is located in the cylinder, but on the back side of the plug, as shown in Figure 3.4. The pistons regulate the pressure of the gas contained in the segments of the cylinder on either side of the plug. The plug allows gas molecules to pass slowly, but the rate of transfer through the plug is too slow to allow an equilibration in pressure between the two sides of the cylinder. Let us assume that we can apply a pressure P_1 on the left piston and a pressure P_2 on the right piston and that whatever mechanical device we use, it will precisely maintain those pressures as gas molecules pass through the plug.

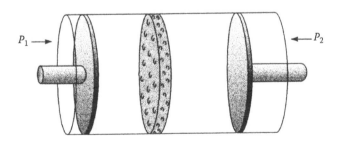

FIGURE 3.4
An apparatus for the Joule–Thomson expansion of a gas. The apparatus consists of a cylinder with pistons on both ends. Between the pistons is a porous plug which allows gas molecules to pass slowly from one side to the other. By external mechanical means, the pressure on the left is maintained at some value P_1, and the pressure on the right is P_2, with $P_1 > P_2$. The whole apparatus is insulated and does not exchange heat with the surroundings.

There will be a slow change in the volume of the gas on the right and on the left of the plug as gas passes through the plug in response to the difference in pressure applied by the pistons. The change in volume implies work done by one piston and done on the other piston. When after some period of time the change in volume is ΔV_1 on the left and ΔV_2 on the right, the net work done is simply the sum of pressure–volume work on the left side, $-P_1\Delta V_1$, and on the right side, $-P_2\Delta V_2$.

$$w = -(P_1\Delta V_1 + P_2\Delta V_2) \tag{3.49}$$

Notice that if the choice of piston pressures is such that gas moves from the left to the right, then $\Delta V_1 < 0$ and $\Delta V_2 > 0$. Since the process is carried out adiabatically, $q = 0$, and therefore, $\Delta U = w$. The enthalpy change for the entire system is zero; this is an isenthalpic (constant enthalpy) process.

$$\Delta H = \Delta H_1 + \Delta H_2$$

$$= \Delta(U_1 + P_1V_1) + \Delta(U_2 + P_2V_2)$$

$$= \Delta U_1 + \Delta U_2 + P_1\Delta V_1 + P_2\Delta V_2$$

$$= w + P_1\Delta V_1 + P_2\Delta V_2 = 0$$

Now, when the temperature on the left, T_1, is measured and the temperature on the right, T_2, is measured, we can evaluate at such instant in the process the quantity $(T_2 - T_1)/(P_2 - P_1)$. If the experiment is repeated with a decreasing difference between the pressure on the left and the pressure on the right, this quantity will approach $\partial T/\partial P$. The variable that is held constant in this derivative quantity is H because the setup insures an isenthalpic process. This derivative is called the **Joule–Thomson coefficient**:

$$\mu_{JT} = \left(\frac{\partial T}{\partial P}\right)_H \tag{3.50}$$

Taking P and T as the independent variables of the enthalpy function [i.e., $H = H(P,T)$], a rule of partial differentiation (see Appendix A) leads to the following relation among partial derivatives.

$$\left(\frac{\partial H}{\partial P}\right)_T = -\left(\frac{\partial T}{\partial P}\right)_H \left(\frac{\partial H}{\partial T}\right)_P \tag{3.51}$$

The partial derivatives on the right side are the Joule–Thomson coefficient and the constant pressure heat capacity.

For ideal gases, the partial derivative on the left side of Equation 3.51 is zero which means that the Joule–Thomson coefficient for an ideal gas is zero. We can see this by differentiating $H = U + PV$ with respect to P at constant T.

$$\left(\frac{\partial H}{\partial P}\right)_T = \left(\frac{\partial U}{\partial P}\right)_T + P\left(\frac{\partial V}{\partial P}\right)_T + V \tag{3.52}$$

The internal energy of an ideal gas depends only on temperature, and thus, at constant temperature, the derivative of U on the right side of Equation 3.52 is zero. Furthermore, with the ideal gas equation $PV = nRT$, the remaining two terms on the right side of Equation 3.52 cancel because $(\partial V/\partial P)_T = -nRT/P^2 = -V/P$.

For real gases, the Joule–Thomson coefficient is not zero but rather is a value that tends to vary with temperature and also with pressure since by Equation 3.51

$$\mu_{JT} = -\frac{1}{C_P}\left(\frac{\partial H}{\partial P}\right)_T$$

For most real gases, the Joule–Thomson coefficient at room temperature is a positive value. That means the gases cool on expansion. At around 300 K and 1 bar, argon, nitrogen, carbon dioxide, and air all cool on expansion. This effect is commonly used to liquefy gases. A gas that has been subjected to high pressure is allowed to freely expand by venting into a lower-pressure chamber. As the gas is vented, cooling may lead to liquefaction of some of it. The fraction that is vented is repressurized before starting the process again. Often through several cycles, the gas can be cooled to the point of liquefaction or solidification. An example that may be familiar is a fire extinguisher with pressurized carbon dioxide. The cooling from its expansion when the extinguisher is used can lead to the formation of some solid carbon dioxide (dry ice), and it may cool the surrounding air enough to condense water vapor (fog).

The dependence of the Joule–Thomson coefficient on temperature and pressure is such that for a given pressure, there is a temperature at which μ_{JT} goes to zero and beyond which it changes sign. For a pressure of 1 bar, this temperature is called the **inversion temperature**. Hydrogen and helium have inversion temperatures well below those of most other gases, and hence, liquefaction of hydrogen and helium requires precooling of the gas below the inversion temperature.

3.7 Conclusions

Reversibility means that equilibrium is maintained throughout a change of state. Gases can expand and do work on their surroundings and can be compressed. Expansion and compression can occur reversibly or irreversibly. Gases can exchange heat with the surroundings (unless isolated or adiabatic), and both heat and work correspond to a *transfer* of energy. Both also have the same molecular basis, an exchange of momentum between the gas particles and particles in the surroundings via collisions.

The total of heat and work is the change in internal energy, ΔU. U is a state function, its differential is exact, and it does not depend on the path for the change of state. Work and heat are inexact differentials and do depend on the path. (Thus, the efficiency of engines that convert heat to mechanical work can be adjusted by altering the pathways.) The first law of thermodynamics is the conservation of energy: $\Delta U = q + w$.

The differential of entropy for a change of state is the differential of the heat transfer in a reversible process divided by the temperature. For an irreversible process, the entropy change can be obtained by identifying an equivalent reversible process because S is a state function.

The second law of thermodynamics says that the change in entropy for an isolated system is always zero or greater than zero. For a reversible process, the entropy change of an isolated system is zero, whereas for an irreversible process it is greater than zero. Such a process may take place spontaneously as the system moves to a state of greater entropy.

The third law of thermodynamics states that absolute entropy approaches zero as temperature approaches zero.

Heat capacities and Joule–Thomson coefficients are properties of substances. For a monatomic ideal gas, the heat capacity at constant pressure is $5nR/2$ and the heat capacity at constant volume is $3nR/2$; the Joule–Thomson coefficient is zero. Real gases have more complicated behavior as a consequence of interparticle interactions.

Point of Interest: Heat Capacities of Solids

The study of the heat capacities of solids has sparked the discovery of a number of fundamental ideas. The subject starts with the work of Dulong and Petit who in 1819 published an empirical rule stating that the heat capacity per gram atom is the same for the solids of all elements. Pierre Louis Dulong was born in Rouen, France, in the middle of February 1785, and Alexis Therese Pettit was born in Vesoul, France, on October 2, 1791. They began collaborating in 1815, working first on comparing air and mercury thermometers in order to measure temperature accurately over a wider range than had previously been possible. Later, they measured specific heats of finely powdered metals and of sulfur. (Specific heat, the energy required to raise 1 g of a substance 1°C, is proportional to the heat capacity of a substance.) They found that the product of the specific heat and the atomic weight was nearly the same for all the substances they examined. This added support for Dalton's atomic theory, and it even resolved a few disputes over the atomic weights of certain elements. Petit died of tuberculosis in 1820. Dulong continued with work on the specific heats of gases and on heats of reactions. He died in Paris in 1838.

The rule of Dulong and Petit was challenged very early. At room temperature, light elements were found to have heat capacities below those predicted by the rule. At higher temperatures, the predicted values were surpassed. In 1871, Boltzmann showed that if atoms in a solid were vibrating, the product of the heat capacity and the atomic mass should be constant as in the rule of Dulong and Petit. This provided a more fundamental basis for the rule but did not account for all observations.

By the late 1800s, study of heat capacities had led to the perplexing observation that the heat capacities of solids at quite low temperatures were very much below those expected from the Dulong–Petit rule. Some measurements gave heat capacities only 1% of the predicted value. About 1907, Einstein tied quantum behavior to heat capacities. He showed that if the vibrational energies of atoms in a solid were quantized, heat capacity would diminish sharply at low temperature. The high-temperature limit of Einstein's theory was the result achieved earlier by Boltzmann. Einstein's theory proved not as quantitatively accurate in its application to the heat capacity curves of solids as it was to those of diatomic gases.

A more complete, more accurate theory for solids was developed by Peter Debye and reported in 1912. Debye was drawn into thoroughly analyzing vibrations in crystalline materials and showed that the vibrations were not those of isolated atoms at a single frequency but rather vibrations with a broad spectrum of frequencies. He developed a good approximation to the distribution of the frequencies, and with that distribution formula, heat capacity variation with temperature in solids became quantified and predictable. Debye's treatment of vibrations in solids has applications to thermal, mechanical, and even optical properties of materials.

Peter Joseph William Debye was born in the Netherlands on March 24, 1884. He received a degree in electrical engineering in 1905 and then went on to obtain a PhD in physics from the University of Munich in 1908. Three years later he succeeded Einstein at the University of Zürich in the position of professor of theoretical physics. He held several professorships at leading universities in Germany, the Netherlands, and Switzerland in the years thereafter until coming to the United States shortly after the start of World War II. From 1940 until becoming professor emeritus in 1950, he was professor and head of the chemistry department at Cornell University in Ithaca, New York. Debye died in Ithaca on November 2, 1966.

Debye's first major area of work involved the temperature dependence of the dielectric constant. He proposed that molecules could possess a permanent electric dipole. Molecules with permanent dipoles would tend to be aligned by an external electric field, but this would be offset in part by thermal motion. His kinetic analysis of the thermal motion yielded a formula relating the dielectric constant to the size of the permanent dipole and to the temperature. His formula resulted in three developments: (1) quantitative accounting of the temperature dependence of dielectric constants, (2) establishing that molecules may possess permanent dipoles, and (3) providing a means for determining molecular dipole moments by measuring the dielectric constant of a pure liquid. Today, the Debye is a standard unit for measuring the size of molecular dipole moments.

Following the discovery in 1912 by William Henry Bragg and his son William Lawrence Bragg that x-rays were diffracted by crystals, Debye reported that the thermal vibrations of the atoms in crystals affected the x-ray patterns. Based on his work on heat capacities (specific heats), he determined the temperature dependence of the intensity of the diffraction pattern. This led him to consider the x-ray diffraction of randomly oriented molecules,

Peter Joseph William Debye

and with Paul Scherer, he established the basis for x-ray structural analysis of colloidal systems, crystal powders, and polycrystalline metals. Debye received the Nobel Prize in chemistry in 1936, and in his honor, the American Chemical Society gives the annual Peter Debye Award in recognition of contemporary contributions to physical chemistry.

Heat capacities, the subject of one section of this chapter, have been studied for almost two centuries. These simple properties have had a role in a number of significant developments in physics and chemistry. The Dulong–Petit rule put heat capacity measurements in support of the modern atomic theory that was taking hold at the time. Had Dulong and Petit not attempted to find that simple relation from their data, the exceptions to their rule—the problems of heat capacities—could not have drawn the same attention in the decades following their work. Their rule provided a target for criticism and a reason for further study of heat capacity data. In the hands of Boltzmann and then Einstein, this problem was one trigger of the quantum revolution of the twentieth century. Debye completed the rigorous theory for heat capacities of solids and in the process developed key insight for x-ray diffraction of powders, a very important structural tool in chemistry. Heat capacity seems like a minor issue in molecular science, but with the unexpected turns of scientific discovery, it has become like a slow-growing tree with many fruitful branches.

Exercises

3.1 Calculate the pressure for a cylindrical piston system such as depicted in Figure 3.1 given that the mass, m, is 1.0 kg and the area of the piston is 0.2 m^2.

3.2 If a cylindrical piston system (Figure 3.1) were in balance with a pressure of 1.0 bar from 1.0 mol of an ideal gas, how much mass has to be loaded on top of the piston in order to make the height of the piston change by 2.0 cm given that the cross-sectional area of the cylinder is 25.0 cm^2? What will the new pressure be?

3.3 For 1.0 mol of an ideal gas, calculate the work done by the gas on the surroundings if it undergoes a reversible isothermal expansion at $T = 400$ K whereby the pressure changes from 1.0 to 0.5 bar. Repeat the calculation if the gas then goes from 0.5 to 0.25 bar, and then again if it goes from 0.25 to 0.125 bar.

3.4 Calculate the work done by 2.5 mol of an ideal gas to expand reversibly and isothermally to five times its initial volume at 373°C.

3.5 0.50 mol of an ideal gas is compressed from 54.6 to 20.0 L isothermally and reversibly at room temperature (25°C). What is the amount of work done on the gas?

3.6 If an ideal gas expands reversibly and adiabatically to three times its initial volume, write an expression for the final pressure in terms of the initial pressure and the volumes.

3.7 If 1.0 mol of a gas that obeys the equation of state $PV = n(RT + bP)$ is initially at a temperature of 300 K, a volume of 22.0 L, and a pressure of 1.0 bar, what is the value of the parameter b? If the pressure is reduced to 0.5 bar in an isothermal expansion, what is the volume of the gas?

3.8 How much work is expended by 1.0 mol of an ideal monatomic gas that is initially at 400 K and 1 bar pressure as it expands adiabatically against a fixed external pressure of 0.1 bar?

3.9 What is the temperature of the gas in Exercise 3.8 after the expansion?

3.10 If 1.0 mol of an ideal monatomic gas sample absorbs 10 kJ of heat on isothermal expansion at $T = 400$ K, what are the initial and final volumes of the gas sample?

3.11 Consider two states of 1.0 mol of an ideal gas: $P_1 = 1$ bar, $V_1 = 20$ L, and $P_2 = 10$ bar, $V_2 = 1$ L. Select a realistic step or steps to go from one state to the other irreversibly (i.e., steps wherein a piston changes the volume realistically, the temperature is maintained by a heat bath, and so on). Next, describe a hypothetical step or steps such that this change of state may be accomplished reversibly.

3.12 Find the entropy change for the gas expansion in Exercise 3.8 by finding a set of reversible steps that lead to the same final state.

3.13 Evaluate the heat and entropy change for the three expansion processes for the ideal gas in Exercise 3.3.

3.14 Using the data in Table 3.1, find the amount of heat absorbed by 2.0 mol of carbon monoxide in an isobaric (constant-pressure) temperature change from 400 to 1000 K.

3.15 Find $(\partial S/\partial T)_V$ for an ideal gas.

3.16 Derive an expression for the work done in a reversible isothermal expansion from volume V_1 to volume V_2 for 1.0 mol of a gas that obeys the van der Waals gas equation (Equation 2.31). For an expansion from 20.0 to 40.0 L at $T = 300$ K, compare the work expended by an ideal gas to that expended by argon, taking argon to obey the van der Waals gas equation and using coefficients in Table 2.2.

3.17 Use Equation 3.13 to show that the efficiency of the Carnot cycle depends only on the temperatures, T_1 and T_2, of the heat reservoirs.

3.18 For the four steps of a Carnot cycle, make a sketch of (1) how temperature varies with pressure, (2) how temperature varies with entropy, and (3) how internal energy varies with entropy.

3.19 Consider the following four steps in a reversible engine cycle:

1. Isobaric increase in volume
2. Adiabatic expansion
3. Isobaric decrease in volume
4. Adiabatic compression

On a graph of pressure versus volume, make a sketch of the changes for the four steps. Do the same for a graph of volume versus temperature and for a graph of entropy versus temperature.

3.20 Using the data in Table 3.1 and the heat capacity expression $\delta q = C_p dT$, find the entropy change for a sample of 1.0 mol of oxygen that undergoes an isobaric reversible temperature change from 300 to 400 K.

3.21 Find the Joule–Thompson coefficient for a van der Waals gas in terms of the van der Waals gas parameters a and b.

3.22 Using Equation 3.40, find $(\partial U/\partial V)_T$ for a Berthelot gas (Equation 2.30).

3.23 Using $C_P - C_V$ for a gas that obeys the equation of state in Equation 3.32.

3.24 Explain which would be better as a refrigerant gas in a household device, hydrogen or nitrogen. (Consider the Joule–Thompson coefficients.)

Bibliography

Klotz, I. M. and R. M. Rosenberg, *Chemical Thermodynamics* (Wiley, New York, 1994).

Kuhn, T. S., *Black-Body Theory and the Quantum Discontinuity, 1894–1912* (Oxford University Press, New York, 1978). This provides a fascinating and critical account of the development of the origins of quantum mechanics. In particular, it challenges the account of many texts about the origin of the quantum hypothesis.

Servos, J. W., *Physical Chemistry from Ostwald to Pauling* (Princeton University Press, Princeton, NJ, 1990).

4

Phases and Multicomponent Systems

An ideal gas exists only as a gas because there are no intermolecular forces between the particles. However, intermolecular forces do exist in real substances with the result that real substances condense into liquids and freeze into solids. Liquids and solids are different phases than gases, and their thermodynamic description introduces added complexity to the machinery of thermodynamics. This chapter is concerned with the generalization of the thermodynamic principles for application to systems with more than one phase and more than one component. It also explains some of the important phenomena associated with phase changes.

4.1 Phases and Phase Diagrams

In the first three chapters, we considered only *pure substances*, those that contain one chemically distinct species. A pure substance may exist in different **phases**, and these are different homogeneous regions in the system. Materials in different phases are mechanically separable. An ice cube is homogeneous, and when placed in a glass of water, it is a different region than the surrounding (homogeneous) liquid. The ice cube consists of water in the solid phase and can be mechanically separated from the liquid phase water in the glass. Some species exist in more than the three most obvious phases of solid, liquid, and gas because there are different solid or liquid state phases. A phase is recognized as a state of matter that is uniform in its physical features and chemical composition.

The number of phases that coexist at equilibrium and the number of chemical species combined to make up a thermodynamic system affect the variations that can be freely made to the system. To understand this, let us return to the ideal gas for which $PV = nRT$. This relation among the three thermodynamic state functions, P, V, and T, couples their particular values. For some sample of gas, we are free to mechanically set the volume and the temperature, for instance, but then the pressure is dictated by the gas law. For a gas, real or ideal, of one chemically distinct substance, there are two independent **degrees of freedom**, two thermodynamic state variables that can be set independently while maintaining the system in a state of equilibrium; $f = 2$ with f designating the degrees of freedom. Now consider the difference arising from multiple phases. For an ice cube in water to remain at equilibrium, there must be no net melting or freezing. If this system were held at 1 bar pressure, the temperature would have to be 273 K (the melting point of ice at 1 bar) to insure that there is no net change in the amount of ice and amount of liquid. Were the pressure to be varied, then we should expect some small change in volume in accord with relations among P, V, and T for liquid and solid water, just as we have for gases. However, unless the temperature was changed in a specific way dictated by the appropriate PVT relations, the pressure change could lead to melting or freezing since the melting point varies with pressure. Maintaining equilibrium in the water–ice system allows for only one degree of freedom.

There is a clear difference between a single-phase system and a multiphase system. Whereas we can specify the pressure and the temperature of pure water vapor, which means there are two degrees of freedom, there is only one degree of freedom for liquid water and ice to remain in equilibrium. An arbitrary change in pressure must be accompanied by a specific change in temperature and volume. This means the number of degrees of freedom is reduced by the number of phases, p, that are in equilibrium. Therefore, $f = 3 - p$ for a pure substance.

Since there is a limit on the number of degrees of freedom of a system with different phases existing in equilibrium, it is possible to use a two-dimensional plot (e.g., an x–y plot) to show the possible behavior. In the beginning of this text, P–V isotherms were drawn as two-dimensional plots of the allowed behavior of a gas. We could instead have drawn V–T isobars to show the relation between volume and temperature at specific pressures. To understand and follow phase behavior, it is a plot of pressure versus temperature that is normally the most useful. This is because temperature and pressure, not volume and entropy, are most easily varied in the laboratory. Volume is not independent when the variables of interest are pressure and temperature, and so we could consider drawing "constant-volume" curves on a P–T plot. More interesting to follow is the pressure and temperature at which phase equilibrium is maintained, regardless of the volume.

A **phase diagram** is a P–T plot of **phase equilibrium curves** for all phases of some substance. Each line or curve on a phase diagram represents a phase boundary. On one side a system at equilibrium exists entirely in one phase, and on the other it exists in another phase. Along the line, the two phases coexist. A phase boundary line gives temperature as function of pressure (or pressure as a function of temperature) for which this phase equilibrium is maintained. The strict relation between temperature and pressure in following a phase boundary line means that one degree of freedom is eliminated. If temperature were varied, pressure would have to change according to the phase boundary line in order for the two phases to continue to coexist.

Let us consider the form of the phase diagram of water, shown in Figure 4.1, and the information that is incorporated into it. There are three phase equilibrium lines corresponding to the coexistence of liquid and solid, of liquid and vapor, and of solid and vapor. These curves meet at a single point where all three phases coexist, and it is called the **triple point**. This point is the one specific pressure, temperature, and (implicitly) volume at which this occurs. For water, the triple point occurs at a very low pressure. We will see that phase diagram information is related to equations of state, which in turn are related to intermolecular interactions. Thus, laboratory measurement of phase behavior data provides information on certain intrinsic molecular properties related to intermolecular interaction.

Normal phase transition temperatures (i.e., freezing point, boiling point) are those temperatures at which two phases coexist with the pressure at 1 bar. A horizontal line drawn on a phase diagram corresponding to a pressure of 1 bar crosses the phase equilibrium curves at the normal phase transition temperatures.

The pressure–volume (P–V) isotherms of a gas that can liquefy show certain qualitative differences from those of an ideal gas, which cannot be liquefied. Generally, the volume of a liquid changes very slowly with pressure. Thus, P–V isotherms of a liquid tend to be almost vertical lines. At some temperature and pressure, this has to change to the P–V isotherm behavior we have already considered for gases. The change must be continuous since our experience indicates no pressure and temperature where a substance instantaneously changes its volume. Volume is a continuous, rather than discontinuous, function of temperature and pressure regardless of the number of phases. As a result, there is one P–V isotherm with an inflection point, and this is illustrated in Figure 4.2. At the inflection

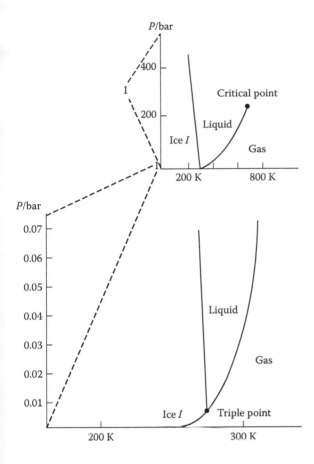

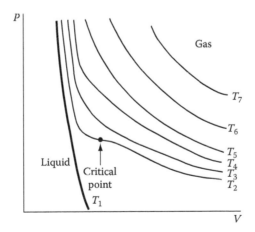

FIGURE 4.1
The phase diagram of water drawn on two different scales covering different temperature–pressure ranges. The triple point is shown on the expanded scale on the bottom drawing.

FIGURE 4.2
P–V isotherms ($T_7 > T_6 > \cdots > T_2 > T_1$) of a hypothetical pure substance that can exist as a gas and as a liquid. The isotherms for the gas phase are similar to those of an ideal gas. Deviations from ideal behavior become apparent at lower temperatures. The one isotherm (at T_1) shown for the liquid state has the typical pressure–volume behavior of a liquid, a small change in volume corresponding to a large change in pressure. To change smoothly from one type of behavior to the other, there must be a *P–V* isotherm (T_2) with an inflection (critical) point.

point, the first derivative of pressure with respect to volume is zero since the curve is flat at that point. As well, the second derivative is zero because the curvature is neither upward nor downward. The implications of this inflection point are quite interesting.

The mathematical conditions for an inflection point in the P–V isotherms of a pure substance are

$$\left.\frac{\partial P}{\partial V}\right|_{T_c} = 0 \quad \text{and} \quad \left.\frac{\partial^2 P}{\partial V^2}\right|_{T_c} = 0 \tag{4.1}$$

where the subscript c designates that this condition holds at the specific point of inflection which is referred to as the critical point. T_c, the critical point temperature, is the temperature of the P–V isotherm that includes the inflection point. When we differentiate the ideal gas equation of state ($PV = nRT$), we find that these conditions cannot be satisfied, a result consistent with the fact that an ideal gas exists only as a gas and not in a condensed phase:

$$\frac{\partial P}{\partial V} = \frac{-nRT}{V^2} \neq 0 \quad \text{for } T > 0$$

$$\frac{\partial^2 P}{\partial V^2} = \frac{2nRT}{V^3} \neq 0 \quad \text{for } T > 0$$

However, the conditions are satisfied for real gases because of how their intermolecular interactions give rise to more complicated equations of state. This may be illustrated for a gas whose behavior is reasonably well represented by the van der Waals gas equation (Equation 2.31).

$$P = \frac{nRT}{V - nb} - \frac{an^2}{V^2}$$

Again, a and b are constants specific to a given gas in the van der Waals equation. The first and second derivatives of the pressure with respect to volume are the following.

$$\frac{\partial P}{\partial V} = -\frac{nRT}{(V - nb)^2} + \frac{2an^2}{V^3} \tag{4.2}$$

$$\frac{\partial^2 P}{\partial V^2} = \frac{2nRT}{(V - nb)^3} - \frac{6an^2}{V^4} \tag{4.3}$$

The critical point, if it exists, is the specific state of temperature, pressure, and volume (i.e., T_c, P_c, and V_c) at which these two derivatives become zero. That produces two equations in two unknowns:

$$-\frac{nRT_c}{(V_c - nb)^2} + \frac{2an^2}{V_c^3} = 0 \tag{4.4}$$

$$\frac{2nRT_c}{(V_c - nb)^3} - \frac{6an^2}{V_c^4} = 0 \qquad (4.5)$$

The unknowns are V_c and T_c. Algebraic solution yields values for the critical temperature and critical volume.

$$V_c = 3nb \qquad (4.6)$$

$$T_c = \frac{8a}{27bR} \qquad (4.7)$$

Substitution of V_c and T_c into the equation of state yields the value of P_c.

Critical phenomena, behavior near or beyond the critical point, was first reported in 1869. Above the critical point temperature, a substance exists at all pressures as one pure phase, a critical fluid phase. Above the critical point temperature, the system is homogeneous with no appearance of a discontinuity, and the density is uniform even throughout a process that takes the substance from very low to very high pressure; no pressure increase can bring about a phase change in this region.

While many pure chemical species exhibit three phases, vapor, liquid, and solid, there are other phases that can occur, mostly in the solid state. For instance, water can freeze into different crystalline forms. Because they are distinct crystal structures, they correspond to different solid phases. Phase equilibrium curves have been determined such as those shown in Figure 4.3.

Elemental carbon is another interesting example because of the sharply different properties of its two solid phases, graphite and diamond. The equilibrium between these is at very high temperatures and pressures. Crossing that equilibrium is a way to synthetically produce diamonds, and this has been done industrially for decades. Of course, there is a chemical bonding difference between graphite and carbon, something that we take as analogous to the clear-cut bonding difference between oxygen atoms in O_2 versus O_3. Thus, the phase equilibrium between diamond and graphite can as well be considered a reaction equilibrium. Likewise, C_{60} or buckminsterfullerene might be taken as another solid phase of carbon, though it corresponds to still another type of bonding between carbon atoms. On the other hand, in the different solid phases of water, water molecules maintain the same chemical bonding.

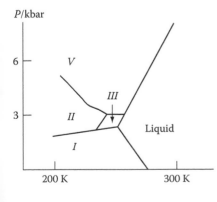

P/kbar

6 — V

III

3 — II

I

Liquid

200 K   300 K

FIGURE 4.3
The phase diagram of water showing the different solid phases designated ice *I*, ice *II*, and so on.

4.2 The Chemical Potential

It is sometimes convenient to refer to thermodynamic values for precisely 1 mol of a substance. The enthalpy, internal energy, entropy, and so on can be expressed per mole. An overbar is used to identify such molar values, and the general definition is

$$\overline{X} = \left(\frac{\partial X}{\partial n}\right)_{T,P} \quad \text{for } X \in \{A, G, H, U, V, S\} \tag{4.8a}$$

Thus, $\overline{H}$, the derivative of the enthalpy with respect to the number of moles of a substance, is the **molar enthalpy**. For systems involving different phases of one or of several substances, a set of variables, the **composition variables**, $\{n_1, n_2, ...\}$ give the number of moles of each substance in a phase, and then the definition of a molar value becomes

$$\overline{X}_i = \left(\frac{\partial X}{\partial n_i}\right)_{T,P,n_j \neq n_i} \tag{4.8b}$$

Equation 4.8a is simply Equation 4.8b applied to a system with one pure substance; however, the values defined by Equation 4.8b are usually referred to as *partial* molar values. It follows from this definition that a total thermodynamic value X for a system can be expressed as a sum of products of partial molar values and the corresponding n_i values.

$$X = \sum_i \overline{X}_i n_i \tag{4.9}$$

For example, the internal energy of ice and water in equilibrium is the sum of the partial molar internal energy of ice times the number of moles of ice plus the partial molar internal energy of liquid water times the number of moles of liquid.

The **partial molar Gibbs energy**, $\overline{G}_i$, has special utility such that it is usually given a special name, the **chemical potential**, and a distinct symbol, which herein is g_i ($g_i = \overline{G}_i$). Equation 4.9 for G is

$$G = \sum_i g_i n_i \tag{4.10}$$

We may think of each of the chemical potentials as building blocks for the total Gibbs energy of a system.

Systems of chemical interest often are not chemically pure substances but are mixtures of two or more chemically distinct species. The mixture may be a solution if it is uniform at the molecular level or a dispersion if it is uniform on a much larger scale, such as that of small droplets of oil dispersed in liquid water. A system of one or more chemically distinct species may exist in different phases, and the number of **components** of a system is the number of independent species needed to specify the composition of all the phases in a system. For instance, a closed system of water vapor over a dilute water solution of copper sulfate is a two-component system. This is because the liquid and vapor of water are in equilibrium and only the total amount of water is adjustable, while the amount of copper sulfate in the solution is independently adjustable. To specify the composition of all the phases in this system requires specifying the amount of water and the amount of copper sulfate.

For multicomponent and multiphase systems, there are more degrees of freedom than for a one-component single-phase system. In addition to two independent variables, such as P and T, or S and V, to describe the state of a system, there are the composition variables; however, there are constraints on these variables. For a system with c components, there are c composition variables for a given phase, but in terms of intensive variables, it is the concentrations (mole fractions) that are adjustable; there are $c - 1$ of these for each phase. If there are p phases, then there are $p(c - 1)$ adjustable concentrations. With p phases present in equilibrium, there are $p - 1$ constraints for each component in order for the phases to maintain coexistence. Thus, there are $c(p - 1)$ constraints. The number of degrees of freedom, f, is the sum of 2 (P and T) and $p(c - 1)$ less the $c(p - 1)$ constraints.

$$f = 2 + p(c-1) - c(p-1) = 2 - p + c \tag{4.11}$$

This relation, which is due to Gibbs, is called the **phase rule**. f is the number of parameters we can adjust and still maintain equilibrium conditions in the system.

Allowing for variable composition of a system means that the thermodynamic state functions are functions not only of two variables such as T and P, but also of the composition variables; for example, $G = G(T, P, n_1, n_2, \ldots)$. The differential of G is

$$dG = \left(\frac{\partial G}{\partial P}\right)_{T,n_j} dP + \left(\frac{\partial G}{\partial T}\right)_{P,n_j} + \sum_i \left(\frac{\partial G}{\partial n_i}\right)_{T,P,n_j \neq n_i} dn_i$$

Using the definition of the chemical potential and two thermodynamic compass relations, this equation is written

$$dG = VdP - SdT + \sum_i g_i dn_i \tag{4.12}$$

We can also let $G = G(g_1, g_2, \ldots, n_1, n_2, \ldots)$ as implied by Equation 4.10 and express dG as

$$dG = \sum_i (g_i dn_i + n_i dg_i) \tag{4.13}$$

This is often called the **Gibbs–Duhem equation**. Subtracting Equation 4.13 from Equation 4.12 yields

$$dG - dG = 0 = VdP - SdT - \sum_i n_i dg_i \tag{4.14}$$

This is another form of the Gibbs–Duhem equation, and it is simply a relationship involving the differential of pressure, temperature, and the component chemical potentials, but *not* G.

Notice that under conditions of constant temperature ($dT = 0$) and constant pressure ($dP = 0$), the Gibbs–Duhem equation leads to

$$\sum_i n_i dg_i = 0 \quad (dP = 0, dT = 0) \tag{4.15}$$

In Chapter 3, we found that the Gibbs energy is a convenient energy quantity for constant-temperature and constant-pressure processes, since it is dG that indicates whether a change will occur spontaneously. Another way of saying this is that at constant temperature and pressure, a system is at equilibrium (does not change spontaneously) if $dG = 0$. Consider a two-phase system at equilibrium such as ice in liquid water. With $dG = 0$ in Equation 4.13, and with Equation 4.15, we have

$$\overset{phases}{\underset{i}{\sum}} g_i dn_i = 0 \Rightarrow g_{solid} dn_{solid} + g_{liquid} dn_{liquid} = 0 \tag{4.16}$$

Since the system is closed, an increase in the amount of one phase leads to a decrease in the other phase; that is, $dn_{solid} = -dn_{liquid}$. Using this relation between the differentials in Equation 4.16, the chemical potentials of the two phases are seen to be equal, $g_{solid} = g_{liquid}$. This is a general point; for a system at equilibrium, the chemical potentials of like substances in different phases are the same. Consider a system of several components using a numerical subscript to designate a substance and a second subscript (α, β, γ,...) to designate the phase. Thus, $n_{2\beta}$ refers to the number of moles in the β-phase for component 2. The differential of G given in Equation 4.13 becomes

$$dG = \sum_i \sum_{\mu=\alpha,\beta,...} (n_{i\mu} dg_{i\mu} + g_{i\mu} dn_{i\mu}) \tag{4.17}$$

and at equilibrium with $dP = 0$ and $dT = 0$, Equation 4.16 becomes

$$\sum_i \sum_{\mu=\alpha,\beta,...} g_{i\mu} dn_{i\mu} = 0 \tag{4.18}$$

Taking the system as closed means the change in the total amount of any one component is zero. That change is the sum of the changes in the amount of the component in each of its phases, and so the following must hold for every ith component.

$$\sum_{\mu=\alpha,\beta,...} dn_{i\mu} = 0 \tag{4.19}$$

Equations 4.18 and 4.19 are satisfied if and only if the following hold.

$$g_{1\alpha} = g_{1\beta} = \cdots = g_1$$

$$g_{2\alpha} = g_{2\beta} = \cdots = g_2 \tag{4.20}$$

$$g_{3\alpha} = g_{3\beta} = \cdots = g_3$$

and so on.

The chemical potentials are identical for phases of a given component.

4.3 Clapeyron Equation

We have already seen that phase diagrams are a concise representation of the conditions under which different phases of a pure substance are found to exist in equilibrium. Along the phase diagram lines that separate different phase regions, the chemical potentials of the phases on each side of the boundary are equal. Conditions at which two phases are found coexisting but at which the chemical potentials are not equal, such as ice floating in a glass of water at room temperature, are not equilibrium conditions. Under those conditions, spontaneous phase change takes place depleting material in the phase with the greater chemical potential, as in ice melting in a glass of water at room temperature.

We can relate the derivatives of chemical potentials to derivatives of G by differentiation of Equation 4.10, $G = \sum_j g_i n_j$.

$$\left(\frac{\partial G}{\partial P}\right)_{T,n_j} = \sum_j \left(\frac{\partial g_j}{\partial P}\right)_{T,n_i \neq n_j} n_j \tag{4.21}$$

From Equation 4.12, we find that the partial derivative on the left side of Equation 4.21 is V. Using the definition of partial molar volumes yields

$$V = \sum_i \overline{V}_i n_i = \sum_i \left(\frac{\partial g_i}{\partial P}\right)_{T,n_j \neq n_i} n_i$$

Therefore, a relation exists for chemical potentials and partial molar volumes that is analogous to the relation between G and V obtained from Equation 4.12

$$\left(\frac{\partial g_i}{\partial P}\right)_{T,n_j \neq n_i} = \overline{V}_i \tag{4.22}$$

Repeating this analysis for differentiation with respect to T instead yields

$$\left(\frac{\partial g_i}{\partial T}\right)_{P,n_j \neq n_i} = -\overline{S}_i \tag{4.23}$$

Let us consider two phases, designated α and β, of some pure substance. Along the phase equilibrium line on the phase diagram, the equality of their chemical potentials given by Equation 4.20 (i.e., $g_\alpha = g_\beta$) implies an equality of differential changes in each.

$$\left(\frac{\partial g_\alpha}{\partial P}\right)_T dP + \left(\frac{\partial g_\alpha}{\partial T}\right)_P dT = \left(\frac{\partial g_\beta}{\partial P}\right)_T dP + \left(\frac{\partial g_\beta}{\partial T}\right)_P dT \tag{4.24}$$

With Equations 4.22 and 4.23, this becomes

$$\overline{V}_\alpha dP - \overline{S}_\alpha dT = \overline{V}_\beta dP - \overline{S}_\beta dT \tag{4.25}$$

Rearrangement gives

$$\left(\overline{V_\alpha} - \overline{V_\beta}\right) dP = \left(\overline{S_\alpha} - \overline{S_\beta}\right) dT \tag{4.26}$$

On the left side of the equation is the molar volume change $\Delta \overline{V_{tr}} = \overline{V_\alpha} - \overline{V_\beta}$ for the phase transition. Likewise, on the right side is the molar entropy change for the transition.

In terms of molar changes, Equation 4.26 is written

$$\Delta \overline{V_{tr}} dP = \Delta \overline{S_{tr}} dT \tag{4.27}$$

(The overbar becomes superfluous here; the form of the equation is valid for any number of moles.) Since $\Delta G_{tr} = 0$, the enthalpy change of a phase transition is $\Delta H_{tr} = T\Delta S_{tr}$. And so

$$T\Delta V_{tr} dP = \Delta H_{tr} dT \tag{4.28}$$

This is one form of the **Clapeyron equation** (Benoit-Paul Emile Clapeyron, France, 1799–1864).

The significance of the Clapeyron equation is most apparent if it is rearranged:

$$\frac{dT}{dP} = \frac{T\Delta V_{tr}}{\Delta H_{tr}} \quad \text{or} \quad \frac{dP}{dT} = \frac{\Delta H_{tr}}{T\Delta V_{tr}} \tag{4.29}$$

In this form, we can see that the slope of a phase equilibrium line at some specific temperature depends on the molar transition enthalpy and molar transition volume. This equation predicts the linear dependence of temperature and pressure on each other while phase equilibrium is maintained. We can see from this relation that for an increase in pressure ($dP > 0$), the temperature at which the transition takes place increases or decreases according to the signs of the transition volume and the transition enthalpy.

$\Delta V_{tr} > 0$ and $\Delta H_{tr} > 0$ or $\Delta V_{tr} < 0$ and $\Delta H_{tr} < 0$

- Pressure increase leads to transition temperature increase ($dT > 0$)
- Temperature increase leads to transition pressure increase ($dP > 0$)

$\Delta V_{tr} > 0$ and $\Delta H_{tr} < 0$ or $\Delta V_{tr} < 0$ and $\Delta H_{tr} > 0$

- Pressure increase leads to transition temperature decrease ($dT < 0$)
- Temperature increase leads to transition pressure decrease ($dP < 0$)

For example, since ΔV_{tr} for vaporization of water is positive (volume increases) and ΔH_{tr} is also positive (heat is required), increasing the pressure raises the boiling point temperature. In contrast, increasing the pressure *lowers* the melting point of water because ΔV_{tr} is negative. Ice happens to be less dense (greater molar volume) than liquid water, which is unusual. The phase diagram for carbon dioxide in Figure 4.4 shows the more typical situation of the slope of the solid–liquid phase boundary line being positive. Then, increasing pressure leads to increasing transition temperature.

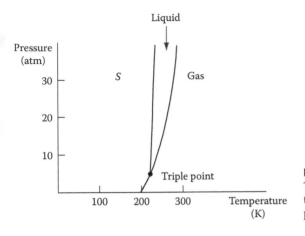

FIGURE 4.4
The phase diagram of carbon dioxide. Notice that the slopes of the phase equilibrium curves are positive.

How much is a boiling point or melting point temperature affected by a particular change in pressure? The precise answer requires the information represented by the entire phase equilibrium curve. However, the Clapeyron equation provides an approximation. Letting ΔT be the change in the transition temperature and ΔP be the specified change in pressure and taking Equation 4.28 to hold approximately for other than infinitesimal changes in the vicinity of the transition temperature yields

$$T_{tr}\Delta V_{tr}\Delta P = \Delta H_{tr}\Delta T \tag{4.30}$$

or

$$\Delta T = \Delta P \left(\frac{T_{tr}\Delta V_{tr}}{\Delta H_{tr}} \right) \tag{4.31}$$

The accuracy of Equation 4.31 depends in part on how close to linearity the phase equilibrium curve is over the range T_{tr} to $T_{tr} + \Delta T$. Avoiding the approximation calls for integrating Equation 4.28, as shown in Example 4.1.

A special approximate version of the Clapeyron equation is available for transitions from solid to vapor (sublimation) or from liquid to vapor. For these two types of transitions, the volume change is approximately the volume of the gas. This is because the volume of 1 mol of gas is typically very much more than the volume of 1 mol of the substance in its solid or liquid phase.

$$\Delta\overline{V} = \overline{V}_g - \overline{V}_s \approx \overline{V}_g \tag{4.32}$$

At the same time, if we approximate the PVT behavior of the gas as ideal, then $\overline{V}_g = RT/P$. This approximation of Equation 4.28 yields

$$RT^2 \left(\frac{dP}{P} \right) = \Delta\overline{H_{tr}}dT \tag{4.33}$$

This is the **Clausius–Clapeyron equation** (Rudolf Julius Clausius, Germany, 1822–1888). Taking the enthalpy of the transition to be constant over some range of temperature and pressure (a further approximation) and then integrating Equation 4.33 yields a useful expression.

$$\int_{P_1}^{P_2} d(\ln P) = \Delta\overline{H}_{tr} \int_{T_1}^{T_2} \frac{dT}{RT^2}$$

$$\ln\left(\frac{P_2}{P_1}\right) = -\frac{\Delta\overline{H}_{tr}}{R}\left(\frac{1}{T_2} - \frac{1}{T_1}\right)$$

(4.34)

This expression implies that by measuring the equilibrium vapor pressure at the two temperatures, T_1 and T_2, the enthalpy of vaporization can be calculated. Also, taking the exponential of Equation 4.34 shows a feature of comparing two or more species.

$$\frac{P_2}{P_1} = e^{\Delta\overline{H}_{tr}(T_2-T_1)/RT_1T_2}$$

(4.35)

A given set of temperatures, T_1 and T_2, corresponds to a ratio of pressures, P_2 to P_1, that varies exponentially from one substance to the next with the phase transition enthalpy.

Example 4.1: The Solid–Liquid Phase Boundary

Problem: What is the relation between temperature and pressure expected for fusion or melting? What is the form of the phase diagram line for the boundary between a liquid and a solid phase?

Solution: The Clapeyron equation, Equation 4.28, provides the relation between temperature and pressure for a phase transition. We need to consider how it applies in the case of a solid-to-liquid (or liquid-to-solid) phase transition. The approximation of Equation 4.32 to the molar volume for a phase transition is based in part on the fact that for a solid or a liquid, the molar volume is largely independent of temperature, as well as being a value much less than the molar volume of a gas. This means that for fusion or melting, the change in the molar transition volume in Equation 4.28 is usually well approximated as a constant (independent of T). The following steps are carried out with this approximation, as well as with the assumption that the transition enthalpy is independent of temperature.

$$dP = \frac{\Delta H_{tr}}{\Delta V_{tr}}\frac{dT}{T}$$

$$\int_{P_1}^{P_2} dP = \frac{\Delta\overline{H}_{tr}}{\Delta\overline{V}_{tr}} \int_{T_1}^{T_2} \frac{dT}{T}$$

$$P_2 - P_1 = \frac{\Delta\overline{H}_{tr}}{\Delta\overline{V}_{tr}} \ln\left(\frac{T_2}{T_1}\right)$$

For the process of fusion, the enthalpy change (solid → liquid) is a positive value. (See Table 4.1 for a compilation of several transition temperatures and enthalpies.) It is

TABLE 4.1

Molar Transition Enthalpies and Transition Temperatures[a]

	T_m (°C)	$\Delta\bar{H}_{fusion}$ (kJ)	T_b (°C)	$\Delta\bar{H}_{vaporization}$ (kJ)
AgCl	455	13.2	1547	199
Al	660.32	10.789	2519	294
Ar	−189.36	1.18	−185.85	6.43
Au	1064.18	12.72	2856	324
Cl_2	−101.5	6.40	−34.04	20.41
F_2	−219.66	0.51	−188.12	6.62
H_2	−259.34	0.12	−252.87	0.90
H_2O	0.0	6.01	100.0	40.657
H_2S	−85.5	2.38	−59.55	18.67
N_2	−210.0	0.71	−195.79	5.57
N_2O	−90.8	6.54	−88.48	16.53
NH_3	−77.73	5.66	−33.33	23.33
Ne	−248.61	0.328	−246.08	1.71
CCl_4	−22.62	2.56	76.8	29.82
CO	−205.02	0.833	−191.5	6.04
CH_4	−182.47	0.94	−161.48	8.19
CH_3OH	−97.53	3.215	64.6	35.21
C_2H_6	−182.79	2.72	−88.6	14.69
C_6H_6	5.49	9.87	80.09	30.72

Source: Values from Lide, D.R., *Handbook of Chemistry and Physics*, 86th edn., CRC Press, Boca Raton, FL, 2006.

[a] The enthalpies are defined such that the direction for fusion is from solid to liquid, and the direction for vaporization is from liquid to vapor. The enthalpy of fusion is the value at T_m and 1 bar, and the enthalpy of vaporization is the value at T_b and 1 bar.

typical that the volume change (solid → liquid) is positive, corresponding to the solid being denser than the liquid. With those conditions, the derived relation means that pressure along the phase boundary curve increases with temperature. If the volume change is negative, however, the phase boundary curve will have an opposite slope, as happens for the melting or freezing of water. Try this for yourself for water and ethanol and you will find they have opposite slopes, as water expands when it freezes whereas ethanol does the opposite.

4.4 First- and Second-Order Phase Transitions

What gives rise to condensation and fusion? We know that the hypothetical substance we call an ideal gas does not condense or solidify, and this is because it possesses no intermolecular interactions. The molecules of real gases do interact in a significant way, and that is the fundamental basis for condensation and fusion. At long range, there is an attraction between real gas particles. This attraction covers a considerable range in real gas systems, from near zero up to $40\,\mathrm{kJ\,mol^{-1}}$. As a gas is cooled, collision events occur with less kinetic energy. Lowering the temperature means changing the distribution of particle speeds and slowing the mean speed. The attraction between particles—their tendency to stick together and clump—is ultimately manifested in condensation once the kinetic energies have been sufficiently reduced. Then, the particles in the condensed phase are essentially in contact, though they have sufficient thermal (kinetic) energy to move about. On average, they do not have sufficient energy to escape (vaporize). For this reason, we see a phase separation between vapor and liquid, between those molecules that do not have sufficient energy to overcome the attraction of other molecules in the liquid and those that do. As the temperature is lowered further, thermal movement is reduced until the particles are essentially locked into their closest arrangements with their neighbors. This is the solid state. Because of intrinsic molecular structure, there may be different packing arrangements, including different crystal structures, and these may be manifested in different solid phases.

There are a number of characteristics of the type of phase transitions we have considered so far. In practice, these characteristics are often used to determine data points for phase equilibrium lines. One of the obvious characteristics is a discontinuity in enthalpy as a function of temperature. Consider a sample of ice at $-100°C$ and a pressure of $1\,\mathrm{bar}$. The enthalpy changes smoothly as the temperature is increased until the system reaches $0°C$, the melting temperature. At this point, there is a jump in the enthalpy corresponding to the enthalpy of melting. After all the ice has melted, the temperature can be increased and the enthalpy will be a new but different function of the temperature. Since the heat capacity, C_P, is the derivative of the enthalpy with respect to temperature, C_P as a function of temperature is also discontinuous at the phase transition. Vaporization of liquid water at $100°C$ and 1 bar leads to a sharp increase in volume. Thus, volume is a discontinuous function at a phase transition. The same holds for entropy.

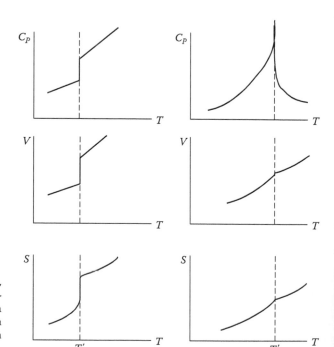

FIGURE 4.5
Typical behavior of heat capacity, volume, and entropy of a pure substance as a function of temperature given that the system undergoes a first-order phase transition (left) versus a second-order phase transition (right) at the temperature T'.

The chemical potential of two phases at equilibrium must be identical, and so the chemical potential of a substance as a function of temperature *is* continuous at the phase transition. The discontinuous functions just mentioned, volume, entropy, enthalpy, and heat capacity, are all related to first derivatives of the chemical potential. Therefore, the type of transitions we have considered so far, the ones to which we are most accustomed, is categorized as having a discontinuity in the first derivative of the chemical potential with respect to temperature at the phase transition. These are called **first-order phase transitions**.

Another type of transition is the **second-order phase transition** in which the first derivative of the chemical potential is continuous while the second derivative is not. This means that enthalpy, volume, and entropy vary continuously with temperature through a second-order phase transition temperature. This behavior is qualitatively different from that of a first-order phase transition, as illustrated in Figure 4.5. Whereas first-order phase transitions occur at a definite temperature for a given pressure, and with separation of the phases, second-order transitions do not exhibit a separation of phases and occur over a range of temperatures. The transition from superfluid helium to normal liquid helium and the transition from being a superconducting metal to being an ordinary conductor are examples of second-order transitions.

4.5 Conclusions

Interactions between atoms and molecules are the reason for deviation from ideal behavior. A critical point, one manifestation of nonideal behavior, is an inflection point on a P–V isotherm. Another manifestation is the existence of different phases. Solids and liquids are condensed phases, and it is not unusual for there to be several solid phases for a pure substance. Each phase is identified as a single homogeneous region of the same substance. Phase diagrams show lines of temperature versus pressure that represent a phase boundary. On one side of a line a system at equilibrium exists entirely in one phase, and on the other, it exists in the other phase. Along the line, the two phases are in equilibrium and the substance may exist in both. For a pure substance in a single phase, there are two adjustable thermodynamic functions, or two degrees of freedom, such as temperature and pressure. However, the relation between temperature and pressure along a phase boundary line means that one degree of freedom is eliminated if the system is to be maintained so that the two phases coexist. Temperature might be varied, but then pressure must follow the phase boundary line. A triple point is found where phase boundary lines intersect, and at that specific PVT point, three phases coexist.

The nature of phase boundary curves is understood from relations derived from the equality of the chemical potentials of a substance in different phases at equilibrium. The Clapeyron equation relates temperature changes to pressure changes (along a phase boundary curve) and to the enthalpy and volume change of the particular transition. The Clausius–Clapeyron equation is an approximation of the Clapeyron equation for transitions from condensed phase to vapor phase. Relations can be generated for other types of phase transitions, such as those involving only condensed phases. These relations show that laboratory measurement of temperature and pressure values along a phase equilibrium curve provides a means for obtaining enthalpies of phase transitions.

A system consisting of more than one pure substance is one that has some particular number of components. Components are chemically distinct species. The thermodynamic degrees of freedom (f) of a multicomponent or of a single-component system are given by the Gibbs phase rule, $f = 2 - p + c$, where c is the number of components and p is the number of phases that coexist. The chemical potential is the partial derivative of G with respect to the number of moles of a component in a given phase. At equilibrium, chemical potentials of a component in its different phases are equal.

Point of Interest: Josiah Willard Gibbs

Josiah Willard Gibbs

Josiah Willard Gibbs was a master of thermodynamics, and his work has had lasting impact. He was born in New Haven, Connecticut, on February 11, 1839, the only son among the five children in his family. He graduated from Yale College in 1858, where his father was a professor of sacred literature. He continued his studies at Yale, receiving a PhD in engineering—the first awarded at Yale and one of the first in the United States—in 1863. His endeavors as an engineer led to a patent for an improved railway car brake and to a new design for a steam-engine governor.

Gibbs' interests changed from engineering to more fundamental issues not long after he received his PhD, though it may have been more of a direct evolution than a change. He spent a 3 year period studying physics and mathematics in Paris, Berlin, and Heidelberg. Then, in 1871, he was appointed as a professor of mathematical physics at Yale. He received no salary for 9 years in this position, but that changed when he received and declined an offer for a salaried position at Johns Hopkins University. Gibbs continued teaching at Yale until his death on April 28, 1903.

Gibbs was known to value geometrical reasoning. His first scientific paper, published in 1873, was titled Graphical Methods in the Thermodynamics of Fluids. For decades, thermodynamic processes had been analyzed with pressure–volume graphs. Gibbs showed that there were other useful choices for the coordinate axes, such as temperature and entropy. In the process he did much to clearly explain the significance of entropy. In his next paper, he considered thermodynamic relations in terms of a geometrical surface in three dimensions. He used volume, entropy, and internal energy as the orthogonal axes and showed that temperature and pressure were then determined by the plane tangent to the surface at a given point. Maxwell was very much influenced by Gibbs' geometrical arguments and began to incorporate them into his own work, but with others Gibbs' ideas tended to be slow to take hold. They were often presented in an austere, terse form that made them difficult to grasp.

Gibbs is responsible for the criterion of thermodynamic equilibrium and for a general theory of thermodynamic equilibrium, for which his primary application of interest was chemistry. He introduced the chemical potential and devised the phase rule (Equation 4.11) using his thermodynamic surfaces to account for the coexistence of different phases. By 1878, he had completed a major publication that collected most of his work on thermodynamics. It was titled *On the Equilibrium of Heterogeneous Substances*, and it greatly expanded the applications of thermodynamics by including electromagnetic and electrochemical phenomena, surfaces, and other problems.

There was a continuing interchange of thinking between Maxwell and Gibbs. For instance, in the 1880s, Gibbs became interested in optics and in Maxwell's electromagnetic theory of light. He published several papers on optical dispersion. Gibbs had also studied Maxwell's and Boltzmann's work on the kinetic theory of gases. This prompted him to start presenting a course on statistical mechanics in the 1890s, and in 1902, his book *Elementary Principles in Statistical Mechanics Developed with Special Reference to the Rational Foundation of Thermodynamics* appeared. The methods he presented proved to be more general than those in use, and they came to dominate the field of statistical mechanics. Gibbs did not see the widespread acceptance of his ideas on statistical mechanics, however, because he died after an illness the year following the publication of the book.

Exercises

4.1 Verify by substitution that Equations 4.6 and 4.7 satisfy Equations 4.4 and 4.5. Then find the critical pressure for a van der Waals gas in terms of the constants a and b.

4.2 Using the data in Table 2.2, calculate values for the critical volume and critical temperature of the following gases: hydrogen, water, methane, and benzene.

4.3 A liquid has a vapor pressure of 25.6 mm Hg at 25°C and a normal boiling point of 356°C. Calculate the heat of vaporization of this liquid.

4.4 Write the van der Waals gas equation in terms of the critical temperature and critical volume instead of the constants a and b by using Equations 4.6 and 4.7 to substitute for a and b in Equation 2.31.

4.5 The volume change for a phase transition at some temperature can be obtained from the densities of each phase at the given temperature. (The volume is the number of moles divided by the molar density.) Given that the density of ice and liquid water at 273 K and 1 bar are, respectively, 0.915 and 1.000 g cm^{-3}, find the molar volume change for melting at 273 K and 1 bar.

4.6 An aqueous solution contains the salts NaCl, KCl, KBr, and KI. How many components are there in the solution?

4.7 How many components are there in the following systems?

    a. Solid ammonium chloride in equilibrium with vapor phase NH_3 and HCl

    b. The products of the reaction of solid sodium with an excess of chlorine gas

    c. A solution of vinegar in water plus olive oil (salad dressing)

4.8 Find the number of degrees of freedom, f, for the following systems.

    a. A pure sample of water existing in vapor, liquid, and gas together (equilibrium)

    b. An unopened 12 oz can of a carbonated beverage

    c. The mercury inside a thermometer

4.9 An equilibrium has been established between gaseous water, oxygen, methane, and carbon dioxide. How many degrees of freedom exist in this system?

4.10 Use the Gibbs–Duhem equation, Equation 4.14, applied to a *one*-component system to show the analogy between chemical potentials and Gibbs energies. (Rewrite Equation 4.14 for one component. What does this say?)

4.11 Using the molar volume change for melting from Exercise 4.5 and the enthalpy of fusion for water of 6.01 kJ mol^{-1}, estimate the change in the melting point if the pressure is changed from 1.0 to 1.1 bar.

4.12 Calculate the vapor pressure of water at 150°C using the Clausius–Clapeyron equation assuming the heat of vaporization is independent of temperature.

4.13 For a substance with an enthalpy of vaporization of 10.0 kJ mol^{-1} at 300 K and 1 bar, estimate the change in the boiling temperature if the pressure is decreased by 20%.

4.14 Calculate the boiling point of water on a mountain top where the pressure is 0.876 atm.

4.15 Calculate the boiling point of water in a pressure cooker that can maintain a pressure of 135 kPa.

4.16 Give a reasonable basis for the fact that many recipes for baking a cake suggest a different oven temperature for high altitudes. Should that temperature be higher or lower than the baking temperature at sea level?

4.17 Determine if any of the equations of state in Table 2.1 describe gases that would exhibit critical points. Find the critical temperature, pressure, and volume for any that do. Is there any type of functional feature in an equation of state that goes along with the existence of a critical point?

4.18 [Spreadsheet problem] Enter two columns of values that are the a and b coefficients in Table 2.1 for all diatomic and triatomic molecules. Evaluate V_c, T_c, and P_c for each gas.

4.19 [Spreadsheet problem] Assuming that the van der Waals equation of state holds for ammonia, and using the coefficients for ammonia in Table 2.2, find T_c Equation 4.7). Enter a column of data corresponding to 20 values in a range of volumes for 1 mol of ammonia: 0.01, 0.02, 0.03 L, and so on. Evaluate the pressure that satisfies the equation of state at the critical temperature for each of these pressures. Use the data to make a plot of the critical temperature P–V isotherm.

4.20 Prove the following:

$$\left(\frac{\partial G}{\partial n_i}\right)_{P,T,n_j \neq n_i} = \left(\frac{\partial U}{\partial n_i}\right)_{S,V,n_j \neq n_i} = \left(\frac{\partial H}{\partial n_i}\right)_{P,S,n_j \neq n_i} = \left(\frac{\partial A}{\partial n_i}\right)_{V,T,n_j \neq n_i}$$

4.21 Following Example 4.1, approximate $\ln(T_2/T_1)$ by rewriting it as $\ln(1 + x)$ with $x = (T_2 - T_1)/T_1$ and truncating the power series expansion (see Appendix A) to the first term in x. Does this indicate that the phase boundary curves between liquids and solids can be approximated as straight lines? If so, for what conditions does this hold?

4.22 Measurement of two sets of equilibrium temperature–pressure points for a gas–liquid equilibrium can give a value for the enthalpy of the phase change according to Equation 4.35. Explain how measurement of a series of temperature–pressure data points would confirm or contradict the assumption that the enthalpy change is not temperature dependent.

4.23 Consider a transition enthalpy in Equation 4.27 that had an explicit temperature dependence of the form

$$\Delta \overline{H_{tr}}(T) = A + BT + CT^2$$

where A, B, and C are constants. Rearrange and integrate to obtain a relation between two temperature pressure points (T_1, P_1) and (T_2, P_2) as in Example 4.1.

4.24 Offer a molecular explanation of the small differences between standard melting and boiling temperatures for the two rare gases in Table 4.1 in contrast with the larger differences for F_2 and Cl_2.

4.25 For the nonmetallic species in Table 4.1, make a plot of the molar enthalpy of vaporization versus boiling temperature. Suggest a reason for this rough correlation.

4.26 [Spreadsheet problem] With $P_1 = 1$ bar and $T_1 = T_b$ (boiling temperature), let $T_2 = T_1 + 10$, $T_1 + 20, \ldots, T_1 + 100$, and evaluate P_2 in Equation 4.34 for the vaporization of H_2O and of H_2S using data in Table 4.1. Compare the two sets of results.

Bibliography

Engel, T. and P. Reid, *Thermodynamics, Statistical Mechanics, and Kinetics* (Prentice-Hall, New York, 2009).

Klotz, I. M. and R. M. Rosenberg, *Chemical Thermodynamics* (Wiley, New York, 1994).

5

Activity and Equilibrium of Gases and Solutions

The condition of equilibrium is important in the analysis of systems undergoing reaction, phase change, or solvation. This chapter explores equilibrium thermodynamics with applications to gases, solutions, and mixtures. Activity is an important device that is introduced to help understand equilibrium and how equilibrium can be shifted. We begin by considering standard states and the means for collecting and using extensive thermodynamic data.

5.1 Activities and Fugacities of Gases

All thermodynamic values may be referenced to some chosen value. For instance, when a Chicago weather report on a January day says that the temperature is "ten degrees below freezing," it means the temperature is 10°F below a certain reference point temperature, 32°F, the normal freezing point temperature of water. Referencing values is useful in thermodynamic analysis because often it is changes, not absolute values, that are obtained from measurements. **Standard state** conditions refer to a generally accepted thermodynamic state that serves as a common reference point. For condensed phases, the standard state is the most stable phase of a pure substance at a pressure* of 1 bar. For a gas, the standard state is roughly a pressure of 1 bar; however, somewhat later we will consider the gas phase definition of a standard state more carefully. A superscript $^\circ$ identifies a value as a standard state value.

The chemical potential, g_i, of some species i in a thermodynamic system can be referenced to its standard state value, g_i°. A measure of the difference from the reference value is a quantity called the **chemical activity**, a_i, defined by the following relation for constant temperature.

$$a_i \equiv e^{(g_i - g_i^\circ)/RT} \tag{5.1}$$

Taking the logarithm and rearranging yields an equivalent expression.

$$g_i = g_i^\circ + RT \ln a_i \tag{5.2}$$

Notice that if the chemical potential on the left side of Equation 5.2 happens to be that of the reference state, the corresponding activity value used on the right-hand side of the expression (i.e., a_i°) must be identically 1.0 for the equality to hold. Thus, at a given

---

* For some time, precisely 1.0 atm has been used as the reference point for defining a standard state of a substance, but since an atmosphere is not an S.I. unit (see Appendix F), the reference point for a standard state was changed to 1 bar, a pressure roughly 1% less than 1 atm. For high precision use of tabulated thermodynamic data, it is important to determine if 1 atm or 1 bar was adopted for the standard state condition.

temperature, the activity varies exponentially with the chemical potential, relative to the chosen standard state at which the activity is precisely 1.0.

If temperature is held constant for some process, the differential expression that follows from Equation 5.2 is

$$dg_i = RTd(\ln a_i) \tag{5.3}$$

For a one-component system with n moles, $ng = G$ by the definition of the chemical potential, and then $ndg = dG = VdP - SdT$. For an isothermal process, that is, with $dT = 0$, Equation 5.3 becomes

$$dg = RTd(\ln a) = \overline{V}dP \tag{5.4}$$

Rearrangement yields

$$d(\ln a) = \frac{\overline{V}}{RT}dP \tag{5.5}$$

Let us designate $P°$ as the standard state pressure of the gas and then integrate Equation 5.5 from the standard state, with activity of 1.0, to some unspecified state at which the pressure is P' and the activity is a'.

$$\frac{1}{RT}\int_{P°}^{P'} \overline{V}dP = \int_{\ln 1.0}^{\ln a'} d(\ln a) = \ln a' \tag{5.6}$$

To apply this equation to a specific gas, we need to know the molar volume as a function of pressure. For an ideal gas, $\overline{V} = V/n = RT/P$. Inserting RT/P on the left side in Equation 5.7 produces

$$\frac{1}{RT}\int_{P°}^{P'} \frac{RT}{P}dP = \ln \frac{P}{P°} = \ln a' \tag{5.7}$$

Thus, for an ideal gas

$$\frac{P}{P°} = a' \tag{5.8}$$

The activity of the ideal gas is simply the ratio of the pressure to the reference pressure, and this can be substituted into Equation 5.2.

$$g = g° + RT \ln \frac{P}{P°} \tag{5.9}$$

The common choice for the standard state pressure of an ideal gas is $P° = 1$ bar (100 kPa).

In principle, we can use a nonideal equation of state to carry out the integration in Equation 5.6. However, there is a practical idea that avoids the complicated forms of

Equation 5.9 that would result from such a calculation. It is to make Equation 5.9 apply to all gases by using an effective pressure function specific to each gas in place of the pressure in Equation 5.9. The effective pressure is called the **fugacity**, and its symbol is f. The definition of the fugacity of a gas is

$$f = f° e^{(g - g°)/RT} \tag{5.10a}$$

which is the same as

$$g = g° + RT \ln \frac{f}{f°} \tag{5.10b}$$

The common choice for the standard state of real gases is that for which $f° = 1$ bar. As pressure is diminished, the fugacity approaches the true pressure. In the limit of zero pressure, the effective pressure and the true pressure become the same, and this is expressed in the following way.

$$\lim_{P \to 0} \frac{f}{P} = 1 \tag{5.11}$$

This is because all gases approach ideal behavior in the limit of low pressure. The ratio of fugacity to pressure is the **fugacity coefficient**, γ.

$$\gamma = \frac{f}{P} \tag{5.12}$$

For ideal gases $\gamma = 1$ and $f = P$. Comparing Equation 5.2 with Equation 5.10b shows the relation between activity and fugacity.

$$a = \frac{f}{f°} \tag{5.13}$$

The fugacity may be calculated from relations of the activity and the equations of state.

We may consider the fugacity of a real gas in terms of the deviation from ideal behavior. We begin by integrating Equation 5.5 between two limits.

$$\int_{\ln a''}^{\ln a'} d(\ln a) = \int_{P''}^{P'} \frac{\overline{V}_{real}}{RT} dP \tag{5.14}$$

$$RT \ln \frac{a'}{a''} = \int_{P''}^{P'} \overline{V}_{real} dP$$

From Equation 5.13, we rewrite Equation 5.14 in terms of fugacities.

$$RT \ln \frac{f'}{f''} = \int_{P''}^{P'} \overline{V}_{real} dP \tag{5.15}$$

Since Equation 5.15 is general, it applies to an ideal gas. In that case, the fugacities are the pressures. Therefore,

$$RT \ln \frac{P'}{P''} = \int_{P''}^{P'} \overline{V}_{ideal} dP \tag{5.16}$$

The next step is to subtract Equation 5.16 from Equation 5.15.

$$RT \left(\ln \frac{f'}{f''} - \ln \frac{P'}{P''} \right) = \int_{P''}^{P'} (\overline{V}_{real} - \overline{V}_{ideal}) dP \tag{5.17}$$

We now let P'' approach zero and apply Equation 5.11.

$$\ln \frac{f'}{f''} - \ln \frac{P'}{P''} = \ln \left(\frac{P''}{f''} \frac{f'}{P'} \right)$$

$$\lim_{P'' \to 0} \ln \left(\frac{P''}{f''} \frac{f'}{P'} \right) = \ln \frac{f'}{P'} \tag{5.18}$$

$$\therefore RT \ln \frac{f'}{P'} = \int_{0}^{P'} (\overline{V}_{real} - \overline{V}_{ideal}) dP$$

From Equation 5.12, notice that $\ln(f/P) = \ln \gamma$, and thus, Equation 5.18 is an expression for the fugacity coefficient.

Equation 5.18 serves to compare real behavior to ideal behavior. For certain real gases, the integral is negative. For instance, attractive forces among real gas molecules may lead to a molar volume smaller than that of an ideal gas whose particles are non-interacting. A negative value for the integral implies a fugacity that is less than the pressure, whereas a positive integral corresponds to a gas with a fugacity greater than the pressure.

As an example of the use of Equation 5.18, we will find ln γ for a gas that obeys the following equation of state.

$$\overline{V} = \frac{RT}{P} + b + cP \tag{5.19}$$

Subtracting the ideal gas molar volume and inserting in Equation 5.18 yields

$$\overline{V} - \overline{V}_{ideal} = b + cP \tag{5.20}$$

$$RT \ln \gamma = \int_{0}^{P'} (b + cP) dP = bP' + \frac{cP'^2}{2} \tag{5.21}$$

Forming the exponential of this expression and suppressing the prime on P, now that we have integrated over P and no longer need to distinguish it, yields

$$\gamma = e^{(2bP + cP^2)/2RT} \tag{5.22}$$

Notice that for low pressure and/or high temperature, the fugacity is nearly equal to 1.0. It is for these conditions that nonideal equations of state approach the ideal gas equation of state.

From the equation of state, we can derive an expression such as Equation 5.22 for the fugacity coefficient of a real gas. From that we obtain the fugacity via Equation 5.12 and the activity via Equation 5.13. Then, via Equation 5.2, we obtain the Gibbs energy (or chemical potential) for the gas relative to its standard state value. In this way, the activity provides an important connection between the equation of state, a PVT equation, and one of the thermodynamic energies. Now, continuing with the example of the gas with Equation 5.19 as its equation of state, application of Equation 5.2 yields the following expression for the chemical potential (Gibbs energy for a pure substance).

$$\overline{G} - \overline{G}^{\circ} = RT \ln P + bP + \frac{cP^2}{2} \tag{5.23}$$

Using the relations among the Gibbs energy, the entropy, the enthalpy, and so on, each thermodynamic function can be expressed relative to its standard state value.

Example 5.1: Fugacity and Activity of a Real Gas

Problem: Find the fugacity coefficient and determine the behavior of G at high temperatures and low pressures for a gas for which the activity is found to vary with pressure and temperature according to the equation

$$a = e^{cP(T^2 - d)/RT^3}$$

where c and d are constants specific to this gas.

Solution: Equations 5.12 and 5.13 relate the activity to the fugacity coefficient:

$$\gamma = \frac{af^{\circ}}{P} = \frac{f^{\circ}}{P} e^{cP(T^2 - d)/RT^3}$$

The logarithm of the activity expression is

$$\ln a = \frac{cP}{RT^3}(T^2 - d)$$

The differential of this expression can be used with Equation 5.3 to find dG (with $dT = 0$).

$$dG = RTd\ln a = \frac{c}{T^2}(T^2 - d)dP$$

This dG is a molar quantity which we can also designate as dg. We now integrate on the left side from the standard state to an unspecified state, and on the right side from a reference state pressure, P°, to the pressure, P, of the unspecified state.

$$\overline{G} - \overline{G}^{\circ} = c\left(1 - \frac{d}{T^2}\right)(P - P^{\circ})$$

At high temperatures and low pressures, the right-hand side of this expression approaches a constant.

5.2 Activities of Solutions

We began to analyze solutions in Chapter 4, and we know that solutions can be formed for gas, liquid, and solid phases. **Dalton's law of partial pressure** is an observation that to a good approximation the pressure in a multicomponent gas is the sum of pressures, the partial pressures, that each component would have if it were the only substance in the given volume. Other thermodynamic functions have a more complicated dependence on composition. For instance, at room temperature, a water solution of ethanol has a slightly smaller volume than the sum of the volumes of the water and the ethanol prior to mixing. The interaction of water and ethanol molecules results in a density of the solution that is greater than the density anticipated from those of the pure components assuming no attractive interaction.

Partial molar functions are derivatives of thermodynamic functions with respect to the number of moles of a component. Thus, for some thermodynamic function R, we define the following.

$$\overline{R}_i = \left(\frac{\partial R}{\partial n_i} \right)_{T,P,n_j \neq n_i} \tag{5.24}$$

With this definition, the differential in R at constant temperature and pressure can be expressed as

$$dR = \sum_i \overline{R}_i dn_i \tag{5.25}$$

From this, we can see that the chemical potential in a multicomponent system is the partial molar Gibbs energy. The partial molar volume, partial molar entropy, and so on are all defined by Equation 5.24. These functions are useful in analyzing solutions and mixtures.

The activities of species in solutions can be related to activities of gases under conditions of gas–liquid equilibrium because at equilibrium the chemical potentials are equal. Via Equation 5.2, the following holds for each substance in a system.

$$g_i^\circ(gas) + RT \ln a_i(gas) = g_i^\circ(liquid) + RT \ln a_i(liquid) \tag{5.26}$$

The reference point for the chemical potentials, the standard state values, can be eliminated by comparison of two systems with the same standard states.

A useful system for comparison consists of (I) a pure substance i present in liquid and vapor at the same total pressure as in (II) the vapor over a solution that has substance i as one component. In I, the vapor consists of pure i plus some nondissolving inert gas that brings the pressure of the otherwise pure i substance system up to the same pressure as that of the solution system. Here is a representation of the two.

System I	System II
Vapor	*Vapor*
i + inert gas	$A + B + \cdots + i + \cdots$
Pressure fixed at P' by inert gas	Pressure $= P'$
Liquid	*Liquid*
i(pure)	$A + B + \cdots + i + \cdots$

The equality of chemical potentials, as in Equation 5.26, for the two systems gives

$$g_i^\circ (liquid) + RT \ln a_i^I (liquid) = g_i^\circ (gas) + RT \ln a_i^I (gas)$$

$$g_i^\circ (liquid) + RT \ln a_i^{II} (liquid) = g_i^\circ (gas) + RT \ln a_i^{II} (gas)$$

Subtracting the second equation from the first eliminates the reference chemical potentials, and then dividing by RT yields

$$\ln a_i^I (liquid) - \ln a_i^{II} (liquid) = \ln a_i^I (gas) - \ln a_i^{II} (gas) \tag{5.27}$$

The antilogarithm of this expression gives

$$\frac{a_i^I (liquid)}{a_i^{II} (liquid)} = \frac{a_i^I (gas)}{a_i^{II} (gas)} \tag{5.28}$$

This ratio is a relation between gas and liquid activities.

A good approximation to Equation 5.28 is that the activity of the pure liquid, $a_i^I (liquid)$, is very nearly 1. This follows from Equation 5.6 which relates $\ln a$ to an integral of volume over pressure. For most liquids, until very great pressures are reached, the volume change is almost negligible over a considerable pressure range. Thus, for the pure liquid, the activity is very nearly 1.

$$a_i^{II} (liquid) = \frac{a_i^{II} (gas)}{a_i^I (gas)} \tag{5.29}$$

The activity of the liquid is given approximately by the ratio of the activities of the gas.

We can interpret this result by considering a gas that behaves ideally. Then, the gas phase activities in Equation 5.29 are simply the partial pressures.

$$a_i^{II} (liquid) = \frac{P_i^{II}}{P_i^I} (ideal) \tag{5.30}$$

This means that the activity of the ith component of the liquid solution (system II) is the ratio of the partial pressure of the ith component in the vapor above the liquid (system II) divided by the vapor pressure of i over pure i in a system with the same total pressure (system I). Dalton's law states that the partial pressures of components of a gas are proportional to their **mole fractions**, X_i, where

$$X_i = \frac{n_i}{\sum_j n_j} \tag{5.31}$$

Thus, the activity in Equation 5.30 can be expressed in terms of the total pressure and mole fraction. Given that the total pressure is the same in system I as in system II, we have

$$a_i^{II} (liquid) = \frac{X_i^{II} (gas)}{X_i^I (gas)} \tag{5.32}$$

Therefore, via Equation 5.30, we obtain the activity of a component of a solution from the ratio of the partial pressures of the component over the solution and over its pure liquid *at the same pressure*, or via ratios of the mole fractions of the gases in the two systems (Equation 5.32).

5.3 Vapor Pressure Behavior of Solutions

The vapor pressure of a binary mixture at some constant temperature can be measured as a function of mole fractions of the components of the mixture. Let us consider some hypothetical mixture of species A and B at room temperature. The vapor pressure of pure A is found to be 100 millibars (mb) and the vapor pressure of pure B is found to be 200 mb. These values are the end points of the curve that gives the vapor pressure of the mixture as a function of the mole fraction of either A or B, as illustrated in Figure 5.1. We might anticipate that the behavior of the vapor pressure between these end points will be that of a straight line. For instance, at a mole fraction of 0.5, we might expect the vapor pressure of the mixture to be simply the average of the pure liquid vapor pressures, 150 mb. A solution that exhibited this type of behavior would be considered an **ideal solution**. The activity of a component of an ideal solution is simply its mole fraction. Thus, with X_i as the mole fraction of the ith component in a binary (two-component) liquid and P_i^{pure} as the vapor pressure of pure i,

$$P_i = X_i P_i^{pure} \tag{5.33}$$

$$a_i = X_i \tag{5.34}$$

Equation 5.33 is **Raoult's law** (Francois Marie Raoult, France, 1830–1901), an approximate expression for the vapor pressure of the components of a mixture.

The hypothetical system in Figure 5.1 illustrates vapor pressure behavior that is not ideal but representative of real mixtures. The deviation from ideality may be regarded as arising

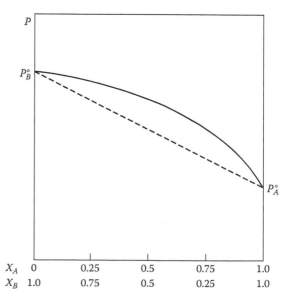

FIGURE 5.1
Vapor pressure of a hypothetical mixture of two species A and B as a function of the mole fraction at constant temperature. The end points give the vapor pressures of the pure liquids. Straight-line behavior is ideal (dashed line). Real systems tend to deviate to higher pressure or to lower pressure with a curved path between the end points (solid curve).

| X_A | 0 | 0.25 | 0.5 | 0.75 | 1.0 |
| X_B | 1.0 | 0.75 | 0.5 | 0.25 | 1.0 |

from interactions between A and B molecules in the liquid. First, we must realize that the $A–A$ interactions affect the vapor pressure of pure A. The stronger the interactions, the lower the vapor pressure. Likewise, $B–B$ interactions affect the vapor pressure of B. Now, when A and B are mixed, it is found for a given A molecule that some of the nearby molecule positions are occupied by B molecules. Then, it is the combination of $A–A$ and $A–B$ interactions that is important in determining the likelihood that an A molecule will escape to the vapor. Therefore, we conclude that the vapor pressure deviates from ideal behavior in the direction of lower pressure when the $A–B$ interaction is more attractive than the $A–A$ and $B–B$ interactions. It deviates toward higher pressure when the $A–B$ interaction is less attractive than the $A–A$ and $B–B$ interactions. The nature of the deviation may be even more complicated than the hypothetical system shown in Figure 5.1. Ideal behavior is approached when the two species being mixed are so similar that the $A–B$ interaction is approximately the same as the $A–A$ and the $B–B$ interactions.

The deviation from ideal solution behavior for a given component can be characterized by defining a parameter, γ, as the ratio of the true activity to the ideal activity (the mole fraction).

$$\gamma_i \equiv \frac{a_i}{X_i} \tag{5.35}$$

This is called the **activity coefficient**, and it is analogous to the fugacity coefficient for gases. For a species that exists as a liquid in its standard state at the temperature of the solution, the activity coefficient approaches 1 as the mole fraction of the substance approaches unity. This is simply a statement that Raoult's law is at its best as an approximation in the regions close to the end points on the vapor pressure curve.

The limiting behavior when a species is dissolved in a liquid is described by an approximation known as **Henry's law** (William Henry, England, 1775–1836). Henry's law says that at very low concentrations of a dissolved gas, the partial pressure of the gas over the solution is directly proportional to the concentration of the gas in the solution, or in other words, the partial pressure varies linearly with the concentration. This law is an approximation because a deviation from this linear relationship is usually found at some pressure. The slope of the line relating the pressure of the gas to the mole fraction is called the **Henry's law constant**. Figure 5.2 is a plot of vapor pressure

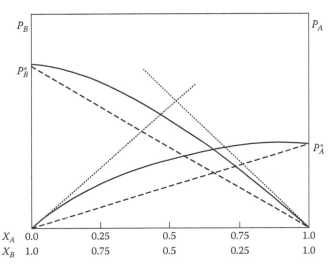

FIGURE 5.2
Vapor pressure curves for an $A–B$ mixture. The dashed lines show the behavior expected if Raoult's law held throughout, and the dotted line shows the behavior if Henry's law held throughout. The solid lines represent the true behavior.

of a binary system that compares Henry's law behavior, essentially low-concentration behavior, and Raoult's law behavior.

Figure 5.2 is a representative illustration of the component partial pressures and the total vapor pressure. Notice that at a mole fraction of 0.5 (equal concentrations), the composition of the vapor is not equally balanced between the two species. Rather, the one with the greater vapor pressure makes the greater contribution to the vapor. This fact is the basis for the process of **distillation**. If we start with a 50/50 liquid mixture of two species A and B, with A the more volatile, the first step is to lower the pressure until vapor and liquid are in equilibrium. The vapor will be richer in A than in B. The vapor is then mechanically separated and condensed to a liquid by increasing the pressure. This liquid is enriched in A. Next, the pressure is again lowered to bring this liquid into equilibrium with vapor, a vapor that is now further enriched in A. That vapor can be mechanically removed, and the process continued again and again until the desired purity of A is achieved. In some cases, mixtures behave as if they were pure substances; they cannot be separated by distillation. These mixtures are called **azeotropes**, and one example is a mixture of 95% ethanol and 5% water.

A more conventional route to distillation is by heating and cooling rather than changing the pressure. The best way to follow this process is with a plot of temperature versus mole fraction, as illustrated in Figure 5.3. We will take pressure to be constant, perhaps 1 bar. The two important lines to draw on such a plot are the boiling temperature line and the vaporization line. The boiling temperature line (the bottom line) gives the temperature, as a function of composition, at which liquid and vapor are in equilibrium. Below this

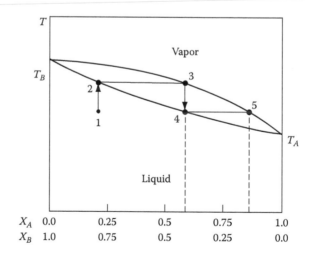

FIGURE 5.3

For a hypothetical mixture of two species A and B, the composition of liquid and vapor is related to the temperature on this plot. At some pressure such as 1 bar, T_A is the boiling point of pure A and T_B is the boiling point of pure B. If the system starts at point 1, heating it to the temperature at point 2 will initiate boiling. Because of different volatilities of A and B, the composition of the vapor differs from that of the liquid at the outset. Of course, if boiling were continued until all the liquid was gone, then the composition would have to be the same as that of the original liquid. However, at the outset, the composition of the vapor corresponds to the intersection of a horizontal line at the given temperature with the upper curve, the curve that gives the temperature at which only vapor will exist. So, if the first vapor from the boiling liquid (point 2) is pumped off to a new chamber at the same temperature, the composition will be that corresponding to point 3. The temperature is then lowered to condense the vapor into liquid. This liquid can then be heated to the boiling point, and the first vapor to come off will have the composition corresponding to point 5. Repeated steps will continue to purify the sample of the most volatile substance.

line, the liquid will not be found to be boiling. The line above indicates the temperature at which there can be no more liquid. (For a single substance, this line is coincident with the phase boundary line between liquid and vapor.) If the temperature of the liquid is raised until it is boiling (point 1 to point 2 in Figure 5.3), the vapor that is produced will have the composition corresponding to the upper line at the same temperature (point 3 in Figure 5.3). So if this vapor is mechanically separated and condensed, one substance will have been enriched. Then, this liquid is heated, and a further enriched vapor is produced, separated, and condensed. This cycling process, called **fractional distillation**, can then be repeated until essentially any desired purity is achieved.

5.4 Equilibrium Constants

Activity has an essential connection with the state of equilibrium in a reactive system, and in turn, the chemical potential is involved. It is useful to realize that the term "reaction" can be applied to almost any type of change, and in that sense we have already considered the reactions of phase change, solution, and mixing. Though true chemical reactions will be considered in detail in the next chapter, the analysis here can be viewed within the context of the types of reactions with which we have already worked. Following is a general statement for what occurs in any reaction.

$$m_1 R_1 + m_2 R_2 + \cdots \rightarrow n_1 P_1 + n_2 P_2 + \cdots \tag{5.36}$$

Notice, for instance, that this reaction statement could perfectly well correspond to a process of mixing m_1 and m_2 moles of *reactant* substances R_1 and R_2, respectively, to produce $n_1 = m_1$ and $n_2 = m_2$ of *product* substances P_1 and P_2, respectively. The change in the Gibbs energy for the general reaction process can be obtained by adding and subtracting the chemical potentials for products and reactants.

$$\Delta G_{rxn} = \sum_i^{prod} n_i g_i - \sum_j^{react} m_j g_j \tag{5.37}$$

Next, we employ Equation 5.2 to write this expression in terms of the activities of the reactants and products.

$$\Delta G_{rxn} = \sum_i^{prod} \left(n_i g_i^\circ + n_i RT \ln a_i \right) - \sum_j^{react} \left(m_j g_j^\circ + m_j RT \ln a_j \right) \tag{5.38}$$

The standard state chemical potentials in Equation 5.38 combine to give the standard state value of ΔG_{rxn}°, and the activities can then be arranged as a quotient,

$$\Delta G_{rxn} = \Delta G_{rxn}^\circ + RT \ln \left(\frac{\prod_i a_i^{n_i}}{\prod_j a_j^{m_j}} \right) \tag{5.39}$$

Notice that at standard state conditions, all activities are unity, the value of the quotient within the parentheses in Equation 5.39 is 1, and with those conditions, the ΔG of the reaction is properly the ΔG_{rxn}°, the standard state value for the free energy change of the reaction.

The quotient in Equation 5.39, which is the product of the activities of the product species (each to a power given by the reaction stoichiometry), divided by a like product for the reactants, is called the **activity quotient** or the **reaction quotient**. It is often designated Q. If the system is at equilibrium, the activity quotient takes on a special value, and to see this, let us first consider equilibrium carefully. In Chapter 1, equilibrium was defined as the arrangement of an isolated system among available configurations that has the maximum entropy. At equilibrium, there is no tendency for the system to spontaneously move away from equilibrium (although it may instantaneously fluctuate around the equilibrium). If temperature is kept constant, as we have been considering in this development, then maintaining a maximum entropy implies that there are two constraints, $dT = 0$ and $dS = 0$. If S and T are used as the variables for G, that is, $G = G(S, T)$, then we can establish that $dG_{eq} = 0$. So long as equilibrium is maintained and temperature is held constant, the energy does not change. This means that at equilibrium ΔG_{rxn} is zero. A very simple application is the phase change of ice melting into liquid water. For equilibrium to be maintained, temperature and pressure must be varied together in order to follow the phase boundary line between solid and liquid water. The condition of equilibrium constrains temperature and pressure. Then, when temperature is held fixed, there is no allowed pressure change that maintains equilibrium. With the differential relation $dG = -SdT + VdP$, the constraints that $dT = 0$ and $dP = 0$ mean that $dG = 0$. Now, in Equation 5.39, setting ΔG_{rxn} equal to zero for equilibrium conditions leads to

$$\Delta G_{rxn}^{\circ} = -RT \ln\left(\frac{\prod_i a_i^{n_i}}{\prod_j a_j^{m_j}}\right)_{eq}$$

(5.40)

The activity quotient at equilibrium is called the **equilibrium constant**, and it is usually designated K or K_{eq}. With these definitions, we can write Equation 5.39 in a very compact form.

$$\Delta G_{rxn} = -RT \ln K + RT \ln Q = RT \ln\left(\frac{Q}{K}\right)$$

(5.41)

Notice that the equilibrium constant and the activity quotient are both dimensionless. If Q/K is greater than 1, ΔG_{rxn} is positive, which means the reaction will not proceed spontaneously toward products. If Q/K is less than 1, ΔG_{rxn} is negative and the products are energetically favored. And, if $Q = K$, the system is at equilibrium.

Le Chatelier's principle (Henry Louis Le Chatelier, France, 1850–1936) says that displacing a system from equilibrium results in a new equilibrium being reached that at least partially offsets the effect of the displacement. Pressure and temperature are two variables that might be adjusted so as to displace a system from equilibrium. Or, we may simply add reactants (or products). When we add reactants, we expect some to be consumed as equilibrium is reestablished. The quantitative information on what happens is contained in Equation 5.41.

From the thermodynamic equivalence that $G = H - TS$, we write that for a reaction carried out at standard state conditions at constant temperature,

$$\Delta G°_{rxn} = \Delta H°_{rxn} - T\Delta S°_{rxn} \tag{5.42}$$

Comparing this with Equation 5.40, we have

$$-RT\ln K = \Delta H°_{rxn} - T\Delta S°_{rxn}$$

$$\ln K = -\frac{\Delta H°_{rxn}}{RT} + \frac{\Delta S°_{rxn}}{R} \tag{5.43}$$

This expression can tell us how the equilibrium constant varies with temperature at constant pressure by taking the first derivative with respect to temperature. From Equations 5.42 and 5.43,

$$\ln K = -\frac{1}{R}\frac{\Delta G°_{rxn}}{T} \tag{5.44}$$

Taking the partial derivative with respect to temperature means

$$\left(\frac{\partial \ln K}{\partial T}\right)_P = -\frac{1}{R}\left(\frac{\partial(\Delta G°_{rxn}/T)}{\partial T}\right)_P \tag{5.45}$$

The right-hand side may be calculated from thermodynamic relations for G since they apply to ΔG functions as well:

$$dG = -SdT + VdP \Rightarrow \left(\frac{\partial G}{\partial T}\right)_P = -S$$

$$G = H - TS \Rightarrow \frac{G}{T} = \frac{H}{T} - S \tag{5.46}$$

$$\therefore \left(\frac{\partial G}{\partial T}\right)_P - \frac{G}{T} = -\frac{H}{T}$$

Differentiation of G/T yields an expression related to the left-hand side of Equation 5.46:

$$\left[\frac{\partial}{\partial T}\left(\frac{G}{T}\right)\right]_P = \frac{1}{T}\left(-\frac{G}{T} + \left(\frac{\partial G}{\partial T}\right)_P\right) \tag{5.47}$$

Combining Equations 5.46 and 5.47 yields the **Gibbs–Helmholtz equation.**

$$\left(\frac{\partial(G/T)}{\partial T}\right)_P = -\frac{H}{T^2} \tag{5.48}$$

Using Equation 5.48 for ΔG and ΔH and then applying Equation 5.45 yields

$$\left(\frac{\partial \ln K}{\partial T}\right)_P = -\frac{\Delta H^\circ_{rxn}}{RT^2} \tag{5.49}$$

This result is the **van't Hoff equation** (Jacobus Henricus van't Hoff, Holland, 1852–1911). For a reaction with an enthalpy change independent of temperature, Equation 5.49 indicates that $\ln K$ diminishes with increasing temperature if the reaction enthalpy change is negative. This means that increasing the temperature pushes the equilibrium toward the reactants for an exothermic reaction. Increasing the temperature favors the products for an endothermic reaction.

Another useful relationship is obtained by invoking Equation 5.43 at two temperatures, T_1 and T_2, with corresponding equilibrium constants, K_1 and K_2, and then taking the difference:

$$\ln K_1 - \ln K_2 = \frac{\Delta H^\circ_{rxn}}{RT_2} - \frac{\Delta H^\circ_{rxn}}{RT_1} + \frac{\Delta S^\circ_{rxn}}{R} - \frac{\Delta S^\circ_{rxn}}{R}$$

$$\ln\left(\frac{K_1}{K_2}\right) = \frac{\Delta H^\circ_{rxn}}{R}\left(\frac{1}{T_2} - \frac{1}{T_1}\right) \tag{5.50}$$

When we know the equilibrium constant at one temperature and the enthalpy of the reaction, we can use Equation 5.50 to predict the equilibrium constant at another temperature. Or, by knowing the equilibrium constant at two temperatures, we can obtain the enthalpy of the reaction.

5.5 Phase Equilibria Involving Solutions

Addition of a solute to a pure liquid affects the phase equilibria. An easily recognized manifestation of this is the freezing point depression of water when sodium chloride is dissolved in it. The first step in understanding phase equilibria of solutions is to examine the changes in thermodynamic state functions that occur on mixing. As already discussed, mixing can be regarded as a reaction, one in which the number of components, c, and the number of moles of each component, for example, n_i, do not change. Thus, the Gibbs energy change for the process can be expressed in terms of chemical potentials of the components after mixing, g_i', and before mixing, g_i.

$$\Delta G_{mix} = \sum_i^c n_i(g_i' - g_i) \tag{5.51}$$

The chemical potentials in Equation 5.51 can be expressed in terms of the standard state chemical potentials and the activities via Equation 5.2, but then the standard state chemical potentials sum to zero as long as the substances, their quantities, and their standard states are the same before and after the mixing process. This yields

$$\Delta G_{mix} = \sum_i^c n_i RT \left(\ln a_i' - \ln a_i \right) \tag{5.52}$$

As previously discussed, a prime designates the after-mixing values.

For each component, we can make a substitution for the activities in Equation 5.52. First, assuming the unmixed components are in their standard states, their activities are identically 1, and so $\ln a_i = 0$. Second, using Equation 5.35, the activities of the mixed components can be expressed in terms of the mole fraction and the component's activity coefficient: $a_i' = X_i \gamma_i$. This simplification of Equation 5.52 leads to

$$\Delta G_{mix} = RT \sum_i^c n_i (\ln X_i \gamma_i) \tag{5.53}$$

We now use the relations among thermodynamic state functions to find the changes in other values on mixing. For instance, $G = H - TS$ implies $\Delta G = \Delta H - \Delta(TS)$. Note that we have been taking the process to be isothermal and isobaric, and so $\Delta(TS) = T\Delta S$. Also, the mole fractions in the mixture are fixed; they do not vary with temperature or pressure, although the activity coefficients might. This means that $d(\ln X\gamma) = d \ln X + d \ln \gamma = d \ln \gamma$. Following are some of the expressions for mixing that are obtained.

$$\Delta V_{mix} = \left(\frac{\partial \Delta G_{mix}}{\partial P} \right)_{T,X} = RT \sum_i^c n_i \left(\frac{d \ln \gamma_i}{\partial P} \right)_{T,X} \tag{5.54}$$

$$\Delta S_{mix} = -\left(\frac{\partial \Delta G_{mix}}{\partial T} \right)_{P,X} = -\sum_i^c n_i R \left[\ln X_i \gamma_i + T \left(\frac{d \ln \gamma_i}{\partial T} \right)_{P,X} \right] \tag{5.55}$$

$$\Delta H_{mix} = \Delta G_{mix} + T\Delta S_{mix} = -RT^2 \sum_i^c n_i \left(\frac{d \ln \gamma_i}{\partial T} \right)_{P,X} \tag{5.56}$$

Notice that if the solutions are ideal, the activity coefficients are 1.0 and the preceding equations simplify considerably. The volume change and the enthalpy change are zero in the limit of ideal behavior.

We have already considered one significant manifestation of mixing that leads to the formation of solution, and that is the effect on vapor pressure. Because of intermolecular interactions, the vapor pressure over a solution of A dissolved in B deviates from the sum of vapor pressures of pure A and pure B (i.e., the Raoult's law vapor pressure). Also because of intermolecular interactions, the boiling temperature of B with A dissolved in it is different from that of pure B. Indeed, the phase equilibrium curves may be displaced, and this is illustrated in Figure 5.4. Let us analyze how this comes about.

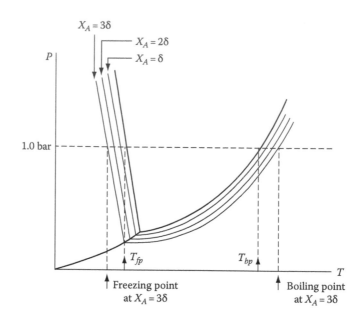

FIGURE 5.4

Progressive effect on phase equilibria of a solvent from increasing the amount of solute. The heavy lines are those of the pure solvent. As the concentration of the solute increases, phase equilibria curves for the liquid–vapor interface and the solid–vapor interface are displaced further and further from those of the pure solvent. This leads to elevation of the boiling point and lowering of the freezing point.

A phase change can be regarded as a reaction. Along a phase equilibrium curve, the reaction is at equilibrium. How is the equilibrium shifted by forming a dilute solution? If we start with pure B and dissolve a small amount of A, the activity of B in the liquid phase changes. (For simplicity, take A as nonvolatile.) Treating the solution as ideal, a good approximation for a low concentration of A in B, the activity of the liquid is the mole fraction, X_B. To understand the effect on the phase equilibrium curve, we ask what the change in activity does at standard state conditions, and we use Equation 5.50 to answer the question. The enthalpy change to use in the equation is that of the phase change. K_1 is the equilibrium constant for the phase change of pure B, and so it is identically 1. (The activities of pure B in two phases at equilibrium are the same.) T_1 is the boiling point of pure B at the given pressure. K_2 is the equilibrium constant when the solution has formed, and it is the ratio of the activity of the gas (1.0) to the activity of the liquid (X_B) for a vaporization reaction. T_2 is the new boiling temperature.

$$\ln\left(\frac{K_1}{K_2}\right) = \ln\left(\frac{1}{1/X_B}\right) = \ln X_B = \frac{\Delta H_{vap}}{R}\left(\frac{1}{T_2} - \frac{1}{T_1}\right) \tag{5.57}$$

$X_B = 1 - X_A$, and $\ln(1 - X_A)$ is approximately $-X_A$ (for $X_A \ll 1$). Thus, the new boiling point, T_2, depends simply on the concentration of A in B.

$$-X_A = \frac{\Delta H_{vap}}{R}\left(\frac{1}{T_2} - \frac{1}{T_1}\right) \quad \text{or} \quad T_2 = \left(\frac{-RX_A}{\Delta H_{vap}} + \frac{1}{T_1}\right)^{-1} \tag{5.58}$$

Notice that $1/T_2$ varies directly with $1/T_1$, which is consistent with the offset phase equilibria curves of the solution (Figure 5.4) following those of the pure solvent. Also, notice that because the enthalpy of vaporization must be positive, the boiling point of the solution (T_2) must be greater than that of the pure solvent (T_1).

The same analysis can be applied to freezing, assuming dilute (ideal) solutions, with the same result, Equation 5.58. Of course, the reaction is to form a solid from a liquid, as opposed to a vapor from a liquid, and therefore, the sign on the enthalpy change is negative. This means that the freezing point is lowered instead of being raised. Figure 5.4 shows this effect clearly.

From this analysis for dilute solutions, that is, when A is at a low concentration in B or when B is at a low concentration in A, we infer that the qualitative behavior of the freezing and boiling point temperatures across the whole range of composition is largely a continuation of the low-concentration behavior. A depresses the freezing point of B, and B depresses the freezing point of A, as shown in Figure 5.5. However, somewhere these curves cross, and this point is called the **eutectic point**. It is the lowest freezing point temperature for any composition of the solution.

A final aspect relating to activities of solutions is **osmotic pressure**. Membranes have been found and developed that are rigid, so that pressure is not transmitted, and selective, so that only certain chemical substances (e.g., water) pass through. If such a membrane is placed between a pure solvent and a solution made from the solvent and something that does not pass through the membrane, there will be a difference in the activities of the solvent on the two sides of the membrane which can be brought into equilibrium by a different pressure on the two sides of the membrane. This is the osmotic pressure. Figure 5.6 shows the type of setup that is used to measure osmotic pressure. If one counteracts the osmotic pressure by applying a stronger external pressure, the system will respond

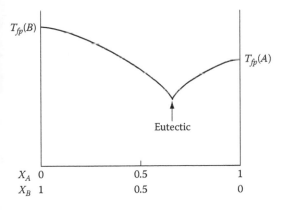

FIGURE 5.5
Freezing point of a hypothetical binary mixture as a function of composition.

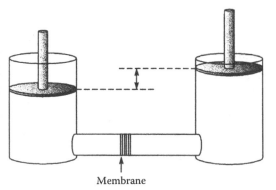

FIGURE 5.6
Osmotic pressure develops in the solution on the right because of the reduced chemical potential of the solvent on the right relative to that of the pure solvent on the left. The pressure is proportional to the height to which the liquid rises in the tube.

by passing solvent molecules from the solution into the pure solvent container. This is the reverse of osmosis, and the process of **reverse osmosis** is a widely used means of purification, especially for water.

The activity, a, of the solvent containing the solute is not 1, and that implies a difference in the chemical potential of the solvent on the right in Figure 5.6 versus the pure solvent on the left, which has an activity of 1, assuming the left container is open to the atmosphere at a pressure of 1 bar. Yet, the chemical potential of the solvent must be the same in the right and left containers because solvent can pass freely through the membrane. Hence, the offsetting change in the chemical potential is the pressure. The incremental contribution to the chemical potential is VdP, and since the volume of a liquid is essentially constant with respect to pressure, the net change is the molar volume of the liquid times the pressure difference, P, between the left and right containers. Therefore, the sum of the effects of the activity difference and the pressure difference on the chemical potential is zero.

$$RT \ln a + \overline{V} \Delta P = RT \ln a + \overline{V} \Pi = 0 \qquad (5.59)$$

This expression can be rearranged to yield the osmotic pressure, P, in terms of the activity. Now, if the solution is very dilute, or if we assume that it is ideal, the activity is the mole fraction, X, of the solvent. Then, as was done in developing Equation 5.58, for a low concentration, we approximate the natural logarithm of X by $-X_{solute}$. That is, $\ln X = \ln(1 - X_{solute}) \approx -X_{solute}$. Equation 5.59 becomes

$$\Pi = \frac{RTX_{solute}}{\overline{V}} \qquad (5.60)$$

Notice that this is almost identical to the ideal gas equation of state, $P = RT/\overline{V}$, if we take Π to be P. So osmotic pressure increases with temperature and with concentration of the solute. In the case where a solute dissociates, then the mole fraction of solute used in Equation 5.60 should be the mole fraction of ions in solution. For instance, this means multiplying X_{solute} by 2 for a substance that completely dissociates into two ions in solution. It is interesting that in some cases the measurement of osmotic pressure by a setup as simple as that in Figure 5.6 can be a very sensitive means for measuring concentration. As you can see, if P is known, Equation 5.60 can be used to obtain X_{solute}.

5.6 Conclusions

This chapter, like the three before it, has considered changes in thermodynamic systems, but it has also been concerned with equilibrium. The condition of equilibrium serves to unify the analysis of chemical reactions, phase changes, mixing, and formation of solutions.

A significant point of the analysis and applications in this chapter is that we mostly need to follow changes in a chemical system in a relative way, meaning with respect to some reference. Standard states are a defined reference point.

The treatment of nonideal systems is helped by the introduction of fugacity for gases and the introduction of activity for all phases. Fugacity is an effective pressure that approaches the true pressure at low pressure, for at low pressures real gases approach ideal behavior.

Activity may be thought of as an effective mole fraction. *RT* times the natural logarithm of the activity of a pure substance is the chemical potential *relative* to the chosen reference (standard state). This means the activity is exactly 1 for a substance in its standard state. For condensed phases, the standard state of a pure substance is its most stable phase. The pressure for standard states is 1 bar for liquids, solids, and the ideal gas. The standard state of a real gas corresponds to its fugacity being 1 bar.

At equilibrium, the activities of the different phases of a given component of a multicomponent system are equal.

The partial vapor pressures of components over a liquid solution do not usually have a perfectly linear dependence on their respective mole fractions in the liquid because of different activities of the components.

Equilibrium is a condition where the activities of the reactants and products are such that $\Delta G = 0$. A system prepared at nonequilibrium conditions will adjust spontaneously toward products if ΔG is less than zero or toward reactants if ΔG is greater than zero. This holds for reactions defined generally to include phase changes, mixing, and so on. From this analysis, we find the basis for an age-old technique, fractional distillation, and we quantitatively understand the phenomena of a solute raising the boiling point and lowering the freezing point of a solvent.

The activity quotient is the product of activities of the products of a reaction divided by the product of activities of the reactants. The value of the activity quotient at equilibrium is the equilibrium constant.

Point of Interest: Gilbert Newton Lewis

Gilbert Newton Lewis

G. N. Lewis had an extraordinary impact on the development of modern chemical thermodynamics, physical chemistry education in the United States, and the development of the chemistry department of the University of California at Berkeley.

Gilbert Newton Lewis was born in Massachusetts on October 25, 1875. His mother and his father, a lawyer, were the source of his education until age 14. His family relocated to Lincoln, Nebraska, and so it was the University of Nebraska where Lewis enrolled in 1891. After 2 years, he transferred to Harvard where he received a BS degree in 1896. A year later, Lewis began graduate work in physical chemistry at Harvard. He completed his PhD thesis on electrochemical potentials in 1899. He held an instructorship at Harvard for several years but eventually chose to move to the Massachusetts Institute of Technology to work with Arthur Amos Noyes (1866–1936). He stayed there for 7 years until accepting an offer from the University of California at Berkeley to become dean of the College of Chemistry in 1912. He died in Berkeley on March 23, 1946.

Lewis' contributions to science stand out prominently in thermodynamics and chemical bonding theory, but they were also quite significant in photochemistry and acid–base theory (from which we have Lewis acids and Lewis bases). Perhaps because of the freedom he experienced in his early education, Lewis had a bent for unorthodox views and an inclination to explore diverse problems. An unpublished paper from his days at Harvard was on blackbody radiation and early elements of the quantum theory of matter and radiation. He postulated that light has a pressure, and using that idea several years later, he derived Einstein's relativistic relationship between mass and energy without invoking the principle of relativity. In the 1920s, he returned once more to this area, seeking paradoxes and unorthodox concepts about space and time—encouraged by the quantum revolution that he saw occurring in physics. However, this proved to be the least successful of his scientific pursuits.

Lewis did his thesis research at Harvard on thermodynamics, and that was the start of one of his most lasting areas of study. He recognized how the whole of existing principles could be applied to more than ideal systems. By 1901, he had proposed the idea of, and coined the term, "fugacity," explaining it as the tendency of a substance to change from one phase to another.

Lewis recognized that the prevailing enthalpic view of chemical reactions did not account for the tendency of reactions to occur and that free energy and entropy were needed for a more predictive picture. He saw fugacity as a fundamental element of the thermodynamics of real systems. However, he never succeeded in obtaining the entropy function from the fugacity. Instead of it being fundamental, fugacity proved more of practical value by converting certain equations for real systems to the form of ideal systems.

In 1907, Lewis' generalization of the concept of fugacity led him to introduce the concept of activity. This has provided the theoretical machinery for treating deviations from ideality, particularly for solutions. Indeed, Lewis' achievements in thermodynamics are very much of a practical sort, advancing applications to nonideal systems. In 1923, writing with Merle Randall, Lewis collected his working knowledge of thermodynamics into a text. With its publication, thermodynamics was increasingly regarded as essential in chemical education. The text was revised in 1961 by Kenneth S. Pitzer and Leo Brewer and remained an important volume.

In 1916, Lewis published a paper arguing that chemical bonds are a shared pairs of electrons. He gave a cubic picture of atoms in which the shared electrons were at adjoining edges of cubes. This led to the octet rule for valence electronic structure, and Lewis used this to account for the structures of a number of molecules that had not been solved. Lewis' scientific work was interrupted by World War I. In 1918, he was in France as part of the Chemical Warfare Service. In the first half of 1919, Irving Langmuir brought sudden attention to Lewis' picture of chemical bonding. He elaborated on Lewis' idea and successfully advanced what came to be called Lewis–Langmuir theory. The cubic picture of atoms as static distributions of electrons became obsolete with the development of quantum mechanics, but the idea of shared electron bonding survived.

Lewis' mark on chemical education comes not only from his thermodynamics text but also from his building of one of the leading chemistry departments in the United States. He was persuaded to move to Berkeley in 1912 by the president of the University of California. He gave up the position of deputy director of the Research Laboratory of Physical Chemistry at MIT but took with him several graduate students and colleagues. Lewis rapidly transformed the chemistry department at Berkeley into a major research center. Under Lewis, Berkeley moved to the top among U.S. institutions in the area of physical chemistry. At the same time, Lewis discouraged divisionalization in administering the department, preferring to see all of his faculty as simply professors of chemistry. The department at Berkeley grew in size during this period and became one of the leading producers of PhD chemists. The percentage of Berkeley undergraduates enrolled in chemistry doubled in 6 years. Lewis' success was noticed and emulated elsewhere, and that helped move physical chemistry from a rather peripheral role to a central role in chemical education. In building his department, Lewis had developed a reputation as a shrewd judge of, and tough competitor for, talented chemists.

Exercises

5.1 Find the volume of 1 mol of an ideal gas at a pressure of 5 bar and a temperature of 500 K. Then, using the van der Waals gas equation (Equation 2.31) and the values in Table 2.1, compute the pressure of 1 mol samples of the following gases if they are at the same volume and temperature: neon, mercury, and benzene.

5.2 Subtract the ideal gas equation of state from each of the following equations of state, Equations 2.29 through 2.31, to give expressions for the difference between real and ideal gas behavior. Then, show that at constant temperature, these expressions go to zero in the limit of pressure going to zero.

5.3 At 300 K, an alcohol's partial pressure over a 0.5 mol water solution is found to be 0.05 bar, whereas over the pure liquid it is 0.08 bar. The partial pressure of water over the solution is found to be 0.02 bar compared to 0.03 bar over pure water. Compare the activities of the liquids in solution with their mole fractions. What are the values of the activity coefficients (ratio of activity to mole fraction)?

5.4 At a temperature of 373 K, two organic solvents are found to have equilibrium vapor pressures that are in the ratio of very nearly 1–3. On that basis, find the mole fractions of the two in the vapor over a solution of the two with the mole fraction of the first component being (a) 0.25, (b) 0.5, and (c) 0.75.

5.5 Using Equation 5.58 and values in Table 4.1, compute the changes in freezing and boiling point temperatures of water, hydrogen sulfide, carbon tetrachloride, methanol, and benzene from dissolving some soluble component A such that its mole fraction X_A is 0.01.

5.6 Apply Equation 5.58 and use values in Table 4.1 to estimate the mole fraction of silver that will be found in a silver–gold solid solution if its melting point is 100 K below that of pure gold.

5.7   The boiling point of 1,2-dichloroethane at 1 bar is 83.5°C. Calculate the molar heat of vaporization and the vapor pressure at 75°C.

5.8   At a certain temperature, the vapor pressure of some pure substance X is 98 mm Hg and the vapor pressure of another some pure substance Y is 41 mm Hg. The vapor pressure of X follows the equation

$$P_X = 98(0.15x + 0.85x^2)$$

and that of Y follows the equation

$$P_Y = 41(1 - 0.18x - 0.82x^2)$$

where x is the mole fraction of X in the solution. Calculate the mole fraction of X in the vapor if the mole fraction in the solution is 0.36.

5.9   What is the solubility (in moles of naphthalene per kg of toluene) of naphthalene in toluene at 20°C? The heat of fusion of naphthalene is 148.3 J g^{-1} and its normal melting point is 80.3°C.

5.10 The vapor pressure of an aqueous solution of 125.0 g of solute in 1.00 L of water at 35.0°C is 37.5 mm Hg. What is the molar mass of the solute? The vapor pressure of pure water at this temperature is 42.2 mm Hg.

5.11 Considering gas equations of state as general power series expansions in P, V, and T, identify a form for a gas equation of state that would not approach ideal behavior in the low-pressure limit. Find one feature of this equation that makes it an unphysical choice for representing the equation of state of a real gas.

5.12 Find the fugacity for a sample of 1 mol of neon at $T = 400\,K$ and $P = 1$ bar if the equation of state is

$$\overline{V} = \frac{RT}{P} + b \quad \text{and} \quad b = 0.017\,L\,mol^{-1}$$

5.13 Find an expression for the activity, a, of a gas in terms of the temperature and pressure if the gas obeys a Berthelot gas equation,

$$P\overline{V} = RT + cP\left(1 - \frac{d}{T^2}\right) \quad \text{where } c \text{ and } d \text{ are constants}$$

5.14 The **compressibility factor** of a gas, Z, is defined by $Z = P\overline{V}/RT$, a quantity that is 1 for an ideal gas but that is a function of P, V, and T for real gases. Find Z for a gas that has Equation 2.29 as its equation of state.

5.15 Using the definition of Z in Exercise 5.14, show that in general the integrand in the fugacity coefficient equation, Equation 5.18, can be replaced by $RT(Z - 1)/P$.

5.16 Find the fugacity at $T = 400\,K$ and $P = 100$ bar of a gas that has Equation 2.32 as its equation of state with the molar constant b having the value 5.0×10^{-2} L mol^{-1}. Repeat the evaluation at the same temperature but at a pressure of 1 bar.

5.17 Derive an expression for the fugacity of a gas that obeys Equation 2.29.

5.18 Derive an expression for the fugacity coefficient of a van der Waals gas. To do this it will be helpful to find dP in terms of dV from the equation of state and then carry out the necessary integration over the molar volume rather than over pressure.

5.19 In a binary liquid solution at 300 K, the partial pressure of component 1 is found to have a quadratic dependence on the mole fraction of component 2:

$$P_1 = aX_1(1 + bX_2^2)$$

Taking a as having the value of 10 bar and b (dimensionless) to be 2.1, make a plot of P_1 versus mole fraction of the solution. Draw lines to indicate the Raoult's law and Henry's law behavior. Determine the activity of component 1 at 0.5 mole fraction.

5.20 When tap water is heated, one often sees bubbles form. When tap water is frozen, bubbles are often seen trapped in the ice. Explain the likely cause(s) of these phenomena and explain how to predict the volume of the bubbles that would form for a given quantity of water.

5.21 Equation 5.50 is approximate unless ΔH is independent of temperature. Consider a temperature-dependent form of the reaction enthalpy change: $\Delta H = a + bT + cT^2$. Integrate Equation 5.49 with this form of the enthalpy change and compare with Equation 5.50.

5.22 Following the development in Equations 5.54 through 5.56, find an expression for ΔU_{mix} and $\Delta C_{P,mix}$. Evaluate these and the three values in Equations 5.54 through 5.56 for a mixing process that produces an ideal solution.

5.23 [Spreadsheet problem] Using values in Table 4.1 for N_2O and Cl_2, apply Equation 5.58 to find the boiling points for when the mole fraction of a solute, X_A, is 0.05, 0.1, 0.15, 0.2,..., 0.95. If a solution were formed with N_2O and Cl_2, at what mole fraction values would the boiling point temperature curves for the two species cross? (Determine the eutectic point by plotting both sets of temperature values against mole fractions.)

5.24 If one wished to use reverse osmosis to "force pure water out" of water containing a solute, what pressure would be needed if the solute concentrations were on the order of parts per thousand? In considering home RO (reverse osmosis) water purification systems, how does this pressure compare with the difference between the water pressure of the water system and atmospheric pressure?

5.25 The vapor pressure of ethanol has the following temperature dependence:

T (°C)	P (mm Hg)
0	11.8
15	32.2
30	78.5
45	173.2
60	351.8

Plot these data points in an appropriate manner and calculate the heat of vaporization and the normal boiling point of ethanol.

Bibliography

Kaufman, M., *Principles of Thermodynamics* (Dekker, New York, 2002).
Klotz, I. M. and R. M. Rosenberg, *Chemical Thermodynamics* (Wiley, New York, 1994).

6

Chemical Reactions: Kinetics, Dynamics, and Equilibrium

From a macroscopic view, a chemical reaction has taken place when one or more substances have changed into one or more substances chemically distinct from the original ones. At the atomic and molecular levels, it is the making and breaking of bonds that signifies the occurrence of a reaction. A reaction implies significant changes in the participating atomic and molecular species. Thus, the energetics and entropics of the bonding changes influence whether or not a reaction takes place. Reactions are also subject to external influences and are affected by changes in thermodynamic state variables such as temperature and pressure. This is manifested in the rates at which reactions take place. This chapter begins with a microscopic view of chemical reactions and the molecular energetics of reactions. It then adds kinetic and thermodynamic analyses to account for the different macroscopic rates at which reactions proceed. It concludes with a discussion of electrochemical reactions and cells in which another external influence, an electrical potential, affects the reaction.

6.1 Reaction of Atoms and Molecules

The formation of molecule Z from the reaction of molecule X with molecule Y is recognized in a laboratory by some means that detects the presence of molecule Z in a reaction vessel that initially contained only molecules X and Y. At the microscopic level, one may ask at what instant the reaction actually takes place. In part, the answer to this question lies in tracking the approach and interaction of molecules X and Y. A conceptual device for that purpose is the **potential energy surface**.

A potential energy surface is simply the energy of a system of atoms expressed or represented as a function of the atomic position coordinates in space (or relative position coordinates). "Surface" is used to refer to this energy function by analogy with considering some function $z(x, y)$ in Cartesian space to be a surface. At each (x, y) point, the height of the surface above the x–y plane is $z(x, y)$. A potential energy surface may be a function of more than two variables, depending on the complexity of the reacting species.

A Cartesian surface, for example, $z(x, y)$, may be represented in a two-dimensional graph by means of contours, curves drawn through (x, y) points for which z is some particular value. The same may done for potential energy surfaces, but if there are more than two variables, all but two must be fixed. This is a **slice** through a multidimensional surface.

Consider the hypothetical reaction of an atom A with a diatomic molecule BC that produces the diatomic AB plus the atom C. The distances from A to B and from B to C are the variables that can be used for the horizontal and vertical axes (x–y axes) in graphically representing the surface. Such a potential energy surface is depicted in Figure 6.1, and we can use it to follow the course of a reaction event. The outset of the process of A approaching BC corresponds to a path along the potential energy surface in Figure 6.1 that starts at the upper left. As the A-to-B

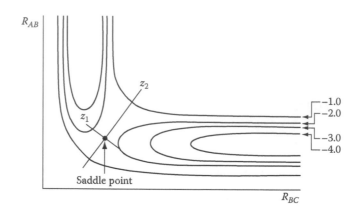

FIGURE 6.1

Contours in arbitrary energy units of the potential energy surface for the hypothetical colinear reaction $A + BC \rightarrow AB + C$. The horizontal axis is the B-to-C separation distance, and the vertical axis is the A-to-B distance. At the saddle point, a special coordinate system is defined by the two axes z_1 and z_2. Taking the saddle point as the origin of this system (i.e., $z_1 = 0$ and $z_2 = 0$), we see that z_1 is the direction in which the energy contours lead downhill away from the saddle point, whereas z_2 is a direction in which the energy is greater away from the saddle point.

distance diminishes, the path leads through the region of the **saddle point**. The saddle point is the "mountain pass" through the energy potential "mountain peaks" on either side. Once beyond the pass, the path becomes downhill in energy and it curves so that the B-to-C distance then changes much more quickly than the A-to-B distance. In this region, the A–B bond has formed and atom C departs from the collective system.

The saddle point is very important in understanding the reaction because if the A, B, and C atoms were arranged so as to be at the saddle point with zero kinetic energy—a hypothetical circumstance that is precluded by quantum mechanics—then there would be no net force toward either reactant or product directions. This suggests choosing the saddle point as a dividing point between reactants and products. Actually, we should regard the *line* along the whole "mountain pass" as the boundary between reactants and products because nothing says that the collision of A with BC always goes precisely through the saddle point. In other words, we use something analogous to a "continental divide" to distinguish reactants from products. The geometrical arrangement of the system at the saddle point is called the **transition state structure**, or sometimes simply the **transition state**. The reaction may occur by passing from one trough to the other through the region surrounding the saddle point.

As shown in Figures 6.1 and 6.2, it is possible to define a coordinate axis, z_1, that passes from one trough to the other directly through the transition state. The orthogonal direction, z_2, follows the "continental divide." The potential energy at steps along z_1 is depicted in Figure 6.3. A convenient choice for this coordinate system origin ($z_1 = z_2 = 0$) is the saddle point, and then the saddle point is identified as the extremum such that

$$\left. \frac{\partial V(z_1, z_2)}{\partial z_1} \right|_{\substack{z_1=0 \\ z_2=0}} = 0$$

$$\left. \frac{\partial V(z_1, z_2)}{\partial z_2} \right|_{\substack{z_1=0 \\ z_2=0}} = 0$$

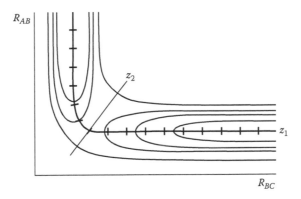

FIGURE 6.2
Minimum energy path (heavy line) superimposed on the contours of the potential energy surface for the hypothetical colinear reaction $A + BC \rightarrow AB + C$, as in Figure 6.1. The minimum energy path follows the bottom of one trough to the transition state structure and then follows the bottom of the other trough. We can consider the potential energy as a function of the position along the minimum energy path, Figure 6.3, by considering the path to be a (curved) coordinate axis with regular intervals to mark distance along the curve.

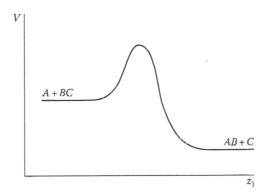

FIGURE 6.3
One-dimensional slice following the z_1 coordinate of the potential energy surface in Figure 6.2. The curve gives the energy following the minimum energy path as a coordinate axis. The highest point on this curve is the saddle point, and the energy of that point relative to the energy of the reactants is the energetic barrier that must be surmounted for the reaction to take place.

The curvature (second derivative), though, is different for the z_1 and z_2 directions. Holding $z_2 = 0$ and moving in the z_1-direction from some point $z_1 < 0$, the potential goes up until $z_1 = 0$. Then it goes down. This means that

$$\left. \frac{\partial^2 V(z_1, z_2)}{\partial z_1^2} \right|_{\substack{z_1 = 0 \\ z_2 = 0}} < 0$$

Holding $z_1 = 0$ and moving in the z_2 direction, the potential goes down as we approach $z_2 = 0$. The curvature in this direction is opposite the curvature in the z_1 direction, and so

$$\left. \frac{\partial^2 V(z_1, z_2)}{\partial z_2^2} \right|_{\substack{z_1 = 0 \\ z_2 = 0}} > 0$$

If there were more than two degrees of freedom for the system, the general defini-tion of the transition state would be the point at which all the first derivatives of the potential energy are zero, and at which the curvature in one and only one direction is negative. The coordinate axis of that direction is called the **reaction coordinate** or **reaction path**.

A line, or generally a potential surface slice, drawn through the saddle point perpendic-ular to the reaction path serves as a boundary between reactants and products, between $A + BC$ and $AB + C$. The basis of an idea called transition state theory is to use this bound-ary to decide at the microscopic level if a reaction has taken place. Simply, if the dynamic path of motion taken by the species crosses the boundary, then a reaction, either forward or backward, is said to have occurred. This basis for deciding if a reaction has taken place avoids complications of using criteria that relate to the bonding. For instance, we might try to say that the reaction has taken place when one bond is broken or when a new bond is formed, but it would be difficult to define such a point. In contrast, the transition state structure is an unambiguous feature of a potential energy surface. The two ideas are not at odds; it is usually the case that chemical bonds are being formed and are being broken in the vicinity of the transition state structure.

There are a number of types of reaction potential surfaces for three atoms. In the hypo-thetical case just considered, one trough leads to an energetic barrier, the "mountain pass," and then down to another trough. Another type of surface we can distinguish is that of a barrierless exoergic reaction. For this type of surface, the reactant trough leads continu-ously downhill to the product trough, and so there is no barrier. Still another type of sur-face is one for which there is an intermediate, a feature manifested on a potential surface as a local minimum. The minimum can be shallow, implying at best a fleeting existence for the species associated with the structure at that minimum, or it can be deep. For instance, in the reaction $H + CN \rightarrow HC + N$ there is some potential energy surface point correspond-ing to the stable triatomic molecule HCN. This point is the bottom of a well in the potential energy surface, a feature depicted in the hypothetical surface in Figure 6.4. Notice that even with the existence of a potential well, transition states occur.

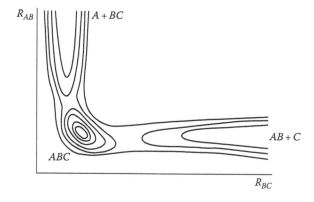

FIGURE 6.4
Contours in arbitrary energy units of the potential energy surface for the hypothetical colinear reaction involv-ing a stable triatomic molecule, ABC. The overall process is $A + BC \rightarrow ABC \rightarrow AB + C$. The horizontal axis is the B-to-C separation distance, and the vertical axis is the A-to-B distance. There is a saddle point between the $A + BC$ trough and the ABC triatomic (potential surface minimum) and a saddle point between the minimum and the $AB + C$ trough.

6.2 Collisions and Transport

A collision event among structureless ideal gas particles may be manifested by a change in momentum. The same holds for certain macroscopic objects. For billiard balls arranged on a pool table, or for air hockey pucks, one indication that a collision has occurred is clear-cut, a deflection of moving objects from their initial straight-line path. These hard-sphere objects happen to collide with a particularly simple interaction potential, one that is zero except for the point of contact when it is infinite or essentially infinite. The absence of a more elaborate potential makes it easy to anticipate the outcome of a collision by analyzing the positions of the objects and the direction of their motions. In Figure 6.5, we see a ball of diameter $2r$ moving toward two other balls designated A and B. From the direction of motion, it is clear that there will be a collision with B but not with A. The moving ball will hit the first ball that extends into the cylinder whose axis is the direction of motion and whose radius is r. Notice that the center of any ball touching the outside surface of the cylinder is at a distance equal to its radius from the cylinder. Thus, the minimum distance for a collision is the sum of the radii of the colliding spheres.

At long range, molecular interaction potentials are weak relative to chemical bonding interactions. The potential surfaces along coordinates that correspond to the separations between molecules have very shallow features. In other words, the detailed potential surfaces discussed in the prior section are important when the interacting particles (atoms and molecules) are quite close to each other. We can analyze the likelihood that the particles will be quite close by ignoring the interaction potential with its shallow features and treating the objects as hard-sphere gas particles. This is truly a long-range analysis, strictly appropriate at the large separation distances where the interactions are nil. We can then improve the picture to account for the short-range features. Thus, the problem of collision rates is one of classical and statistical mechanics.

How often do collisions occur in some interval of time for some distribution of gas particles? Let us assume that the distribution of particles is uniform with a density of D; that is, there are D particles per unit volume. If all the particles have the same radius, r, then we can envision a cylinder with a radius twice that (the sum of the radii of two particles), $R = 2r$, whose volume includes the *centers* of all particles colliding with one given particle. If we multiply the density by the volume of this cylinder for one moving particle, the result will be the number of collisions, N, that take place as the particle moves through the cylinder. The volume of the cylinder is the area of its cross section times its length. The cross-sectional area is πR^2, and the length of the cylinder is the distance the ball moves in the increment of time, Δt,

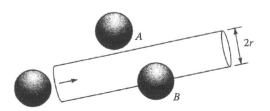

FIGURE 6.5
A solid ball moving in the direction of the arrow will collide with ball B but will miss A. B extends into the cylinder whose axis is the direction of motion and whose diameter is $2r$, where r is the radius of the balls. A is entirely outside the cylinder. The center of a ball just touching the surface of the cylinder will be at a distance $R = 2r$ from the center of the moving ball at the instant of impact.

that we are interested in. That distance is the product of the velocity of the ball, v, and Δt. Therefore, the number of collisions may be expressed as

$$N = \pi R^2 v \Delta t D \tag{6.1}$$

Notice that N is dimensionless. It is the number of collisions, and it is given by the product of an area (length squared), a velocity (length per unit time), a period of time, and a density which is a number per unit volume (length cubed).

From Equation 6.1, the number of collisions for one particle per unit time is simply $N/\Delta t$. The *average* number of collisions in a sample per unit time per particle is $N/\Delta t$, but with the velocity, v, replaced by the mean or *average* relative speed, $\langle s \rangle$, of the distribution of particle speeds. This number of collisions is designated z.

$$z = \pi R^2 \langle s \rangle D \tag{6.2}$$

$\langle s \rangle$ is given by Equation 2.17 with one difference. The analysis for the *relative* speed among particles requires the use of an effective mass or reduced mass, μ, in place of the mass of one particle in Equation 2.17: $\mu_{ij} = m_i m_j / (m_i + m_j)$. If all the particles have the same mass, then $\mu = m/2$. Defining σ to be πR^2, an area referred to as the cross section of a hypothetical cylinder, and substituting for $\langle s \rangle$ yields

$$z = \sigma D \sqrt{\frac{8kT}{\pi \mu}} \tag{6.3}$$

The total number of collisions per unit volume is the number of collisions per particle times the number of particles per unit volume, which is the density, D, divided by 2 to avoid double-counting each collision.

If there are two types of particles, A and B, then there will be a density for each, D_A and D_B. The number of collisions per unit time per unit volume will then include collisions between A and A, B and B, and A and B. Thus, there will be a reduced mass for each type of pair of particles, μ_{AA}, μ_{AB}, and μ_{BB}, and there will be a cross section specific to each pair, σ_{AA}, σ_{AB}, and σ_{BB}. To find the rate of collisions of A with A, we multiply the expression in Equation 6.3 by $D_A/2$. The result is a quantity designated Z_{AA}. (Z is used for the *total* number of collisions, whereas z is used for the number of collisions per particle.)

$$Z_{AA} = \sigma_{AA} D_A \sqrt{\frac{8kT}{\pi \mu_{AA}}} \frac{D_A}{2} \tag{6.4}$$

Likewise, the number of collisions per unit time per unit volume of B with B is

$$Z_{BB} = \sigma_{BB} D_B \sqrt{\frac{8kT}{\pi \mu_{BB}}} \frac{D_B}{2} \tag{6.5}$$

Finally, the number of collisions of A with B is obtained with Equation 6.3, using the density of one species times the density of the other.

$$Z_{AB} = Z_{BA} = \sigma_{AB} D_B D_A \sqrt{\frac{8kT}{\pi \mu_{AB}}} \qquad (6.6)$$

To find the total collision rate, we multiply each of the densities in the expressions in Equations 6.4 through 6.6 by the volume of the sample. Notice that the density times the volume of the sample equals the molar concentration of that species times Avogadro's constant, N_A:

$$D_A V = N_A [A]$$

Therefore, the rate of collision of A and B per unit time in the sample is

$$Z'_{AB} = \sigma N_A^2 [A][B] \sqrt{\frac{8kT}{\pi \mu_{AB}}} \qquad (6.7)$$

Should there be the possibility of a chemical reaction between A and B in some sample, then the reaction cannot occur unless A and B collide. On this elementary basis, we can argue that the reaction rate cannot exceed the collision rate, Z'_{AB}, given by Equation 6.7.

It should be noted that the difference between the collision rate for billiard balls and the collision rate for real molecules is that the cross section for molecules cannot be stated in terms of the radius, r, of a molecule as was done for colliding balls. Nonetheless, there will be a cross section for two species coming together and, in a rough sense, the cross section should be related to the geometries of the two interacting species and perhaps to the close-in potential surface features. But rather than considering that here, we shall regard the cross section as a phenomenological value.

The **mean free path**, λ, is a value that further characterizes the collision events of a system of particles. It is the average distance traveled between collisions, and thus it is found as the average speed $\langle s \rangle$ divided by the collision rate z from Equation 6.3.

$$\lambda = \frac{\langle s \rangle}{z} \qquad (6.8)$$

If the system is a mixture of particles A and B, then there will be mean free paths for A particles, λ_A, and B particles, λ_B, that may be different because of different A and B masses and different diameters.

The next question we need to ask about reaction probabilities is this: Does every collision of molecule A with molecule B lead to an $A + B \rightarrow C$ reaction event? The nature of potential energy surfaces reveals one requirement for a collision event to lead to a reaction: The reactive system must pass any activation barrier that is present. In later chapters, it will be shown that the quantum mechanical picture of atoms and molecules reveals the possibility of "tunneling" through a potential barrier (not going over the mountain pass but going through the mountain). Other than by this means, getting past a barrier means the system must have more energy than the potential energy at the saddle point. (The possibility of quantum mechanical tunneling will not be considered at this point; however, we may keep in mind that tunneling could serve to increase the rate of any reaction we analyze.)

The kinetic energy of two colliding particles is energy that is available for conversion to potential energy; it is energy that may take a system up and over a saddle point.

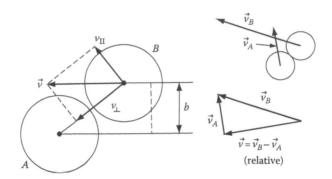

FIGURE 6.6
The collision of two hard spheres with relative velocity $\vec{v}$ can be analyzed at the point of impact in terms of components of the velocity along the line connecting the mass centers and perpendicular to that line. The impact parameter, b, is the projection of the vector between the mass centers onto an axis perpendicular to the velocity vector. On the right is the relative velocity vector in terms of the individual velocity vectors of the colliding particle.

However, not all the kinetic energy is available. The relative directions of the colliding particles are quite important. If you were hitting a baseball high into the air with a bat, which would convert more of the bat's kinetic energy into ball height (overcoming the gravitational potential), "connecting" with the ball or a glancing hit? The glancing hit would deliver only a fraction of the available kinetic energy to the ball. The same holds for two colliding spheres. Figure 6.6 illustrates a glancing collision of two spheres, one fixed and one moving. The component of the velocity of the moving particle at the point of hard-sphere contact has one component directed along the line of centers of the two particles. This is the direction that corresponds to going directly up the "mountain pass" on the potential energy surface. The component of the moving particle's velocity vector in the orthogonal direction, on the other hand, corresponds to motion in a direction orthogonal to the minimum energy path (along the side of the mountain range).

The projection of the vector between the two centers in Figure 6.6 onto an axis perpendicular to the relative velocity vector is called the **impact parameter**. The percentage of the kinetic energy available for a reactive system to cross a potential energy barrier increases as the impact parameter is diminished. Glancing collisions tend to have a smaller likelihood for reaction than collisions with zero impact parameters.

If molecules are involved in a reaction process, the orientation of the molecule or molecules is an important factor in the likelihood of reaction. If atom A can react with molecule BC to produce $AB + C$, whereas $AC + B$ is not energetically favored, then A stands a much better chance of reacting with the BC molecule if it approaches from the B end than from the C end.

If two colliding molecules do not react, they may scatter elastically (no change in their individual internal state energies and no change in total kinetic energy) or inelastically (with changes in internal state energies). The nature of the interaction between the colliding species and the manifold of internal quantum states of the colliding species influences the likelihood of elastic and inelastic scattering.

An atomic-molecular level view of reaction rates draws on the analysis of collision probability, distribution of orientations in collision events, competition with nonreactive scattering, and so on. In principle, computer simulations offer a means of predicting reaction rates by following, and averaging over, a large number of trajectories of the reactant particles, with the initial conditions of the set of trajectories chosen to match properties of a macroscopic sample, for example, temperature. Neglecting quantum mechanical tunneling, the trajectories are paths along the reactive system's potential energy surface dictated by initial conditions (velocities and molecular orientations) and by Newton's laws of mechanics. One trajectory may correspond to a head-on reactive collision, whereas another may correspond to an approach that does not take the system over the saddle point. With a sufficiently large sampling of trajectories, one simulates the distribution of events.

The microscopic picture of reactions is qualitatively consistent with macroscopic observations of the rates of reactions. Laboratory measurements of most reaction systems show a rather simple dependence of the reaction rate on the concentrations, and this is consistent with an overriding requirement for reaction, that the reacting particles collide. Our analysis of collision events shows the dependence of a rate on a concentration; however, the proportionality factor between the rate and the concentration, called the rate constant, depends on temperature and on numerous properties of the reacting species and their interaction potential surface.

A **transport property** is the capacity of a substance to transport or transfer matter, momentum, energy, heat, electric charge, and so on, such transport being accomplished, in part, via collisions. **Diffusion, viscosity,** and **thermal conductivity** are among the many transport properties of a substance. The speed of sound is determined by the ability of a substance to transmit a compression wave. Without giving the derivation, we state that the speed of sound in a gas or a liquid is

$$s_{sound} = \left(\frac{C_P}{C_V} \frac{kT}{m} \right)^{1/2} \tag{6.9}$$

For an ideal monatomic gas, the ratio of the heat capacities is 5/3, and so the speed of sound has a simple relationship to $\langle s \rangle$, the mean speed.

$$\frac{\langle s \rangle}{s_{sound}} = \frac{\sqrt{8kT/\pi m}}{\sqrt{5kT/3m}} = \sqrt{\frac{3(8)}{5\pi}} = 1.2361$$

Flux describes the rate of flow of something across some area or boundary. Flux is expressed in units of the substance that is flowing per unit area per unit time. Thus, electric flux is expressed in charge per cm^2 per second. A phenomenological result, that is, the result of observation and measurement, is that flux tends to be proportional to a gradient in whatever is flowing. To illustrate this, let us consider a gas not presumed to be uniformly distributed in the space of some vessel. The positions of the gas particles are given in an (x, y, z) coordinate system with axes such that the plane $z = 0$ divides the left-hand side of the vessel from the right-hand side. We will try to determine whether gas particles in the system are moving to the left or to the right with respect to the plane at $z = 0$. Let $N(z)$ be the number of particles per unit length at some instant. We may argue that the number of molecules going from left to right at $z = 0$ at that instant varies as the number of molecules that were within the mean free path, λ, on the left, whereas the number going from right to left at $z = 0$ varies as the number at $z = +\lambda$, that is, $N(\lambda)$. That is, particles at a position of $z = -\lambda$ have a chance of crossing the plane $z = 0$ before any further collision event has occurred, and they will be moving to the right. Particles at $z = \lambda$ have a chance of crossing the plane at $z = 0$ and will be moving to the left. The rate at which these particles cross the $z = 0$ plane (the boundary for measuring the particle flux) is the number of particles times the mean speed of the particles. The net flux, J, obeys the following proportionality:

$$J \propto \langle s \rangle N(-\lambda) - \langle s \rangle N(\lambda) \qquad (6.10)$$

power series expansion (see Appendix A) of N about $z = 0$ yields

$$J \propto \langle s \rangle \left(N(0) - \lambda \frac{dN}{dz}\bigg|_{z=0} + \frac{\lambda^2}{2} \frac{d^2N}{dz^2}\bigg|_{z=0} + \cdots \right)$$

$$- \langle s \rangle \left(N(0) + \lambda \frac{dN}{dz}\bigg|_{z=0} + \frac{\lambda^2}{2} \frac{d^2N}{dz^2}\bigg|_{z=0} + \cdots \right)$$

$$= -\langle s \rangle \left(2\lambda \frac{dN}{dz}\bigg|_{z=0} + \frac{\lambda^3}{3} \frac{d^3N}{dz^3}\bigg|_{z=0} + \cdots \right) \qquad (6.11)$$

Often, a good approximation is to truncate this expression after the first term. Doing that shows that the flux is proportional to the mean speed and the concentration gradient (dN/dz).

$$J \propto -\langle s \rangle \left(\lambda \frac{dN}{dz}\bigg|_{z=0} \right) \qquad (6.12)$$

It is convenient to designate this gradient (dN/dz) as N_z. If N_z is zero, as is the case in a uniform distribution, then the net flux of particles is zero. On the other hand, a nonzero gradient gives rise to a flux in the opposite direction, one that eventually may diminish the gradient. A phenomenological approach at this point is to assign a value, D, as a constant of proportionality.

$$J(matter) = -DN_z \qquad (6.13)$$

D is referred to as the **diffusion coefficient** of a substance.

For an ideal monatomic gas, $U = 3nRT/2$, and the flow of energy (i.e., heat) through a substance depends on the temperature gradient, dT/dz. The flux of energy is proportional to this gradient via a phenomenological constant, κ, called the **coefficient of thermal conductivity**.

$$J(energy) = -\kappa \frac{dT}{dz} \qquad (6.14)$$

The **coefficient of viscosity** is the constant in a similar expression for the flux in momentum transfer. It is proportional to the gradient in momentum along a direction orthogonal to the flow. For a substance of uniform composition, this is proportional to the gradient in particle velocities simply because $p = mv$, and the masses are identical.

$$J(momentum) = -\eta \frac{dv_{x,y}}{dz} \qquad (6.15)$$

The analysis leading to Equation 6.12 shows a dependence of the flux on the mean speed. For Equations 6.13 and 6.14, this dependence is rolled into the constant of proportionality, and so thermal conductivity, viscosity, and diffusion coefficients have a $T^{1/2}$ dependence on temperature. These coefficients and their temperature dependence may be important in reaction probabilities for systems in which gradients exist for the densities of the reactants.

6.3 Rate Equations

A macroscopic **reaction rate** is defined as the derivative of the concentration of some species involved in the reaction with respect to time.

$$R = -\frac{d[X]}{dt}\frac{1}{v_X} \tag{6.16}$$

where
   X is the chemical symbol of a reactant species
   $[X]$ denotes its concentration
   v_X is the signed stoichiometric number for species X in the overall atom-balanced reaction

If X is a product species, v_X is negative corresponding to X increasing as the reaction progresses, whereas v_X is positive if X is a reactant. In general, the rate R is some function of the concentrations of the species in the system, and so $R = R([A], [B], [C],...)$. A **rate equation** is a specific equality between the time derivative in Equation 6.16 and the function of concentrations, $R([A], [B], [C],...)$; it is a differential equation in time. The concentrations, for example, $[A]$ and $[B]$, are functions of time because as a reaction progresses, concentrations change. The units for a reaction rate are the units of concentration and inverse time.

Reactions that have rate equations with a linear dependence on only one concentration are said to be **first order reactions**. Reactions that depend on the concentration of only one species to the second power or that depend on the product of the concentrations of two species, each to the first power, are **second order reactions**. If a reaction happens to be independent of concentration, it is a **zero order reaction**. In general, if a reaction rate is given by

$$R = k[A]^{n_A}[B]^{n_B}[C]^{n_C}\cdots \tag{6.17}$$

where k is a constant, then the sum of exponent values,

$$n = n_A + n_B + n_C + \cdots \tag{6.18}$$

is the order of the reaction. Each of the exponents, such as n_A or n_B, is referred to as the order of the reaction with respect to A or to B.

An **elementary reaction** is one in which the reaction occurs in a single collision. A number of elementary reaction steps may be involved in an overall reaction. The **molecularity** of an elementary reaction refers to the number of reactant molecules involved in the reaction step. A spontaneous decomposition reaction, for instance, is a **unimolecular** reaction. A reaction involving two combining species, such as

$$H_2 + Cl \rightarrow HCl + H$$

is a **bimolecular** reaction. And if three reactants are involved in one step, the reaction is **termolecular**. We can normally expect the molecularity and the order of the reaction to be the same for an elementary reaction.

A **complex reaction** is one requiring more than one collision or involving several elementary reaction steps. It is possible that the order of a complex reaction with respect to some species does not turn out to be positive or even that the overall order is not an integer. The order of the reaction, then, is more an observational result of the dependence of the rate on concentrations.

The proportionality factor in a rate equation is usually designated k, and it may be subscripted to indicate a particular reaction step. In an experiment, the determination of the reaction order and of the rate constant may be primary goals. Certainly, their measurement provides key data for understanding the detailed nature of the reaction events.

The rate equation for a zero order reaction is simply

$$R = k[A]^0 = k \tag{6.19a}$$

Taking A to be the reactant, this expression is the same as

$$-\frac{d[A]}{dt} = k \tag{6.19b}$$

Integrating this expression over time, from the initial time $t = 0$ when the concentration of A was the specific value $[A]_0$ to the specific time t' when the concentration was $[A]_{t'}$, yields

$$\int_{[A]_0}^{[A]_{t'}} d[A] = k \int_0^{t'} dt$$

$$[A]_{t'} - [A]_0 = -k(t' - 0) \tag{6.20}$$

$$[A]_{t'} = [A]_0 - kt'$$

This result clearly indicates that the concentration $[A]$ should be diminishing linearly with time from the initial concentration $[A]_0$.

Do the reactant concentrations in other than zero order reactions show the same dependence on time as in Equation 6.20? The answer is no. For a first order reaction, such as a spontaneous unimolecular rearrangement, the rate equation ($\nu_A = 1$) is

$$-\frac{d[A]}{dt} = k[A] \tag{6.21}$$

Division by $[A]$ and integration of Equation 6.21 goes as follows:

$$-\int_{[A]_0}^{[A]_{t'}} \frac{1}{[A]} d[A] = k \int_0^{t'} dt$$

$$\ln\left(\frac{[A]_{t'}}{[A]_0}\right) = -kt' \tag{6.22}$$

$$[A]_{t'} = [A]_0 e^{-kt'}$$

Equation 6.22, the integrated rate expression for a first order reaction, is a statement that the concentration of A diminishes as an exponential in time starting from the initial concentration $[A]_0$. So, if we measure $[A]$ at various times in the course of the reaction and plot that concentration against time, the resulting curve will be qualitatively different from that of a zero order reaction, and as we shall see, different from a higher order reaction. This is shown in Figure 6.7.

A second order reaction involving one reactant, A, has the following rate equation ($\nu_A = 1$):

$$-\frac{d[A]}{dt} = k[A]^2 \tag{6.23}$$

Rearrangement and integration, as in the previous orders, goes as follows:

$$-\int_{[A]_0}^{[A]_{t'}} \frac{1}{[A]^2} d[A] = k \int_0^{t'} dt$$

$$\frac{1}{[A]_{t'}} - \frac{1}{[A]_0} = kt' \tag{6.24}$$

$$[A]_{t'} = \frac{[A]_0}{1 + kt'[A]_0}$$

This expression shows an inverse dependence on time, not the exponential dependence in Equation 6.22.

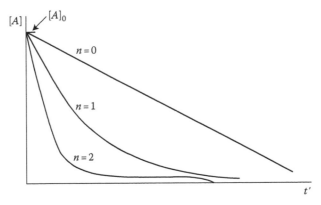

FIGURE 6.7
Plot of the form of the curves that show the dependence of the concentration of a reactant $[A]$ on time for reactions that are zero, first, and second order in A.

The **half-life** of a reaction, τ, is the time required for one-half of the reactants to be consumed. That is, at time $t = \tau$, $[A]_\tau = [A]_0/2$. This time is a characteristic, measurable value of a reaction. By monitoring the concentration of a reactant, one finds τ by noting the time at which the reactant concentration has diminished to one-half its initial value. For a zero order reaction, this condition, applied to Equation 6.20, is

$$[A]_0 = 2k\tau \tag{6.25}$$

For a first order reaction, Equation 6.22 gives the following at time $t = \tau$:

$$\ln 2 = k\tau \tag{6.26}$$

And, for a second order reaction, Equation 6.24 gives

$$\frac{1}{[A]_0} = k\tau \tag{6.27}$$

Notice that the first order reaction half-life is unique in having no dependence on the initial concentration, $[A]_0$. These half-life equations highlight a means for determining reaction order. If t is determined for several different initial concentrations, $[A]_0$, then the data can be used to determine if τ varies linearly with $[A]_0$ as in Equation 6.25 or inversely as in Equation 6.27, or if it is independent as in Equation 6.26. Once the order of the reaction is determined, the measurement of τ and $[A]_0$ means that the value of the rate constant, k, is known via one of these three equations or by a corresponding equation for a higher order reaction.

Example 6.1 illustrates the corresponding analysis for an $A + B$ reaction.

6.4 Rate Laws for Complex Reactions

Rate equations become more complicated when an overall reaction process occurs in a sequence of steps. Let us analyze this possibility by considering a hypothetical two-step sequence. In the first step, A changes to B, and in the second, B changes to C.

$$A \to B \to C$$

The rate of depletion of species A is

$$R_1 = -\frac{d[A]}{dt} = k_1[A] \tag{6.28}$$

Example 6.1: Integrated Rate Expression for an $A + B$ Reaction

Problem: Find an expression for the concentration of reactants as a function of time for a bimolecular reaction $v_A A + v_B B \to v_C C$.

Solution: The differential rate expression for a second order reaction of A with B is

$$-\frac{1}{v_A}\frac{d[A]}{dt} = -\frac{1}{v_B}\frac{d[B]}{dt} = k[A][B]$$

At any instant of time, the amount of A consumed relative to some initial concentration is $[A] - [A]_0$, and likewise for B, the change is $[B] - [B]_0$. The reaction stoichiometry requires that these changes divided by the stoichiometry numbers are equal:

$$\frac{[A]-[A]_0}{\nu_A} = \frac{[B]-[B]_0}{\nu_B}$$

Rearrangement followed by using "c" to represent a constant term gives

$$\nu_B[A] = \nu_A[B] - (\nu_A[B]_0 - \nu_B[A]_0) = \nu_A[B] - c$$

With this relation, we substitute for $[B]$ in the product $k[A][B]$ in the differential rate expression and obtain an expression involving only the concentration variable $[A]$. We can also substitute for $[A]$ and obtain an expression for $[B]$.

$$-\frac{1}{[A]}\frac{d[A]}{dt} = k(\nu_B[A] + c)$$

$$-\frac{1}{[B]}\frac{d[B]}{dt} = k(\nu_A[B] - c)$$

Converting the left-hand side of these equations to the logarithmic equivalent and then subtracting the second from the first yields

$$-\frac{d\ln[A]}{dt} + \frac{d\ln[B]}{dt} = k(\nu_B[A] - \nu_A[B] + 2c)$$

The right-hand side of this expression simplifies to kc because from the definition of c and the preceding stoichiometric requirement on concentration changes, $\nu_B[A] - \nu_A[B] + c = 0$. We now integrate from time 0 to time t'.

$$-\int_{[A]_0}^{[A]_{t'}} d\ln[A] + \int_{[B]_0}^{[B]_{t'}} d\ln[B] = kc\int_0^{t'} dt$$

$$-\ln\frac{[A]_{t'}}{[A]_0} + \ln\frac{[B]_{t'}}{[B]_0} = kct'$$

$$\frac{[A]_{t'}}{[B]_{t'}} = \frac{[A]_0}{[B]_0}e^{-kct'}$$

Thus, the ratio of concentrations varies exponentially with time. However, if the initial concentrations of the reactants could be chosen so that $c = 0$, the ratio would be constant throughout the course of the reaction.

However, the rate of change in the concentration of B is more complicated because it is an intermediate in the conversion of A to C. One step produces B, while the other removes B.

$$\frac{d[B]}{dt} = R_1 - R_2 \tag{6.29}$$

where R_2 is the rate for the second reaction step, that is,

$$R_2 = \frac{d[C]}{dt} = k_2[B] \qquad (6.30)$$

Since the solution of the first rate equation is

$$\frac{[A]_t}{[A]_0} = e^{-k_1 t} \qquad (6.31)$$

then the second equation, Equation 6.29, is a more complicated differential equation.

$$\frac{d[B]}{dt} = k_1[A]_0 e^{-k_1 t} - k_2[B] \qquad (6.32)$$

We can verify by substitution that the general solution to this differential equation is

$$[B]_t = \frac{k_1[A]_0}{k_2 - k_1}(e^{-k_1 t} - e^{-k_2 t}) \qquad (6.33)$$

From Equations 6.31 and 6.33, an expression for $[C]_t$ may also be developed since the sum of the concentrations, $[A]$, $[B]$, and $[C]$, at any instant of time, is the same as the initial concentration of reactant, $[A]_0$. Therefore,

$$[C]_t = [A]_0 - [A]_t - [B]_t \qquad (6.34)$$

(At time $t = 0$, $[B] = 0$ and $[C] = 0$.)

Let us consider Equation 6.33 under the condition that the second step of the reaction sequence happens to be much faster than the first step; that is, $k_2 \gg k_1$. Then, as t increases, $e^{-k_1 t}$ becomes increasingly greater than $e^{-k_2 t}$. For long-time behavior, we can neglect the second exponential term in Equation 6.33, and therefore,

$$[B]_t = \frac{k_1}{k_2}[A]_0 e^{-k_1 t} = \frac{k_1}{k_2}[A]_t \qquad (6.35)$$

This situation is one in which A converts to B relatively slowly, but then B converts to C quickly thereafter. Thus, it is not surprising that Equation 6.35 reveals that the concentration of species B is much less than the concentration of the reactant species, A. Now, if we combine Equation 6.35 with Equation 6.34, we obtain the time dependence of the concentration of species C.

$$[C]_t = [A]_0 - [A]_0 e^{-k_1 t} - \frac{k_1}{k_2}[A]_0 e^{-k_1 t}$$

$$= [A]_0 \left(1 - \left[1 + \frac{k_1}{k_2}\right] e^{-k_1 t}\right)$$

$$= [A]_0 (1 - e^{-k_1 t})$$

Notice that this is the expression we would obtain for a reaction where species A changes directly to species C at the rate k_1. We recognize from this that the first step, the slow step, is the important step in determining the rate at which species C is produced. We say, then, that the slow step is the **rate-determining step** of the overall process.

When a fast reaction step follows a slow reaction, as in the preceding case, another way of analyzing the kinetics is to assume that the concentration of the intermediate is unchanging in time. With the conditions we have been using, we can argue that the concentration of species B in the sample is at a steady-state level, or is almost unchanging in time, because as soon as some A converts to B the rapid step 2 converts the B to C. Invoking this **steady-state approximation** with Equation 6.29 yields

$$\frac{d[B]}{dt} = R_1 - R_2 = k_1[A] - k_2[B] \approx 0 \Rightarrow [B]_t = \frac{k_1}{k_2}[A]_t \qquad (6.36)$$

This result is identical to Equation 6.35; the steady-state approximation introduces the approximations made earlier in a different way.

That reaction events can be regarded as trajectories that take a system over a saddle point region on a potential energy surface brings up the idea that trajectories starting from the product side might cross the saddle point region going in the opposite direction. A reaction can be reversible. From a microscopic point of view, all reactions can be regarded as reversible. However, if the product potential energy trough is substantially below the reactant trough, there may be little likelihood of reversibility. So far, this has been an implicit assumption in our analysis of rate expressions. With such a difference in potential energies, products would be produced with a significant amount of kinetic energy and this might be so rapidly dissipated as to preclude the reverse reaction.

When there is appreciable reversibility in a reaction system, the rate expressions are different. Consider the following system of elementary reactions:

$$A \rightarrow B + C$$

$$B + C \rightarrow A$$

The second reaction is simply the reverse of the first. However, the rates for the two reactions are intrinsically different. The first reaction is spontaneous and first order. The second reaction is bimolecular and second order. A convention is to refer to one reaction as the forward reaction, and to the other as the reverse reaction. The forward reaction can be the one we intuitively take as the most favorable under the initial conditions of the system. Using k_f as the rate constant for the first reaction and k_r for the second reaction in this simple example, the differential rate expression is

$$\frac{d[A]}{dt} = -k_f[A] + k_r[B][C] \qquad (6.37)$$

With somewhat more involved calculus than in previous cases, one can obtain the integrated rate expression from Equation 6.37. In general, reversible reactions can be analyzed as a reaction sequence, taking the reverse reaction to be another step.

6.5 Temperature Dependence and Solvent Effects

Our examination of collision events led to the conclusion that the rate of two-body collisions has a second order dependence on the concentration(s) of the species (Equation 6.7). We realize that a two-body collision event corresponds to an elementary second order (bimolecular) process, and hence, the rate law for a bimolecular process has the same second order dependence on concentration. A complete gas kinetic analysis of collisions requires that we recognize that molecules in a gas have a distribution of velocities. Since this distribution is prescribed by the temperature, we can ask what happens when the temperature is changed.

Let us consider temperature dependence starting from the viewpoint of strictly hard-sphere collisions. From Figure 6.5, we realize that in a classical mechanical view, colliding reactants would not overcome the potential barrier or would not pass through a potential surface's transition state unless they had total kinetic energy (translational energy) at least equal to the potential barrier. If the barrier height for a reaction of molecules A and B is V_{AB}, how many pairs of molecules in a sample of A and B molecules would have at least this much energy? The Maxwell–Boltzmann distribution law indicates that the number of species of a certain energy, E, in a sample at thermal equilibrium and temperature T, varies as the exponential of $-E/kT$, where k is Boltzmann's constant. If we sum the number of molecules with energies greater than V_{AB} (integrate the distribution function over E from $E = V_{AB}$ to $E = \infty$), the result is $e^{-V_{AB}/kT}$. This factor times the collision rate of Equation 6.7 is the hard-sphere reaction rate:

$$R = \frac{Z'_{AB}}{N_A} e^{-V_{AB}/kT} \tag{6.38}$$

The factor of N_A is introduced so that the rate is in terms of moles rather than molecules. Notice that the collision rate Z'_{AB} is temperature-dependent (Equation 6.7) through its relation to the average relative speed. We know that the rate expression for an elementary $A + B$ reaction is $R = k_{rxn}[A][B]$, where the subscript rxn distinguishes the rate constant from Boltzmann's constant. Thus, the hard-sphere approximation for the rate constant is

$$k_{rxn} = \frac{Z'_{AB} e^{-V_{AB}/kT}}{N_A[A][B]} \tag{6.39}$$

Again, an important feature is that the rate constant, k_{rxn}, varies with temperature.

Laboratory determination of an exponential dependence of reaction rate constants on temperature was accomplished over 100 years ago by Svante Arrhenius (Sweden, 1859–1927). His work provided an expression that can be applied in studying most chemical reactions.

$$k_{rxn}(T) = A e^{-E_{act}/RT} \tag{6.40}$$

E_{act} stands for the **activation energy** which is the molar equivalent of the barrier height designated V_{AB} in this discussion. The value A in Equation 6.40 is sometimes called a **frequency or pre-exponential factor**. It is somewhat dependent on temperature, varying as $\sqrt{T}$ in the hard-sphere picture because of the collision rate's dependence on temperature. This factor can often be approximated as a constant over limited ranges of temperature.

Equation 6.40 indicates that the activation energy of a reaction can be found by plotting the logarithm of k_{rxn} against $1/T$, taking A to be constant. The slope of the line is proportional to E_{act}. For elementary reactions, the activation barrier has a positive value, and thus, the rate of the reaction increases with increasing temperature via the exponential dependence of $-1/T$. For complex reactions, it is possible to have an effective E_{act} that is negative, and this is manifested in a rate constant that decreases with increasing temperature.

If one could obtain the activation energies for the forward and reverse directions of a reaction, one would have the energy differences between the transition state energy and the reactant energies in one case (forward), and between the transition state energy and the product energies in the other (reverse). The difference between these two activation barriers is the energy difference between the reactants and the products.

Reactions that take place in solution may have complicated mechanisms because of increased intermolecular interactions in the condensed phase and reactions with solvent molecules. In solutions with polar solvents, ions are often involved in the mechanisms. For instance, if acetaldehyde, CH_3CHO, were to react with some other compound in a water solution, the reaction might take place with protonated or hydrated forms of acetaldehyde that form in solution: CH_3CHOH^+, $[CH_3CO \cdot H_2O]$, and $[CH_3CO \cdot H_3O^+]$. The overall reaction would then involve elementary reaction steps such as $[CH_3CO \cdot H_2O] + H_3O^+ \leftrightarrow [CH_3CO \cdot H_3O^+] + H_2O$.

The analysis of gas phase collision rates carried out earlier in this chapter can help us understand certain of the complexities of solutions. We might argue that molecules have the same number of degrees of freedom in either fluid (gas or liquid), that the Maxwell–Boltzmann distribution law applies to the molecular energies in both phases, and that the mean free path length differs only because of the difference in densities (concentrations). On that basis, two reactive species A and B might be expected to have the same collision rates in the gas and solution phases provided their concentrations were set to be the same in both phases. However, the close proximity of solvent molecules in the condensed phase affects the collision rate in a way that is not incorporated into the gas phase analysis. Whereas A and B will collide, perhaps react, and then separate in the gas phase, the presence of solvent molecules in the solution tends to keep A and B together for some time after their collision. That time, though short, allows A and B to collide again and again before finally separating. Thus, a key difference between solution and gas phase reactions is the solvent's effect on enhancing the collision frequency, often quite sizably.

Another possible effect of a solvent amounts to changing the potential energy surface of the reaction. Consider the one-dimensional potential energy surface slice in Figure 6.3 to be that of some hypothetical gas phase bimolecular reaction. If placed in a solvent, the species in this reaction would experience intermolecular interactions with the solvent molecules. We should always expect some attractive (stabilizing) interaction if a solution forms at all, and therefore, in the solvent, the potential energy slice, as depicted in Figure 6.3, and the whole potential energy surface are shifted downward. However, as depicted in Figure 6.8, that shift need not be uniform. The solvent molecules may have a more favorable interaction with products than with reactants, or with the transition state structure than with products, and so on. Therefore, the solvent may change the energy difference between reactants and products and/or it may change the activation barrier, and that in turn, would affect the rate of reaction. One way to observe this is to measure reaction rates in a series of solvents that display different interactions with the reactive species.

In addition to the solvent providing a "cage" that enhances the effective collision frequency, the solvent's transport properties affect the probability that A and B will be brought into proximity. In other words, a solution rate of reaction depends on the solvent's viscosity, or diffusion coefficient. If every collision of A with B leads to a reaction, the reaction

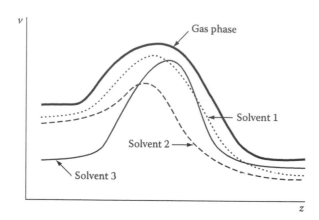

FIGURE 6.8
Possible solvent effects on the potential energy surface slice that follows the minimum energy path for a hypothetical bimolecular reaction. The vertical axis is the potential energy, and the horizontal axis is the reaction coordinate. Solvent 1 has an attractive interaction that is essentially constant along the reaction coordinate. Solvent 2 gives rise to a greater energy lowering at the transition state than elsewhere, and so it increases the rate of reaction. Solvent 3 shows the greatest energy lowering for the reactants and thereby increases the activation energy.

rate will be controlled by the viscosity of the solvent. The more viscous the solvent, the slower the reaction rate. These reactions are termed **diffusion-controlled** reactions. On the other hand, if A and B do not react on every collision, then the solvent's ability to cage an AB pair—also dependent on the solvent's transport properties—will enter into the reaction rate, along with the elements specific to isolated A and B that were considered for gas phase reactions, for example, the activation barrier.

6.6 Reaction Thermodynamics

In the simplest picture, the energy of the transition state structure on a potential energy surface is the activation energy of reaction. The difference between the energies of the reactants and products, ΔU_{rxn}, is the difference between the energy at the bottom of the surface trough corresponding to noninteracting reactants and the energy at the bottom of the trough corresponding to noninteracting products. Let us connect this molecular level information with the thermodynamics of reactions.

Enthalpy is related to internal energy via $H = U + PV$. For a process at constant pressure (1 bar), $\Delta H = \Delta U + P\Delta V$. The standard state change in volume that occurs in the course of a reaction is determined by the standard state molar volumes of the reactants and products.

$$\Delta V_{rxn} = \overset{products}{\sum_{i}} n_i \overline{V}_i^\circ - \overset{reactants}{\sum_{j}} n_j \overline{V}_j^\circ = -\overset{r,p}{\sum_{i}} \nu_i \overline{V}_i^\circ \tag{6.41}$$

where
    n_i is the number of moles of product i
    n_j is the number of moles of reactant j (unsigned stoichiometric numbers)
    ν_i is a signed stoichiometric number as in Equation 6.16

The standard state molar volume of many liquids at room temperature is around 0.1 L. If all reactants and products are liquids, the change in volume will be small, considerably less than the typical molar volume. A more sizable ΔV_{rxn} is found if there is a net change in the number of moles of gas. Since the typical standard state molar volume of a gas at room temperature is around 25 L, a reaction that converts 1 mol of liquid to 1 mol of gas has a ΔV_{rxn} of around 25 L. If 25 L is multiplied by a pressure of 1 bar, the resulting $P\Delta V$ contribution to ΔH in this case is 2.5 kJ mol^{-1}. Reaction enthalpies are often much greater, on the order of 100 kJ mol^{-1}. In view of that, it is reasonable to use an approximation to this term. If the change in molar volumes of all condensed phase reactants and products is taken to be negligible, and if all the molar volumes of all gas phase reactants and products are approximately represented as the ideal value (i.e., $P\overline{V}_i = RT$ for all i),

$$P\Delta V_{rxn} \equiv \Delta n_{gas} RT \tag{6.42}$$

where Δn_{gas} is the net change in the number of moles of gas for the reaction. In the limit of low pressure (low temperature), ΔH_{rxn} approaches ΔU_{rxn} and then the enthalpy in the limit of $T = 0$ can be associated with the energy difference between the reactant and product troughs on the potential surface. Even at higher temperatures, the enthalpy of the reaction usually serves as a good approximation to ΔU_{rxn}, and vice versa.

A consequence of the first law of thermodynamics, the conservation of energy, is that we can combine standard enthalpies of reactions to produce the standard enthalpy of another reaction. If a reaction can be accomplished in a set of steps, each a reaction with a known standard enthalpy, the sum of the standard enthalpies of the steps is the standard enthalpy of the overall reaction. This statement is often called **Hess's law** (Germain Henri Hess, Russia, 1802–1850). Standard enthalpies of formation are available in published tables, and a few selected values are listed in Table 6.1. Enthalpies of formation are reaction enthalpies for which the reactants are pure elemental substances. Thus, if we can develop a set of reaction steps where the reactants in each step are elemental substances, the overall reaction enthalpy will be a combination of enthalpies of formation. Here is an example:

Reaction 1	$H_2 + \frac{1}{2}O_2 \rightarrow H_2O$	$\Delta \overline{H}_f^\circ (H_2O)$
Reaction 2	$H_2 + C + \frac{1}{2}O_2 \rightarrow H_2CO$	$\Delta \overline{H}_f^\circ (H_2CO)$
Reaction 3	$2H_2 + C \rightarrow CH_4$	$\Delta \overline{H}_f^\circ (CH_4)$
Overall reaction	$H_2CO + H_2O \rightarrow CH_4 + O_2$	$\Delta \overline{H}_{rxn}^\circ$

Reaction 3 produces one product that is not in elemental form, whereas reactions 1 and 2 produce reactants. The enthalpy of the overall reaction is the enthalpy of reaction 3 (product) less the enthalpies of reactions 1 and 2 (reactants).

$$\Delta \overline{H}_{rxn}^\circ = \Delta \overline{H}_f^\circ(CH_4) - \Delta \overline{H}_f^\circ(H_2O) - \Delta \overline{H}_f^\circ(H_2CO) = 398 \text{ kJ mol}^{-1} \text{ (evaluated at 298 K)}$$

This particular reaction is endothermic, or "uphill" in energy. The enthalpy change for a reverse reaction is the negative of the enthalpy change in the forward direction, and so the

TABLE 6.1

Standard State Thermodynamic Properties[a]

Substance (kJ)	$\Delta \overline{H}^\circ_{f,298K}$ (kJ)	$\Delta \overline{G}^\circ_{f,298K}$ (JK^{-1})	$\overline{S}^\circ_{298K}$ (JK^{-1})	$\overline{C}^\circ_{p,298K}$
Ag (s)	0	0	42.6	25.4
Al (s)	0	0	28.3	24.4
Ar (g)	0	0	154.8	20.8
Au (s)	0	0	47.4	25.4
B (s)	0	0	5.9	11.1
C (graphite)	0	0	5.7	8.5
C (diamond)	1.9	2.9	2.4	6.1
CO (g)	−110.5	−137.2	197.7	29.1
CO$_2$ (g)	−393.5	−394.4	213.8	37.1
CH$_4$ (g)	−74.6	−50.5	186.3	35.7
CH$_3$OH (l)	−239.2	−166.6	126.8	81.1
C$_2$H$_2$ (g)	227.4	209.9	200.9	44.0
C$_2$H$_4$ (g)	52.4	68.4	219.3	42.9
C$_2$H$_6$ (g)	−84.0	−32.0	229.2	52.5
Ca (s)	0	0	41.6	25.9
CaCl$_2$ (s)	−795.4	−748.8	108.4	72.9
CaO (s)	−634.9	−603.3	38.1	42.0
CaCO$_3$ (s, calcite)	−1207.6	−1129.1	91.7	83.5
Cl$_2$ (g)	0	0	223.1	33.9
Co (s)	0	0	30.0	24.8
Cr (s)	0	0	23.8	23.4
Cu (s)	0	0	33.2	24.4
CuO (s)	−157.3	−129.7	42.6	42.3
Cu$_2$O (s)	−168.6	−146.0	93.1	63.6
F$_2$ (g)	0	0	202.8	31.3
Fe (s)	0	0	27.3	25.1
H$_2$ (g)	0	0	130.7	28.8
H$_2$O z(l)	−285.8	−237.1	70.0	75.3
H$_2$O$_2$ (l)	−187.8	−120.4	109.6	89.1
HCl (g)	−92.3	−95.3	186.9	29.1
H$_2$S (g)	−20.6	−33.4	205.8	34.2
He (g)	0	0	126.2	20.8
Hg (l)	0	0	75.9	28.0
Li (s)	0	0	29.1	24.8
LiH (s)	−90.5	−68.3	20.0	27.9
Li$_2$O (s)	−597.9	−561.2	37.6	54.1
Mg (s)	0	0	32.7	24.9
MgO (s)	−601.6	−569.3	27.0	37.2
MgS (s)	−346.0	−341.8	50.3	45.6
N$_2$ (g)	0	0	191.6	29.1
NO$_2$ (g)	33.2	51.3	240.1	37.2
N$_2$O (g)	81.6	103.7	220.0	38.6
N$_2$O$_4$ (l)	−19.5	97.5	209.2	142.7

TABLE 6.1 (continued)

Standard State Thermodynamic Properties[a]

Substance (kJ)	$\overline{\Delta H}^{\circ}_{f,298K}$ (kJ)	$\overline{\Delta G}^{\circ}_{f,298K}$ (JK^{-1})	$\overline{S}^{\circ}_{298K}$ (JK^{-1})	$\overline{C}^{\circ}_{p,298K}$
NH$_3$ (g)	−45.9	−16.4	192.8	35.1
NH$_4$Cl (s)	−314.4	−202.9	94.6	84.1
NF$_3$ (g)	−132.1	−90.6	260.8	53.4
O$_2$ (g)	0	0	205.2	29.4
O$_3$ (g)	142.7	163.2	238.9	39.2
S (s, rhombic)	0	0	32.1	22.6
SiH$_4$ (g)	34.3	56.9	204.6	42.8
SiO$_2$ (s, α-quartz)	−910.7	−856.3	41.5	44.4

Source: Values from Lide, D.R., *Handbook of Chemistry and Physics*, 86th edn., CRC Press, Boca Raton, FL, 2006.

[a] Values are for a temperature of 298.15 K and pressure of 1 bar.

standard state enthalpy change for the exothermic reaction of methane and oxygen to produce water and formaldehyde is −398 kJ mol^{-1} at 298 K. A general statement of Hess's law is

$$\Delta H^{\circ}_{rxn} = - \overset{products}{\underset{i}{\sum}} \nu_i \overline{\Delta H}^{\circ}_f(i) - \overset{reactants}{\underset{j}{\sum}} \nu_j \overline{\Delta H}^{\circ}_f(j) \tag{6.43}$$

where the ν's are the signed stoichiometric numbers as in Equation 6.16 (negative values for products and positive values for reactants). Note that while we can directly calculate the enthalpy change *were* a reaction to occur, whether the enthalpy change is uphill versus downhill does not determine *if* the reaction will occur.

Enthalpies of formation, such as those in Table 6.1, have been tabulated for standard states at certain temperatures, and that allows us to predict other reaction enthalpies at the same temperatures. At other than those temperatures, heat capacities are used to calculate a reaction enthalpy. When H is expressed as a function of T and P, the differential enthalpy for an individual reactant or product, i, is

$$dH_i = \left(\frac{\partial H_i}{\partial T}\right)_P dT + \left(\frac{\partial H_i}{\partial P}\right)_T dP \tag{6.44}$$

Our analysis using standard state values implies $dP = 0$, and under that constraint, the second term in Equation 6.44 is zero. The remaining partial derivative is the constant-pressure heat capacity of the ith substance, $C_{P,i}$. To evaluate the enthalpy (or other thermodynamic state functions) at a temperature, call it T_2, different from a temperature T_1 for which there are known enthalpies, we invoke Hess's law, treating the temperature changes of reactants and products as reaction steps. Here is an example:

Reaction 1	$A(T_2) + B(T_2) \rightarrow A(T_1) + B(T_1)$
Reaction 2	$A(T_1) + B(T_1) \rightarrow C(T_1) + D(T_1)$
Reaction 3	$C(T_1) + D(T_1) \rightarrow C(T_2) + D(T_2)$
Overall reaction	$A(T_2) + B(T_2) \rightarrow C(T_2) + D(T_2)$

The enthalpies of reactions 1 and 3 are obtained by integrating the heat capacities over the temperature difference. The enthalpy of reaction 2 is obtained from the tabulated data. The sum of the enthalpies of reactions 1–3 is the enthalpy of the overall reaction of interest since it can be accomplished stepwise via reactions 1–3. In general,

$$\Delta H_{rxn}(T_2) = \Delta H_{rxn}(T_1) - \int_{T_1}^{T_2} \sum_i \nu_i C_{P,i} dT \tag{6.45}$$

using the signed stoichiometric numbers (ν_i positive for reactants and negative for products) with the summation going over all reactants and products.

The Gibbs energy is a thermodynamic state function, and so, as for volume (Equation 6.41) and enthalpy (Equation 6.43),

$$\Delta G_{rxn}^\circ = \overbrace{\sum_i n_i \Delta \overline{G}_f^\circ(i)}^{products} - \overbrace{\sum_j n_j \Delta \overline{G}_f^\circ(j)}^{reactants} = -\sum_i^{r,p} \nu_i \Delta \overline{G}_f^\circ \tag{6.46}$$

with signed stoichiometric numbers ν_i and a collective sum over reactants and products on the far right and unsigned stoichiometric numbers n_i in the middle. Table 6.1 includes standard state values for the Gibbs energy of formation for a number of substances from which ΔG° values for other reactions can be calculated. ΔG° values are related to equilibrium constants, Equation 5.44, and in some cases concentration measurements at equilibrium standard state conditions provide these values. The thermodynamic relationship $G = H - TS$ means that for an isothermal process,

$$\Delta G = \Delta H - T\Delta S \quad \text{or} \quad \Delta \overline{G}_{rxn}^\circ = \Delta \overline{H}_{rxn}^\circ - T\Delta \overline{S}_{rxn}^\circ \tag{6.47}$$

Example 6.2: Temperature Dependence of a Reaction Enthalpy

Problem: Find the molar enthalpy of reaction for the standard state formation of ammonia from hydrogen and nitrogen at 298, 400, 600, and 800 K (1) assuming constant heat capacities for reactants and products and (2) using $29.75 + 0.025T - 150{,}000/T^2$ for the heat capacity (J mol^{-1}) of ammonia.

Solution: The reaction to produce 1 mol of ammonia is

$$\tfrac{1}{2}N_2 + \tfrac{3}{2}H_2 \rightarrow NH_3$$

At 298 K, the enthalpy of the reaction is the enthalpy of formation of ammonia given in Table 6.1, -45.9 kJ mol^{-1}. (1) With constant heat capacities, the change in this value on carrying out the reaction at another temperature involves simply multiplying the heat capacities by the temperature difference. We break the reaction into steps, the first being the lowering of the reactants' temperature from 400 K (later 600 and 800 K) to 298 K. The second is the formation reaction at 298 K, and the third is raising the temperature of the product to 400 K. The enthalpy change for the first step is the temperature difference times the reactant heat capacities.

$$(298 - 400)\left[\tfrac{1}{2}C_P(N_2) + \tfrac{3}{2}C_P(H_2)\right] = -102\left[57.79\,\text{J}\right] = -5.89\,\text{kJ}$$

For the third step, the temperature difference is the opposite.

$$(400 - 298) \left[C_P(NH_3) \right] = 3.58 \text{ kJ}$$

Therefore, $\Delta \overline{H}^{\circ}_{rxn, 400K} = -45.9 - 5.89 + 3.58 = -48.2 \text{ kJ}$

Similarly, $\Delta \overline{H}^{\circ}_{rxn, 600K} = -52.7 \text{ kJ}$ and $= -57.3 \text{ kJ}$

(2) If a heat capacity depends on temperature, the enthalpy changes for the first and third steps in the overall process are found by integration over temperature; that is, $\int C_p dT$. For the given conditions of this problem, this needs to be done only for the product, ammonia, that is, for the third step; however, a more complete treatment would, by the same means, take into account the temperature dependence of the heat capacities of nitrogen and hydrogen.

We use the analysis from part 1 for the enthalpy change in the first step.

$$\int_{298}^{400} \left(29.75 + 0.025T - \frac{150,000}{T^2} \right) dT = 29.75(400 - 298) + \frac{0.025T^2}{2} \Big|_{298}^{400} + \frac{150,000}{T} \Big|_{298}^{400}$$

$$= 3.035 + 0.890 - 0.128 = 3.797 \text{ kJ}$$

This value is quite close to the value obtained assuming a constant heat capacity, 3.58 kJ. The reaction enthalpy using this value for the third step is then −48.0 kJ. At higher temperatures, the integrated temperature-dependent heat capacity yields results with more sizable differences than those of part 1,

$$\int_{298}^{600} C_P(T) dT = 12.122 \text{ kJ} \quad \text{versus} \quad 10.588 \text{ kJ with constant } C_P$$

$$\int_{298}^{800} C_P(T) dT = 21.509 \text{ kJ} \quad \text{versus} \quad 17.600 \text{ kJ with constant } C_P$$

Thus,

$$\Delta \overline{H}^{\circ}_{rxn, 298K} = -45.9 \text{ kJ mol}^{-1}$$

$$\Delta \overline{H}^{\circ}_{rxn, 400K} = -48.0 \text{ kJ mol}^{-1}$$

$$\Delta \overline{H}^{\circ}_{rxn, 600K} = -51.3 \text{ kJ mol}^{-1}$$

$$\Delta \overline{H}^{\circ}_{rxn, 800K} = -53.4 \text{ kJ mol}^{-1}$$

Notice that the enthalpy varies with temperature by about 10% over this rather sizable temperature range.

The entropy change is

Example 6.3: ΔS and ΔG of a Reaction

Problem: Find the change in entropy, enthalpy, and Gibbs energy using Hess's law and the data in Table 6.1 for the combustion of methane to produce carbon dioxide and water at 298 K. Show how the entropy change for the reaction could have been obtained from the enthalpies and Gibbs energies.

Solution: The reaction is $CH_4 + 2O_2 \rightarrow CO_2 + 2H_2O$. Combining values from the table according to the reaction stoichiometry (factors for molar quantities in parentheses) yields the desired quantities.

	(−1) CH_4	+	(−2) O_2	+	CO_2	+	(2) H_2O	Equals
$\Delta H^\circ_{rxn,298K}$	74.6		0		−393.5		−571.6	−890.5 kJ
$\Delta G^\circ_{rxn,298K}$	50.5		0		−394.4		−474.2	−818.1 kJ
ΔS°_{rxn}	−186.3		−410.4		213.8		140.0	−242.9 J K^{-1}

ΔS°_{rxn} can also be obtained as

$$\frac{(\Delta H^\circ_{rxn,298K} - \Delta G^\circ_{rxn,298K})}{T} : \frac{(-890.5 + 818.0)}{298} = -0.2433 \text{ kJ K}^{-1} = 243.3 \text{ J K}^{-1}$$

$$\Delta S^\circ_{rxn} = \sum_i^{products} n_i \bar{S}^\circ_i - \sum_j^{reactants} n_j \bar{S}^\circ_j = -\sum_i^{r,p} v_i \bar{S}^\circ_i \tag{6.48}$$

Again, entropy is a thermodynamic state function, and thus, Equation 6.48 amounts to a Hess's law for entropy. Note the similarity of Equations 6.41, 6.43, 6.46, and 6.48. The molar entropies of reactants and products in Equation 6.48 are defined on an absolute basis rather than on the relative basis used for enthalpies and Gibbs energies (Δ values). However, from our fundamental view of entropy, we can recognize that certain contributions to the entropy, such as the nuclear configurations, will not change in a chemical process. Thus, a workable choice for the zero of the absolute entropy scale ignores such contributions. With this, the entropy of a perfect crystalline substance at 0 K is zero. From using $G = G(T,P)$ and relations we can extract from the thermodynamic compass, the constant-pressure entropy change in a reaction is related to the Gibbs energy.

$$\left(\frac{\partial \Delta G^\circ_{rxn}}{\partial T} \right)_P = -\Delta S^\circ_{rxn} \tag{6.49}$$

Equation 5.43 allows us to relate calculated enthalpy and entropy changes to equilibrium constants:

$$\ln K = -\frac{\Delta H^\circ_{rxn}}{RT} + \frac{\Delta S^\circ_{rxn}}{R}$$

This tells us that the temperature dependence of the equilibrium constant is primarily the $1/T$ dependence in the enthalpy term, though ΔH and ΔS may exhibit temperature dependence. At very high temperatures, the enthalpy term is diminished in importance, and the entropy term may dominate. Thus, $\ln K$ asymptotically approaches $\Delta S^\circ_{rxn}/R$ with increasing temperature.

6.7 Electrochemical Reactions

An **electrolyte** is a substance that forms ions in solution, and electrolyte solutions are conductors of electric charge. The formation of ions in solution, such as that from dissolving salt in water, is a reaction.

$$\text{NaCl (s)} \rightarrow \text{Na}^+ \text{(aq)} + \text{Cl}^- \text{(aq)} \tag{6.50}$$

We can carry out measurements to obtain the ΔG and ΔH values for this reaction as for any other reaction. We could also obtain these values at standard state conditions if we had standard state formation enthalpies and Gibbs energies not only for NaCl but also for the product ions. The standard state formation values for the ions correspond to reactions that involve an electron along with the ion of interest:

$$\text{Na (s)} - e^- \rightarrow \text{Na}^+ \text{(aq)} \tag{6.51}$$

$$\tfrac{1}{2}\text{Cl}_2 \text{(g)} + e^- \rightarrow \text{Cl}^- \text{(aq)} \tag{6.52}$$

Attempting to identify "standard state formation" values for an electron would amount to a particular way of setting a reference point for the standard states of ions. Instead, the reference point is chosen such that the enthalpy, entropy, and Gibbs energy for the following reaction are zero:

$$\tfrac{1}{2}\text{H}_2 \text{(g)} - e^- \rightarrow \text{H}^+ \text{(aq)} \quad \Delta \overline{H}^\circ_f \equiv 0 \quad \Delta \overline{G}^\circ_f \equiv 0 \quad \overline{S}^\circ_f \equiv 0 \tag{6.53}$$

If we add the reactions in Equations 6.52 and 6.53, we obtain

$$\tfrac{1}{2}\text{Cl}_2 \text{(g)} + \tfrac{1}{2}\text{H}_2 \text{(g)} \rightarrow \text{H}^+ \text{(aq)} + \text{Cl}^- \text{(aq)} \tag{6.54}$$

With the definition of a reference in Equation 6.53, Hess's law means that the enthalpy, free energy, and entropy for reaction in Equation 6.52 are the same as in Equation 6.54. The reference choice allows us to assign values to ion formation reactions using values for reactions that do not involve a net addition or subtraction of an electron.

An **electrochemical cell** refers to an assembly in which a reaction system experiences an electrical potential that is applied across two electrodes. The electrodes can be metal or graphite or another conductor. The potential is externally controlled and is, therefore, another thermodynamic degree of freedom. We note this by adding 1 to the phase rule expression (Equation 4.11) for the degrees of freedom:

$$f_{el.cell} = 3 - p + c \qquad (6.55)$$

The potential across the electrodes is, then, a variable to be treated along with pressure and temperature. A schematic diagram of an electrochemical cell is given in Figure 6.9.

There are two types of electrochemical cells. A **galvanic or voltaic cell** produces electric current from a spontaneous reaction, whereas an **electrolytic cell** has an external current source, such as a battery, to drive a nonspontaneous reaction. Reactions in electrochemical cells are **redox reactions** since they involve a transfer of electrons from a reducing agent to an oxidizing agent. The reduction reaction withdraws electrons from one electrode, the positive electrode or **cathode**. The oxidation reaction supplies electrons to the other electrode, the negative electrode or **anode**. Current flows via an external circuit from the anode to the cathode.

For any redox reaction, the number of moles of electrons transferred depends on the stoichiometry of the reaction. For example, if copper is placed in a solution of silver nitrate, the following overall reaction implies two half-reactions, the reduction and oxidation reactions.

$$Cu\ (s) + 2AgNO_3\ (aq) \rightarrow 2Ag\ (s) + Cu(NO_3)_2\ (aq)$$

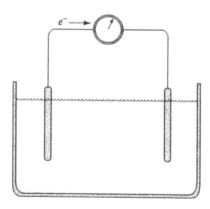

FIGURE 6.9

Schematic representation of an electrochemical cell. The electrodes are conducting materials connected by a conductor (e.g., copper wire). A voltage across the electrodes can be measured by a potentiometer inserted into the circuit. It is common to place the anode on the left and the cathode on the right in these types of schematic drawings. Negative charge flows away from the anode (left) into the external part of the circuit. This means that oxidation, a loss of electrons, occurs at or in the vicinity of the anode. The electrons lost from the surrounding material flow into the anode; it is the negative electrode. At the cathode, or the positive electrode, reduction takes place as electrons flow from the external circuit to the substance in the cell. This flow of electrons corresponds to the cell voltage being measured as a positive value ($E > 0$). When it is negative, the reaction is spontaneous in the reverse direction. As well, introduction of an external voltage source (e.g., a battery) in the circuit may drive the reaction in one direction or the other.

$$\text{Oxidation: Cu (s)} \rightarrow \text{Cu}^{2+} \text{(aq)} + 2e^-$$

$$\text{Reduction: Ag}^+ \text{(aq)} + e^- \rightarrow \text{Ag (s)}$$

The number of moles of electrons transferred in the overall reaction, n, is 2, which is the same as the stoichiometric number for silver nitrate.

To analyze energy in electrochemical cells, let us first consider a system consisting of a pair of oppositely charged conductive plates held in parallel in a vacuum. There is an electrical potential difference, E, between these plates, and their geometric arrangement insures that the electric field between them is uniform. Classical electrostatic theory gives the energy of moving some small incremental amount of charge, δZ, from one plate to the other to be $-E\,\delta Z$. This implies the following relation between charge and potential energy:

$$dU = -EdZ \tag{6.56}$$

An electrochemical cell is a more complicated setup than charged parallel plates in a vacuum, and thus, the field may not be uniform throughout the material between the electrodes. However, if E is the voltage between electrodes, Equation 6.56 serves to give the energy change for an infinitesimal amount of charge flowing from one electrode to the other.

The differential of the electrical energy in Equation 6.56 implies a corresponding term in the differential of the Gibbs energy, Equation 4.12:

$$dG = VdP - SdT + \sum_i g_i dn_i - EdZ \tag{6.57}$$

Operating conditions of most interest for electrochemical cells are constant pressure and constant temperature. Constant composition is approached in the limit of negligible current flow. Under these conditions,

$$dG = dU = -EdZ \tag{6.58}$$

With n as the number of moles of electrons transferred in an overall reaction of an electrochemical cell,

$$Z = Fn \tag{6.59}$$

where F is the charge of 1 mol of electrons and is called **Faraday's constant** (Michael Faraday, England, 1791–1867). The value of F is 9.648534×10^4 C mol^{-1} (Appendix F). Let us define $d\eta$ to be a dimensionless measure of reaction progress such that the changes in composition given by the differentials of the composition values, dn_i, for the substances in the reaction are $d\eta$ times the stoichiometric number, v_i. For the number of moles of electrons transferred, this means that $dn = nd\eta$. Therefore, with Equations 6.58 and 6.59,

$$dZ = Fnd\eta \tag{6.60}$$

$$dG = -EFnd\eta \tag{6.61}$$

We also have that

$$dG = \Delta \bar{G}_{rxn} d\eta \tag{6.62}$$

Equating the expressions for dG in Equations 6.61 and 6.62 and removing the common factor of $d\eta$,

$$\Delta \bar{G}_{rxn} = -nEF \tag{6.63}$$

Usually, the redox reaction is taken to be balanced with stoichiometric numbers corresponding to the transfer of 1 mol of electrons ($n = 1$).

Equation 5.41, which related the change in Gibbs energy for a reaction to the equilibrium constant, K, and the activity quotient, Q, can be combined with Equation 6.63:

$$E = -\frac{RT}{nF} \ln\left(\frac{Q}{K}\right) \tag{6.64}$$

This is one form of the **Nernst equation** (Walther Hermann Nernst, Germany, 1864–1941). Among other things, this equation shows that if the system is at equilibrium ($Q = K$), the voltage will be zero. Another form of the Nernst equation comes about from recognizing that we can express Equation 6.63 under standard state conditions. Then, the standard state change in the Gibbs energy is related to a voltage designated $E°$, the **standard cell voltage**. Under standard state conditions, $Q = 1$, and so

$$E° = \frac{RT}{nF} \ln K \tag{6.65}$$

If this is subtracted from Equation 6.64, the result is an equivalent form of the Nernst equation. (It contains the same information.)

$$E = E° - \frac{RT}{nF} \ln Q \tag{6.66}$$

The Nernst equation implies that measuring an electrochemical cell voltage can be used to find the equilibrium constant and thereby $\Delta G°_{rxn}$. The measurement of the voltage across two electrodes is properly found by balancing an opposing voltage such that no current flows. An adjustable external voltage source is connected between the two electrodes along with a current meter. When adjustment reduces the current to zero, then the cell's voltage is the negative of the voltage of the external source. This **zero-current voltage** is E. To fully appreciate the utility of such a measurement, we need to realize that there are several ways of physically separating the reactions occurring at the anode and at the cathode. Figure 6.10 illustrates certain possibilities.

A combination reaction of two gases, such as $\frac{1}{2}H_2 + \frac{1}{2}Cl_2 \rightarrow HCl$, can be accomplished in an electrochemical cell. To do so, molecular hydrogen needs to be reduced to form solvated H^+ ions and to give up electrons to the anode, whereas molecular chlorine needs to form Cl^- ions in solution while accepting electrons from the cathode. If the two gases are mixed, the energy released by the reaction may end up as heat rather than electric current. So, as shown in Figure 6.10, a gas electrode serves to keep the reactant gases apart. Each electrode must be in its own compartment. Gas is bubbled near a catalytic electrode surface

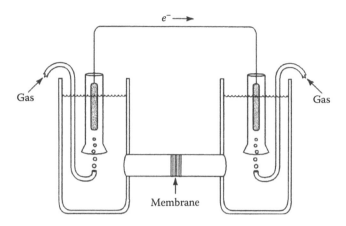

FIGURE 6.10
Electrochemical cell with an ion-selective membrane separating the two halves. For reactions involving gas phase reactants, a cell can be arranged so that the gases are kept separate as shown. These two, and other approaches, make it possible to physically separate reactions at the anode of a cell from reactions at the cathode.

with any unreacted gas escaping or being captured. Each gas is physically isolated, and so we can be sure the reduction of one gas takes place at one electrode while the oxidation of the other gas takes place at the other electrode. Other devices can accomplish a physical separation. An ion-selective membrane may be used to form a physical boundary between the two solutions (Figure 6.10). It allows certain ions to pass, and so electrical neutrality can be maintained in both compartments, as it would if they were directly connected. A salt bridge is still another device that selectively passes ions from one side to the other.

Electrochemical cells can be used and operated in many ways. **Fuel cells** generate voltage by the separate oxidation and reduction of two substances flowing into the compartments of a cell as in Figure 6.10. H_2 and O_2 are useful in fuel cells because the product of their reaction H_2O is easily handled. An external voltage greater than E of the cell can reverse the direction of a reaction. For instance, molecular hydrogen and molecular oxygen can be produced from water electrolytically. Because the cell voltage depends on reactant activities, essentially concentrations, voltage in an electrochemical cell can be used to measure concentrations. pH meters have operated using this principle for at least three-quarters of a century.

Example 6.4: ΔG of an Electrochemical Cell

Problem: Using reduction potentials, find ΔG°_{rxn} for the reaction of molecular hydrogen and molecular fluorine to produce HF.

Solution: The overall reaction can be written in terms of half-reactions that may take place at the electrodes in a cell of the type shown in Figure 6.10.

$$\text{Reduction: } F_2\,(g) + 2e^- \rightarrow 2F^-$$

$$\text{Oxidation: } H_2\,(g) \rightarrow 2H^+ + 2e^-$$

$$\text{Overall: } H_2\,(g) + F_2\,(g) \rightarrow 2HF$$

Note that the overall reaction, the sum of the first two, is said to produce HF, not H^+ and F^-. The two ions in the half-reactions are produced in aqueous solution. The overall reaction produces HF in aqueous solution, which is the same under the assumption of complete ionization of HF (an approximation).

From Table 6.2, the potential for the first half-reaction is 2.866 V. The second half-reaction is the reference (relative to which the first is measured), and so its potential is zero. Thus, an electrochemical cell that generates HF from hydrogen and fluorine gas generates a potential of 2.866 V under standard state conditions. The Gibbs energy change under standard state conditions is obtained directly via Equation 6.63.

$$\overline{\Delta G}_{rxn} = -nEF = -2(9.648534 \times 10^4)(2.866) = 5.531 \times 10^5 \, J = 553.1 \, kJ$$

Note that the number of moles of electrons transferred in this reaction is 2. Therefore $n = 2$ when using Equation 6.63.

Voltages are assigned to electrochemical half-reactions by choosing a reference standard, that is, choosing a particular other half-reaction to have zero voltage. Several of these values are given in Table 6.2. By adding standard voltages of half-reactions, one obtains the voltage of a cell, as in Example 6.4.

Equilibrium of a reaction can be shifted by a change in pressure or temperature; the equilibrium constant depends on temperature and pressure. Because of the Nernst relation, such a change results in a voltage change in an electrochemical cell. Example 6.5 illustrates this.

TABLE 6.2

Reduction Potentials for Reactions in Water at 298 K Relative to a Hydrogen Standard

Reaction	Potential (V)
$Li^+ + e^- \rightarrow Li \, (s)$	−3.0401
$K^+ + e^- \rightarrow K \, (s)$	−2.931
$Ca^{2+} + 2e^- \rightarrow Ca \, (s)$	−2.868
$Na^+ + e^- \rightarrow Na \, (s)$	−2.71
$Mg^{2+} + 2e^- \rightarrow Mg \, (s)$	−2.372
$Pb^{2+} + 2e^- \rightarrow Pb \, (s)$	−0.1262
$2H^+ + 2e^- \rightarrow H_2$	0.0 (reference standard)
$AgCl + e^- \rightarrow Ag \, (s) + Cl^-$	0.22233
$Cu^{2+} + 2e^- \rightarrow Cu \, (s)$	0.3419
$Cu^+ + e^- \rightarrow Cu \, (s)$	0.521
$Hg_2^{2+} + 2e^- \rightarrow 2Hg \, (l)$	0.7973
$Ag^+ + e^- \rightarrow Ag \, (s)$	0.7996
$2NO_3^- + 4H^+ + 2e^- \rightarrow N_2O_4 + 2H_2O$	0.803
$Cl_2(g) + 2e^- \rightarrow 2Cl^-$	1.35827
$F_2(g) + 2e^- \rightarrow 2F^-$	2.866
$XeF + e^- \rightarrow Xe + F^-$	3.4

Source: Values from Lide, D.R., *Handbook of Chemistry and Physics*, 86th edn., CRC Press, Boca Raton, FL, 2006.

Example 6.5: Effect of Temperature on Cell Voltage

Problem: For the cell reaction of hydrogen gas and chlorine gas to produce HCl in aqueous solution (dissociated into ions), how does temperature affect the cell voltage given that the standard state molar entropy for the products is $56.3 \, J \, K^{-1} \, mol^{-1}$?

Solution: Voltage is related to ΔG, and so temperature dependence comes about through the temperature dependence of ΔG. At constant pressure, we have

$$\Delta G_{rxn} = -nEF = \Delta H_{rxn} - T\Delta S_{rxn}$$

Taking the enthalpy and entropy changes to be constant, the derivative with respect to temperature yields

$$\left(\frac{\partial E}{\partial T} \right)_P = \frac{1}{nF} \Delta S_{rxn}$$

The next step is to find ΔS_{rxn}. From Table 6.1, we obtain the molar entropies for hydrogen (130.7) and chlorine (223.1) gas. The balanced chemical reaction required to produce 1 mol of HCl requires ½ mol of H_2 and ½ mol of Cl_2, $\frac{1}{2}H_2 + \frac{1}{2}Cl_2 \rightarrow H^+ + Cl^-$. The entropy of the product was given, and thus,

$$\Delta S_{rxn} = 56.3 - 0.5(130.7) - 0.5(223.1) = -120.6 \, J \, K^{-1} \, mol^{-1}$$

Therefore,

$$\left(\frac{\partial E}{\partial T} \right)_P = \frac{-120.6}{(1)(9.648534 \times 10^4)} = -1.25 \times 10^{-3} \, V \, K^{-1}$$

Since the voltage for this cell at 298 K is 1.35827 V, the temperature dependence is relatively slight. For instance, a change from a temperature of 298 to 398 K would change the voltage by 0.125 V, or by less than 10%. For cell reactions that involve only one phase (no gas reactants), ΔS_{rxn} would likely be smaller. That reduces voltage sensitivity to temperature.

Notice that just as measuring the cell voltage affords a means for determining ΔG_{rxn}, measuring the temperature dependence of the voltage affords a means of measuring ΔS_{rxn}.

Another feature of cell behavior can be recognized by using the thermodynamic relation $q_{rev} = T\Delta S_{rxn}$. The sign of the derivative of voltage with respect to temperature is the sign of ΔS_{rxn} and is therefore the sign of q_{rev}. So, in this example, the cell evolves heat (negative sign of q_{rev}) as it operates.

6.8 Conclusions

Elementary reaction steps are those for which products are produced from reactants as the result of single collision events. They are found to be unimolecular, bimolecular, or termolecular. A microscopic view of an elementary reaction develops from the concept of a potential energy surface. A key feature of the potential energy surface for a reactive system is the transition state structure or the saddle point. This point is the minimum potential that must be overcome to get from reactants to products (and back, perhaps).

Since reactions occur through collisions, the rates for elementary reactions depend directly on the concentrations of the reactants, and an expression relating rate to concentrations is a rate equation. The proportionality factor in a rate equation is the rate constant, k. This constant is a function of temperature because the thermal distribution of reactant velocities plays a role in the likelihood that reactant species will have sufficient energy to overcome the potential barrier. Measuring the temperature dependence of k affords a means for finding the potential barrier or activation energy. Other factors affect the likelihood of a reaction taking place in a given collision event, and these include the relative orientations of interacting species, quantum mechanical tunneling through a potential barrier, the relative directions of colliding species, and for reactions in solution, the solvent's ability to temporarily "cage" reactants.

Complex reactions are a sequence of elementary reaction steps. Their overall rate laws are more complicated but develop out of the rate equations of the elementary steps. The goal of experimental investigation of reaction kinetics is often to elucidate the sequence of individual steps that make up a complex reaction, that is, to find the mechanism of a complex reaction. Contemporary experimental and theoretical work on chemical reactions include efforts to achieve such detailed information that all the factors affecting reaction rates can be quantified and fundamentally accounted for.

Thermodynamic state functions obey the same relations whether applied to a process such as expansion of a gas or to a chemical reaction. Thus, we can sum molar enthalpies, molar volumes, molar entropies, and molar free energies of products, subtract from them the corresponding sum over reactants, and obtain the given quantity for the reaction. ΔU corresponds to the difference in reactant and product energies on the potential energy surface (i.e., differences in bond energies or molecular stabilities) and is approximately equal to the constant-pressure change in energy ΔH. For a reaction to proceed without input of energy requires that the Gibbs energy change, ΔG, be negative (downhill). The rate, at which the reaction goes, versus whether or not it goes at all, depends on activation barriers and other features that do not enter ΔG.

"Reaction" can be used as a global term in that the thermodynamic and kinetic analyses that have been discussed can apply not only to ordinary chemical reactions but also to electrochemical cells, to phase changes, to the expansion of a gas, and so on. In prior chapters, we examined the thermodynamic features of a phase change such as the melting of ice. Viewing that change as a reaction allows us to consider not only the conditions required for ice to melt but also the rate at which melting takes place. At the molecular level, the melting process requires breaking weak water–water bonds, bonds that hold an individual water molecule in an ice lattice. That process has an intrinsic activation barrier just as a chemical reaction does. Thus, we may say that all the molecular phenomena of change are unified at the macroscopic level under thermodynamics. As well, thermodynamics connects with more detailed, molecular pictures via kinetics and via two subjects for subsequent chapters, quantum mechanics and statistical mechanics. In the end, physical chemistry takes the fundamental laws of physics, applied to electrons and nuclei, and carries them to such everyday phenomena as the formation of ice. Though this body of knowledge is already massive, discovery continues as we learn not only how to explain and account for phenomena but how to control them.

Point of Interest: Galactic Reaction Chemistry

The mass contained in our galaxy is not only that of stars and their satellites. By the 1930s, the existence of an interstellar medium was recognized, and today this medium is thought to hold about 25% of the mass of our galaxy. Interstellar space contains electrons, atoms, ions, and molecules in varying densities. The regions of greatest density are referred to as interstellar clouds, and though some of them have as many as a million atomic/molecular particles per cubic centimeter, the most typical density is about one particle per cm^3. This is an extremely low density compared to atmospheric conditions on earth where we find about 1019 particles per cm^3. The immense size of the interstellar clouds accounts for their large fraction of the galaxy's mass. The interstellar medium of the Milky Way has a characteristic thickness—its dimension perpendicular to the galactic plane—of about 600 light years, and it extends, with varying density across the galactic plane.

Knowledge of the composition of the interstellar medium has been growing rapidly in the last 25 years. Spectroscopic techniques, including some that will be considered in Chapter 10, can sometimes reveal the "fingerprints" of molecules in the electromagnetic radiation received from interstellar space. In 1963 the existence of OH was reported, and in 1968 Charles Townes and coworkers identified the first polyatomic molecule, ammonia, in a cloud associated with the constellation Sagittarius. The next year water was identified, and Lewis Snyder found the first organic molecule, formaldehyde. Today, there are over 100 molecules and molecular ions that have been detected in space.

Many of the interstellar molecules are exotic. Several were thought to exist in space while awaiting the terrestrial synthesis and laboratory fingerprinting that confirmed their existence (i.e., laboratory spectra that matched signals received from space). For a few others, the case is compelling that the molecules exist in space even though there is no terrestrial workup (i.e., these molecules have not been made on Earth). Some of the exotic species, particularly ions, exist for microseconds in a laboratory apparatus, but because of the low cloud densities and temperatures are thought to exist for days in space. Collision frequencies are relatively low in the interstellar medium.

Because interstellar chemistry offers the prospect of studying unique molecules, unique conditions, and a reaction system that is truly galactic, the field has attracted a good number of investigators. Astronomers use radiotelescopes to search for new molecules, to identify new clouds, and even to draw maps of the densities of certain molecular species. Discovering how molecules have come to exist in interstellar space, and predicting what interesting molecules may be sitting out there undetected involves the use of reaction kinetics and thermodynamics.

The analysis of interstellar reaction chemistry presumes elementary bimolecular reactions because of the very low density of atomic, molecular, and ionic particles in space. Because these reactants are cold ($<100\,K$), their kinetic energy may be insufficient to overcome high activation barriers, and as a result, reactions involving ions are believed to have the biggest role. (The ions may arise in the first place when a molecule or atom is excited by cosmic rays.) Though these assumptions narrow the set of reactions that may be taking place in clouds, the number of possible reactions is still sizable. Investigators are left dealing with hundreds and perhaps thousands of bimolecular reactions involving hundreds of reactants. It is a complex reaction system, and yet it is amenable to rate law studies based on the kinetic arguments we discussed in this chapter. Unfortunately, an added complication is that laboratory measurements of the reaction rates, reaction enthalpies, reaction barriers, and certain other information are not available for all the ion-molecule reactions that are considered candidates for leading roles in interstellar chemistry. These studies may ultimately point to the existence of more complicated molecules or to special requirements for the formation of observed species, such as dark grains of matter.

Exercises

6.1 Calculate mean free path lengths for pure helium and for pure argon at a pressure of 1 bar and a temperature of $300\,\text{K}$, using $R = 1.2\,\text{Å}$ for helium and $R = 1.9\,\text{Å}$ for argon.

6.2 Calculate the number of collisions per atom of Neon at 350°C assuming it is an ideal gas at $1.5\,\text{atm}$. The radius of a Neon atom is $38\,\text{pm}$.

6.3 Assume ideal gas heat capacities for N_2 and then calculate the speed of sound for an atmosphere of pure nitrogen at $250\,\text{K}$ (a very cold outside temperature) and at $310\,\text{K}$ (a very warm outside temperature).

6.4 Using Equation 6.40, calculate E_{act} for some reaction where it is known that the rate constant doubles on changing the temperature

   a. From 10 to $20\,\text{K}$

   b. From 100 to $110\,\text{K}$

   c. From 300 to $310\,\text{K}$

   d. From 1000 to $1010\,\text{K}$

6.5 Using the data in Table 6.1 find the standard state enthalpy of reaction at $T = 298\,\text{K}$ for the following:

   a. $CH_4 + 2O_2 \rightarrow CO_2 + 2H_2O$

   b. $C_2H_6 + 3.5O_2 \rightarrow 2CO_2 + 3H_2O$

   c. $2N_2O + 3O_2 \rightarrow 2N_2O_4$

   d. $SiH_4 + 2H_2O_2 \rightarrow SiO_2 + 2H_2O + 2H_2$

6.6 Find $\Delta \overline{H}^{\circ}_{rxn}$ at $T = 298\,\text{K}$ for the following reactions using values given in Table 6.1:

   a. $2CH_3OH + 3O_2 \rightarrow 4H_2O + 2CO_2$

   b. $2CH_4 + C_2H_2 \rightarrow C_2H_4 + C_2H_6$

6.7 Using the data in Table 6.1 and Equation 5.43, calculate the equilibrium constant for the conversion of graphite to diamond at $T = 300\,\text{K}$ and at $T = 10^5\,\text{K}$.

6.8 If a piece of solid lead were placed in an aqueous solution of the salt of one of the other metals listed in Table 6.2, for which would the lead be oxidized?

6.9 Using the reduction potentials in Table 6.2, find $\Delta \overline{G}^{\circ}_{rxn,298K}$ for the reaction $Na(s) + \frac{1}{2}Cl_2(g) \rightarrow Na^+(aq) + Cl^-(aq)$.

6.10 Design a cell for which the overall reaction is

$$H_2O(l) \rightarrow H^+(aq) + OH^-(aq)$$

Calculate ΔG° at $298\,\text{K}$ for this cell. What is the equilibrium constant for this reaction?

6.11 Repeat the analysis leading to Equation 6.7 to find an expression for Z'_{ABC}, the rate for a three-body collision of A, B, and C, in terms of temperature and concentration.

6.12 Follow the steps leading to Equation 6.24 for a third order reaction.

6.13 Show by substitution that Equation 6.33 is a solution of Equation 6.32.

6.14 Find the integrated rate expression for an elementary reaction where $4A \rightarrow B$.

6.15 Find an expression for the half-life of the third-order reaction $3A \rightarrow B$.

6.16 Find the integrated rate expression for the reaction $2A + B \rightarrow$ products assuming that the concentration of B does not change. This might occur where collision with B is necessary for two A molecules to react but where B is not consumed in the reaction.

6.17 Find the integrated rate expression for a reaction that is empirically determined to be of order 3/2 in one reactive species.

6.18 Consider a situation where a reactant A reacts to form two different products (a parallel reaction). The reaction can be visualized as

$$A \xrightarrow{\ k_1\ } B$$

$$A \xrightarrow{\ k_2\ } C$$

Determine the integrated rate law expression for the concentration of B and C at any time t.

6.19 Consider a trimerization sequence in which

$$A + A \rightarrow A_2$$

$$A + A_2 \rightarrow A_3$$

Find the differential rate expression for the change in the concentration of A.

6.20 The kinetics of hydrolysis of an ester can be studied by titration of the acid by-product with base. For the following set of data

Time (min)	Volume of Base Titrated (mL)
0	0
15	18.7
30	30.1
45	36.9
∞	50.0

a. Determine the order of the reaction

b. Calculate the half-life

6.21 Compute the fraction of F_2 molecules in a gas sample that would have kinetic energy in excess of $100 \, \text{kJ mol}^{-1}$ at the following temperatures: 10, 300, and 600 K. Repeat this evaluation for H_2 in order to assess the role of particle mass on the energy available for overcoming a reaction barrier.

6.22 Enzymes are biochemical catalysts. They are part of a reaction but are not consumed in the course of a reaction. This means their concentration is constant. Consider the following overall reaction where E is an enzyme, S is a substrate (reactant), and P is the product:

$$E + S \rightarrow E + P$$

Determine the differential rate equations with the following assumptions:

a. The reaction is elementary with no likelihood of the reverse reaction.

b. The forward and reverse reactions are elementary reactions with different rate constants.

c. The reaction is a complex reaction whereby an intermediate species I is formed by the reaction between E and S. I can then undergo a unimolecular reaction to yield P and E. Both elementary steps can occur in the reverse direction.

Can a series of experimental measurements be designed that would distinguish among the three possible mechanisms?

6.23 Find the range of values given in Table 6.1 for the heat capacities of substances that are gases in their standard state. Find the average of these values. How well would that average serve as an approximate value for real gases? How does the average compare with the heat capacity of an ideal gas? Do the same for solid metals.

6.24 Find $\Delta \overline{H}^{\circ}_{rxn,T}$ and $\Delta \overline{G}^{\circ}_{rxn,T}$ at $T = 200$, 400, and $800\,K$ for the reaction HCCH + $H_2 \rightarrow$ H_2CCH_2 using values given in Table 6.1.

6.25 Calculate the equilibrium constant for the reaction $CO + \frac{1}{2}O_2 \rightarrow CO_2$ under standard state conditions at $T = 1000\,K$ using data in Table 6.1. Compare with the equilibrium constant calculated for the same temperature assuming that the heat capacities of the three substances vary linearly with temperature and using these values:

	CO	O_2	CO_2
$\overline{C}^{\circ}_{P,298K}$ (J K^{-1})	29.141	29.378	37.135
$\overline{C}^{\circ}_{P,1200K}$ (J K^{-1})	34.169	35.683	56.354

(Hint: Begin by finding the heat capacity of each substance as a linear function of temperature.)

6.26 Using the temperature dependent heat capacities obtained in Exercise 6.21, find $\Delta \overline{H}^{\circ}_{rxn,600K}$ and $\Delta \overline{G}^{\circ}_{rxn,600K}$ for the reaction $CO + \frac{1}{2}O_2 \rightarrow CO_2$.

6.27 Devise an experiment to measure a reaction rate using an electrochemical cell, a potentiometer, and a precision timer. (Can reaction kinetics be studied for a reaction in a cell?)

6.28 Using data in Table 6.2, identify half-reactions that could take place in an electrochemical cell to yield each of the following overall reactions. Determine ΔG°_{rxn} for each and indicate the direction of the change in cell voltage (if any) were pressure to be doubled.

a. $Mg(s) + Cl_2(g) \rightarrow MgCl_2$

b. $Li(s) + K^+ \rightarrow Li^+ + K(s)$

6.29 Find ΔG°_{rxn} for an electrochemical cell where lead metal is being oxidized at one electrode and chlorine gas is reacting to form chloride ions at the other electrode. Predict the qualitative effect on the voltage of this cell if the pressure is increased.

6.30 A hydrogen–oxygen fuel cell is constructed to be operated at $350\,K$ and a total pressure of 10.0 bar. What is the voltage produced by this cell under its operating parameters and under standard conditions assuming the gases are behaving ideally?

Bibliography

Bard, A. J. and L. R. Faulkner, *Electrochemical Methods: Fundamentals and Applications*, 2nd edn. (Wiley, New York, 2000).

Billing, G. D. and K. V. Mikkelsen, *Introduction to Molecular Dynamics and Chemical Kinetics* (Wiley, New York, 1996).

Hecht, C. E., *Statistical Thermodynamics and Kinetic Theory* (Freeman, New York, 1990).

Houston, P. L., *Chemical Kinetics and Reaction Dynamics* (Dover Publications, New York, 2006).

Moore, J. W. and R. G. Pearson, *Kinetics and Mechanism*, 3rd edn. (Wiley, New York, 1981).

Steinfeld, J. I., J. S. Francisco, and W. L. Hase, *Chemical Kinetics and Dynamics*, 2nd edn. (Prentice Hall, Englewood Cliffs, NJ, 1998).

7

Vibrational Mechanics of Particle Systems

With atoms and molecules taken to be single particles, earlier chapters have followed gas kinetic analysis of collisions, gas pressure, and transfer of energy as heat and work. However, the internal structure and mechanics of molecules—that they are not single point masses—can play a role in thermodynamic behavior and reaction energetics. This chapter focuses on the mechanics of vibration, an internal motion exhibited by all molecules. Though we start by using classical mechanics, it turns out to be an incomplete theory in that it fails to correctly describe very small, very low-mass particle systems. To go beyond classical pictures calls for us to invoke quantum mechanical ideas which are introduced here. The contrast and the correspondence between the classical and quantum pictures of the vibrational motion of molecules is a primary objective of this chapter.

7.1 Classical Particle Mechanics and Vibration

Differential equations, whose solutions describe the motion of a particle or system of particles, are called **equations of motion**. In the mechanics of Isaac Newton (England, 1642–1727), the equations of motion include one of Newton's laws: The total force acting on an object equals the mass of the object times the time rate of change of the object's velocity, which is the acceleration, that is, $\vec{a} = d\vec{v}/dt$. Letting $\vec{F}$ be the vector of force and m the mass of the object, Newton's relation

$$\vec{F} = m\vec{a} \tag{7.1}$$

is an equation of motion. If one finds or knows $\vec{F}$, this equation ultimately describes everything about the object's position and motion. Solution of this differential equation yields the position vector as a function of time, that is, $\vec{r}(t)$. This means we may be able to predict at every instant of time the position of the object and, via the time derivatives of $\vec{r}(t)$, the object's velocity and acceleration. This description of position, velocity, and acceleration is a feature of a **classical mechanics** picture. It is consistent with our everyday experience with moving objects; however, we will find that it does not hold for very small, very low-mass particles.

Newton's formulation is not the only way in which classical equations of motion can be formulated. Lagrange (Joseph Louis Lagrange, France, 1736–1813), Hamilton (William Rowan Hamilton, Ireland, 1805–1865), and others developed different means, and it is the formulation of Hamilton that has proven the most useful framework for developing the mechanics of quantum systems. It is important to realize that Newtonian, Lagrangian, and Hamiltonian mechanics offer equivalent descriptions of classical systems.

The $\vec{r}(t)$ found with one formulation is the same as that found with another, even if the route to that function appears different.

At this point, we shall restrict attention to the mechanics of systems of point-mass particles. This means that particles are taken to have no volume. The mass of each particle is treated as if enclosed in an infinitesimally small region of space, a point. And mostly, we will restrict our attention to **conservative** systems, which are those for which the potential energy has no explicit dependence on time (i.e., there are no potentials changing with time). These restrictions are not an aspect of the particular mechanical formulation; they are used here as a convenience that is in keeping with the types of systems of most immediate interest in chemistry.

The classical kinetic energy, T, of any particle is the square of the momentum divided by twice the particle's mass. Momentum is a vector, and so, in Cartesian space, there is an x-component, a y-component, and a z-component. Thus,

$$T = \frac{\left|\vec{p}\right|^2}{2m} = \frac{p_x^2 + p_y^2 + p_z^2}{2m} = \frac{p \cdot p}{2m} \tag{7.2}$$

where two standard forms of notation for a vector quantity have been used, namely, $\vec{p}$ and p. If there is a system of N different particles, a subscript serves to distinguish the masses and momenta of the particles, for example, m_i and p_i, where $i = 1, \dots, N$. The kinetic energy for the system is just the sum of the kinetic energies of each of the particles:

$$T = \sum_{i=1}^{N} \frac{p_i \cdot p_i}{2m_i} \tag{7.3}$$

Written in this way, the kinetic energy appears as an explicit function of the momentum components of each of the particles; that is, $T = T(p_1, p_2, \dots, p_N)$.

The potential energy, V, of a conservative system of particles depends only on the position coordinates of the particles. For example, the potential energy of two electrically charged particles, one with charge Q_1 and the other with charge Q_2, has the form $Q_1 Q_2 / r$, where r is the separation distance between the particles. If a Cartesian coordinate system is used to specify particle positions, the first particle's position can be given by (x_1, y_1, z_1) and the second one's position by (x_2, y_2, z_2). The potential energy in this case is a function of these six coordinates:

$$V = V(x_1, y_1, z_1, x_2, y_2, z_2) = \frac{Q_1 Q_2}{\sqrt{(x_2 - x_1)^2 + (y_2 - y_1)^2 + (z_2 - z_1)^2}}$$

It is not required that position coordinates be Cartesian coordinates, though this is often the most convenient. In some important examples, spherical polar coordinates will be used. It is a good idea, then, to carry out further mechanical developments without specifying the type of coordinate system, merely requiring that it be sufficient to specify the positions of all the particles in the system. This is usually described as using a **generalized coordinate system**, and q (or q_i) is often used as the symbol for a generalized position coordinate. The number of degrees of freedom is independent of the choice of coordinate system, and so for a single particle the three possible degrees of freedom (directions in three-dimensional space) can be denoted by the coordinates q_1, q_2, and q_3. Each generalized position coordinate defines a direction in which there is a corresponding, or conjugate, component of momentum; for every q_i there is a p_i.

For conservative systems in which the potential energy does not depend on momenta, the **Hamiltonian**, H, is defined to be simply the sum of T and V expressed as a function of position and momentum coordinates. For such systems, the Hamiltonian is the total energy, E, of the system.

$$H = T + V = E \tag{7.4}$$

For nonconservative systems, the potential has an explicit dependence on time, and then the Hamiltonian has to be a function of time as well as of the other coordinates.

Hamilton determined that for a generalized coordinate system, the equations of motion can be found from H:

$$\frac{\partial H}{\partial q_i} = -\frac{dp_i}{dt} = -\dot{p}_i \tag{7.5}$$

$$\frac{\partial H}{\partial p_i} = \frac{dq_i}{dt} = \dot{q}_i \tag{7.6}$$

(A dot over a variable or functions signifies the first derivative with respect to time. A second dot would signify the second derivative with respect to time, and so on.) For each degree of freedom (each direction), the partial derivative of the Hamiltonian with respect to a position coordinate is equal to the negative of the time derivative of the corresponding, or conjugate, momentum coordinate. The partial derivative of the Hamiltonian with respect to a momentum coordinate equals the time derivative of the conjugate position coordinate. Simultaneous solution of these differential equations yields a description of the mechanics of the system.

An important application of Hamilton's classical equations of motion is the vibrational mechanics of systems of particles. We shall begin with application to an idealized, simple problem, the harmonic oscillator. The harmonic oscillator is a special model problem consisting of a mass able to move in one direction and connected by a spring to an infinitely heavy wall, as shown in Figure 7.1. The spring is special because it is harmonic, which means that the restoring force is linearly proportional to the value of the coordinate that

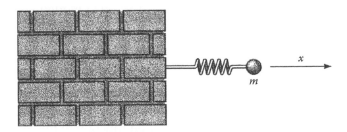

FIGURE 7.1
The harmonic oscillator is a special model problem in mechanics. It consists of a particle of mass m attached by a harmonic spring to an unmovable wall. In the one-dimensional harmonic oscillator problem, the particle has only one degree of freedom corresponding to motion to and away from the wall. In the discussion, this is the x-direction.

gives the displacement from its equilibrium. An equivalent statement is that the potential energy for stretching or compressing the spring varies quadratically with the displacement from the equilibrium length. If x is chosen to be the position coordinate of the mass, m, and x_0 is the equilibrium length of the spring (length of the spring when the system is at rest), the harmonic potential experienced by the mass in Figure 7.1 has the following mathematical form:

$$V(x) = \frac{1}{2}k(x - x_0)^2 \tag{7.7}$$

The constant, k, in this expression is called the **spring constant** or the **force constant**.

Since there is but one degree of freedom for the one-dimensional harmonic oscillator, the kinetic energy is simply $p^2/2m$. Thus, the Hamiltonian for this system is

$$H(x,p) = \frac{p^2}{2m} + \frac{1}{2}k(x - x_0)^2 \tag{7.8}$$

Differentiating this Hamiltonian function with respect to x and using Equation 7.5 yields

$$k(x - x_0) = -\dot{p} \tag{7.9}$$

Simply differentiating H with respect to p and using Equation 7.6 yields

$$\frac{p}{m} = \dot{x} \tag{7.10}$$

Equations 7.9 and 7.10 are the equations of motion of the harmonic oscillator. Equation 7.10 is also the definition of the momentum that would be invoked in Newtonian mechanics; that is, multiplying the equation by m gives p equal to the mass times the velocity, $p = m\dot{x}$. In general, solving the equations of motion can be done by any standard means of solving differential equations. In this particular example, the first step is to take the time derivative of Equation 7.10 after first multiplying through by m.

$$\dot{p} = m\ddot{x}$$

This relates the time derivative of the momentum to the acceleration or the second time derivative of the position function, $\ddot{x}$. But the first equation of motion also tells us that the time derivative of the momentum is $-k(x - x_0)$. Therefore, we equate these to give

$$-k(x - x_0) = m\ddot{x} \tag{7.11}$$

This expression is the differential equation that determines the position function, $x(t)$.

Hamilton's formulation, applied to the harmonic oscillator problem, has yielded a differential equation, Equation 7.11, that relates the position function to the second time derivative of the position function. This differential equation has a general solution that can be expressed in either of two ways. They are mathematically equivalent:

$$x(t) = x_0 + a\sin(\omega t) + b\cos(\omega t) \tag{7.12a}$$

$$x(t) = x_0 + Ae^{-i\omega t} + Be^{i\omega t}$$

(7.12b)

$$i \equiv \sqrt{-1}$$

The constants in these functions are a, b, and ω in the first and A, B, and ω in the second. x_0 is the position at which the potential is zero, the equilibrium length of the spring. The constant ω must have units of inverse time, and it is a frequency. The other constants must have units of length because $x(t)$ is the position of the particle. These equations are tested by using them in Equation 7.11, which calls for calculating the second derivative of either Equation 7.12a or b. With either choice,

$$\ddot{x}(t) = -\omega^2 \left[x(t) - x_0 \right]$$

(7.13)

Comparing this with the differential equation derived for the harmonic oscillator, Equation 7.11, one can see that the frequency is simply related to the force constant and the particle's mass.

$$\omega = \sqrt{\frac{k}{m}}$$

(7.14)

Equation 7.14 indicates how the vibrational frequency is affected by changes to the oscillator. For instance, if the mass were heavier, then the frequency of oscillation would be lower. In a molecule, isotope substitution might accomplish such a change if it led to increased particle mass. If the force constant, k, were to become larger, that is, if the spring were stiffer, Equation 7.14 indicates that the frequency would be increased. In a molecular system, increasing the bond order, as in changing from a carbon–carbon single bond to a carbon–carbon double bond usually leads to a larger force constant.

The constants a and b, or A and B, in Equation 7.12 are found from initial conditions. For instance, if it is known that at time $t = 0$ the position is x_0, that is, $x(0) = x_0$, the following will be obtained:

$$x(t) = x_0 + a\sin(0) + b\cos(0) = x_0 \Rightarrow b = 0$$

(7.15)

That is, the condition $x(0) = x_0$ is enforced by setting the constant b to zero. Notice that this condition merely amounts to choosing the time at which "the clock is started" to be some instant when the particle is at its equilibrium position. The remaining unknown constant, a, is given by the maximum displacement from equilibrium that the particle reaches. In other words, $x_0 + a$ and $x_0 - a$ are the furthest positions from equilibrium that the particle reaches in the course of its vibrational motion.

At maximum displacement, the velocity of the particle goes to zero, meaning that at that instant the kinetic energy is zero; all the energy of the oscillator is potential energy at that point. This is the point where the particle is at one of its **turning points** (i.e., where it turns from going in one direction to going in the opposite direction), and again, this is either the point $x_0 + a$ or the point $x_0 - a$. Notice that a is in no way restricted to any particular value. It may be zero, meaning the oscillator is not moving in time. It may be 1.0 or 1.0234 or 120,000.1. This statement just reinforces everyday experience that we can stretch a spring to almost *any* length desired and release it in the classical world, at least so long as the spring does not break. The significance of this is that we can adjust the energy of the oscillating spring system in a continuous manner. It will turn out that this applies only to the *classical* harmonic oscillator and does not hold in the quantum world.

7.2 Vibration in Several Degrees of Freedom

Molecular problems are problems of many particles, and as a result, there are many degrees of freedom. To illustrate how Hamilton's formulation is used with more than one degree of freedom, consider the double oscillator problem depicted in Figure 7.2. This is a system with two degrees of freedom, the positions along the x-axis of the two particles. Subscripts 1 and 2 distinguish the particles, whereas subscripts a and b distinguish the two springs. The particles have masses m_1 and m_2, and their positions are x_1 and x_2. The harmonic springs have force constants k_a and k_b and equilibrium lengths x_a and x_b. The classical Hamiltonian function for this system is the sum of the kinetic energies of the two particles and the potential energies from stretching the two springs. We first analyze spring b by noting that its length is $x_2 - x_1$. Its displacement from equilibrium is this length less x_b, assuming that $x_2 > x_1$ always. The potential energy stored in spring b is this displacement squared, times $k_b/2$. The potential energy in spring a is the square of its displacement from equilibrium, $x_1 - x_a$, times $k_a/2$. Adding to these the kinetic energy for the particles gives the Hamiltonian function, H, for this problem.

$$H(x_1, x_2, p_1, p_2) = \frac{p_1^2}{2m_1} + \frac{p_2^2}{2m_2} + \frac{1}{2}k_a(x_1 - x_a)^2 + \frac{1}{2}k_b(x_2 - x_1 - x_b)^2 \qquad (7.16)$$

The equations of motion, by Hamilton's formulation, follow from applying Equations 7.5 and 7.6 with this Hamiltonian.

$$\frac{p_1}{m_1} = \dot{x}_1 \qquad (7.17a)$$

$$k_a(x_1 - x_a) - k_b(x_2 - x_1 - x_b) = -\dot{p}_1 \qquad (7.17b)$$

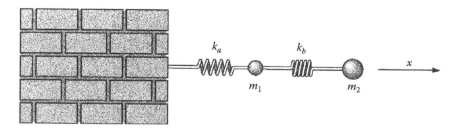

FIGURE 7.2
Double harmonic oscillator. This consists of two particles, 1 and 2, that move only in the direction of the x-axis. The masses of 1 and 2 are m_1 and m_2, respectively, and their positions are x_1 and x_2. Spring a is connected to an unmovable wall and to particle 1. Spring b connects the two particles.

$$\frac{p_2}{m_2} = \dot{x}_2 \tag{7.18a}$$

$$k_b(x_2 - x_1 - x_b) = -\dot{p}_2 \tag{7.18b}$$

These are **coupled** differential equations that reflect the fact that the two springs are physically coupled. They are connected at particle 1. Solving this type of problem or this set of coupled differential equations is considered briefly later in this chapter. For now, the important feature is that systems of several degrees of freedom, in general, have coupled equations of motion.

There are circumstances where the equations of motion in several degrees of freedom are not coupled. In these cases, the Hamiltonian has a special form referred to as a **separable** form. Separability arises when a particular Hamiltonian can be written as *additive, independent* functions:

$$H(q_1, q_2, q_3, ..., p_1, p_2, p_3, ...) = H'(q_1, p_1) + H''(q_2, q_3, ..., p_2, p_3, ...) \tag{7.19}$$

In Equation 7.19, H' is a function only of the first position and momentum coordinates, while H'' is a separate function involving only the other coordinates. When we apply Equations 7.5 and 7.6 to H in order to find the equations of motion for this system, the equations for q_1 and p_1 are found to be distinct, or mathematically unrelated to any of the equations that involve any of the other coordinates. In fact, the equations for q_1 and p_1 could just as well have been obtained by taking H' to be the Hamiltonian for the motion in the q_1-direction, while the other equations of motion could have been obtained directly from the H'' Hamiltonian. Thus, the additive, independent terms in the Hamiltonian can be separated in order to deduce the equations of motion. Furthermore, since the equations of motion are separable, the actual physical motions of the H' system and the H'' system are unrelated and independent. It is possible that a Hamiltonian can be separable in all variables; that is, it may be that

$$H(q_1, q_2, q_3, ..., p_1, p_2, p_3, ...) = H_1(q_1, p_1) + H_2(q_2, p_2) + H_3(q_3, p_3) + \cdots$$

then all the motions are independent.

In the double-oscillator system in Figure 7.2, the potential energy of spring b was dependent on the position coordinates of particle 1 (x_1) and of particle 2 (x_2). Thus, the Hamiltonian $H(x_1, x_2, p_1, p_2)$ was not separable in x_1 and x_2. To contrast with that situation, we will next consider a system that displays separability in its Hamiltonian. It is the oscillator problem in Figure 7.3. The Hamiltonian is

$$H(x_1, x_2, p_1, p_2) = \frac{p_1^2}{2m_1} + \frac{p_2^2}{2m_2} + \frac{1}{2}k_b(x_2 - x_1 - x_b)^2$$

Because the potential energy term involves both x_1 and x_2, this appears to be an inseparable Hamiltonian function. In fact, in terms of x_1 and x_2, the differential equations of motion are coupled. However, a certain change of coordinates, or more precisely a **coordinate transformation**, leads to a separable form. This coordinate transformation comes about

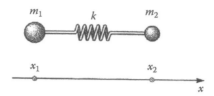

FIGURE 7.3
Oscillator system where two particles, each constrained to move along the x-axis, are attached by a harmonic spring with force constant k. The positions of the two particles along the x-axis are designated x_1 and x_2. The equilibrium length of the spring is the constant x_0.

by defining two new variables. First, we define r to be a coordinate that gives the displacement from the equilibrium length of the spring, and second, we define s to be the position of the center of mass of the whole system:

$$r = x_2 - x_1 - x_0$$

$$s = \frac{m_1 x_1 + m_2 x_2}{m_1 + m_2}$$

By inspection, the potential energy now has the simple form $kr^2/2$. That is, in the $\{r, s\}$ coordinate system, the potential energy is independent of s, the position of the system's mass center.

The kinetic energy of this system can be expressed in terms of momentum coordinates that are conjugate to the position coordinates, r and s. First, Equation 7.5 is used in the $\{x_1, x_2\}$ system to write

$$p_1 = m_1 \dot{x}_1 \quad p_2 = m_2 \dot{x}_2$$

Next, the kinetic energy is expressed in terms of the time derivatives of the position coordinates by using these expressions for the momenta.

$$T = \frac{1}{2}\left(m_1 \dot{x}_1^2 + m_2 \dot{x}_2^2\right)$$

Next, the time derivatives of r and s are substituted for the time derivatives of x_1 and x_2. The relation between velocities in the two coordinate systems comes about from taking the time derivative of the equations that relate r and s with x_1 and x_2:

$$\dot{r} = \dot{x}_2 - \dot{x}_1$$

$$\dot{s} = \frac{m_1 \dot{x}_1 + m_2 \dot{x}_2}{m_1 + m_2}$$

By rearrangement, we obtain from these two equations

$$\dot{x}_1 = \dot{s} - \frac{m_2}{m_1 + m_2}\dot{r}$$

$$\dot{x}_2 = \dot{s} + \frac{m_1}{m_1 + m_2}\dot{r}$$

Using these expressions in the equation for T yields

$$T = \frac{1}{2}\left[(m_1 + m_2)\dot{s}^2 + \frac{m_1 m_2}{m_1 + m_2}\dot{r}^2\right]$$

At this point, the quantity M is introduced to designate the total mass of the system, that is, $M = m_1 + m_2$, and the symbol μ is used for the quantity $m_1 m_2/m_1 + m_2$. μ is called the **reduced mass** of a two-body system. Therefore,

$$T = \frac{1}{2}(\mu \dot{r}^2 + M\dot{s}^2)$$

Notice that this is similar in form to the kinetic energy expressed in the original coordinates x_1 and x_2. The similarity suggests a definition of the conjugate momenta in the $\{r,s\}$ coordinate system:

$$p_r = \mu\dot{r}$$

$$p_s = M\dot{s}$$

In fact, it is Hamilton's equations of motion (Equation 7.5) that insure that these are correct definitions.

The Hamiltonian for the oscillator in Figure 7.3 can now be written fully transformed to the $\{r,s\}$ coordinate system.

$$H(r,s,p_r,p_s) = \frac{p_s^2}{2M} + \frac{p_r^2}{2\mu} + \frac{1}{2}kr^2$$

This is recognized to be a separable Hamiltonian because the first term is independent of the remaining two terms. The first term is interpreted as the kinetic energy of translation of the whole system. The second term is the kinetic energy for the *internal* motion of vibration, while the last term is the potential energy for vibration. In this system, separability arises because the translational motion of the whole system is independent and unrelated to the internal motion of vibration.

As discussed previously, separability of a Hamiltonian means that the equations of motion could also have been developed from the independent terms if they had been treated as individual Hamiltonians. For the example in Figure 7.3, the equations of motion obtained from $H(r, s, p_r, p_s)$ are the same as those obtained for the separate problems of internal motion, where the Hamiltonian is

$$H'(r,p_r) = \frac{p_r^2}{2\mu} + \frac{1}{2}kr^2$$

and of translational motion, for which the Hamiltonian is

$$H''(s,p_s) = \frac{p_s^2}{2M}$$

In other words, the four equations of motion obtained for the entire system using H turn out to be the same as the two equations of motion obtained from the H' function and the two equations of motion obtained from the H'' function. This comes about because the vibration of the system, which is motion in the r-direction, does not affect and is not affected by the translation of the system, which is motion in the s-direction. Therefore, H' alone is sufficient if only internal motion mechanics are of interest; translational motion of the system is uncoupled or separated from the internal motion of vibration.

Comparison of the internal vibrational Hamiltonian for the two-body oscillator in Figure 7.3 with the Hamiltonian for the one-body oscillator in Figure 7.1 shows that the differences are only in the "names" of the masses and their meanings, m versus μ, and the names of the displacement coordinates, x versus r. On this basis, we can say that vibration of the two-body system is mechanically equivalent to vibration of a single body of mass m attached to an unmovable wall. **Mechanical equivalence** means two systems have equations of motion, or Hamiltonians, of the same form.

The examples given so far have involved particles that were restricted to move only in a straight line, along the x-axis. A more general analysis is necessary, of course, since atoms and molecules do not have to experience such constraints. A free atom can move in the x-direction, the y-direction, or the z-direction. Should any other type of coordinate system be used to give the location of the atom, there will still be three and only three independent coordinates. For a system composed of N atoms or N particles, whatever the type, a total of $3N$ independent coordinates will be required to specify the positions of all N particles. There will be $3N$ spatial degrees of freedom for the system.

We have seen that the overall translational motion of a system of two particles along the x-axis is separable from the vibrational motion. In a three-dimensional picture of the system, translational motion is also separable, but the coordinate transformation is different. In three Cartesian dimensions, the positions of the two particles can be specified as (x_1, y_1, z_1) and (x_2, y_2, z_2). The separation distance between the two particles, r, is then

$$r = \sqrt{(x_2 - x_1)^2 + (y_2 - y_1)^2 + (z_2 - z_1)^2} \tag{7.20}$$

The center of mass of the system is given by three coordinate values (X, Y, Z):

$$X = \frac{1}{M}(m_1 x_1 + m_2 x_2) \tag{7.21a}$$

$$Y = \frac{1}{M}(m_1 y_1 + m_2 y_2) \tag{7.21b}$$

$$Z = \frac{1}{M}(m_1 z_1 + m_2 z_2) \tag{7.21c}$$

Two other new coordinates, θ and φ, complete this particular transformation. They give the orientation of the two-particle system about the center of mass and with respect to a fixed $\{x, y, z\}$ axis system as depicted in Figure 7.4. r, θ, and φ are **spherical polar coordinates**.

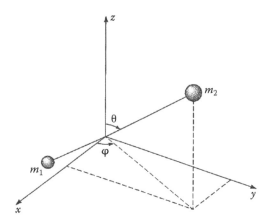

FIGURE 7.4
Spherical polar coordinates for a two-particle system. r is the distance between m_1 and m_2. The center of mass is positioned at the origin of the (x, y, z) coordinate system. θ is the angle between the z-axis and the line connecting the two masses. θ ranges from 0 to π. φ measures the rotation about the z-axis of the line connecting the masses. It ranges from 0 to 2π. (Also see Figure 2.4.)

$$x_2 - x_1 = r\sin\theta\cos\varphi$$

$$y_2 - y_1 = r\sin\theta\sin\varphi \qquad (7.22)$$

$$z_2 - z_1 = r\cos\theta$$

The overall coordinate transformation is $\{x_1, y_1, z_1, x_2, y_2, z_2\} \rightarrow \{X, Y, Z, r, \theta, \varphi\}$. Notice that the number of independent coordinates, six, remains the same.

The transformation to spherical polar coordinates makes the potential energy dependent on only the variable, r, rather than on all six coordinates. The kinetic energy has a contribution associated with translational motion of the whole system,

$$T^{translation} = \frac{1}{2M}\left(p_X^2 + p_Y^2 + p_Z^2\right)$$

It also has a contribution associated with the vibrational motion and a contribution associated with rotation of the two bodies about their center of mass. Thus, the Hamiltonian consists of independent terms associated with motion in the X-direction (i.e., translation of the system along the x-axis), in the Y-direction, and in the Z-direction, and with vibrational and rotational motion. This describes the separability of the problem. There are three translational degrees of freedom that are separable from the rest.

Had we analyzed a system of N particles with $N > 2$, there would still be only three translational degrees of freedom for the system as a whole. And this motion would be separable from the other motions. Thus, there would be $3N - 3$ degrees of freedom for other than translational motion. In the case of two point-mass particles, there were two rotational degrees of freedom corresponding to the two angles θ and φ. This is the case for any linear system, and so there are $(3N - 3) - 2 = 3N - 5$ vibrational degrees of freedom. For nonlinear collections of particles, there are three independent rotations that are rotations about the x-axis, about the y-axis, and about the z-axis. Thus, for nonlinear systems, $3N - 6$ is the number of degrees of freedom remaining for vibrational motion.

It may not seem clear why there should be a different number of rotational degrees of freedom for linear and for nonlinear systems. The subtle difference has to do with the fact that we are considering particles to be point masses, entities without volume or size. If some number of these masses were arranged along the *x*-axis, it would be easy to visualize turning the system about the *y*-axis or turning it about the *z*-axis. On the other hand, turning it about the *x*-axis would be equivalent to doing nothing. It would be the rotation of an infinitesimally thin rod, an operation or motion that does not exist. Only if one or more particles were off-axis would such motion be defined, but then that would be a nonlinear arrangement.

7.3 Quantum Phenomena and Wave Character

Numerous mechanical devices in our everyday world assert the validity of the classical mechanical analyses that have helped in the design of such devices. However, we cannot beforehand be assured that the mechanics of a clock pendulum, for instance, are the same when the size of the pendulum system is reduced (or increased) well beyond the limits of our experience. The mass of an atom may be as little as 10^{-24} g, and this is a very different "size" than that of objects we see and touch. Mechanical behavior in the world of such very tiny particles is, in fact, different.

The first of what are now called quantum phenomena was noticed or detected around the beginning of the twentieth century. The results of numerous experiments that had been performed could not be explained by the electromagnetic and mechanical laws known or accepted at the time. For example, it was found that a beam of monoenergetic electrons could be diffracted in the same way as a monochromatic beam of light impinging on two slits. The diffraction of light was known to be a manifestation of its wave character, and thus, to explain the diffraction of electrons, an abstract type of wave character had to be attributed to small particles. Of course, this was entirely outside the paradigm of classical mechanics, which holds that particles are exactly what we sense and observe them to be in our everyday world: They are distinct objects that exist at a specific place at any instant of time, traveling with a well-specified velocity, unlike electro-magnetic radiation which is distributed as a wave in space and time.

It was no doubt unsatisfying to many scientists during the early part of the century that basic physical laws worked out over almost two centuries did not hold for the tiny world of electrons and atoms. Today, we see the classical picture as a special case of broader theories of matter and radiation.* It took studies of the tiny-particle world and the introduction of several new concepts to develop the more encompassing theories. This was because key features of the quantum world are manifested in ways not directly perceptible in our everyday world.

Particle diffraction was an important stepping stone in the understanding of tiny-particle mechanics. The particles used for a diffraction experiment are free particles, meaning they are not subject to a potential. Electrons, for instance, would first have to be stripped away from some atoms or molecules. Observing diffraction requires that all the particles in the

---

* A question that sometimes comes to mind on first learning of the quantum mechanical revolution is whether a new theory could someday replace quantum mechanics. This may be more a matter for the philosophy of science and how science accepts or rejects new theories, for the successes of quantum ideas have withstood almost 100 years of challenges. Some future scientific revolution that affects quantum theory is less likely to replace quantum theory than to build on it or encompass it in the same way that quantum mechanics ultimately was found to connect with and encompass classical mechanics.

sample move at the same velocity or momentum, and this can be accomplished by the manner in which the particle beam is formed. The quantitative result of the diffraction experiment is a wavelength, λ, which is obtained from measurement of spacings in the diffraction pattern left on a sensitive plate that is etched where electrons impinge. Measured for different particle momenta, p, the wavelength is found to vary inversely with p. This is called the **de Broglie relation**:

$$p = \frac{h}{\lambda} \tag{7.23}$$

λ is the **de Broglie wavelength** (Louis Victor de Broglie, France, 1892–1987). The proportionality constant of this relation, h, is a fundamental constant of nature, Planck's constant (Max Planck, Germany, 1858–1947). The value of h is 6.626069×10^{-34} J s.

Let us find the de Broglie wavelength for an object from our day-to-day world, a bowling ball rolling at 36 km h^{-1}. If the mass of the ball is 8 kg, then the linear momentum of this "particle" is

$$8\,\text{kg} \times 36\,\text{km}\,\text{h}^{-1} = 8\,\text{kg} \times 10\,\text{m}\,\text{s}^{-1} = 80\,\text{kg}\,\text{m}\,\text{s}^{-1}$$

Substituting this linear momentum into the de Broglie relation yields the wavelength:

$$\lambda = \frac{6.626 \times 10^{-34}\,\text{J}\,\text{s}}{80\,\text{kg}\,\text{m}\,\text{s}^{-1}} = 8.28 \times 10^{-36}\,\text{m}$$

This is an incredibly tiny wavelength compared to the size of the bowling ball and a bowler, and it is not surprising that effects associated with something this size are not within the perception of the bowler. However, the de Broglie wavelength of a free electron moving at the same velocity as the bowling ball is much longer:

$$\lambda = \frac{6.626 \times 10^{-34}\,\text{J}\,\text{s}}{(9.1 \times 10^{-31}\,\text{kg})(10\,\text{m}\,\text{s}^{-1})} = 7.3 \times 10^{-5}\,\text{m}$$

The wavelength of the electron, because of its small mass, is significant relative to the dimension of its atomic world, whereas the wavelength of the bowling ball is negligible relative to its dimensions. At least on this basis, it appears a valid approximation to treat the mechanics of the bowling ball classically and neglect its wave character. The wave character of the free electron, though, is not ignorable as evidenced by the phenomenon of electron diffraction.

Other observations made at the beginning of the twentieth century not only showed unexpected phenomena but also seemed to involve the new constant, h, in many ways. The photoelectric effect, which is the emission of electrons from a metal surface on irradiation with light, was explained by Einstein with a nonclassical picture of electromagnetic radiation. He argued that the energy delivered by radiation comes in discrete amounts, called quanta, and that the energy of a quantum is proportional to the frequency of the radiation, ν:

$$E_{radiation} = h\nu \tag{7.24}$$

Notice that the proportionality constant is Planck's constant. As more problems were explained by the unusual quantum and wave notions, it became certain that classical physics was not complete and could not be applied at the subatomic level, at least not without introducing ad hoc conditions and assumptions. The laws of quantum mechanics that emerged were worked out by clever and sometimes indirect lines of thought.

If it is necessary for a quantum-level description that we associate a wavelength with a free particle, then we should have some simple oscillating function associated with the particle. This would be a sine or cosine function (or equivalently, a complex exponential), for example,

$$A(x) = A_0 \cos\left(\frac{2\pi x}{\lambda}\right) \tag{7.25}$$

The constant, A_0, is the maximum amplitude of $A(x)$. A salient feature of this type of function is that we can derive the wavelength from the function as an **eigenvalue**. Notice that when $A(x)$ is differentiated twice, the result is a constant, the eigenvalue C, times $A(x)$.

$$\frac{d^2}{dx^2} A(x) = -\left(\frac{2\pi}{\lambda}\right)^2 A_0 \cos\left(\frac{2\pi x}{\lambda}\right) = CA(x) \tag{7.26}$$

In such cases, the function is called an **eigenfunction** of the operation of second differentiation. One may take "of one's own" or "same" as a loose translation of the German word *eigen*. It is one's own function, or the same function, that is produced on the operation of second differentiation. The constant, C, that shows up multiplying the original function is the eigenvalue. In contrast, $A(x)$ is not an eigenfunction of the operation d/dx since the first order differentiation of the cosine function produces a different function, the sine function.

At this point we might accept the notion that the function $A(x)$ or something similar is connected with the mechanics of a moving free particle. However, we do not have an interpretation of $A(x)$ to make such a connection; its meaning is not apparent. One thing, though, is that $A(x)$ has a wavelength embedded in it, and that wavelength appears in a particular eigenvalue of the function. It was deduced by Erwin Schrödinger (Germany, 1887–1961), Werner Heisenberg (Germany, 1901–1976), and others that eigenvalues have special significance and in particular that they give measurable information about a system, such as the de Broglie wavelength of a free particle. The eigenfunctions themselves came to be known as **wavefunctions** partly because it was the apparent wave character of particles that demanded a new mechanics.

If there were particular operations, such as second differentiation, for which all proper wavefunctions must be eigenfunctions, there would be a general basis for finding wavefunctions for any system. That general basis would be to solve an eigenequation rather than to guess at something like $A(x)$. This focuses attention on the operations that may enter the eigenequation, and from that, specific mechanical elements have come to be associated with specific mathematical operations. One of these can be demonstrated using the de Broglie relation and the free particle example. If the operation $(d/dx)^2$ is scaled (multiplied) by the negative of the square of the constant $\hbar = h/2\pi$ ($\hbar$ is usually called the reduced Planck's constant) and applied to $A(x)$, by Equation 7.26 the result is the square of the momentum times $A(x)$:

$$-\left(\frac{h}{2\pi}\right)^2 \frac{d^2}{dx^2} A(x) = \frac{h^2}{\lambda^2} A(x)$$

This suggests an operation that in quantum mechanics corresponds to the square of the dynamical variable of momentum. We might anticipate that this is useful since the classical energy, at least, is a function of momentum and energy is of key importance in a particle world where energy is quantized.

These several notions about the quantum world form part of a theoretical basis that is usually referred to as the set of **quantum mechanical postulates**. These postulates are ideas or statements that in themselves may or may not be true. However, if predictions that follow naturally from these ideas are proven true, then the ideas are accepted as true even if there is no direct proof.

Postulate I: *For every quantum mechanical system, there exists a wavefunction that contains a full mechanical description of the system.* The wavefunction is a function of position coordinates and possibly time. This is quite different from classical mechanics where the full mechanical description would be a function that gives the position and momentum at any instant in time, for example, $\vec{x}(t)$.

Postulate II: For every dynamical variable, there is an associated mathematical operation. Furthermore, if there is an eigenvalue for such an operation and a particular wavefunction, that eigenvalue is the result that would be obtained from measuring that dynamical variable for the system having this particular wavefunction. This explains, in part, how we can glean a picture of a quantum mechanical system from its wavefunction. An example, so far, is that the operation associated with the square of the momentum has an eigenvalue of $(h/\lambda)^2$ for the free particle wavefunction, $A(x)$. We extract $(h/\lambda)^2$ from the wavefunction by applying the associated operator.

Postulate III: *The wavefunction of a system must be an eigenfunction of the operation associated with the classical Hamiltonian of the system.* The corresponding eigenvalue is the energy of the system since energy is the quantity given by the classical Hamiltonian. Energy is measurable or observable, and so this statement is similar to the second postulate that tells what to make of eigenvalues of wavefunctions. However, the third postulate is a stronger statement than the second postulate in that it requires the wavefunction to be an eigenfunction of one specific operator, the Hamiltonian.

The dynamical variables in the classical Hamiltonian are position and momentum coordinates. As long as wavefunctions are developed* as functions of position coordinates, as was the case for $A(x)$, the mathematical operation associated with a position coordinate is just multiplication by the coordinate. The operation associated with the momentum variable p_i that is conjugate to a Cartesian position variable q_i is

$$-i\hbar \frac{\partial}{\partial q_i}$$

where $i \equiv \sqrt{-1}$. Notice that the square of this operation (i.e., applying it twice) is the same as that used previously for the square of the momentum. The various mathematical operations are best referred to as **operators**, and operators may be distinguished from variables by a "hat" over the character. Thus, the statement $\hat{x} = x$ means that the operator associated with x is the same as multiplication by the variable x, whereas $\hat{p} = -i\hbar(\partial/\partial x)$ means that the

---

* Quantum mechanics can also be developed so that the functions describing the system are functions of momentum coordinates, not position coordinates. This is termed a momentum representation, and in this representation, position and momentum operators take on a different form. The picture of a quantum system, though, is equivalent, and the choice between representations is largely one of convenience.

operator associated with the momentum is $-i\hbar$ times the operation of partial differentiation with respect to x. Operators can be combined and applied repeatedly to form new operators. Thus, if the position and momentum operators are known, an operator can be constructed to correspond with the classical Hamiltonian. This operator is the **Hamiltonian operator**, $\widehat{H}$, or just the Hamiltonian if the quantum mechanical context is clear. The most common symbol used for an unspecified quantum mechanical wavefunction is ψ, and so Postulate III is simply the equation

$$\widehat{H}\psi = E\psi \tag{7.27}$$

This equation is called the **Schrödinger equation**. E is the energy eigenvalue. Solution of the Schrödinger equation is central to all quantum mechanical problems.

7.4 Quantum Mechanical Harmonic Oscillator

The problem of the quantum mechanical description of a harmonic oscillator is our first example of applying the postulates of quantum mechanics. It also provides a valuable comparison with the classical description considered earlier in this chapter. The picture of the system is the same as that in Figure 7.1, a mass m connected to an unmovable wall by a harmonic spring with force constant k. The steps for quantum mechanical treatment of this problem, as well as any other problem, are the following:

1. Express the classical Hamiltonian function for the system.
2. By replacing variables in the classical Hamiltonian with corresponding quantum mechanical operators, develop the quantum mechanical Hamiltonian operator and express the Schrödinger equation specific to the problem.
3. Find the wavefunction from the Schrödinger equation, which means finding the eigenfunction or eigenfunctions of the Hamiltonian operator.

From the earlier analysis of the classical harmonic oscillator we have the Hamiltonian function.

$$H(x,p) = \frac{p^2}{2m} + \frac{1}{2}k(x-x_0)^2$$

For convenience, let us now define the x-coordinate origin to be the location of the mass when the system is at equilibrium. This means choosing the origin such that $x_0 = 0$. Thus,

$$H(x,p) = \frac{p^2}{2m} + \frac{1}{2}kx^2$$

The quantum mechanical Hamiltonian is constructed formally by making x, p, and H operators.

$$\widehat{H} = \frac{\widehat{p}^2}{2m} + \frac{1}{2}k\widehat{x}^2$$

The explicit form of the momentum operator involves second differentiation with respect to x, and the position coordinate operator is just multiplication by x. Thus, the explicit quantum mechanical Hamiltonian is

$$\hat{H} = -\frac{\hbar^2}{2m}\frac{d^2}{dx^2} + \frac{1}{2}kx^2 \tag{7.28}$$

Solution of the Schrödinger equation, $\hat{H}\psi(x) = E\psi(x)$, is the next step.

The Schrödinger equation for the harmonic oscillator happens to be a well-studied differential equation that mathematicians had solved long before the quantum mechanical problem was formulated. There are an infinite number of functions of x that turn out to be valid eigenfunctions of the Hamiltonian operator. A subscript n is used, therefore, to distinguish one solution from another. Each eigenfunction has its own energy eigenvalue. The differential equation at hand is

$$-\frac{\hbar^2}{2m}\frac{d^2\psi_n(x)}{dx^2} + \frac{1}{2}kx^2\psi_n(x) = E_n\psi_n(x) \tag{7.29}$$

All the eigenfunctions, or solutions, can be expressed as

$$\psi_n(x) = \psi_n\left(\frac{z}{\beta}\right) = \frac{N}{\sqrt{2^n n!}}h_n(z)e^{-z^2/2} \quad \text{where } z \equiv \beta x \tag{7.30}$$

These wavefunctions have been written as a function of z for conciseness, where z is simply x scaled by a constant, β. The constant β is related to the force constant and the particle mass:

$$\beta^2 \equiv \frac{\sqrt{km}}{\hbar} \tag{7.31}$$

and

$$N = \left(\frac{\beta^2}{\pi}\right)^{1/4} \tag{7.32}$$

The subscript n has zero as its smallest allowed value and can be continued to infinity. The functions $h_n(z)$ are simple polynomials of z (or of βx) that are called **Hermite polynomials** after the mathematician Charles Hermite (France, 1822–1901) who first identified them. The function e^{-cx^2} is a standard exponential function called a **Gaussian function** after Carl Friedrich Gauss (Germany, 1777–1855). Thus, the eigenfunctions in Equation 7.30 can be described as products of Hermite polynomials, a Gaussian function, and a constant term. At $z = 0$ (i.e., at $x = 0$), the Gaussian function is at its maximum value, 1. As z increases or decreases, the Gaussian function diminishes quickly. For $z \gg 0$ or $z \ll 0$, the Gaussian function becomes vanishingly small, and so the asymptotic limit for each of the wavefunctions is zero. Restated in terms of the position variable x instead of z, the Gaussian function embedded in the wavefunctions is

$$e^{-z^2/2} = e^{-(\beta x)^2/2}$$

Because β increases with the mass of the particle and with the spring constant as in Equation 7.31, a heavier mass or a stiffer spring makes the value of the Gaussian function smaller for any given displacement in x. For the classical harmonic oscillator, we expect a heavier mass or a stiffer spring to allow for smaller excursions from equilibrium. In the quantum mechanical picture, the wavefunction spreads less with a heavier mass or stiffer spring.

Hermite polynomials are very simple polynomials. It turns out that they can be generated in several ways, such as by the following formula:

$$h_n(z) = (-1)^n e^{z^2} \frac{d^n}{dz^n} e^{-z^2} \tag{7.33}$$

Or, they can be generated by a recursive procedure.* The first several are

$$h_0(z) = 1 \qquad\qquad\qquad h_0(\beta x) = 1$$

$$h_1(z) = 2z \qquad\qquad\qquad h_1(\beta x) = 2\beta x$$

$$h_2(z) = 4z^2 - 2 \qquad\qquad\quad h_2(\beta x) = 4\beta^2 x^2 - 2$$

$$h_3(z) = 8z^3 - 12z \qquad\qquad h_3(\beta x) = 8\beta^3 x^3 - 12\beta x$$

$$h_3(z) = 8z^3 - 12z \qquad\qquad h_3(\beta x) = 8\beta^3 x^3 - 12\beta x$$

$$h_4(z) = 16z^4 - 48z^2 + 12 \qquad h_4(\beta x) = 16\beta^4 x^4 - 48\beta^2 x^2 + 12$$

$$h_5(z) = 32z^5 - 160z^3 + 120z \qquad h_5(\beta x) = 32\beta^5 x^5 - 160\beta^3 x^3 + 120\beta x$$

Notice that h_0 is a constant. h_1 has one node, meaning that it has one point where it changes sign, namely, $z = 0$. Generally, h_n has n nodes, and odd functions (i.e., $n = 1,3,5,...$) have one of those at the origin ($x = z = 0$).

A crucial point is to see that these wavefunctions are eigenfunctions of the harmonic oscillator Hamiltonian and to determine the eigenvalues associated with each. For ψ_0,

$$\hat{H}\psi_0 = -\frac{\hbar^2}{2m}\frac{d^2\psi_0}{dx^2} + \frac{1}{2}kx^2\psi_0$$

$$= -\frac{\hbar^2}{2m}\beta^2(\beta^2 x^2 - 1)\psi_0 + \frac{1}{2}kx^2\psi_0$$

$$= \frac{\hbar^2}{2m}\beta^2\psi_0 + \left(\frac{k}{2} - \frac{\beta^2\hbar^2}{2m}\right)x^2\psi_0$$

---

* The recursive relation shows how to generate the $n + 1$ polynomial from the n-order polynomial given that $h_0(z) = 1$: $h_{n+1}(z) = 2zh_n(z) - dh_n(z)/dz$.

A rearrangement of terms has been made between the second and third steps. This rearrangement leads not only to one term that is just a constant times the wavefunction but also to a second term that is a different function entirely, $x^2\psi_0$. However, if the expression for b from Equation 7.31 is used, then the factor that multiplies $x^2\psi_0$ is found to be zero. Thus, ψ_0 is indeed an eigenfunction of the Hamiltonian. The eigenvalue, E_0, of the Schrödinger equation, $H\psi_0 = E_0\psi_0$, is

$$E_0 = \frac{\hbar^2\beta^2}{2m} = \frac{\hbar^2}{2m}\frac{\sqrt{km}}{\hbar} = \frac{1}{2}\hbar\sqrt{\frac{k}{m}} = \frac{1}{2}\hbar\omega$$

where ω is the intrinsic frequency of the harmonic oscillator, the same frequency as in the classical description. If the Hamiltonian is applied to the ψ_1 wavefunction from Equation 7.30, the following results:

$$\hat{H}\psi_1 = -\frac{\hbar^2}{2m}\frac{d^2\psi_1}{dx^2} + \frac{1}{2}kx^2\psi_1$$

$$= -\frac{\hbar^2}{2m}\beta^2(\beta^2x^2 - 3)\psi_1 + \frac{1}{2}kx^2\psi_1$$

$$= 3\frac{\hbar^2}{2m}\beta^2\psi_1 + \left(\frac{k}{2} - \frac{\beta^4\hbar^2}{2m}\right)x^2\psi_1$$

$$= 3\frac{\hbar^2}{2m}\beta^2\psi_1 = \frac{3}{2}\hbar\omega\psi_1$$

The energy eigenvalue for this state is precisely $\hbar\omega$ more than the energy of ψ_0. And if the next state, ψ_2, were so tested, its eigenenergy would be $\hbar\omega$ more than the energy of ψ_1. In fact, the energy eigenvalues of the harmonic oscillator can be expressed generally by using the index n that distinguishes the different states.

$$E_n = \left(n + \frac{1}{2}\right)\hbar\omega \tag{7.34}$$

n will be referred to as a quantum number from now on. The state with the lowest energy is always called the ground state, and for the harmonic oscillator, this is the $n = 0$ state.

There are two very important features of Equation 7.34 that are typical of many quantum systems. First, the allowed energies of the system are not continuous. Whereas the classical harmonic oscillator can be made to have any energy at all just by stretching it to a suitable length, the quantum mechanical oscillator can have only the particular energies specified by Equation 7.34. Any energy added to the oscillator or removed from it should come in chunks (quanta) of the amount that separates the allowed energies, that is, integer multiples of $\hbar\omega$. For an oscillator in our everyday world, $\hbar$ is so small that the allowed energies are essentially continuous; this is the correspondence between the classical and the quantum descriptions. The second important feature is that the ground state energy is not zero. There exists no state of the system for which the energy is zero, and so the quantum

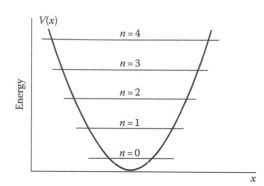

FIGURE 7.5
The energy levels of the quantum mechanical harmonic oscillator depicted as horizontal lines intersecting the potential energy function for the oscillator.

mechanical oscillator can never be completely at rest! The lowest energy of an oscillator system, the energy of the ground state, is referred to as the **zero point energy**.

Figure 7.5 shows a helpful way of displaying the quantum mechanical information about the harmonic oscillator or about other simple problems. The position coordinate, x, labels the horizontal coordinate axis of a two-coordinate graph. The vertical axis is in units of energy, and the first item drawn is the potential, $V(x)$. As shown in Figure 7.5, $V(x)$ for the harmonic oscillator is a parabola that opens upward. The next step in the display is to draw straight horizontal lines at energies that correspond to the eigenenergies of the system. Each is labeled with the quantum number, n. Notice the regular spacing of these lines as dictated by Equation 7.34. The horizontal lines designate the energy levels of the system. The particular points where these lines cross the parabola, $V(x)$, are points at which the

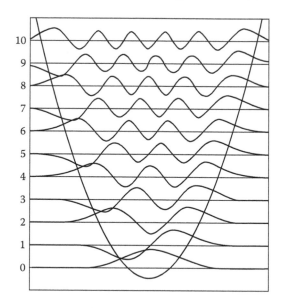

FIGURE 7.6
Wavefunctions of low-lying levels of the quantum mechanical harmonic oscillator. The functions are drawn so that the baseline is the energy level line as in Figure 7.5, and that means that the function is zero-valued or has a node at any point where it crosses the energy level line. Since it is the qualitative form that is usually of interest, the vertical axis for the wavefunctions is in arbitrary units. Notice that for the highest levels drawn, the wavefunction has maximum amplitude close to the classical turning points. A truly classical oscillator is moving at its slowest speed as it changes direction at a turning point, and so it is more likely to be found around the turning points than in the middle.

potential energy is the same as the energy of the oscillator. Were this a classical problem, these points would be those at which the particle has no kinetic energy; it has stopped and is ready to turn back. Such points are called classical turning points, though the quantum mechanical particle does not have to turn back at them.

The horizontal lines in Figure 7.5 that identify the energy levels serve as baselines for drawing the wavefunctions. Typically, the vertical scale for the wavefunctions is arbitrary since it is the qualitative form of each that is of most interest. Figure 7.6 is a drawing of this sort, and one of the things we can see is that the number of nodes of each wavefunction equals the quantum number of the wavefunction.

7.5 Harmonic Vibration of Many Particles

The oscillator problem in Figure 7.2 yielded a nonseparable Hamiltonian because the two springs were connected to one of the masses and this coupled the stretching and contracting of the springs. The corresponding mathematical feature is a cross-term in the potential energy expression, a term that involves both of the position coordinates, x_1 and x_2. This problem is representative of the complexity of the problems of molecular vibration since we could imagine the chemical bonds of a molecule behaving much like springs connecting atoms. The vibrational potentials of some molecule considered in this way would have plenty of cross-terms of the atomic displacement coordinates. Pulling a bit on one atom would obviously affect the other atoms. All the motions are coupled through the network of springs, and were the Hamiltonian to be written in terms of atomic position coordinates, it would not be separable. However, the oscillator problem in Figure 7.3 also had a cross-term involving the particle position coordinates, and yet that problem was finally separated into internal vibration and the translation of the whole system. A coordinate transformation was the key to obtaining that Hamiltonian in separable form. So, we may ask, can a coordinate transformation lead to separability for a general type of problem including the one illustrated in Figure 7.7? In fact, there is a specific coordinate transformation that can accomplish just that. However, the process of making that transformation assumes that all the springs are harmonic. Therefore, when we later apply this process to molecules, it will be as an approximation of the behavior of chemical bonds. The potentials for chemical bonds are always at least somewhat anharmonic, meaning that there is a

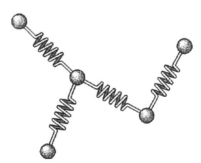

FIGURE 7.7
Vibrating system of particles connected by springs. The potential energy of this system is a sum of the potential energies of the springs. Such energies have at least a quadratic dependence on the mass position coordinates, and that is the mathematical basis for the motions of the masses to be coupled.

cubic, quartic, or higher-order dependence of the potential energy on the extent of extension or compression of the "spring" (bond).

If separability of the harmonic vibrations is achieved by a coordinate transformation, the Hamiltonian in the final coordinate system will have the following general form:

$$H = \sum_i \left(\frac{1}{2}\alpha_i q_i^2 + \frac{p_i^2}{2\beta_i} \right) \tag{7.35}$$

In other words, separability means that each degree of freedom has an independent term in the Hamiltonian. Just the form of this Hamiltonian tells much about the vibrations of the system, and this can be seen by considering a system that is mechanically equivalent. That system would consist of a set of different harmonic oscillators, each of the type shown in Figure 7.1 and none connected to another. For the ith oscillator of this set, there would be a displacement coordinate, q_i, a spring constant which could be called α_i, and the particle's mass which could be called β_i. One could initiate vibration by stretching and releasing any one of these oscillators, and there would be no effect on the others. By mechanical equivalence, the same is true for the coupled harmonic oscillator problem (e.g., Figure 7.7); however, the q_i-coordinates are more complicated because they turn out to be linear combinations of the many different particle displacement coordinates. A given q_i could correspond to a displacement of all the particles at once.

The coordinates $\{q_i\}$ for which the vibrational Hamiltonian takes on the separable form of Equation 7.35 are referred to as **normal coordinates**. Motion that follows along the direction of these coordinates is referred to as **normal mode** vibration. For each normal mode, there is a vibrational frequency and, by mechanical equivalence arguments, its value is

$$\omega_i = \sqrt{\frac{\alpha_i}{\beta_i}} \tag{7.36}$$

The important quantities α_i and β_i depend on the masses of the particles and the force constants of the springs, and they do so in a way that will have to be worked out through coordinate transformations.

Figure 7.8 shows another oscillator problem that involves two particles. When the particles are of the same mass and the three springs happen to have the same force constant, the two normal coordinates have a simple qualitative form. One corresponds to the two particles moving together; at the same time, both are moving to the left or both are moving to the right. The other normal coordinate corresponds to the two particles moving

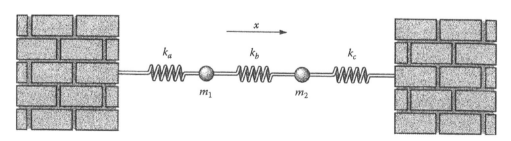

FIGURE 7.8
A vibrating system of two particles constrained to move in the x-direction and connected by springs to each other and to two unmovable walls.

oppositely; when one is moving to the right, the other is moving to the left. These two normal modes of vibration have different frequencies of oscillation. This fact is physically reasonable because in the parallel motion the length of the central spring does not change. Only two springs are stretched or contracted. In the opposed motion, when the two outer springs are squeezed, the inner one is stretched. The frequency of the latter type of motion is greater than the frequency of the parallel motion.

With the example in Figure 7.8 in mind, we can understand several features of a normal mode description of a classical system of masses and harmonic springs. If the particles in the system are displaced from equilibrium by an amount proportional to one normal coordinate, releasing the particles will result in the system vibrating *only* in that mode. In the example in Figure 7.8, with equal masses and like springs, displacement from equilibrium of one particle to the left and the other to the right by the same amount is displacement proportional to one of the normal coordinates. Displacement of both particles to the right (or left) by the same amount is displacement along the other normal coordinate.

When a system vibrates in a pure mode, all the particles in the system reach their point of maximum displacement at the same instant and all pass through their equilibrium positions at the same instant. They all vibrate with the same frequency and phase. In carbon dioxide, for instance, the pure symmetric stretching vibrational mode is one that equivalently stretches both carbon–oxygen bonds simultaneously. In this vibrational mode, the carbon atom remains at the midpoint of the line between the oxygen atoms as they oscillate back and forth with the same frequency. A system such as that in Figure 7.8 can be found to vibrate in a more complicated manner than simple normal mode motion even if the springs are perfectly harmonic. However, any such motion can be represented as a superposition of the different normal mode motions.

The quantum mechanical information that follows from a normal mode analysis must reveal the same mechanical equivalence to a set of disconnected oscillators as the classical analysis. Each such oscillator (normal mode of vibration) can exist in any of the states possible for a one-dimensional harmonic oscillator. Each has its own contribution to the energy of the system, and thus, the Hamiltonian in Equation 7.35 corresponds to a quantum mechanical energy level expression

$$E_n = \sum_i \left(n_i + \frac{1}{2} \right) \hbar \omega_i \tag{7.37}$$

Because each vibrational mode is independent, a different vibrational quantum number is needed for each. A state of the system is specified by a particular set of choices for all the n_i's.

7.6 Conclusions

In a classical description of systems of particles, equations of motion are the differential equations whose solution is predictive of the positions and velocities of every particle at every instant of time. Very low-mass particles, such as atoms and electrons, behave according to different mechanical laws, the laws of quantum mechanics. In the quantum mechanical world, particles may display character associated with waves and have other behavior that does not follow from classical mechanics. The quantum mechanical equivalent of solving classical equations of motion is solving the Schrödinger equation, $H\psi = E\psi$. H is the Hamiltonian, a differential operator associated with the energy. E is the energy eigenvalue, and ψ is the wavefunction, which in some way contains the mechanical description of the system.

The harmonic oscillator is a model system in both quantum and classical mechanics. Classically, it is a particle moving back and forth at a frequency, $\sqrt{k/m}$, where k is the force constant and m is the particle mass. Its quantum mechanical description shows the system existing only with certain allowed energies, each separated by Planck's constant times the oscillator's frequency. A system of many particles connected by harmonic springs can exhibit normal mode vibration, whereby all the particles move with the same frequency. For a system of N particles, there are $3N - 6$ normal modes, or $3N - 5$ if the system is linear.

Point of Interest: Molecular Force Fields

Molecular force fields and molecular dynamics molecular force fields are explicit potential energy functions for the stretching, bending, and twisting of molecules. The simplest have a functional form that is harmonic in the displacement coordinate for each vibrational degree of freedom as in the potential in the Hamiltonian in Equation 7.35. The force constants for harmonic terms and the constants used in any other terms comprise a set of parameters. To choose the parameter values for a molecule is to choose its force field. The first use of force fields involved vibrations of polyatomic molecules; the parameters in a harmonic potential were selected empirically to yield normal mode frequencies as close as possible to experimental vibrational frequencies. Chemical intuition played a strong hand as it became clear that force field parameters could be *transferred* from one molecule to another for fragments with the same chemical bonding. For example, the force constant for stretching a carbon–carbon triple bond is roughly the same in many different molecules with carbon–carbon triple bonds. Thus, if one obtains an empirical force constant for a particular chemical bond using spectroscopic data for a chosen molecule, that force constant can be taken as a reasonably good value for the same type of chemical bond but in a different molecule than the one analyzed. A force field constructed for a molecule by transferring parameter values from other molecules can be used to predict its vibrational frequencies.

In the last two decades or so, transferable force fields have been used as a molecular structure tool. The various parameter sets and functional forms have now been so carefully developed that they can be applied to predict bond lengths and bond angles of large molecules for which experimental determination of the structure may not be available. The minimum of a molecule's vibrational potential corresponds to the equilibrium structure of a molecule. A search through different geometries of a molecule to find the lowest energy according to the chosen force field is a search for an equilibrium structure of that molecule. This procedure for calculating molecular structures, called molecular mechanics (MM), is encoded in a number of commercial software packages.

Within the last 20 years or so, MM potentials have been used to simulate dynamic behavior of large molecules, often in biophysical problems, via a procedure called molecular dynamics (MD). Quantum effects are entirely neglected in MD, and instead the dynamics of atoms (point masses) are obtained by integrating the classical equations of motion over numerous, very short time steps. Such computer simulation of dynamic behavior gives a numerical representation of atomic positions as a function of time. Provided quantum effects are relatively unimportant—as they may be for certain issues involving dynamics—an MD simulation can be useful in following conformational changes, determining the effect of solvents on molecular structure, or finding changes in structure with variation in temperature.

Molecular mechanics and molecular dynamics (MD) are related but different. The main purpose of MD is modeling of dynamic molecular motion. However, MM can also provide important dynamic parameters, such as energy barriers between different conformers or steepness of a potential energy surface around a local minimum. MD and MM are usually based on the same classical force fields, but MD may also be based on quantum chemical methods like DFT.

There are several ways of defining the environment surrounding the molecule or molecules of interest in molecular mechanics. A system can be simulated in vacuo with no surrounding environment at all, but this is usually not desirable because it can introduce significant errors in the molecular geometry. One way to solvate a system is to place water molecules in the simulation box with the molecules of interest and treat the water molecules explicitly using the same parameters as those in the molecule. A variety of water models exist with increasing levels of complexity, representing water as a simple hard sphere, as three separate particles with fixed bond angles, or even as four or five separate interaction centers to account for unpaired electrons on the oxygen atom. Of course, the more complex the water model, the more computationally intensive the simulation. A compromise approach can be used in which the explicitly represented water molecules are replaced by a mathematical expression that reproduces the average behavior of water molecules (or any other solvent of interest). This method is useful for minimizing errors that arise from gas-phase simulations and reproduces bulk solvent properties well, but cannot reproduce situations in which individual water molecules have interesting interactions with the molecules being studied.

Exercises

7.1 Verify Equation 7.13 with both Equation 7.12a and b by forming the second time derivative of $x(t)$.

7.2 What force constant for a harmonic spring gives rise to a vibrational frequency, ω, of $10\,s^{-1}$ when a mass of $1\,kg$ is attached? What would be the frequency if such a spring were attached to a particle with the mass of (a) an electron, (b) a hydrogen atom, (c) a baseball?

7.3 Calculate the de Broglie wavelength for a free electron with a velocity 10% of the speed of light. What would be the speed of a $0.01\,kg$ marble if it had the same de Broglie wavelength?

7.4 Calculate the de Broglie wavelength for a helium atom with a speed equal to the mean speed for helium at $T = 300\,K$.

7.5 Consider a $0.01\,kg$ mass attached to a spring with a force constant such that the oscillation frequency is 1 cycle s^{-1}. This is a system that we might encounter in our macroscopic world, but treating it quantum mechanically, what is the energy separation between different allowed energy states and what is the zero-point energy? Is the quantization of energy consistent with our perceptions of continuous allowed energies for classical (macroscopic) oscillators? What would be the n quantum number of this oscillator if it had an energy of about 10^{-10} J?

7.6 Compute the reduced mass of $H^{35}Cl$ and $D^{35}Cl$ in kg. Calculate their ratio.

7.7 Calculate the force constants (in N m^{-1}) for the molecules in Exercise 7.6 given that their vibrational frequencies are 2886 and 2096 cm^{-1}, respectively.

7.8 For a hypothetical harmonic oscillator with a mass equal to the mass of a hydrogen atom and a force constant such that the frequency of vibration is 1.0×10^{14} cycles s^{-1}, find the classical turning points for the $n = 0, 1, 2$, and 3 levels.

7.9 A mass of 100.0 g is on a harmonic spring with a force constant of 15.0 N m^{-1}. The spring moves through a distance of 15.0 cm during this motion. What is the frequency of this oscillation?

7.10 What is the wavelength of radiation that must be absorbed for a harmonic oscillator with a frequency of 5.0×10^{13} Hz to cause the oscillator to move from the $n = 6$ to the $n = 7$ level?

7.11 Set up the classical Hamiltonian for a particle of mass m that can move in the three directions, x, y, and z, and experiences a gravitational potential of the form $V(x, y, z) = mgz$. Then, using Equations 7.5 and 7.6, write the equations of motion for this particle.

7.12 Write the explicit form of the Hamiltonian for a system of two charged particles connected by a massless harmonic spring with force constant k. Use Q_1 and Q_2 for the charges of the particles and m_1 and m_2 for their masses.

7.13 From the exponential relation that for any real number α, $e^{i\alpha} = \cos\alpha + i\sin\alpha$, find the constants A and B in Equation 7.12b so that this $x(t)$ is the same as the $x(t)$ in Equation 7.12a for the special case when $b = 0$ and $a = 1$. Next, generalize your result for the constant a left unspecified. This amounts to writing A and B as a function of a.

7.14 Write the potential energy of the following system of point-mass particles and harmonic springs as a function of the position coordinates of the particles. Assume that they are constrained to move only in the x–y plane.

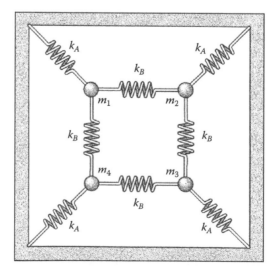

Let l be the length of each side of the square, let s_A be the equilibrium length of the springs with force constant k_A, and let s_B be the equilibrium length of the springs with force constant k_B.

7.15 Construct the Hamiltonian function for a system of N identical point masses constrained to move only along the x-axis, each connected to the next by a harmonic spring with force constant k. Set up the differential equations of motion for this system using Hamilton's prescription. [Hint: You may wish to try this for specific cases such as $N = 3$ and $N = 4$, and then generalize your formulas using summations over position/momentum coordinates as necessary.]

7.16 Develop Hamilton's equations of motion for the system in Exercise 7.14.

7.17 Set up the Hamiltonian for a particle of mass m that is connected to a harmonic spring with force constant k. The other end of the spring is connected to a pivot that allows the spring-and-mass assembly to turn about freely in any direction. The location of the pivot is fixed in space and does not move. Show that this system is mechanically equivalent in vibrational and rotational motion to a system of two particles connected by the same spring, assuming the particles can move in all three dimensions.

7.18 In what qualitative manner would a symmetric stretch mode of the system in Exercise 7.11 differ in having $m_1 = m_2 = m_3 = m_4$ and in having $m_1 = 2m_2 = m_3 = 2m_4$?

7.19 What would be the qualitative forms of the normal modes of vibration of the oscillator in Figure 7.6 if there were not two but three equally massive particles connected one to another and to the walls by harmonic springs with the same force constants? And if there were four particles?

7.20 If the two particles in Figure 7.8 were considered more realistically, there would be a y- and a z-degree of freedom for each particle to move. This would mean four more vibrational degrees of freedom, and they would be associated with the bending modes of the system. Qualitatively, what would they look like? Should any of them have the same normal mode frequency?

7.21 Starting from Equation 7.30, write out the first five harmonic oscillator wavefunctions explicitly in terms of x.

7.22 Verify that ψ_0 and ψ_1 for the harmonic oscillator are normalized and orthogonal to each other.

7.23 Apply the harmonic oscillator Hamiltonian to ψ_2 and to ψ_3 and verify that they are eigenfunctions.

7.24 Use the recursion relation for Hermite polynomials to find $h_6(z)$.

7.25 Evaluate $\psi_n(0)$ (i.e., find the value of the nth wavefunction at $x = 0$) for the $n = 0, 1, 2, 3, 4, 5$, and 6 states of the harmonic oscillator.

7.26 Consider a general three dimensional harmonic oscillator with force constants k_x, k_y, and k_z. Write down the total energy expression for the energy levels of this system. What is the ground state energy?

Bibliography

Graybeal, J. D., *Molecular Spectroscopy* (McGraw-Hill, New York, 1993).

8

Molecular Quantum Mechanics

Though quantum mechanical analysis can often be more complicated than classical mechanical analysis, the results offer rich understanding of the detailed basis of chemical phenomena. This chapter focuses on the development of some basic elements of quantum mechanics beginning with formal aspects of using quantum mechanical operators. It examines the treatment of certain model problems used in chemistry, and it introduces the concept of approximate solution of the Schrödinger equation. Finally, two powerful approximation techniques are considered.

8.1 Quantum Mechanical Operators

Because of their important role in quantum mechanical problems, it is useful to understand operators and their workings. An operator is a mathematical device that relates two functions. An operator equation is any equation that includes an operator. Thus, $g(x) = 3f(x)$ can be regarded as an operator equation since the functions $g(x)$ and $f(x)$ are related by the operation of multiplication by 3. The equation can be expressed as $g(x) = \hat{O}(x)$, where $\hat{O} = 3$. Though operators can also be as simple as multiplication by a constant such as 3, they can be multiplication by a variable, such as x, or differentiation with respect to one or more variables. *Anything* that can relate one function to another function can be labeled an operator.

Two operators can be combined by addition or by multiplication to create another operator. The process of operator addition and the process of operator multiplication, though, require definition.

$\hat{c} = \hat{a} + \hat{b}$ means the result of operating with $\hat{c}$ on an arbitrary function, f, yields the same result as operating with $\hat{a}$ on f and adding that to the result of operating with $\hat{b}$ on f.

$\hat{d} = \hat{a}\hat{b}$ means that the result of operating with $\hat{d}$ on an arbitrary function, f, yields the same result as operating on f with $\hat{b}$ and subsequently operating on that result with $\hat{a}$.

With these definitions, operators can be added and multiplied, and in this way algebraic expressions can be written involving just operators.

Multiplication of operators is successive application of two or more operators. However, this is not the same as multiplication of numbers; the order of the operators being multiplied may make a difference. Consider the example where one operator, $\hat{a}$, is multiplication by the variable x, $\hat{a} = x$, and the other operator is differentiation with respect to x, that is, $\hat{b} = d/dx$. To invoke the definition of operator multiplication, let $f(x)$ be an arbitrary function; that is, let it stand for any function whatsoever. The product of the operator multiplication of $\hat{a}$ times $\hat{b}$ is found from applying $\hat{b}$ and then $\hat{a}$ to the arbitrary function. But this result is seen to be different from multiplication of $\hat{b}$ times $\hat{a}$.

$$\widehat{ab}f(x) = x\frac{df(x)}{dx} \Rightarrow \widehat{a} = x\frac{d}{dx}$$

$$\widehat{ba}f(x) = \frac{d}{dx}(xf(x)) = x\frac{df(x)}{dx} + f(x) \Rightarrow \widehat{a} = \left(x\frac{d}{dx} + 1\right)$$

This illustrates that operator multiplication is not necessarily commutative, whereas multiplication of numbers and variables is.

A **commutator** is a special operator constructed from two operators as the difference between the two ways they can be multiplied. The commutator is designated by placing the two operators being multiplied inside square brackets as in the following:

$$[\widehat{A},\widehat{B}] \equiv \widehat{A}\widehat{B} - \widehat{B}\widehat{A} \tag{8.1}$$

If the two operators do commute, which means that the same result is achieved from applying them in either order, then the commutator is equal to zero. On the basis of the multiplication results for the $\widehat{a}$ and $\widehat{b}$ operators just considered, their commutator is $[\widehat{a}, \widehat{b}] = -1$.

A special type of operator equation is the eigenvalue equation, which can be expressed generally as

$$\widehat{O}f = cf \tag{8.2}$$

This equation says that for some particular function f and some operator $\widehat{O}$, operating on f with $\widehat{O}$ yields a constant, c, times f. The Schrödinger equation is, of course, an important example of an eigenvalue equation.

The manipulation of wavefunctions and operators can be presented with a concise notation due to P. A. M. Dirac (England, 1902–1984) called *bra-ket notation*. We shall use a form of this notation restricted to functions of spatial coordinates. With it, the integration symbols, $\int \ldots dx\, dy\ldots$, are replaced with angle brackets, $\langle \cdots \rangle$. Anything placed within these brackets is meant to be integrated over all spatial coordinates of the system. Between the brackets there is a vertical bar that separates functions, as in $\langle f|g \rangle$. It is implicit in this notation that the complex conjugate is to be used for anything between the left bracket and the vertical bar, and so $\langle f|g \rangle = \int f^*(x)\, g(x)\, dx$. (The complex conjugate of a number or function is that number or function with $-i$ replacing i [the square root of -1] wherever it appears.) Should we happen to write a function to the left of the bar that is explicitly the complex conjugate of some function, it remains *implicit* to take its complex conjugate: $\langle f^*|g \rangle = \int f(x)g(x)\, dx$ since the complex conjugate of the complex conjugate is the original function.

An operator acting on a function on the left side of the vertical bar does not act on anything to the right of the vertical bar. An operator to the right of the vertical bar acts only on functions to its right, as would be natural from operator algebra. Thus, the integral expression $\int \Psi^* \widehat{O}\Psi dx$ involving the wavefunction Ψ and the operator $\widehat{O}$ is written $\langle \Psi|\widehat{O}\Psi \rangle$.

Likewise, $\langle O\Psi|\Phi \rangle \equiv \int \Phi(\widehat{O}^*\Psi^*)dx$ because the operator acts only on the function Ψ by this notation. (Sometimes a second vertical bar is placed to the right of the operator, thereby sandwiching it between two bars, for example, $\langle \Psi | \hat{O} | \Psi \rangle$, with no change in meaning.)

An operator, $\hat{O}$ is said to have the important property of being **Hermitian** for some set of functions if for every pair of functions, Ψ_i and Ψ_j, from that set the following equivalence exists:

$$\int \Psi_i^* \hat{O} \Psi_j dx = \int \Psi_j [\hat{O}\Psi_i^*] dx \qquad (8.3a)$$

The difference between the left and right integrals is simply that instead of the operator being applied to Ψ_j, the complex conjugate of the operator is applied to the other function, Ψ_i. Equation 8.3a is written as

$$\left\langle \Psi_i \middle| \hat{O}\Psi_j \right\rangle = \left\langle \hat{O}\Psi_i \middle| \Psi_j \right\rangle \qquad (8.3b)$$

in Dirac notation. There are a number of important consequences of the Hermitian property.

Theorem 8.1: Eigenvalues of a Hermitian Operator Are Real Numbers

Given: A wavefunction ψ that is an eigenfunction of Hermitian operator $\hat{G}$ with associated eigenvalue g.

$$\hat{G}\psi = g\psi \quad \text{given}$$

$$\psi^* \hat{G}\psi = \psi^* g\psi = g\psi^* \psi \quad \text{from left-multiplying by } \psi^*$$

$$\left\langle \psi \middle| \hat{G}\psi \right\rangle = g \left\langle \psi \middle| \psi \right\rangle \quad \text{from integrating}$$

$$\left\langle \psi \middle| \hat{G}\psi \right\rangle = \left\langle \hat{G}\psi \middle| \psi \right\rangle \quad \text{because } \hat{G} \text{ is Hermitian}$$

$$\left\langle \hat{G}\psi \middle| \psi \right\rangle = \left\langle g\psi \middle| \psi \right\rangle = g^* \left\langle \psi \middle| \psi \right\rangle \quad \text{from using } \hat{G}\psi = g\psi$$

$$\therefore g \left\langle \psi \middle| \psi \right\rangle = g^* \left\langle \psi \middle| \psi \right\rangle \quad \text{or} \quad g = g^*$$

A number that is equal to its complex conjugate can be only a real number, and thus, g is real.

Two functions are said to be **orthogonal** over some region of space if over that region the integral of the product of one function with the complex conjugate of the other function is zero. That is, $\langle \psi_i | \psi_j \rangle = 0$ is a statement of orthogonality, that ψ_i and ψ_j are orthogonal. An analogous use of the term is that two geometric vectors are orthogonal if their dot product is zero. The angle between two such vectors would be 90°, and in a similar, though abstract, sense orthogonal functions may be thought of as being in completely different "directions" in a "space of functions."

Theorem 8.2: Eigenfunctions of a Hermitian Operator with Different Associated Eigenvalues Are Orthogonal Functions

Given: A set of wavefunctions $\{\psi_i\}$ that are each eigenfunctions of a Hermitian operator $\hat{G}$ with associated eigenvalues g_i, none of which are equal.

$$\left\langle \psi_i \middle| \hat{G}\psi_j \right\rangle = \left\langle \psi_i \middle| g\,\psi_j \right\rangle = g\left\langle \psi_i \middle| \psi_j \right\rangle \quad \text{from given information}$$

$$\left\langle \psi_i \middle| \hat{G}\psi_j \right\rangle = \left\langle \hat{G}\psi_i \middle| \psi_j \right\rangle \quad \text{because } \hat{G} \text{ is Hermitian}$$

$$\left\langle g_i\psi_i \middle| \psi_j \right\rangle = g_i\left\langle \psi_i \middle| \psi_j \right\rangle \quad \text{from given information}$$

Since it is given that $g_i \neq g_j$ then $\left\langle \psi_i \middle| \psi_j \right\rangle = 0$.

This theorem shows that every eigenfunction of a Hermitian operator is orthogonal to every other eigenfunction with a different eigenvalue. Furthermore, it is possible to prove that eigenfunctions of a Hermitian operator can be constructed to be orthogonal even if some have like eigenvalues. This is a corollary to the prior theorem which we consider without a detailed proof.

Corollary 8.2.1

A set of eigenfunctions of a Hermitian operator with the same eigenvalues can be transformed (linearly combined) into orthogonal functions while remaining eigenfunctions of the operator.

Hamiltonian operators are Hermitian, and therefore, this corollary provides for all the solutions of the Schrödinger equation to be an orthogonal set.

In a number of types of problems, there may be two or more Hermitian operators of interest, and it is possible that there may be a set of functions that are simultaneously eigenfunctions of all the operators. For this to be found, the operators must commute with each other.

Theorem 8.3

If a set of functions, $\{f_i\}$, are eigenfunctions of two different operators, $\hat{A}$ and $\hat{B}$, with associated sets of eigenvalues, $\{a_i\}$ and $\{b_i\}$, respectively, the operators commute for that set of functions

$$\hat{A}f_i = a_i f_i \quad \text{and} \quad \hat{B}f_i = b_i f_i \quad \text{from given conditions}$$

$$\hat{A}\hat{B}f_i = \hat{A}(b_i f_i) = b_i \hat{A}f_i = b_i a_i f_i \quad \text{applying the operators}$$

$$\hat{B}\hat{A}f_i = \hat{B}(a_i f_i) = a_i \hat{B}f_i = a_i b_i f_i \quad \text{applying the operators}$$

$$\hat{A}\hat{B}f_i - \hat{B}\hat{A}f_i = a_i b_i f_i - b_i a_i f_i = 0 \quad \text{by subtraction}$$

$$\therefore \hat{A}\hat{B} - \hat{B}\hat{A} = 0$$

Corollary 8.3.1

If two operators commute, it is possible to find a set of functions that are simultaneously eigenfunctions of both.

8.2 Information from Wavefunctions

The operator theorems in the last section hint at an important aspect of wave mechanical descriptions: The eigenvalue of some eigenequation can be formally isolated by the process of multiplying by the complex conjugate of the wavefunction and then integrating over the physical space of the system. In the case of the eigenvalue of the Schrödinger equation, we have

$$\hat{H}\psi = E\psi$$

$$\langle\psi|\hat{H}\psi\rangle = E\langle\psi|\psi\rangle \quad \text{or} \quad \int\psi^*\hat{H}\psi dx = E\int\psi^*\psi dx$$

$$E = \frac{\langle\psi|\hat{H}\psi\rangle}{\langle\psi|\psi\rangle} \quad \text{or} \quad E = \frac{\int\psi^*\hat{H}\psi dx}{\int\psi^*\psi dx}$$

If an operator for which the wavefunctions were not eigenfunctions were of interest, the same type of expression could be constructed. For example, the wavefunctions of the harmonic oscillator are not eigenfunctions of the position operator, x, and yet we might ask about the significance of the value we shall designate $\langle x\rangle$.

$$\langle x\rangle = \frac{\langle\psi|x\psi\rangle}{\langle\psi|\psi\rangle} = \frac{\int\psi^*x\psi dx}{\int\psi^*\psi dx} \tag{8.4}$$

To interpret this expression, consider a hypothetical setup where an enormous (infinite) number of measurements were made of the position of a particle of a harmonic oscillator. Because of the particle's mechanical behavior, the probability per unit length of finding the particle at any one position may vary from position to position. That is, the probability per unit length is some function of x, $P(x)$. Now, the probability that on any one measurement a specific value x' is obtained is $P(x')\delta x'$ (probability per unit length at the point x' times an increment of length around x'). To average the enormous number of position measurements in this hypothetical situation, measured values, x', times the probability $P(x')\delta x'$ are accumulated (summed). Likewise, the net probabilities $P(x')\delta x'$ are accumulated. Then, the average value for the position measurement is the accumulation of the probabilities times position divided by the accumulated probabilities. Since x is a continuous variable, the accumulation process corresponds to integration.

$$\text{Average position} = \frac{\sum_i x_i P(x_i) \delta x_i}{\sum_i P(x_i) \delta x_i} \rightarrow \frac{\int x P(x) dx}{\int P(x) dx} \tag{8.5}$$

(If $P(x)$ is **normalized**, the integral in the denominator on the right is 1.) Comparing Equations 8.4 and 8.5 suggests that if $\psi^*(x)\psi(x)$, called the *absolute square* or the *square* of the wavefunction, is a probability density, then Equation 8.4 will give the average value of x for the quantum mechanical system described by ψ.

In a classical picture of electromagnetic radiation, it is the square of the wave amplitude that gives the energy density. Hence, there is a physical significance to the square of the wave associated with radiation. Early in the development of quantum mechanics, the significance of the square of the wavefunction was recognized as being the probability per unit volume, or the **probability density**, of finding the particles of a system to be at specific locations in space. Designating the probability density as ρ, this says,

$$\rho(q_1, q_2, \ldots) \equiv \psi^*(q_1, q_2, \ldots)\psi(q_1, q_2, \ldots) \tag{8.6}$$

The probability density function may be as close as one can get to attaching physical significance to a wavefunction. Consider the ground state wavefunction of the harmonic oscillator. The probability density function is

$$\rho_0 = \frac{\beta}{\sqrt{\pi}} e^{-\beta^2 x^2}$$

This function has its maximum value at $x = 0$, which is the position at equilibrium, and thus, the probability per unit length (length instead of volume since this is a one-dimensional system) of finding the particle is greatest at $x = 0$. Loosely, the most likely spot to find the particle as it moves about in its ground state is in the vicinity of $x = 0$.

The probability density functions for a number of the lowest energy states of the harmonic oscillator are depicted in Figure 8.1. These functions are sketched using the associated energy level lines as the baseline in the same way that the wavefunctions are displayed in Figure 7.6. The nodes of the wavefunction are those where the probability density is zero. Notice that as the quantum number increases, the probability density becomes relatively greater in the turning point regions than around $x = 0$. Classical mechanical treatment of a harmonic oscillator shows that it spends the most time per unit length in the vicinity of the turning points because at those points, the velocity of the particle goes to zero and the particle reverses direction. Nowhere else does it move as slowly. Thus, the quantum mechanical description begins to resemble the classical description as the energy of the oscillator becomes much greater than the separation between the quantum energy levels.

For some one-particle quantum mechanical systems, such as the harmonic oscillator, the probability of finding the particle in a particular region of space depends on the integral of the probability density over that region. For example, the net probability of finding the ground state harmonic oscillator particle between $x = 0.0$ and $x = 0.1$ is the integral of ρ_0 from 0.0 to 0.1, divided by the integral of ρ_0 over all space:

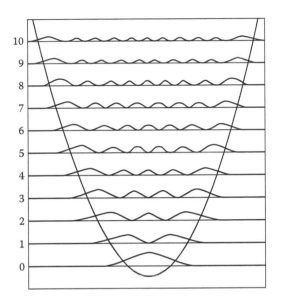

FIGURE 8.1

Probability density functions of low-lying levels of the quantum mechanical harmonic oscillator. These functions are the squares of the wavefunctions. They are drawn so that the baseline, or point of zero probability density, is the energy level line as in Figure 7.6. Notice that as n increases, the probability density at $x = 0$ diminishes. Furthermore, at levels such as $n = 10$ the points of maximum probability density are close to the classical turning points, unlike the situation for the lower states.

$$\frac{\int_{0.0}^{0.1} \rho_0 \, dx}{\int_{-\infty}^{\infty} \rho_0 \, dx}$$

The value in the denominator is the probability of finding the particle *somewhere* or *anywhere*. The condition of **normalization** means that we choose this probability integral—the sum of probabilities of finding the particle at all positions—to be unity. Normalization is not implicit in the solution of the Schrödinger equation. Notice that when ψ_0 is an eigenfunction of some Hamiltonian H, then the function $\Gamma = 3\psi_0$, for example, is just as much an eigenfunction of the Hamiltonian as ψ_0, and it has the same eigenvalue:

$$\hat{H}\Gamma = \hat{H}(3\psi_0) = 3\hat{H}\psi_0 = 3E_0\psi_0 = E_0(3\psi_0) = E_0\Gamma$$

If ψ_0 were a normalized wavefunction, that is, if $\langle \psi_0 | \psi_0 \rangle = 1$, the integral of $\Gamma^* \Gamma$ would be the integral of $(3\psi_0^*)(3\psi_0)$, and therefore $\langle \Gamma | \Gamma \rangle$ would be 9, not 1. Now, if Γ happened to be the function found on solving the Schrödinger equation, we could multiply it by the inverse square root of $\langle \Gamma | \Gamma \rangle$ (i.e., $1/\sqrt{9}$) and thereby generate the normalized wavefunction, ψ_0. A wavefunction for which the integral of the probability density over all space is 1, is said to *normalized*. If a solution of the Schrödinger equation is not normalized, it can be made so by multiplying it by a constant called a **normalization factor**. The normalization factor is obtained from the integral of the square of the unnormalized wavefunction. Thus, if Γ were an unnormalized wavefunction, then $\Gamma' = N\Gamma$ would be a normalized wavefunction provided

$$N \equiv \frac{1}{\langle \Gamma | \Gamma \rangle} \tag{8.7}$$

If Γ happens to be normalized already, this expression will give $N = 1$.

Postulate II says that a measurement of a dynamical variable will yield an eigenvalue *if* the wavefunction is an eigenfunction of the operator associated with that variable. For the case of the harmonic oscillator, we have considered what would result from measuring the position of the particle, x, even though none of the wavefunctions are eigenfunctions of the operator x. Individual measurements may differ from one to the next according to the quantum mechanical probability of the particle being at any particular point. In this circumstance, the most meaningful measurement of the particle's position is the average of many measurements. Because the square of the normalized wavefunction is the probability density, an expression for that average for the system in its nth state follows from Equation 8.4:

$$\langle x \rangle_n = \int_{-\infty}^{\infty} \psi_n^*(x) x \psi_n(x) dx \tag{8.8}$$

In general, the average or mean value of measuring the quantity associated with some operator $\widehat{T}$ for a system described by a normalized wavefunction Ψ is

$$\langle T \rangle = \int \Psi^* \widehat{T} \Psi d\tau \tag{8.9}$$

where $d\tau$ is the volume element for the coordinate system in which Ψ is given. For example, $d\tau = dx\, dy\, dz$ for a single-particle system that exists in three dimensions. This average value is also called the **expectation value** of T. If subscripts are attached to this symbol, they are used to indicate the quantum number(s) of the particular wavefunction used in the integral.

For the harmonic oscillator, it turns out* that for all n, $\langle x \rangle_n = 0$. For all states of the harmonic oscillator, the expectation value of the position of the particle is zero, which is the middle of the potential well. This is a consequence of the fact that for every state, the probability density at some x' is exactly the same as at the point $-x'$. The probability density function is symmetric with respect to the middle of the well, and there is the same chance of being a certain amount to the left of the middle as being to the right by that amount. An average of many measurements gives zero, although any one measurement need not yield zero. In fact, a series of measurements may give many different values that are distributed in some range.

---

* The function in the integrand, $f(x) = \psi_n^*(x) x \psi_n(x)$, is an odd function because it is the product of an odd function, x, and an even function, $\psi_n^*(x)\psi_n(x)$. This means for all x, $f(x) = -f(x)$. When an odd function is integrated from $-\infty$ to $+\infty$, the result is zero.

The **standard deviation** is a statistical characterization of the range of most measurements in some series of trials. It is the root mean square (rms) deviation from the average value. Let's say a particular set of measurements of some quantity yielded the five values 1.2, 2.4, 2.0, 1.8, and 1.6, the average of these being 1.8. The deviations of each measurement from this average are −0.6, 0.6, 0.2, 0.0, and −0.2, respectively. The deviations squared are 0.36, 0.36, 0.04, 0.0, and 0.04, and the mean of these values is 0.16. The rms deviation is $\sqrt{0.16} = 0.4$. Three of the five measurements happen to be in the range 1.8 ± 0.4.

The same type of deviation in measurement from an average value can be evaluated from the wavefunction in a quantum mechanical description of a system. First, there is an operator that corresponds to "measuring the deviation from the average" of some dynamical variable. When that variable is T, this operator is simply $\hat{T} - \langle T \rangle$, the T operator less the numerical value that is the expectation value of T. The square of this operator, $(\hat{T} - \langle T \rangle)^2$, corresponds to measuring the square of the deviation from the average. Thus, the mean square of the deviation is the expectation value of this operator,

$$(\Delta T)^2 \equiv \int \Psi^* \left(\hat{T} - \langle T \rangle \right)^2 \Psi d\tau \tag{8.10}$$

and the root mean square deviation is the square root of this number, ΔT. It is called the **uncertainty** in the measurement of the variable.

From Equation 8.10, another expression can be developed for ΔT.

$$\int \Psi^* \left(\hat{T} - \langle T \rangle \right)^2 \Psi d\tau = \int \Psi^* \left(\hat{T}^2 - 2\hat{T} \langle T \rangle + \langle T \rangle^2 \right) \Psi d\tau$$

$$= \int \Psi^* \hat{T}^2 \Psi d\tau - 2\langle T \rangle \int \Psi^* \hat{T} \Psi d\tau + \langle T \rangle^2 \int \Psi^* \Psi d\tau$$

$$= \langle T^2 \rangle - 2\langle T \rangle^2 + \langle T \rangle^2 = \langle T^2 \rangle - \langle T \rangle^2$$

Therefore,

$$\Delta T = \sqrt{\langle T^2 \rangle - \langle T \rangle^2} \tag{8.11}$$

The uncertainty in T is the square root of the difference between the expectation value of the operator T^2 and the square of the expectation value of T.

Heisenberg's uncertainty relation states that the product of the measurement uncertainties of two conjugate variables, such as position and momentum, is a number on the order of Planck's constant or larger. So if position were measurable with a small uncertainty, Heisenberg's principle would imply a quite large uncertainty for a measurement of momentum. This is another point of difference from classical mechanics in which we can know *both* the position and momentum exactly at any instant of time. Of course, there is a correspondence between the two pictures. Recall that $\hbar$ is a very tiny value relative to the sizes of objects in our macroscopic everyday world, thereby making the uncertainty dictated by quantum mechanics imperceptible.

Example 8.1: Position Uncertainty for the Harmonic Oscillator

Problem: Obtain an expression for Δx for the ground state of the one-dimensional harmonic oscillator.

Solution: To obtain Δx, we must find the expectation values of x and of x^2, according to Equation 8.11. From Equation 7.29 we can write the explicit form of the ground state wavefunction for the harmonic oscillator.

$$\psi_0(x) = \left(\frac{\beta^2}{\pi}\right)^{1/4} e^{-\beta^2 x^2/2}$$

Then, the expectation values are obtained by integration, for which one can consult integral tables (Appendix D).

$$\langle x\rangle_0 = \frac{\beta}{\sqrt{\pi}} \int_{-\infty}^{\infty} x e^{-\beta^2 x^2/2} dx$$

$$= 0$$

$$\langle x^2\rangle_0 = \frac{\beta}{\sqrt{\pi}} \int_{-\infty}^{\infty} x^2 e^{-\beta^2 x^2/2} dx$$

$$= \frac{\beta}{\sqrt{\pi}} \frac{1}{2} \frac{\sqrt{\pi}}{\beta^3} = \frac{1}{2\beta^2}$$

Therefore,

$$\Delta x_0 = \sqrt{\langle x^2\rangle_0 - \langle x\rangle_0^2}$$

$$= \frac{1}{\beta\sqrt{2}} = \sqrt{\frac{\hbar}{2}} \, (km)^{-1/4}$$

The last step follows from the definition of the harmonic oscillator constant β in Equation 7.30. The result indicates that the position uncertainty for the ground state of a harmonic oscillator diminishes as the mass is increased and as the force constant is increased.

If an operator is Hermitian, the expectation values are real, and this can be proved by following steps similar to those in the proof of Theorem 8.1. Consequently, observables always have associated operators that are Hermitian. We do not, for instance, measure length as having a real and imaginary component.

When a wavefunction is an eigenfunction of an operator of interest, the uncertainty is identically zero, which is consistent with Postulate II. For instance, when the operator $\hat{T}$ applied to some function Γ yields the eigenvalue t times Γ, the following result is obtained:

$$\hat{T}^2\Gamma = \hat{T}(\hat{T}\Gamma) = \hat{T}t\Gamma = t^2\Gamma \Rightarrow \langle T^2\rangle = t^2 = \langle T\rangle^2$$

This means ΔT is zero. There is no uncertainty in the measurement of T in this case. Every measurement yields the same value, namely, t. Then, according to Heisenberg's uncertainty relation, the uncertainty in the variable conjugate to T is infinite.

8.3 Multidimensional Problems and Separability

An example of a multidimensional quantum mechanical problem is the two dimensional oscillator shown in Figure 8.2. The Hamiltonian for this problem is

$$\hat{H} = \frac{\hat{p}_x^2}{2m} + \frac{k_x x^2}{2} + \frac{\hat{p}_y^2}{2m} + \frac{k_y y^2}{2} \tag{8.12}$$

This is a separable Hamiltonian, and it has been written so as to make that apparent. The first two terms involve only the x-coordinate, and the last two terms involve only the y-coordinate. As was done for separable, classical Hamiltonians, the two sets of terms can be freely designated independent Hamiltonians:

$$\hat{H}_x = \frac{\hat{p}_x^2}{2m} + \frac{k_x x^2}{2}$$

$$\hat{H}_y = \frac{\hat{p}_y^2}{2m} + \frac{k_y y^2}{2}$$

$$\hat{H} = \hat{H}_x + \hat{H}_y$$

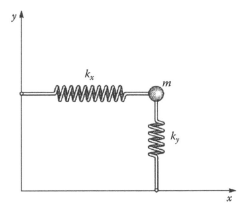

FIGURE 8.2
Two-dimensional oscillator where a particle of mass m is attached to two harmonic springs with force constants k_x and k_y. Each spring is connected to an unmovable rod, but the connection points can slide freely along the rods (i.e., along the x-axis or along the y-axis). Thus, the potential energy for the k_x spring depends on only the x-coordinate of the particle, while the potential energy of the k_y spring depends on only the y-coordinate. The equilibrium lengths of the springs are taken to be zero in the analysis of this example.

The wavefunction for this system is a function of x and y, $\psi = \psi(x, y)$, and the Schrödinger equation

$$(\hat{H}_x + \hat{H}_y)\psi(x, y) = E\psi(x, y)$$

is a separable differential equation. Separability is demonstrated by assuming that the wavefunction, ψ, takes on a product form; that is, the wavefunction is assumed to be a product of independent functions of x- and y-coordinates. These independent functions will be designated $X(x)$ and $Y(y)$, and so, the assumption is made that

$$\psi(x, y) = X(x)Y(y)$$

Now, if the Hamiltonian is applied to this assumed wavefunction, the component Hamiltonians act on only the respective coordinate functions.

$$\hat{H}_x X(x)Y(y) = Y(y)\hat{H}_x X(x)$$

$$\hat{H}_y X(x)Y(y) = X(x)\hat{H}_y Y(y)$$

Thus,

$$Y(y)\hat{H}_x X(x) + X(x)\hat{H}_y Y(y) = EX(x)Y(y)$$

Multiplying by $1/[X(x)Y(y)]$ yields

$$\frac{\hat{H}_x X(x)}{X(x)} + \frac{\hat{H}_y Y(y)}{Y(y)} = E \tag{8.13}$$

This has the same form as the equation $f(x) + g(y) = c$. It is a sum of a function of the x-coordinate (only) and a function of the y-coordinate (only), and that sum is equal to a constant. Since x and y are independent variables, the only way for this to be true is if the function of x is constant and the function of y is also constant. Therefore, on naming the two constants E_x and E_y, we find that Equation 8.13 implies

$$\frac{\hat{H}_x X(x)}{X(x)} = E_x \implies \hat{H}_x X(x) = E_x X(x) \tag{8.14a}$$

$$\frac{\hat{H}_y Y(y)}{Y(y)} = E_y \implies \hat{H}_y Y(y) = E_y Y(y) \tag{8.14b}$$

where $E_x + E_y = E$. There are now two Schrödinger equations. Recall that in a classical analysis of a mechanically separable problem, separate equations of motion are obtained.

The procedure for working with the Schrödinger equation is general for all multidimensional problems that are separable. Separability implies two things:

1. The wavefunction is a product of independent coordinate function of the separable variables.
2. The total energy eigenvalue is a sum of energy terms associated with each separable coordinate.

If separability in a problem is not immediately apparent, a product form of the wavefunction can be tested just the same. If application of the Hamiltonian to the product form [e.g., $X(x)Y(y)$] yields an equation of the same type as Equation 8.13, the problem is separable. If the problem is not separable, the Schrödinger equation ends up as a coupled differential equation, and solution is usually more difficult.

Separability is as advantageous in quantum mechanics as in classical mechanics. For instance, the Schrödinger equations in Equation 8.14 can be recognized as simple, one-dimensional, harmonic oscillator Schrödinger equations. Thus, the harmonic oscillator solutions given in Chapter 7 can be used directly. First, though, it should be recognized that the intrinsic vibrational frequencies of these one-dimensional oscillators are not necessarily the same since they depend on their respective spring constants. There are two frequencies for this system, ω_x and ω_y.

$$\omega_x = \sqrt{\frac{k_x}{m}} \quad \text{and} \quad \omega_y = \sqrt{\frac{k_y}{m}} \tag{8.15}$$

Furthermore, there are two independent quantum numbers because of the independence of vibrational motions in the x- and y-directions. We can call these quantum numbers n_x and n_y. From the one-dimensional Schrödinger equation solutions, the energy eigenvalues are

$$E_x = \left(n_x + \frac{1}{2} \right)\hbar\omega_x \quad \text{and} \quad E_y = \left(n_y + \frac{1}{2} \right)\hbar\omega_y$$

On this basis, the energy of the system as a whole depends on both of the quantum numbers since it is the sum of the x- and y-motion energies. It is appropriate to subscript the total energy, E, with the quantum numbers, and then the following can be said about the energy levels of the system:

$$E_{n_x n_y} = \left(n_x + \frac{1}{2} \right)\hbar\omega_x + \left(n_y + \frac{1}{2} \right)\hbar\omega_y \tag{8.16}$$

The two quantum numbers, being those of a harmonic oscillator problem, can be any positive integer or zero.

The wavefunction, ψ, for the system as a whole is a product, and it must be labeled by the two quantum numbers in order to distinguish the different Schrödinger equation solutions.

$$\psi_{n_x n_y}(x, y) = X_{n_x}(x)Y_{n_y}(y) \tag{8.17}$$

The functions X_{n_x} and Y_{n_y} are simply one-dimensional harmonic oscillator wavefunctions. These product functions describe the **states** of the system as a whole.

An interesting situation arises when this two-dimensional oscillator is **isotropic**, that is, when it is the same in all directions, or specifically when $k_x = k_y$. In that case, $\omega_x = \omega_y$, and the energy expression, Equation 8.16, becomes

$$E_{n_x n_y} = (n_x + n_y + 1)\hbar\omega \quad \omega = \omega_x = \omega_y \tag{8.18}$$

From this expression, the energies of various states can be tabulated by going through the sequence of all the allowed values of the two quantum numbers:

$n_x = 0$	$n_y = 0$	$E_{n_x n_y}/\hbar\omega$
0	0	1
1	0	2
0	1	2
1	1	3
2	0	3
0	2	3
2	1	4
1	2	4
3	0	4
0	3	4
2	2	5
3	1	5
1	3	5
4	0	5
0	4	5

Notice that there are states, such as those with quantum numbers (1, 0) and (0, 1), that have exactly the same energy. They are degenerate. The lowest energy state is the ground state, and in this case it is nondegenerate. It happens to be the only nondegenerate state for this problem. Also notice that at each higher energy level in the problem, the degeneracy, meaning the number of degenerate states, increases.

Quantum mechanical degeneracy often has interesting manifestations in molecular spectroscopy. Of the normal modes of vibration of a linear triatomic molecule such as carbon dioxide, there are two that correspond to a bending motion with the same frequency. When the molecule is aligned with some hypothetical x-axis, the force constant for bending in the x–y plane is the same as for bending in the x–z plane because of the molecule's symmetry. However, since these two bends are in perpendicular planes, they correspond to independent directions. Thus, the energy level expression for the bending vibrations is like Equation 8.18 to the extent to which the bending potential is harmonic.

8.4 Particles with Box and Step Potentials

A special model problem in quantum mechanics is that of a particle-in-a-box. It is the problem of a hypothetical particle of mass m that can move in one dimension, along the x-axis, and that experiences a potential, $V(x)$, with a very simple form. $V(x)$ is zero in some region,

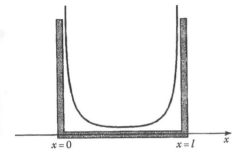

FIGURE 8.3
Particle-in-a-box potential, which is an impenetrable potential outside $x = 0$ and $x = l$ and zero inside, might serve as an approximation of the smoothly changing potential represented by the solid line.

$0 < x < l$, and is infinite elsewhere. The region from $x = 0$ to $x = l$ is inside the "box." This is not a realistic situation by any means because realistic potentials do not become infinite throughout a region and do not change abruptly from zero to infinite. However, certain potential functions may be relatively flat over a region and then turn sharply upward. In such a case, a boxlike $V(x)$, as shown in Figure 8.3, might serve as an approximation to the true potential.

We assume that not even quantum particles are found in regions where the potential is infinite. Then, $V(x)$ really is a box, in that the particle is found only in the region $0 < x < l$. Within that region, the potential is zero, and the Schrödinger equation is quite simple,

$$-\frac{\hbar^2}{2m}\frac{d^2\psi(x)}{dx^2} = E\psi(x) \tag{8.19}$$

The general solution of this differential equation can be written two ways:

$$\psi(x) = Ae^{-i\sqrt{2mEx}/\hbar} + Be^{i\sqrt{2mEx}/\hbar} \tag{8.20a}$$

$$\psi(x) = a\sin\left(\frac{\sqrt{2mEx}}{\hbar}\right) + b\cos\left(\frac{\sqrt{2mEx}}{\hbar}\right) \tag{8.20b}$$

The constants, A and B, or a and b, and the energy eigenvalues do not end up being specified in the course of finding the general solution. However, the Schrödinger equation in Equation 8.19 is valid only in a certain region, the confines of the box, and information about the wavefunction elsewhere remains to be taken into account.

Example 8.2: Degenerate Energy Levels

Problem: Find the degeneracy of the first four energy levels of the harmonic oscillator described by the following Hamiltonian:

$$H = \frac{p_x^2 + p_y^2 + p_z^2}{2m} + \frac{1}{2}kx^2 + \frac{1}{2}(4k)y^2 + \frac{1}{2}(9k)z^2$$

Solution: To find the degeneracy requires that we have solved the Schrödinger equation and know the energies of the states of the system. We notice that this Hamiltonian is separable and that each of the separated Schrödinger equations is a harmonic oscillator equation. We then identify the frequency for each as the square root of the force constant divided by the mass and use the harmonic oscillator solution to write the energies.

$$H_x = \frac{p_x^2}{2m} + \frac{1}{2}kx^2 \qquad \omega_x = \sqrt{\frac{k}{m}} \qquad E_{n_x} = \left(n_x + \frac{1}{2}\right)\hbar\sqrt{\frac{k}{m}}$$

$$H_y = \frac{p_y^2}{2m} + \frac{1}{2}(4k)y^2 \qquad \omega_y = \sqrt{\frac{4k}{m}} \qquad E_{n_y} = \left(n_y + \frac{1}{2}\right)2\hbar\sqrt{\frac{k}{m}}$$

$$H_z = \frac{p_z^2}{2m} + \frac{1}{2}(9k)z^2 \qquad \omega_z = \sqrt{\frac{9k}{m}} \qquad E_{n_z} = \left(n_z + \frac{1}{2}\right)3\hbar\sqrt{\frac{k}{m}}$$

The energy of the system is the sum of the separated energies.

$$E = E_{n_x} + E_{n_y} + E_{n_z} = (n_x + 2n_y + 3n_z + 3)\hbar\sqrt{\frac{k}{m}}$$

The state energies and the degeneracies of the levels can now be obtained by systematically allowing for each of the three quantum numbers to increase from their lowest allowed value of zero and tabulating the corresponding values of the expression ($n_x + 2n_y + 3n_z + 3$).

n_x	n_y	n_z	($n_x + 2n_y + 3n_z + 3$)
0	0	0	3
1	0	0	4
0	1	0	5
2	0	0	5
0	0	1	6
3	0	0	6
1	1	0	6

Thus, the degeneracies for the first four levels are 1, 1, 2, and 3.

The particle-in-a-box potential is discontinuous at $x = 0$ and at $x = l$. This presents a complication in the way we approach the problem because we cannot write one Hamiltonian that is correct for the region $x < 0$, for the region $0 < x < l$, and for the region $x > l$. In general, discontinuous potentials require breaking up the problem into regions. (The boundaries of the regions are wherever the potentials are discontinuous.) Separate Schrödinger equations can be solved for each region, but there are overriding conditions that must be met in order for us to interpret the squares of wavefunctions as probability densities. The conditions are

1. That the wavefunctions are single-valued everywhere
2. That the wavefunctions are continuous
3. That the first derivatives of the wavefunctions are continuous or that the wavefunctions are smooth

Within each region, these conditions are automatically satisfied. At the boundaries between regions, these conditions have to be imposed.

In the particle-in-a-box problem, the wavefunction must be zero-valued at the edges and outside the confines of the box. Thus, the allowed solutions within the box, Equation 8.20, are acceptable as wavefunctions only if they become zero-valued at the edges of the box. This is expressed as boundary conditions:

$$\psi(0) = 0 \quad \text{and} \quad \psi(l) = 0$$

Using the trigonometric form of the solutions in the box, the first of these conditions forces the constant b to be zero. The second condition leads to the following result:

$$\psi(l) = a \sin\left(\frac{l\sqrt{2mE}}{\hbar}\right) = 0 \Rightarrow \frac{l\sqrt{2mE}}{\hbar} = n\pi \tag{8.21}$$

$$|n| = 0,1,2,3,\dots$$

Rearranging this result shows it to be a condition on the energy:

$$E = \frac{n^2\pi^2\hbar^2}{2ml^2} = \frac{n^2h^2}{8ml^2} \tag{8.22}$$

The allowed energies, then, are not continuous because of the boundary potential. In fact, particles that are bound or trapped by potentials have discrete energy levels. Notice that the separation of adjacent energy levels increases with n because of the quadratic dependence of the energy eigenvalues on n in Equation 8.22. In the harmonic oscillator, the bound states were separated by an equal amount; the eigenenergies varied linearly with the quantum number.

Using the relation in Equation 8.21 allows us to express the wavefunctions as

$$\psi_n(x) = a \sin\left(\frac{n\pi x}{l}\right) \tag{8.23}$$

At this point, we reject the possibility of $n = 0$ as unphysical. It would correspond to a wavefunction that is zero-valued everywhere and has a zero probability density everywhere. Also, we exclude the negative integers as possible quantum numbers. Even though they are allowed by Equation 8.21, they do not generate unique wavefunctions. Instead, they correspond to a wavefunction with a sign or phase change since

$$\sin\left(\frac{-n\pi x}{l}\right) = -\sin\left(\frac{n\pi x}{l}\right)$$

Two wavefunctions that differ only in sign or phase have the same probability density, and so they are not distinct states. Thus, the allowed values of n are 1, 2, 3, and so on. Finally, the normalization condition determines the constant a, which turns out to be independent of n:

$$\int_0^l \psi_n^* \psi_n dx = \int_0^l a^2 \sin^2\left(\frac{n\pi x}{l}\right) dx = \frac{a^2 l}{2} = 1$$

a is chosen so that this integral value is 1, a unit probability of finding the particle some-where. The sign of *a* (or the phase of the wavefunction) is arbitrary, and a positive value is chosen. The wavefunctions for the particle-in-a-box problem are

$$\psi_n(x) = \sqrt{\frac{2}{l}}\sin\left(\frac{n\pi x}{l}\right)$$
(8.24)

with the wavefunction being zero outside the box.

The Schrödinger equation for a particle free to move in one dimension with $V(x) = 0$ everywhere is the same as Equation 8.19 with no boundaries. The general solutions of this differential equation are those of Equation 8.20, and let us consider the form given in Equation 8.20a. The constants in the exponential of this equation can be collected into one constant, *k*, called the **wave constant** or **wave number** for the particle.

Example 8.3: Particle in a Three-Dimensional Box

Problem: Find an expression for the allowed energies of a particle of mass *m* found to exist in a three-dimensional box defined by the potential that is zero in the region $0 < x < l, 0 < y < 2l, 0 < z < 3l$ and infinite elsewhere (outside the box region). Determine if degeneracies in energy levels can be found for this problem.

Solution: The Hamiltonian for this problem is simple and separable because it consists of additive terms involving *x*, involving *y*, and involving *z*.

$$\widehat{H} = -\frac{\hbar^2}{2m}\left(\frac{\partial^2}{\partial x^2} + \frac{\partial^2}{\partial y^2} + \frac{\partial^2}{\partial z^2}\right)$$

The separability means the energies are obtained as a sum of energies corresponding to a particle in an *x*-direction box of length *l*, in a *y*-direction box of length *2l*, and in a *z*-direction box of length *3l*. Each direction has an associated quantum number, and then from Equation 8.22, the energy expression is

$$E_{n_x n_y n_z} = \frac{n_x^2 \pi^2 \hbar^2}{2ml^2} + \frac{n_y^2 \pi^2 \hbar^2}{2m(2l)^2} + \frac{n_y^2 \pi^2 \hbar^2}{2m(3l)^2} = \frac{\pi^2 \hbar^2}{2ml^2}\left(n_x^2 + \frac{n_y^2}{4} + \frac{n_z^2}{9}\right)$$

The degeneracies of the levels can now be obtained by systematically allowing for each of the three quantum numbers to increase from their lowest allowed value of 1 and tabulating the corresponding values of

$$\left(n_x^2 + \frac{n_y^2}{4} + \frac{n_z^2}{9}\right)$$

Or, we can simply look to see if there is more than one set of quantum numbers for which the system becomes degenerate. One choice of quantum numbers (n_x, n_y, n_z) that satisfy this condition are (1, 4, 3) and (1, 2, 6). Another pair of degenerate states are those with quantum numbers (3, 4, 3) and (3, 3, 6).

$$k = \frac{\sqrt{2mE}}{\hbar} \tag{8.25}$$

$$\psi(x) = Ae^{-ikx} + Be^{ikx} \tag{8.26}$$

There is no restriction on the energy E or on the value of k because the particle is free: $V(x) = 0$. When the free particle exists in three dimensions, the Schrödinger equation is separable, and the wavefunctions are products of independent functions of x, y, and z that have the form given in Equation 8.26.

$$\Psi(x,y,z) = \left(A_x e^{-ik_x x} + B_x e^{ik_x x}\right)\left(A_y e^{-ik_y y} + B_y e^{ik_y y}\right)\left(A_z e^{-ik_z z} + B_z e^{ik_z z}\right) \tag{8.27}$$

The wave constants k_x, k_y, and k_z are not necessarily the same in each direction. Together, they comprise the **wave vector** $\vec{k}$.

$$\vec{k} = (k_x, k_y, k_z) \tag{8.28}$$

The energy is partitioned in the three directions because of the separability of the Schrödinger equation. Using the symbol del squared,

$$\nabla^2 = \frac{\partial^2}{\partial x^2} + \frac{\partial^2}{\partial y^2} + \frac{\partial^2}{\partial z^2}$$

application of the Hamiltonian to the wavefunction yields

$$\hat{H} = -\frac{\hbar^2}{2m}\nabla^2\Psi(x,y,z) = -\frac{\hbar^2}{2m}\left(k_x^2 + k_y^2 + k_z^2\right)\Psi(x,y,z) = (E_x + E_y + E_z)\Psi(x,y,z) \tag{8.29}$$

This is analogous to expressing the kinetic energy of a free classical particle as a sum of three terms involving the momentum components: $\left(p_x^2 + p_y^2 + p_z^2\right)/2m$.

Returning to the free particle in one dimension, we note that if either A_x were zero or B_x were zero in Equation 8.27, the wavefunction would be an eigenfunction of p_x.

$$\hat{p}_x\left(A_x e^{-ik_x x}\right) = -i\hbar A_x \frac{\partial}{\partial x}e^{-ik_x x} = -\hbar k_x\left(A_x e^{-ik_x x}\right)$$

The eigenvalue for the momentum in this case is $-\hbar k_x$. If instead A_x were zero, the eigenvalue would be $\hbar k_x$. The positive eigenvalue implies that the momentum is that associated with motion in the $+x$ direction, whereas the negative eigenvalue is associated with motion in the opposite direction. This can be generalized for a particle in three dimensions. Thus, the complex exponential forms of the wavefunctions of a free particle can be associated with motion to the right or to the left along each coordinate axis. Then the set of elements termed the wave vector, $\vec{k}$, in Equation 8.28 seems appropriately named as we now see $\vec{k}$ as a vector that identifies the direction that a free particle is moving in three-dimensional space.

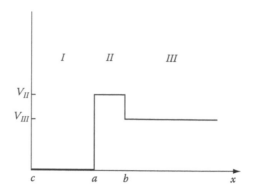

FIGURE 8.4
Hypothetical step potential for a one-dimensional particle. To understand this system, it is helpful to break it into regions. In the first region (*I*), there is a potential well that is almost like the potential of the particle-in-a-box problem. The second region (*II*) has a constant but not infinite potential. The third region (*III*) has a flat potential, and this continues to infinity.

The wavefunctions of the free particle are said to form a **continuum of energy states** because a continuous range of positive energies (or wave vectors) is possible. Unlike the situation for the harmonic oscillator or the particle in the box, there is no restriction on what energies are allowed. This is a typical feature of a particle that is **unbound**, meaning the particle's kinetic energy is sufficient to surmount any potential energy barrier.

An interesting type of problem is found when a potential has a well in one region and approaches a constant (i.e., it is flat) in another region. One example is in Figure 8.4. In this situation, there can be bound states, unbound states, and quantum mechanical tunneling. Recall that a feature of the quantum mechanical harmonic oscillator is that its probability density is not zero beyond the classical turning points, though it is small. Since the classical turning points are the points at which the particle's total energy equals the potential energy, this is where the system exhibits tunneling. For a particle with energy E in region *I* in the potential in Figure 8.4 and $V_{III} < E < V_{II}$, we might expect some tunneling from region *I* into region *II*. Then, as long as the wavefunction does not diminish completely to zero as it continues to the right in region *II*, there is some possibility that the particle will enter region *III*. There it will behave as a free particle since its energy will be greater than the potential, $E > V_{III}$. This tunneling occurrence will make a trapped particle free, something that in the classical world would be the same as walking through a brick wall. Tunneling through a barrier is an important physical process. It explains spontaneous nuclear fission and beta decay, and it explains why certain chemical reactions take place even when the reactants have insufficient energy to surmount the activation barrier. Let us consider certain details of the system in Figure 8.4, taking it as a model barrier problem.

As in the treatment of the particle in a box, we begin with free particle wavefunction forms but do so for each region. In other words, the general solutions of the Schrödinger differential equations for each of the regions treated separately are

$$\psi_I(x) = A_I e^{-ik_I x} + B_I e^{ik_I x} \quad \text{where } k_I = \frac{\sqrt{2mE}}{\hbar} \tag{8.30}$$

$$\psi_{II}(x) = A_{II} e^{-ik_{II} x} + B_{II} e^{ik_{II} x} \quad \text{where } k_{II} = \frac{\sqrt{2m(E - V_{II})}}{\hbar} \tag{8.31}$$

$$\psi_{III}(x) = A_{III}e^{-ik_{III}x} + B_{III}e^{ik_{III}x} \quad \text{where } k_{III} = \frac{\sqrt{2m(E - V_{III})}}{\hbar} \tag{8.32}$$

In order to interpret the square of a wavefunction as a probability density function, conditions exist on the allowed forms of wavefunctions. As mentioned earlier, every wavefunction must be smooth, continuous, and single-valued when the potential has no infinite jumps. In the case of the step potential in Figure 8.4, having these conditions hold at the boundaries between regions implies constraints on the wavefunctions, specifically on the undetermined constants in Equations 8.30 through 8.32. For the wavefunction to be continuous and single-valued the following must hold:

$$\psi_I(a) = \psi_{II}(a) \tag{8.33}$$

$$\psi_{II}(b) = \psi_{III}(b) \tag{8.34}$$

For the wavefunction to be smooth it must have a continuous first derivative everywhere, and so the following conditions must be satisfied:

$$\left.\frac{d\psi_I}{dx}\right|_{x=a} = \left.\frac{d\psi_{II}}{dx}\right|_{x=a} \tag{8.35}$$

$$\left.\frac{d\psi_{II}}{dx}\right|_{x=b} = \left.\frac{d\psi_{III}}{dx}\right|_{x=b} \tag{8.36}$$

Also, as in the particle-in-a-box problem, the wavefunction must vanish at $x = c$ because the potential is infinite at that point:

$$\psi_I(c) = 0 \tag{8.37}$$

Therefore, we have five conditions that relate the six constants A_I, A_{II}, A_{III}, B_I, B_{II}, and B_{III}.

The first boundary to consider is at $x = b$, the boundary between regions II and III. Since a coordinate origin has not been specified, we can choose it to be at this boundary; that is, $b = 0$. The continuity conditions of Equations 8.34 and 8.36 are now

$$A_{III} + B_{III} = A_{II} + B_{II} \quad \text{(from Equation 8.34)}$$

$$-k_{III}A_{III} = -k_{II}A_{II} + k_{II}B_{II} \quad \text{(from Equation 8.36)}$$

Defining $g = k_{III}/k_{II}$ and rearranging, we can obtain A_{II} and B_{II} in terms of A_{III} and B_{III}.

$$A_{II} = \frac{1}{2}(1 + g)A_{III} + \frac{1}{2}(1 - g)B_{III} \tag{8.38}$$

$$B_{II} = \frac{1}{2}(1 - g)A_{III} + \frac{1}{2}(1 + g)B_{III} \tag{8.39}$$

From our analysis of the free particle wavefunctions, we can associate the A_{II} term in ψ_{II} with motion to the left and the B_{II} term with motion to the right. Therefore, at any point in region *II*, the normalized probability that a measurement of the momentum will yield a negative value (motion to the left) is

$$P_{II}^{left} = \frac{A_{II}^* A_{II}}{A_{II}^* A_{II} + B_{II}^* B_{II}} \tag{8.40}$$

The probability of motion in the opposite direction is

$$P_{II}^{right} = \frac{B_{II}^* B_{II}}{A_{II}^* A_{II} + B_{II}^* B_{II}} \tag{8.41}$$

The probability of motion in either direction, the sum of these two values, is properly 1.

Let us now make an assumption, one that may prove drastic under certain conditions but one that will help us extract a qualitative understanding of the effect of a barrier. We shall assume that we are able to control the generation of particles approaching from the far right such that $A_{III} \gg B_{III}$; that is, the likelihood of finding a particle moving to the left in region *III* is much greater than that of finding a particle moving to the right. This leads to the following approximations for Equations 8.38 through 8.41:

$$A_{II} \cong \frac{1}{2}(1+g)A_{III} \quad \text{and} \quad B_{II} \cong \frac{1}{2}(1-g)A_{III}$$

$$P_{II}^{left} \cong \frac{(1+g)^2}{(1+g)^2 + (1-g)^2} = \frac{(1+g)^2}{2(1+g^2)}$$

$$P_{II}^{right} \cong \frac{(1-g)^2}{2(1+g^2)}$$

Under this assumption, we can regard P_{II}^{left} as a measure of the likelihood of **transmission**, *T*, of particles entering from the far right and P_{II}^{right} as a measure of the likelihood of **reflection**, *R*. Substituting for g and then for k_{II} and k_{III} yields

$$P_{II}^{left} = T = \frac{1}{2} + \frac{\sqrt{(E-V_{II})(E-V_{III})}}{2E - V_{II} - V_{III}} \tag{8.42}$$

$$P_{II}^{right} = R = \frac{1}{2} - \frac{\sqrt{(E-V_{II})(E-V_{III})}}{2E - V_{II} - V_{III}} \tag{8.43}$$

This reveals that the likelihood of reflection and transmission at a step potential boundary depends on the difference between the particle's energy and the heights of the two potential steps.

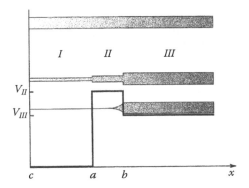

FIGURE 8.5
Transmission of a beam of particles coming from the right and encountering the potential in Figure 8.4 is represented by three horizontal lines corresponding to three specific particle energies. The width of each line is a representation of the transmission probability as a function of x. In the highest energy case, there is a negligible diminishment at the boundary between regions *II* and *III*, but at lower energies, there is a noticeable diminishment.

To understand the implications of Equations 8.42 and 8.43, we shall consider certain specific choices of the potentials and the particle's energy.

1. If V_{II} were equal to V_{III}, the rightmost terms in Equations 8.42 and 8.43 would reduce to 1/2. Then, we would have $T = 1$ and $R = 0$. This means that if there were no potential step (i.e., no change in potential), there would be complete transmission.

2. If the particle's energy, E, were much, much greater than both V_{II} and V_{III}, the rightmost terms would be very nearly equal to 1/2. Again, that would lead to $T = 1$ and $R = 0$. Classically, the difference, $E - V_{II}$ (or $E - V_{III}$), is the particle's kinetic energy in the region. So if it has a lot of kinetic energy, the step in the potential is not too important; mostly the particle is likely to continue ahead. This is illustrated in Figure 8.5.

3. If E were much, much greater than V_{III} but only 10% bigger than V_{II}, an approximation of rightmost terms in Equations 8.42 and 8.43 is

$$\frac{\sqrt{(E-V_{II})(E-V_{III})}}{2E-V_{II}-V_{III}} \cong \frac{\sqrt{(E-V_{II})E}}{2E-V_{II}} = \frac{\sqrt{(1.1V_{II}-V_{II})1.1V_{II}}}{2(1.1V_{II})-V_{II}} = \frac{\sqrt{0.11}}{1.2} = 0.28$$

Then, $T = 0.5 + 0.28 = 0.78$ and $R = 0.5 - 0.28 = 0.22$, and in contrast to the prior specific case, there is a noticeable probability that the incoming particle will be reflected back. This is at odds with the experience in our macroscopic, classical mechanical world where an object encountering a potential will not be stopped or reflected if it has energy in excess of the potential barrier. In the quantum world, there is a real probability of being reflected even if a particle's energy is more than the potential barrier.

Next consider the case where the energy happens to be such that $V_{III} < E < V_{II}$. k_{II} is now imaginary by Equation 8.31, and thus, the general solution to the Schrödinger equation for region *II* is best expressed as

$$\psi_{II}(x) = A_{II}e^{-\alpha_{II}x} + B_{II}e^{\alpha_{II}x} \tag{8.44}$$

$$\alpha = \frac{\sqrt{2m(V_{II}-E)}}{\hbar}$$

This wavefunction shows a nonvanishing probability density in region *II*. Were this a classical particle, there would be no chance of finding it there, and hence this is tunneling through a barrier or in a classically forbidden region.

The model problems involving step potentials illustrate several features of quantum systems that are outside our experience in the macroscopic world. (1) A particle trapped in a potential has discrete allowed energies. The basic examples are the harmonic oscillator and the particle in a box. If the particle is not trapped, there is a continuum of allowed energy states. (2) A particle can tunnel into a region where it is classically forbidden. The likelihood of tunneling diminishes as the difference in the potential and the particle's energy increases. (3) The likelihood that a particle will pass through a barrier region diminishes with increasing width of the region. (4) There is a probability that a particle will be reflected by a barrier even if its energy is sufficient to surmount the barrier.

8.5 Rigid Rotator and Angular Momentum

A body rotating freely about its center of mass exhibits angular momentum. In low-density gases, molecules rotate and possess rotational angular momentum. The importance of angular momentum analysis in many quantum mechanical systems arises from its quantization. In the absence of certain external influences, the total angular momentum of a system takes on only certain values. In the earliest days of quantum theories, the Bohr model (Niels Bohr, Denmark, 1885–1962) of the hydrogen atom had some success in explaining atomic spectra by imposing a quantization condition on the electron's orbital angular momentum. This was a presumption, but the ultimate explanation of the mechanics of the hydrogen atom showed that the electron's angular momentum is indeed quantized. Today, many features of molecular spectra are understood in terms of angular momentum properties, and reaction probabilities have even been found that are dependent on the angular momenta of the colliding atoms or molecules.

The fundamental development of angular momentum in quantum systems draws on classical mechanical theory, and there one finds a notable correspondence between rectilinear and rotational motions. Rectilinear force has an analog in the torque about an axis, and in the appropriate equations of motion, the mass of a straight-moving body plays the same role as a rotating body's moment of inertia. In all respects, analysis of the mechanics of rotating systems is largely a generalization of the types of coordinates to include angular coordinates.

Angular momentum is a vector quantity, and for a moving point mass it is the cross-product of a vector $\vec{r}$, the vector from the axis of about which the particle is rotating to the particle, and the linear momentum vector $\vec{p}$.

$$\vec{L} = \vec{r} \otimes \vec{p} \Rightarrow L_x = yp_z - zp_y, \quad L_y = zp_x - xp_z, \quad L_z = xp_y - yp_x \tag{8.45}$$

For a system of several point-mass particles, the total angular momentum is the vector sum of the angular momenta of each of the particles.

Quantum mechanical operators corresponding to the components of an angular momentum vector can be found directly from the familiar rectilinear position and momentum operators, for example,

$$\hat{L}_x = \widehat{y}\widehat{p}_z - \widehat{z}\widehat{p}_y \tag{8.46}$$

Therefore,

$$\hat{L}_x = -i\hbar\left(y\frac{\partial}{\partial z} - z\frac{\partial}{\partial y} \right)$$

$$\hat{L}_y = -i\hbar\left(z\frac{\partial}{\partial x} - x\frac{\partial}{\partial z} \right) \tag{8.47}$$

$$\hat{L}_z = -i\hbar\left(x\frac{\partial}{\partial y} - y\frac{\partial}{\partial x} \right)$$

With these explicit forms for the operators, it is straightforward to show that the commutator of any pair of them produces $i\hbar$ times the third one:

$$\left[\hat{L}_x, \hat{L}_y \right] = i\hbar\hat{L}_z \tag{8.48}$$

$$\left[\hat{L}_y, \hat{L}_z \right] = i\hbar\hat{L}_x \tag{8.49}$$

$$\left[\hat{L}_z, \hat{L}_x \right] = i\hbar\hat{L}_y \tag{8.50}$$

In practice, it is better to work with angular momentum in an angular coordinate system rather than in a rectilinear system. This means using the spherical polar coordinate system defined as in Figure 7.4 and transforming the operators in Equations 8.48 through 8.50. The coordinate transformation to and from spherical polar coordinates is

$$x = r\sin\theta\cos\phi \quad r^2 = x^2 + y^2 + z^2$$

$$y = r\sin\theta\sin\phi \quad \theta = \arccos\left(\frac{z}{r}\right)$$

$$z = r\cos\theta \qquad \phi = \arctan\left(\frac{y}{x}\right)$$

This transformation enables us to write the angular momentum component operators in either system. Chain rule differentiation provides the substitution for a differential operator in Equation 8.47. For instance,

$$\frac{\partial}{\partial x} = \frac{\partial r}{\partial x}\frac{\partial}{\partial r} + \frac{\partial\theta}{\partial x}\frac{\partial}{\partial\theta} + \frac{\partial\phi}{\partial x}\frac{\partial}{\partial\phi}$$

and from the transformation equations,

$$\frac{\partial r}{\partial x} = \frac{x}{r} = \sin\theta\cos\phi$$

If this is worked through completely, the following operator expressions are obtained:

$$\hat{L}_x = -i\hbar\left(-\sin\phi\frac{\partial}{\partial\theta} - \frac{\cos\phi}{\tan\theta}\frac{\partial}{\partial\phi}\right)$$

$$\hat{L}_y = -i\hbar\left(\cos\phi\frac{\partial}{\partial\theta} - \frac{\sin\phi}{\tan\theta}\frac{\partial}{\partial\phi}\right) \tag{8.51}$$

$$\hat{L}_z = -i\hbar\frac{\partial}{\partial\phi}$$

These can be applied to wavefunctions that are expressed in spherical polar coordinates.

Another useful operator can be found from the angular momentum component operators. The square of the angular momentum of a classical system, or of a quantum mechanical system, is a scalar quantity, the dot product of $\vec{L}$ with itself. The dot product of any vector with itself equals the square of the length of the vector. Thus, $\vec{L}\cdot\vec{L}$ is the square of the magnitude of the angular momentum vector. The quantum mechanical operator that corresponds to this dot product is designated $\hat{L}^2$, and its explicit form can be derived from the component expression for a dot product. Thus,

$$\hat{L}^2 = \hat{L}_x^2 + \hat{L}_y^2 + \hat{L}_z^2$$

In spherical polar coordinates, the following is obtained with Equation 8.51:

$$\hat{L}^2 = -\hbar^2\left(\frac{1}{\sin\theta}\frac{\partial}{\partial\theta}\sin\theta\frac{\partial}{\partial\theta} + \frac{1}{\sin^2\theta}\frac{\partial^2}{\partial\phi^2}\right) \tag{8.52}$$

This operator is related to the Laplacian operator, ∇^2 (del squared), in spherical polar coordinates. From the definition of ∇^2 in Cartesian coordinates, and from the coordinate transformation given previously,

$$\nabla^2 = \frac{\partial^2}{\partial x^2} + \frac{\partial^2}{\partial y^2} + \frac{\partial^2}{\partial z^2} \tag{8.53a}$$

$$= \frac{2}{r}\frac{\partial}{\partial r} + \frac{\partial^2}{\partial r^2} - \frac{\hat{L}^2/\hbar^2}{r^2} \tag{8.53b}$$

The Laplacian is used in the kinetic energy operator for a particle in three dimensions since $\hat{p}^2/2m = -\hbar^2\nabla^2/2m$. Thus, the square of the angular momentum is intimately connected with the amount of kinetic energy of a moving particle.

$\hat{L}^2$ commutes with each component operator, a result that can be derived using the Cartesian forms of the operators or their spherical polar form.

$$\left[\hat{L}^2, \hat{L}_x\right] = 0$$

$$\left[\hat{L}^2, \hat{L}_y\right] = 0 \tag{8.54}$$

$$\left[\hat{L}^2, \hat{L}_z\right] = 0$$

Each of these four operators is Hermitian. By Theorem 8.3, these commutation relations mean that there are functions that are simultaneously eigenfunctions of $\hat{L}^2$ and of any one component of the angular momentum; however, because the component operators do not commute with each other, a mutually commuting set of operators including $\hat{L}^2$ and two (or three) component operators cannot be formed.

Spherical harmonic functions are the functions that are simultaneously eigenfunctions of the two operators $\hat{L}^2$ and $\hat{L}_z$. These functions are usually designated $Y(\theta, \phi)$, and the whole set is obtained by solving the two differential eigenequations,

$$\hat{L}^2 Y = \alpha Y$$

$$\hat{L}_z Y = \beta Y$$

There are an infinite number of solutions to these equations, and two integers are needed to label and distinguish the different spherical harmonic functions. The conventional choices for these two integers are l and m. The θ dependence of these functions is expressed with a special set of polynomials called the **associated Legendre polynomials.**

Associated Legendre polynomials for a variable z, designated $P_l^{|m|}(z)$, can be generated with the following two formulas:

$$P_l^0(z) = \frac{1}{2^l l!} \frac{d^l}{dz^l}\left[(z^2 - 1)^l\right] \tag{8.55}$$

$$P_l^m(z) = (1 - z^2)^{m/2} \frac{d^m}{dz^m} P_l^0(z) \tag{8.56}$$

Notice that if $m > l$, these formulas lead to a function that is zero. Certain of these polynomials* are given explicitly in Table 8.1.

In the spherical harmonic functions, the "variable" for the Legendre polynomials is really a function, $\cos\theta$. This means that after establishing the polynomials explicitly for z using Equations 8.55 and 8.56, the desired equations in terms of θ are obtained by substituting $\cos\theta$ for z throughout. Trigonometric identities are used to simplify the expressions,

---

* The polynomials for $m = 0$, that is, those generated with only Equation 8.55, are generally referred to as the Legendre polynomials, and then the superscript is suppressed. The polynomials that follow from Equation 8.56 are the associated Legendre polynomials.

TABLE 8.1

Associated Legendre Polynomials through $l = 4$

	$P_l^{|m|}(z)$	$P_l^{|m|}(\cos\theta)$
$l = 0$	$P_0^0 = 1$	
$l = 1$	$P_1^0 = z$	$P_1^0 = \cos\theta$
	$P_1^1 = \sqrt{1-z^2}$	$P_1^1 = \sin\theta$
$l = 2$	$P_2^0 = (3z^2 - 1)/2$	$P_2^0 = (3\cos^2\theta - 1)/2$
	$P_2^1 = 3z\sqrt{1-z^2}$	$P_2^1 = 3\sin\theta\cos\theta$
	$P_2^2 = 3(1-z^2)$	$P_2^2 = 3\sin^2\theta$
$l = 3$	$P_3^0 = (5z^3 - 3z)/2$	$P_3^0 = \cos\theta(5\cos^2\theta - 3)/2$
	$P_3^1 = 3(5z^2 - 1)\sqrt{1-z^2}/2$	$P_3^1 = 3\sin\theta(5\cos^2\theta - 1)/2$
	$P_3^2 = 15z(1-z^2)$	$P_3^2 = 15\cos\theta\sin^2\theta$
	$P_3^3 = 15(1-z^2)^{3/2}$	$P_3^3 = 15\sin^3\theta$
$l = 4$	$P_4^0 = (35z^4 - 30z^2 + 3)/8$	$P_4^0 = (35\cos^4\theta - 30\cos^2\theta + 3)/8$
	$P_4^1 = (35z^3 - 15z)\sqrt{1-z^2}/2$	$P_4^1 = \sin\theta\cos\theta(35\cos^2\theta - 15)/2$
	$P_4^2 = (105z^2 - 15)\sqrt{1-z^2}/2$	$P_4^2 = \sin^2\theta(105\cos^2\theta - 15)/2$
	$P_4^3 = 105z(1-z^2)^{3/2}$	$P_4^3 = 105\sin^3\theta\cos\theta$
	$P_4^4 = 105z(1-z^2)^2$	$P_4^4 = 105\sin^4\theta$

and these forms are also presented in Table 8.1. The associated Legendre polynomials are orthogonal functions for the range $z = -1$ to $z = 1$. This corresponds to the range $\cos\theta = -1$ to $\cos\theta = 1$, which is $\theta = \pi$ to $\theta = 0$. Over this range the functions are normalized by a factor involving l and m:

$$\sqrt{\frac{(2l+1)(l-|m|)!}{2(l+|m|)!}}$$

The symbol Θ is used to designate a normalized associated Legendre polynomial in θ; that is,

$$\Theta_{lm}(\theta) = \sqrt{\frac{(2l+1)(l-|m|)!}{2(l+|m|)!}} P_l^{|m|}(\cos\theta) \tag{8.57}$$

Notice that the absolute value of m is used in Equation 8.57, and so the Θ functions are the same for m and $-m$.

Spherical harmonics have an exponential dependence on the angle φ via functions designated Φ_m.

$$\Phi_m(\varphi) = \frac{e^{im\varphi}}{\sqrt{2\pi}} \tag{8.58}$$

The spherical harmonics are products of Φ and Θ functions.

$$Y_{lm}(\theta, \varphi) = \Theta_{lm}(\theta)\Phi_m(\varphi)(-1)^{\lfloor m+|m| \rfloor/2} \tag{8.59}$$

The factor of $(-1)^{\lfloor m+|m| \rfloor/2}$ is an arbitrary phase factor that is introduced to conform to common conventions; its value is always 1 or −1. Notice that the Φ functions are not dependent on the l integer. They are orthogonal since they are eigenfunctions of a Hermitian operator, and they are normalized for the range $\varphi = 0$ to $\varphi = 2\pi$. Thus, the spherical harmonic functions are a set of orthonormal functions over the usual ranges of angles in the spherical polar coordinate system:

$$\langle Y_{lm} | Y_{l'm'} \rangle = \int_0^{2\pi} \int_{0lm}^{\pi} Y_{lm}^*(\theta, \varphi) Y_{l'm'}(\theta, \varphi) \sin\theta \, d\theta \, d\varphi = \delta_{ll'}\delta_{mm'} \tag{8.60}$$

With explicit forms of the spherical harmonic functions, it is possible to show that their eigenvalues with the operators $\hat{L}^2$ and $\hat{L}_z$ are given in terms of l and m.

$$\hat{L}^2 Y_{lm} = l(l+1)\hbar^2 Y_{lm} \tag{8.61}$$

$$\hat{L}_z Y_{lm} = m\hbar Y_{lm} \tag{8.62}$$

l must be zero or a positive integer; smooth, single-valued solutions do not exist for other values. m is restricted because if its absolute value were greater than l, the associated Legendre polynomial would be zero. This would give rise to a zero-valued spherical harmonic function which would not serve as a wavefunction since it would correspond to zero probability density everywhere. Therefore, $|m| \le l$, or $m = -l, -l+1, \ldots, l-1, l$.

An important model problem involving angular momentum is the rigid rotator. Two masses, m_1 and m_2, are taken to be connected by a massless rod that is absolutely rigid. Thus, the separation distance between the masses is fixed at some value, R, the length of the rod. Figure 7.4 serves to depict this problem, provided we take the radial spherical polar coordinate, r, to have the fixed value R (i.e., $r = R$). The moment of inertia of this system is $I = \mu R^2$, where μ is the reduced mass, $m_1 m_2/(m_1 + m_2)$. After separating the motion corresponding to translation of the center of mass of the system, the quantum mechanical kinetic energy operator for the system in Figure 7.4 is

$$\hat{T} = -\frac{\hbar^2}{2\mu}\nabla^2$$

Using Equation 8.53b but taking r to be fixed at the value R, we have

$$\hat{T} = -\frac{\hat{L}^2}{2\mu R^2} = \frac{\hat{L}^2}{2I}$$

In this model situation, there is no potential of any sort, and thus the kinetic energy operator and the Hamiltonian are one and the same.

The Schrödinger equation for the rigid rotator must be

$$\frac{\hat{L}^2}{2I}\psi = E\psi$$

Because the spherical harmonics are eigenfunctions of the $\hat{L}^2$ operator, they are also eigenfunctions of the rigid rotator's Hamiltonian, $\hat{L}^2/2I$. Therefore, the Y_{lm} functions are the wavefunctions for the states of the rigid rotator. The eigenenergies are obtained from Equation 8.61:

$$\frac{\hat{L}^2}{2I}Y_{lm} = l(l+1)\frac{\hbar^2}{2I}Y_{lm} \Rightarrow E_{lm} = l(l+1)\frac{\hbar^2}{2I} \tag{8.63}$$

Notice that the energies increase quadratically with the quantum number l. Also, each energy level has a degeneracy of $2l + 1$ because there are $2l + 1$ choices of the m quantum number for any particular l.

The classical picture of the rigid rotator is that of a dumbbell whirling about its center of mass, and any direction and magnitude of the angular momentum is possible. In the quantum mechanical picture, there are restrictions on the angular momentum because of its quantization. The allowed states are those given by the allowed values of the two quantum numbers:

$$l = 0, 1, 2, 3, \ldots$$

$$m = -l, \ldots, l$$

The wavefunctions are eigenfunctions of the Hamiltonian, and also of $\hat{L}^2$ and of $\hat{L}_z$. The eigenvalues of $\hat{L}^2$ are as given in Equation 8.61. On the basis of the postulates, we should expect that a measurement of the square of the angular momentum will produce one of these values. The square root of such a measurement is interpreted as the magnitude or length of the angular momentum vector. For a given Y_{lm} state, this quantity is $\sqrt{l(l+1)}\hbar$. Thus, the only allowed lengths of the angular momentum vector are 0, $\sqrt{2}\hbar$, $\sqrt{6}\hbar$, $\sqrt{12}\hbar$, $\sqrt{20}\hbar$, and so on.

The eigenvalue of $\hat{L}_z$ for a given Y_{lm} function is $m\hbar$. This is the value of the z-component of the angular momentum vector. Thus, for each state of the rigid rotator, we know the length of the angular momentum vector and its z-component. However, we do not know the x- or y-component precisely since the wavefunctions are not eigenfunctions of the operators $\hat{L}_x$ and $\hat{L}_y$. The information we do have can be depicted by vectors indicating the possible orientations of the angular momentum vector with respect to the z-axis for the different possible lengths of the vector; this is shown in Figure 8.6. So, our whirling quantum dumbbell can "spin" in only fixed or discrete amounts and then only with the angular momentum vector at certain fixed angles.

The three operators, $\hat{L}^2$, $\hat{L}_z$, and $\hat{H}$, commute in this model problem, and as allowed by Theorem 8.3, the wavefunctions describing this system are simultaneously eigenfunctions of all three. One could also form mutually commuting sets by substituting in this group $\hat{L}_x$ or $\hat{L}_y$ for $\hat{L}_z$; however, with the unique transformation from Cartesian coordinates to spherical polar coordinates, either of these choices would make certain mathematical steps more cumbersome. The Hamiltonian must commute with $\hat{L}^2$ since it is proportional to $\hat{L}^2$. This can

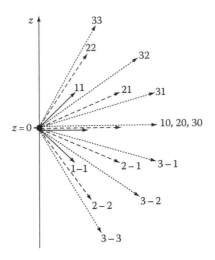

FIGURE 8.6
Angular momentum vectors of the Y_{lm} states of the rigid rotator for $l = 1$ (solid lines), $l = 2$ (dashed lines), and $l = 3$ (dotted lines). The arrows represent the orientation with respect to the z-axis for the particular (l, m) state. The orientations with respect to the x-axis and the y-axis are not determined, and so this picture represents a planar slice through three-dimensional space such that the slice includes the z-axis and is perpendicular to the x–y plane. The angle between this planar slice and the x-axis or the y-axis, though, is arbitrary.

be seen by considering the commutator of some arbitrary operator $\hat{G}$ and that operator scaled by a constant g:

$$[\hat{G}, g\hat{G}] \equiv \hat{G}(g\hat{G}) = g\hat{G}\hat{G} - g(\hat{G}^2 - \hat{G}^2) - 0$$

As indicated earlier, the commutator of $\hat{L}^2$ with any of the angular momentum component operators happens to be zero, which means that $\hat{L}^2$ also commutes with $\hat{L}_x$, $\hat{L}_y$, and $\hat{L}_z$. The component operators, though, do not commute with each other. Therefore, the largest set of mutually commuting operators for the rigid rotator problem consists of three, the Hamiltonian, the $\hat{L}^2$ operator, and any one of the component operators.

It is significant that all the component operators cannot be part of the mutually commuting set. If the component operators, the Hamiltonian, and $\hat{L}^2$ were all mutually commuting, then the corollary to Theorem 8.3 would say that a set of functions can be found that are simultaneously eigenfunctions of the Hamiltonian, of $\hat{L}^2$, and of all three components. In turn, that would say that we can measure each component of the angular momentum with no uncertainty, which means that for any state of the rigid rotator we can know exactly where the angular momentum vector points. That the component operators are *not* mutually commuting means that the wavefunctions of the system can be eigenfunctions of only one component operator and that measurements with no uncertainty can be accomplished for only that one component; the other two have nonzero uncertainty.

By convention, we choose the spherical harmonic functions such that they are eigenfunctions of the z-component operator and not the x- or y-component operator. There is nothing special about the z-direction in space. Rather, since we know that there is one direction in which the angular momentum component is quantized (i.e., the wavefunctions are eigenfunctions of *one* component operator), the convention is that that direction, whatever it happens to be, becomes the z-direction.

8.6 Coupling of Angular Momenta

In many chemical problems, there are multiple sources of angular momenta. For instance, electrons orbiting a molecular axis can give rise to angular momentum at the same time that the rotation of the molecule as a whole about its mass center gives rise to angular momentum. Classically and quantum mechanically, angular momenta add vectorially, and the total angular momentum of an isolated, closed system is conserved. In the classical world, we can know the orientation and length of a given angular momentum vector at any instant in time. The total angular momentum is the vector sum of all such angular momentum vectors. In the quantum world, there is uncertainty in the orientation of an angular momentum vector, and as a result, the addition or coupling of momenta calls for a different sort of analysis.

In a situation where some source of angular momentum is quantized, the associated angular momentum operator commutes with the Hamiltonian. The wavefunctions, then, are simultaneous eigenfunctions. With this in mind, we approach problems involving angular momentum and develop rules that are general, not specific to a given problem. In other words, we consider the angular momentum features of quantum states on their own. An extension of the bra-ket notation is an aid in this endeavor. Instead of placing a function designation (e.g., ψ) in a bra or ket, one places angular momentum quantum numbers for all angular momentum operators for which the wavefunctions must be eigenfunctions (e.g., $|lm\rangle$ for a spherical harmonic function instead of $|Y_{lm}\rangle$). In a realistic application, there may be a number of states with the same angular momentum quantum numbers but different energies, and so we will not distinguish among them in this analysis. Another common practice is to use J in the same way that L has already been used, except that J can represent *any* type of angular momentum, whereas L is usually reserved for orbital angular momentum.

Rules for adding angular momentum need be developed only for two sources at a time. The addition of three sources can be accomplished by adding two together and then adding that result to the third angular momentum vector. Consider two independent sources, $\vec{J}_1$ and $\vec{J}_2$. In the absence of a physical interaction between the two sources, the wavefunctions are eigenfunctions of the total angular momentum operators for each source and of the z-component operators for each source. The designation of an angular momentum state in this situation requires four quantum numbers,

$$|j_1 m_1 j_2 m_2\rangle$$

j_1 and j_2 are the total angular momentum quantum numbers for the two independent sources, while m_1 and m_2 are the corresponding z-component quantum numbers. We express the effect of applying source and z-component operators to these wavefunctions in the following way:

$$\hat{J}_1^2 |j_1 m_1 j_2 m_2\rangle = j_1(j_1+1)\hbar^2 |j_1 m_1 j_2 m_2\rangle$$

$$\hat{J}_2^2 |j_1 m_1 j_2 m_2\rangle = j_2(j_2+1)\hbar^2 |j_1 m_1 j_2 m_2\rangle$$

$$\hat{J}_{1z} |j_1 m_1 j_2 m_2\rangle = m_1\hbar |j_1 m_1 j_2 m_2\rangle$$

$$\hat{J}_{2z} |j_1 m_1 j_2 m_2\rangle = m_2\hbar |j_1 m_1 j_2 m_2\rangle$$

The possible values for the quantum number m_1 are $-j_1, ..., j_1$, and by counting those possibilities, we find there are $2j_1 + 1$ of them. This is the **multiplicity** of states from the j_1 angular momentum. The total number of states for the whole system is a product of multiplicities, $(2j_1 + 1)(2j_2 + 1)$.

Physical coupling of angular momenta comes about through an interaction that is brought into the Hamiltonian of the problem. An interaction that couples momenta must involve both angular momentum vectors, and the form of the interaction Hamiltonian is often that of a dot product of vectors, for example, $\vec{j}_1 \cdot \vec{j}_2$. Such an interaction implies an energetic preference for how the two angular momentum vectors combine in forming a resultant vector. The possibilities, though, are restricted by quantization of the total angular momentum and its z-component. For instance, the maximum possible z-component of the total angular momentum cannot be any more than the sum of the maximum possible z-components of the two vectors being combined. From Figure 8.6, the maximum z-component for any angular momentum vector occurs when the associated m quantum number is equal to the j quantum number. Therefore, a rule exists:

Maximum resultant z-component = (maximum value of m_1 + maximum value of m_2)$\hbar$

$$= (j_1 + j_2)\hbar$$

This must also equal the maximum z-component quantum number of the resultant vector, designated M, times $\hbar$. That is,

Maximum value of $M = j_1 + j_2$

Since any z-component quantum number can take on only certain values because of the length of the associated vector (i.e., $M = -J, ..., J$), knowing the maximum value of M means knowing J. Thus, we conclude that one possible value for the resultant J quantum number is $j_1 + j_2$.

There are other possible resultant J quantum numbers. Notice that there are two ways in which a resultant vector can be composed for which $M = j_1 + j_2 - 1$. One way is with the first vector arranged so that $m_1 = j_1 - 1$, while for the second, $m_2 = j_2$. This is illustrated in Figure 8.7. The second way is with $m_1 = j_1$, but $m_2 = j_2 - 1$. One $\hbar$ less in the resultant z-component comes about through addition of one or the other component vectors at an orientation with one $\hbar$ less in the z-component. These two arrangements of the source vectors do not distinctly correspond to resultant angular momentum states, but they do indicate that there are two states for which $M = j_1 + j_2 - 1$. One of those is expected since it represents one of the possible choices of M with $J = j_1 + j_2$ (the possible J value already identified). The fact that there is another state with the same z-component means that there must be a state with a different J value such that its maximum z-component is $(j_1 + j_2 - 1)\hbar$. That is, the next possible J value is $j_1 + j_2 - 1$.

When the process is continued to the next step down in the z-component of the resultant vector, it turns out that there has to be another possible J value, this one being $j_1 + j_2 - 2$. And this pattern continues until one reaches the minimum value, $|j_1 - j_2|$. Thus, the rule for adding angular momenta is that the possible values for the resultant or total J quantum number span a range given by the quantum numbers associated with the source vectors:

$$J = j_1 + j_2, j_1 + j_2 - 1, j_1 + j_2 - 2, ..., |j_1 - j_2| \tag{8.64}$$

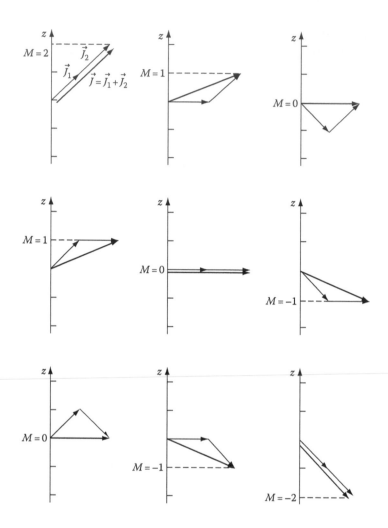

FIGURE 8.7
For two angular momentum vectors, $\vec{J}_1$ and $\vec{J}_2$, with associated quantum numbers j_1 and j_2 both equal to 1, there are $(2j_1 + 1)(2j_2 + 1) = 9$ different combinations of their possible orientations with respect to the z-axis. These 9 ways are shown here, and for each a resultant vector from the sum of $\vec{J}_1$ and $\vec{J}_2$ has been drawn. In each case, the z-component of the resultant vector can be obtained from the sum of the z-component quantum numbers m_1 and m_2.

The resultant states are eigenfunctions of the operators $\widehat{J}^2$, $\widehat{J}_z$, $\widehat{J}_1^2$, and $\widehat{J}_2^2$ can be designated $|JMj_1j_2\rangle$. These functions are not necessarily eigenfunctions of the z-component operators of the individual sources.

Incorporating an angular momentum coupling interaction into a system's Hamiltonian does not change the number of states. The number of states in the absence of any interaction is a product of multiplicities, $(2j_1 + 1)(2j_2 + 1)$. This is equal to the number of states corresponding to the possible resultant J values in Equation 8.64, which is the sum of the multiplicities, $(2J + 1)$, for each possible value of J.

$$(2j_1 + 1)(2j_2 + 1) = \sum_{J=|j_1 - j_2|}^{j_1 + j_2} (2J + 1)$$

This rule is demonstrated with a few examples. We specify values for the source momenta, j_1 and j_2, and find the resultant J's according to Equation 8.64. The sum of the multiplicities for the possible J values is compared with the product of the j_1 and j_2 multiplicities in the accompanying table.

j_1	j_2	$(2j_1+1)(2j_2+1)$	J	$(2J+1)$	$\Sigma(2J+1)$
0	0	1	0	1	1
1	0	3	1	3	3
0	2	5	2	5	5
1	1	9	2	5	
			1	3	
			0	1	9
2	1	15	3	7	
			2	5	
			1	3	15
2	2	25	4	9	
			3	7	
			2	5	
			1	3	
			0	1	25

The J's in this list are the possible values according to Equation 8.64.

An interesting and necessary feature of coupled angular momentum states is that they are eigenfunctions of the coupling interaction operator, $\vec{J}_1 \cdot \vec{J}_2$. This is helpful because a physical coupling interaction may give rise to a term in the Hamiltonian that is proportional to $\vec{J}_1 \cdot \vec{J}_2$. Knowing eigenfunctions of this term can help in finding eigenfunctions of the complete Hamiltonian. It can be shown that the coupled state functions are eigenfunctions of $\vec{J}_1 \cdot \vec{J}_2$ by starting from an expression for the dot product of $\vec{J} = \vec{J}_1 + \vec{J}_2$ and itself.

$$\vec{J} \cdot \vec{J} = \left(\vec{J}_1 + \vec{J}_2\right) \cdot \left(\vec{J}_1 + \vec{J}_2\right) = \vec{J}_1 \cdot \vec{J}_1 + 2\vec{J}_1 \cdot \vec{J}_2 + \vec{J}_2 \cdot \vec{J}_2$$

Rearranging terms yields

$$\vec{J}_1 \cdot \vec{J}_2 = \frac{(\vec{J} \cdot \vec{J} - \vec{J}_1 \cdot \vec{J}_1 - \vec{J}_2 \cdot \vec{J}_2)}{2} = \frac{\left(\hat{J}^2 - \hat{J}_1^2 - \hat{J}_2^2\right)}{2} \tag{8.65}$$

This equation means that the operator corresponding to the dot product of the two source vectors is the same as a combination of the three operators that give the lengths of the three angular momenta. The coupled states, $|JMj_1j_2\rangle$, are eigenfunctions of these three operators, and so they must be eigenfunctions of the combination in Equation 8.65.

$$(\vec{J}_1 \cdot \vec{J}_2)|JMj_1j_2\rangle = \frac{\left(\hat{J}^2 - \hat{J}_1^2 - \hat{J}_2^2\right)}{2|JMj_1j_2\rangle} = \frac{\left[J(J+1) - j_1(j_1+1) - j_2(j_2+1)\right]\hbar^2}{2|JMj_1j_2\rangle} \tag{8.66}$$

This equation will be used later to evaluate the effect of Hamiltonian terms that take the form of a dot product of two angular momentum vectors. Usually, the states that are appropriate in the absence of a coupling interaction are the states labeled $|j_1 m_1 j_2 m_2\rangle$, which are not eigenfunctions of $\vec{J}_1 \cdot \vec{J}_2$.

8.7 Variation Theory

For the most realistic treatment of chemical problems, the Schrödinger equation does not tend to be separable and it is not usually a differential equation easily solved by analytical means. Or, more to the point, the problems of interest are not simple and are not strictly the same as ideal model problems. Powerful techniques have been formulated for dealing with realistic problems. Variation theory and perturbation theory are two such techniques that have been widely employed in understanding many quantum chemical systems.

The **variational principle** is the basis for the variational determination of a wavefunction. This principle states that the lowest eigenvalue of some given operator is at a minimum with respect to small adjustments that might be made to the associated eigenfunction. As a consequence, the expectation value of the operator using some arbitrary function cannot be less than that minimum. In other words, for an arbitrary wavefunction, the expectation value of the operator is greater than or equal to its lowest eigenvalue. Now, the Schrödinger equation calls for finding eigenfunctions of a very specific operator, the Hamiltonian. Hence, the variational principle helps by stating that the expectation value of the Hamiltonian for any wavefunction we guess, choose, or select is greater than or equal to the true ground state energy, the lowest eigenvalue of the Hamiltonian. The expectation value of a chosen or guessed wavefunction amounts to a guess of a system's energy expectation value, and the variation principle indicates that such a guess will be in error on the high side. This has an important implication: A chosen wavefunction can be improved-made more like the true Hamiltonian's eigenfunction-through any adjustment that lowers the expectation value of the energy. This is because the energy cannot be lowered "too much"; the variational principle requires that the expectation value of the energy will never be less than the lowest eigenvalue.

For the variational principle to hold, the chosen wavefunction must satisfy the same conditions we have considered necessary in interpreting the square of a true wavefunction as a probability density. These are the smoothness and continuity and that the wavefunction be normalizable. The function and its first derivative must be continuous in all spatial variables of the system over the entire range of those variables; the function must be single-valued; and the integral of the square of the function over all space must be a finite value. Wavefunctions obtained from analytical solution of a Schrödinger equation and wavefunctions selected for variational treatment should satisfy these conditions.

The process of adjusting chosen wavefunctions, or trial wavefunctions, to minimize the expectation value of the energy is the **variational method**. A useful first example is to return to a problem for which we have analytical solutions, the one-dimensional harmonic oscillator problem, and to see what application of the variational method yields. The Hamiltonian is that in Equation 7.28. The trial wavefunction, Γ, will be taken to be a Gaussian function with an undetermined parameter, α, that is adjustable.

$$\Gamma = \left(\frac{\alpha^2}{\pi}\right)^{1/4} e^{-(\alpha x)^2/2} \tag{8.67}$$

The constant in front of the exponential makes Γ normalized for any choice of α. Other functional forms could be used for the trial wavefunction, for instance $\alpha/(1 + x^2)$. Of course, the chosen trial function must satisfy the conditions of being smooth, continuous, and bounded. We know from the analytical treatment of the harmonic oscillator that Γ in Equation 8.67 is continuous, has a continuous first derivative, is single-valued, and yields a finite value if squared and integrated from $-\infty$ to ∞.

The expectation value, designated W, is obtained by application of the Hamiltonian to the trial function followed by integration with the complex conjugate of the trial function.

$$W = \langle \Gamma | \hat{H} \Gamma \rangle = \left\langle \Gamma \left| -\frac{\hbar^2}{2m}\frac{d^2}{dx^2} + \frac{1}{2}kx^2 \right| \Gamma \right\rangle = \frac{\hbar^2 \alpha^2}{2m} + \left(\frac{k}{4} - \frac{\hbar^2 \alpha^4}{4m}\right)\frac{1}{\alpha^2} = \frac{\hbar^2 \alpha^2}{4m} + \frac{k}{4\alpha^2} \tag{8.68}$$

W turns out to be a function of the parameter α. Since the objective is to adjust the parameter so as to lower W as much as possible, the process needed is that of minimization of $W(\alpha)$, something that can be accomplished using the first derivative.

$$\frac{dW}{d\alpha} = \frac{\hbar^2 \alpha}{2m} - \frac{k}{2\alpha^3}$$

At any point where the first derivative of a function of one variable is zero, the value of the function itself is either a minimum, a maximum, or an inflection point (see Appendix A). Let the point where the first derivative of $W(\alpha)$ is zero be the specific value designated α_{min}.

$$\left.\frac{dW}{d\alpha}\right|_{\alpha=\alpha_{min}} = 0 \Rightarrow \frac{\hbar^2 \alpha_{min}}{2m} - \frac{k}{2\alpha_{min}^3} = 0$$

The expression for the first-derivative function is set equal to zero when the variable is equal to that of the minimum (or maximum) point. This produces the equation on the right for α_{min}, and solving it yields

$$\alpha_{min}^2 = \frac{\sqrt{km}}{\hbar}$$

In certain cases, it may be necessary to test that $W(\alpha = \alpha_{min})$ is a minimum instead of a maximum or an inflection point, especially if there are many solutions to the equation for α_{min}. Assuming a minimum has been found, α_{min} represents the best possible choice in the adjustment of α so as to improve the trial wavefunction. The expectation value of the energy cannot be made any less with any other choice of α. This result is that $\alpha_{min} = \beta$, where β is the constant used in the harmonic oscillator eigenfunctions (Equation 7.31). In this example, variational treatment is capable of yielding the exact ground state wavefunction without analytical solution of the Schrödinger equation. However, this is only because the form of the trial wavefunction happened to be the functional form of the true wavefunction. Had we used $\alpha/(1 + x^2)$ as a trial wavefunction instead, we could not have obtained $e^{-\beta^2 x^2/2}$, the true ground state wavefunction. We would have obtained the closest approximation to it that had the form $1/(1 + x^2)$.

The expectation value of the energy is obtained by inserting the expression for α_{min} in $W(\alpha)$.

$$W(\alpha_{min}) = \frac{\hbar^2 \alpha_{min}^2}{4m} + \frac{k}{4\alpha_{min}^2} = \frac{\hbar}{4}\frac{\sqrt{km}}{m} + \frac{\hbar}{4}\frac{k}{\sqrt{km}} = \frac{\hbar}{2}\sqrt{\frac{k}{m}}$$

This happens to be the true ground state energy of the harmonic oscillator. The power of the variational approach is evident since the exact wavefunction and eigenenergy were obtained without formally solving the differential Schrödinger equation.

To better understand the variational principle, consider a problem with a known set of normalized Hamiltonian eigenfunctions, $\{\psi_i\}$, with corresponding eigenenergies, $\{E_i\}$. It can be proved that any function in the geometrical space of the problem can be represented as a superposition or linear combination of a set of Hamiltonian eigenfunctions. This is much like the idea of a Fourier expansion where a function can be represented by a sum of sine and/or cosine functions over some interval. In this sense, an arbitrary function is an arbitrary linear combination of the eigenfunctions. That is, anything that can serve as a trial function is the same thing as some linear combination of the ψ_i's. If the coefficients, $\{a_i\}$, in this linear combination are considered arbitrary, the following trial function, Γ, is an arbitrary function:

$$\Gamma = N \sum a_i \psi_i$$

So, Γ stands for any trial wavefunction whatsoever, and N is its normalization constant. The expectation value of Γ with the Hamiltonian is easily simplified because the ψ_i functions are eigenfunctions of the Hamiltonian and therefore are orthogonal.

$$W = \left\langle \Gamma \middle| \hat{H}\Gamma \right\rangle = N^2 \left\langle \sum_i a_i \psi_i \middle| H \sum_j a_j \psi_j \right\rangle = N^2 \sum_i \sum_j a_i^* a_j \left\langle \psi_i \middle| \hat{H}\psi_j \right\rangle$$

Example 8.4: Variational Treatment of a Quartic Oscillator

Problem: Use the functional form of the ground state wavefunction of the harmonic oscillator as a trial wavefunction and apply variation theory to find an approximate wavefunction and energy for the quartic oscillator, an oscillator with $V(x) = kx^4$.

Solution: The trial wavefunction has only one adjustable parameter, α, and it is the function given in Equation 8.67. The Hamiltonian is

$$\hat{H} = -\frac{\hbar^2}{2m}\frac{d^2}{dx^2} + kx^4$$

The first step is to evaluate $W(\alpha)$ as was done for the harmonic oscillator in Equations 8.68. (We can take advantage of the fact that the kinetic energy operator is the same, and so we merely have to replace the integral of the operator $kx^2/2$ with that of kx^4.) From consulting integral tables, we have

$$W = \frac{\hbar^2\alpha^2}{4m} + \frac{3}{4}\frac{k}{\alpha^4}$$

The first derivative of W with respect to α is

$$\frac{dW}{d\alpha} = \frac{\hbar^2}{2m}\alpha - 3\frac{k}{\alpha^5}$$

The variational choice of the parameter α for this problem is one that makes this first-derivative function equal to zero.

$$\frac{\hbar^2}{2m}\alpha_{min} - 3\frac{k}{\alpha_{min}^5} = 0 \Rightarrow \alpha_{min}^6 = \frac{6mk}{\hbar^2}$$

Substituting this value in $W(\alpha)$ yields the variational approximation of the eigenenergy for the given trial function.

$$W(\alpha_{min}) = \frac{3}{4}\sqrt[3]{\frac{3\hbar^4 k}{4m^2}}$$

Because of the orthonormality of the ψ_i functions, the value of the integral $\langle\psi_i|\psi_j\rangle$ is 1 if $i = j$ and 0 if $i \neq j$. The symbol δ_{ij} designates this condition. Because of it, the double summation is reduced to a single summation; only summation terms for which $i = j$ contribute to W.

$$W = N^2 \sum_i \sum_j a_i^* a_j E_j \delta_{ij} = \sum_j \left(N^2 a_j^2\right) E_j \equiv \sum_j a_j'^2 E_j$$

The normalization factor, which is

$$N^2 = \left(\sum_i a_i^2\right)^{-1}$$

has been absorbed into the primed coefficients on the last line. The primed coefficients squared must each be less than or equal to 1 because of the following:

$$\langle\Gamma|\Gamma\rangle = 1 = N^2 \sum_i \sum_j a_i^* a_j \langle\psi_i|\psi_j\rangle = N^2 \sum_i a_i^2 = \sum_i a_i'^2$$

The sum of their squares is exactly 1 since Γ is normalized, and so it is impossible for any one coefficient to be greater than 1.

The expectation value W must be greater than or equal to the lowest eigenenergy from the set $\{E_i\}$ because W is a sum of the eigenenergies, each multiplied by a number (e.g., $a_i'^2$) less than 1. To illustrate this reasoning, consider the eigenenergies of the harmonic oscillator, $(n + 1/2)\hbar\omega$, and a trial wavefunction that is a linear combination of the harmonic oscillator functions, with the normalized expansion coefficient for the nth function being a_n'. The expectation value of the energy is

$$W = \frac{\hbar\omega}{2}\left(a_0'^2 + 3a_1'^2 + 5a_2'^2 + 7a_3'^2 + \ldots\right)$$

and the condition on the coefficients is

$$a_0'^2 + a_1'^2 + a_2'^2 + \ldots = 1$$

By inspection, the smallest value for W is $\hbar\omega/2$, which is what we know to be the ground state energy. This is obtained for the value of W only under the condition that $a_0'^2 = 1$ and the other coefficients are zero. Relative to this choice, any change in the coefficients consistent with the normalization constraint makes W a larger number. For instance, with $a_0'^2 = 0.9$, $a_1'^2 = 0.1$, and the other coefficients zero, $W = 3.2\hbar\omega$. This is the variational principle in action; the expectation value of the energy for an arbitrary function cannot be less than the lowest eigenvalue.

8.8 Perturbation Theory

Perturbation theory offers another method for finding quantum mechanical wavefunctions. It is especially suited to problems that are similar to model or ideal situations differing only in some small way. For example, the potential for an oscillator might be harmonic except for a feature such as the small "bump" depicted in Figure 8.8. Because the bump is a small feature, one expects the system's behavior to be quite similar to that of a harmonic oscillator. Perturbation theory affords a way to correct the description of the system, obtained from treating it as a harmonic oscillator, so as to account for the effects of the bump in the potential. In principle, perturbation theory can yield exact wavefunctions and eigenenergies, but usually it is employed as an approximate approach.

In perturbation theory, the Hamiltonian for any problem is partitioned into two or more pieces. The first piece, $\widehat{H}^{(0)}$, is one for which the eigenfunctions and eigenenergies are known. The other pieces of the Hamiltonian represent or constitute the perturbation. This first piece and the associated eigenfunctions and eigenenergies are distinguished in notation by a zero superscript. At the outset of treating a problem using perturbation theory, solutions must already exist for the following Schrödinger equation:

$$\widehat{H}^{(0)}\psi_i^{(0)} = E_i^{(0)}\psi_i^{(0)} \tag{8.69}$$

This is the zero-order perturbation theory Schrödinger equation.

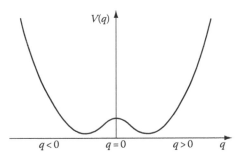

FIGURE 8.8
Harmonic oscillator potential, a parabola, with a small perturbing potential, a bump in the middle. The composite potential is called a **double-minimum potential** or a **double-well potential**. Perturbation theory offers an excellent approach to understanding how the wavefunctions of the unperturbed (harmonic) system differ from the system with a bump in the potential. An example of this type of potential is found for the water molecule. Water is a bent molecule, which means that the minimum of its potential energy as a function of the atomic positions is at a structure that is bent. We can envision a water molecule being straightened into a linear arrangement of the atoms. Of course, that is necessarily a higher-energy arrangement. If the distortion process is continued, the molecule will be bent again but in the opposite sense. Eventually it will look like a mirror image of the original equilibrium structure, and so it will then be at a potential minimum. If the bending angle is taken to be a coordinate, then the potential energy as a function of that angle will have the form of a double-well potential. In the potential curve mentioned earlier, q could be the bending angle. Then, $q = 0$ would correspond to the linear arrangement of the atoms in the water molecule, and to the left and to the right of $q = 0$ would be the two potential minima.

A crucial idea in perturbation theory is the introduction of a parameter, designated here as λ, that smoothly takes the system from that described by the zero-order Hamiltonian to that described by the true Hamiltonian as this parameter is varied from 0 to 1. This means the complete Hamiltonian for the problem is written

$$\hat{H} = \hat{H}^{(0)} + \lambda \hat{H}^{(1)} \tag{8.70}$$

$\hat{H}^{(1)}$ is called the **perturbing Hamiltonian** or the **first-order Hamiltonian**. The role of λ is somewhat abstract. There may be no physical way of smoothly changing a system from some unperturbed model system to the system of interest; however, the parameter λ is a device that accomplishes this in a mathematical sense. When λ is set to zero, Equation 8.70 gives the Hamiltonian for the zero-order or unperturbed situation, whereas when λ is set to 1.0, the Hamiltonian of interest results. With this embedded parameter, the Schrödinger equation with $\hat{H}$ is really a family of equations covering all the choices of λ values. Perturbation theory is a means of dealing with the entire family of equations, even though in the end, one may be interested only in the case where $\lambda = 1$.

Equation 8.70 can be regarded as an expansion of the Hamiltonian operator in a power series in λ. In fact, a general development of perturbation theory would allow for the possibility that, in some way, one might be able to choose a parameter that would give rise to pieces of the Hamiltonian that depend on λ quadratically, or even to higher powers.

$$\hat{H} = \hat{H}^{(0)} + \sum_{n=1}^{\infty} \lambda^n \hat{H}^{(n)} \tag{8.71}$$

$\hat{H}^{(n)}$ is the nth-order Hamiltonian. Since the pieces of the Hamiltonian have factors of 1 to various powers, and since we seek a solution to a family of Schrödinger equations (all choices of λ), we can expect the wavefunction and the eigenenergy to have some dependence on λ as well. The most general way to express such dependence is to allow each to have terms that contain the different powers of λ. Hence, writing the Hamiltonian as a power series implies writing the entire Schrödinger equation as an expansion in λ.

$$\left(\sum_{i=0}\lambda^i\hat{H}^{(i)}\right)\left(\sum_{j=0}\lambda^j\psi_n^{(j)}\right)=\left(\sum_{k=0}\lambda^k E_n^{(k)}\right)\left(\sum_{m=0}\lambda^m\psi_n^{(m)}\right) \qquad (8.72)$$

The subscripts on the wavefunctions and on the energies identify a particular state of interest, the nth state, as in Equation 8.69.

When a particular order of Hamiltonian in Equation 8.72 is applied to a particular order of wavefunction, the result, $\hat{H}^{(i)}\psi_n^{(j)}$, occurs in the Schrödinger equation with a factor of $\lambda^{(i+j)}$. On the right-hand side of Equation 8.72, energies multiply wavefunctions, and these have factors of $\lambda^{(k+m)}$ since k and m are the powers of the individual λ factors on the right-hand side. Terms are found on the left-hand side and on the right-hand side that enter the equation with a factor of λ^0, λ^1, λ^2, and so on. A crucial idea of perturbation theory is that satisfying Equation 8.72 for the family of Schrödinger equations represented by the infinite number of choices of the value of λ implies a mathematical constraint. It is that the terms for any particular power of λ on the left-hand side of Equation 8.72 must equal those with the same particular power of λ on the right-hand side. Notice that the difference between the case of $\lambda = 1$ and of $\lambda = 1/2$ is that a term linear in λ would be diminished by one-half for the latter case, but a quadratic term would be diminished by one-fourth. The various terms enter the Schrödinger equation with a different weighting for every different choice of λ. The only way this can work is for all the terms on the left that depend on λ^1 to equal all those on the right that depend on λ^1, and for all those on the left that depend on λ^2 to equal those on the right with a λ^2 factor, and so on. Following is a list of the first few of these equations. They are found by carrying out the multiplications in Equation 8.72 and collecting terms according to the power of the λ factor.

$$\lambda^0 \text{ terms:} \hat{H}^{(0)}\psi_n^{(0)} = E_n^{(0)}\psi_n^{(0)}$$

$$\lambda^1 \text{ terms:} \hat{H}^{(1)}\psi_n^{(0)} + \hat{H}^{(0)}\psi_n^{(1)} = E_n^{(1)}\psi_n^{(0)} + E_n^{(0)}\psi_n^{(1)}$$

$$\lambda^2 \text{ terms:} \hat{H}^{(2)}\psi_n^{(0)} + \hat{H}^{(1)}\psi_n^{(1)} + \hat{H}^{(0)}\psi_n^{(2)} = E_n^{(2)}\psi_n^{(0)} + E_n^{(1)}\psi_n^{(1)} + E_n^{(1)}\psi_n^{(2)}$$

Instead of one equation, the original Schrödinger equation, there are now an infinite number of equations because the λ power series expansion represents an infinite family of Schrödinger equations.

The equation of λ^0 terms represents the unperturbed or model system, and it is the same as Equation 8.69. Again, its solutions have to be known to apply perturbation theory. Inspection of the λ^1 equation shows that it involves zero-order elements in addition to the first-order elements. The unknowns in this equation are the first-order correction to the wavefunction, $\psi_n^{(1)}$, and the first-order correction to the energy, $E_n^{(1)}$. In the λ^2 equation, zero-order and first-order elements enter in addition to the second-order corrections.

Because of this pattern, the process for solving these equations must proceed from the zero-order equation to the first-order equation to the second-order equation, and so on. In this way, at any order, all the required lower-order corrections will already be known.

Though more and more terms are involved with increasing order of perturbation, N, there are common steps in working with the equations. At each and every order, the energy correction is obtained by multiplying that N-order equation by $\psi_n^{(0)*}$ and integrating over all space. There is a simplification that removes the two terms with $\psi_n^{(N)}$. Notice that on the left-hand side this term is $\hat{H}^{(0)}\psi_n^{(N)}$, while on the right-hand side it is $E_n^{(0)}\psi_n^{(N)}$. These can be grouped together on the left-hand side as $(\hat{H}^{(0)} - E_n^{(0)})\psi_n^{(N)}$.

When this is integrated with the complex conjugate of the zero-order wavefunction, the result is zero.

$$\left\langle \psi_n^{(0)} \left| \left(\hat{H}^{(0)} - E_n^{(0)} \right) \psi_n^{(N)} \right\rangle = \left\langle \left(\hat{H}^{(0)} - E_n^{(0)} \right) \psi_n^{(0)} \left| \psi_n^{(N)} \right\rangle = \left\langle E_n^{(0)}\psi_n^{(0)} - E_n^{(0)}\psi_n^{(0)} \left| \psi_n^{(N)} \right\rangle = 0 \right.\right.\right.$$

The Hamiltonian operator is Hermitian, the operation of multiplication by a constant is Hermitian, and that made it possible to exchange the order in which the operator was applied. With this simplification, the first-order equation yields an expression for the first-order correction to the energy.

$$\left\langle \psi_n^{(0)} \left| \hat{H}^{(1)}\psi_n^{(0)} \right\rangle = \left\langle \psi_n^{(0)} \left| E_n^{(1)}\psi_n^{(0)} \right\rangle = E_n^{(1)} \left\langle \psi_n^{(0)} \left| \psi_n^{(0)} \right\rangle = E_n^{(1)} \right.\right.\right. \tag{8.73}$$

This reveals that the first-order correction to the energy is merely the expectation value of the first-order perturbing Hamiltonian with the zero-order wavefunction. The second-order equation yields an expression for the second-order correction to the energy.

$$\left\langle \psi_n^{(0)} \left| \hat{H}^{(2)}\psi_n^{(0)} \right\rangle + \left\langle \psi_n^{(0)} \left| \hat{H}^{(1)}\psi_n^{(1)} \right\rangle = \left\langle \psi_n^{(0)} \left| E_n^{(2)}\psi_n^{(0)} \right\rangle + \left\langle \psi_n^{(1)} \left| E_n^{(1)}\psi_n^{(1)} \right\rangle \right.\right.\right.\right.$$

$$E_n^{(2)} = \left\langle \psi_n^{(0)} \left| \hat{H}^{(2)}\psi_n^{(0)} \right\rangle + \left\langle \psi_n^{(0)} \left| \left(\hat{H}^{(1)} - E_n^{(1)} \right) \psi_n^{(1)} \right\rangle \right.\right. \tag{8.74}$$

Using this equation, though, requires knowing the first-order correction to the wavefunction.

To find the corrections to the wavefunction, an idea from the previous section is used, namely, that an arbitrary function in some quantum mechanical space can be expressed as a linear combination of eigenfunctions of any Hamiltonian for that space. In perturbation theory, the zero-order equation is presumed to have been solved, and so the prerequisite, a set of eigenfunctions, does exist. Thus, the unknown functions of perturbation theory, the first- or second-order corrections to the wavefunction, and so on, can each be expressed as a linear combination of the zero-order functions. The coefficients in that linear expansion then constitute the unknown information. Let us apply this to the first-order equation, letting c_i be the expansion coefficient for the ith zero-order function.

$$\psi_n^{(1)} = \sum_i c_i \psi_i^{(0)} \tag{8.75}$$

If we group Hamiltonians and energies as was done previously, Equation 8.75 can be used with the first-order Schrödinger equation to give

$$\left(\hat{H}^{(1)} - E_n^{(1)}\right)\psi_n^{(0)} = -\left(\hat{H}^{(0)} - E_n^{(0)}\right)\psi_n^{(1)} = -\left(\hat{H}^{(0)} - E_n^{(0)}\right)\sum_i c_i\psi_i^{(0)} = -\sum_i c_i\left(E_i^{(0)} - E_n^{(0)}\right)\psi_i^{(0)} \qquad (8.76)$$

This result can be used to extract one of the desired coefficients by multiplying by the complex conjugate of one of the zero-order wavefunctions and integrating over all space. Let that one function be $\psi_j^{(0)}$, with $j \neq n$ for now.

$$\left\langle\psi_j^{(0)}\left|\left(\hat{H}^{(1)} - E_n^{(1)}\right)\psi_n^{(0)}\right.\right\rangle = -\sum_i c_i\left(E_i^{(0)} - E_n^{(0)}\right)\left\langle\psi_j^{(0)}\middle|\psi_i^{(0)}\right\rangle$$

$$\left\langle\psi_j^{(0)}\middle|\hat{H}^{(1)}\psi_n^{(0)}\right\rangle - E_n^{(1)}\left\langle\psi_j^{(0)}\middle|\psi_n^{(0)}\right\rangle = -\sum_i c_i\left(E_i^{(0)} - E_n^{(0)}\right)\delta_{ij}$$

$$\left\langle\psi_j^{(0)}\middle|\hat{H}^{(1)}\psi_n^{(0)}\right\rangle = -c_j\left(E_j^{(0)} - E_n^{(0)}\right)$$

$$\therefore c_j = \frac{\left\langle\psi_j^{(0)}\middle|\hat{H}^{(1)}\psi_n^{(0)}\right\rangle}{E_n^{(0)} - E_j^{(0)}}$$

(8.77)

By letting j take on all values except n, we use this equation to find all the coefficients except c_n. (We should not use Equation 8.77 with $j = n$ because the denominator is zero.)

The result presented in Equation 8.77 can be discussed in several ways. We see that a particular c_j will be zero if the integral quantity involving the zero-order wavefunctions for the nth and jth states with the perturbing Hamiltonian is zero. This integral value is what we associate with a coupling between zero-order states. If there is no *coupling* or *interaction* between two states brought about by a perturbation, there will be no mixing of their wavefunctions to first order. Also, if the difference in zero-order energies is large, the extent of mixing will be small because of the denominator in Equation 8.77.

The procedure that has been developed for first-order corrections to the wavefunction can be used at second and higher orders as well. The resulting expressions will be more complicated. There is an unanswered question about first-order corrections at this point: What is the value of the coefficient c_n? The answer comes from another procedure, and it too can be used for all orders. This procedure develops from the normalization condition, and specifically from requiring that the perturbed wavefunction be normalized for any choice of the perturbation parameter, λ.

$$1 = \left\langle\psi_n\middle|\psi_n\right\rangle = \left\langle\psi_n^{(0)} + \lambda\psi_n^{(1)} + \cdots\middle|\psi_n^{(0)} + \lambda\psi_n^{(1)+\cdots}\right\rangle = \left\langle\psi_n^{(0)}\middle|\psi_n^{(0)}\right\rangle + \lambda\left(\left\langle\psi_n^{(1)}\middle|\psi_n^{(0)}\right\rangle + \left\langle\psi_n^{(0)}\middle|\psi_n^{(1)}\right\rangle\right) + \cdots$$

That is, from the power series expansion of the wavefunction used in Equation 8.72, the normalization condition is written as a sum of terms that depend on different orders of λ. The first right-hand-side term is, of course, equal to 1, and so all the remaining terms must add up to zero. Such a result will be true for any choice of λ only if each term is independently zero. The term containing λ to the first power consists of an integral with its complex conjugate. If that integral is zero, so is its complex conjugate. Therefore,

$$0 = \left\langle\psi_n^{(0)}\middle|\psi_n^{(1)}\right\rangle = \left\langle\psi_n^{(0)}\middle|\sum_i c_i\psi_i^{(0)}\right\rangle = \sum_i c_i\left\langle\psi_n^{(0)}\middle|\psi_i^{(0)}\right\rangle = \sum_i c_i\delta_{ni} = c_n$$

This establishes that c_n is zero. For higher order corrections to the wavefunction, the nth coefficient is not necessarily zero; this is a special result at first order.

The perturbation derivations can be carried out to any desired order, but the idea of perturbation theory is that the perturbation, whatever it is, makes a small correction to the zero-order picture. In the typical application, the lowest few orders provide the bulk of the correction needed to make the energies and wavefunctions correct. Of course, *nearly* all is not all; to stop at some low order of perturbation theory (i.e., to truncate the expansion at some order of λ) is to make an approximation. The quality of the approximation that we make, or the order of perturbation theory at which we truncate the process, is our choice when we use this approach.

An illustration of the efficacy of low-order perturbation theory, in fact, of the first order energy corrections, is given in Figure 8.9. The model problem is that of a slightly altered harmonic vibrational potential, the type seen in Figure 8.8. The result of an extensive variational treatment that closely approaches the exact energies and wavefunctions for the lowest several states is shown along with the energy levels obtained from first-order perturbation theory, taking the bump in the potential to be the perturbation of the otherwise harmonic system. The energies the variational and perturbational treatments can be compared with the energies of the unperturbed (harmonic) oscillator. From that, we see that the first-order corrections are similar to the energy changes obtained from a variational treatment. Also, the correspondence between the variational energy and the first-order perturbation theory energy improves as one goes to higher energy levels. This is because for the higher energy states, the relative effect of the bump in the potential is diminished, and so perturbation theory at low order is even more appropriate.

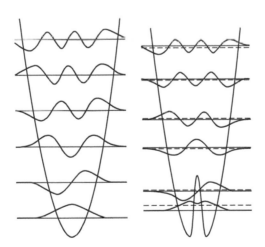

FIGURE 8.9

On the left is a harmonic vibrational potential with the exact energy levels and corresponding wavefunctions shown for the first several states. On the right is the same potential augmented, or perturbed, by a bump in the middle. The solid horizontal lines drawn with this potential are the energy levels obtained from an extensive variational treatment of this problem, whereas the dashed lines are the energy levels obtained from first-order perturbation theory using the system on the left as the zero-order picture and the bump as the perturbation. The wavefunctions on the right for the perturbed system are those obtained from a variational treatment. They demonstrate the detailed changes in the wavefunctions due to the perturbing potential, the most noticeable changes being in the lowest states.

8.9 Conclusions

This chapter has covered the foundations of quantum mechanics that are particularly important for atoms, molecules, and chemistry.

The Schrödinger equation is a prescription; its solutions are the wavefunctions that describe a quantum mechanical system. The square of a normalized wavefunction (the wavefunction multiplied by its complex conjugate) is a probability density function. Wavefunctions must be smooth, continuous functions with a probability density function that is everywhere finite. Wavefunctions for bound states must be normalizable. The probability density may turn out to be nonzero in regions where the energy of the system is less than the potential, an occurrence called tunneling.

Expectation values or mean values correspond to the average of an exhaustive number of measurements. They are obtained from the integral of the associated quantum mechanical operator sandwiched between the complex conjugate of a wavefunction and the wavefunction itself.

Operators that correspond to physical observables are Hermitian operators. Eigenvalues and expectation values of Hermitian operators are real numbers. Eigenfunctions of a Hermitian operator are orthogonal functions. Some pairs of operators commute, and some do not. When a complete set of functions are simultaneously eigenfunctions of a set of operators, then every pair of operators in that set commutes.

Separability in the Schrödinger equation means the Hamiltonian consists of additive, independent terms. Separate Schrödinger equations can then be set up using the independent terms as individual Hamiltonians. The solution of the original Schrödinger equation is a product of the eigenfunctions from the separated Schrödinger equations; the eigenenergy is a sum of the individual eigenenergies.

Degeneracy is the condition in which two or more different solutions of a Schrödinger equation have the same eigenenergy.

This chapter presented two important model systems, the particle in a box and the rigid rotator. The rigid rotator analysis revealed an important feature of angular momentum in quantum mechanical systems. Its magnitude is quantized, not continuous. Commutation relations show that we can know the magnitude of the angular momentum vector and one component exactly (from eigenvalues). Usually, two quantum numbers are used to hold this information, a total angular momentum quantum number (l or J or other designations) and a z-component angular momentum quantum number, m.

Because of the quantization of angular momentum, simple rules exist for the vector sum of two angular momenta. If one source has the quantum number J_1 and the other source J_2, the possible values of J for the composite system range in steps of 1 from $J_1 + J_2$ to $|J_1 - J_2|$.

Two important techniques for finding wavefunctions, or for finding approximations to wavefunctions, were introduced. The use of variation theory involves selecting a trial function with embedded parameters. Adjustment of the parameter values so as to minimize the expectation value of the energy (expectation value with the Hamiltonian operator) brings the function as close to the true wavefunction as possible. The second approach, perturbation theory, is ideally suited to problems that are very much like model problems with known solutions, the difference being in the feature identified as the perturbation. Perturbation theory provides an order-by-order set of corrections to the wavefunctions and eigenenergies of the known, model system. Typically, the higher the order of correction, the better the approximation. The first-order correction to the energy is the expectation value of the perturbing term of the Hamiltonian using the zero-order wavefunction.

Point of Interest: The Quantum Revolution

Photograph of the first conference in 1911 at the Hotel Metropole. *Seated* (L–R): W. Nernst, M. Brillouin, E. Solvay, H. Lorentz, E. Warburg, J. Perrin, W. Wein, M. Curie, and H. Pioncaré. *Standing* (L–R): R. Goldschmidt, M. Planck, H. Rubens, A. Sommerfield, F. Lindemann, M. de Broglie, M. Knudsen, F. Hasenöhrl, G. Hostelet, E. Herzen, J.H. Jeans, E. Rutherford, H. Kamerlingh Onnes, A. Einstein, and P. Langevin.

The Solvay Conference

The development of quantum theory in the early part of the twentieth century was a major scientific revolution. It brought new ideas to the heart of physics, but it was also a revolution in chemistry. Quantum mechanics proved to be the crucial understanding that sparked the evolution of atomic and molecular spectroscopy.

The notion of quantization of energies probably had several origins. Max Karl Ernst Ludwig Planck is often credited with the first expression of the quantum hypothesis. Planck was born in Kiel, Germany, on April 23, 1858. He studied at the University of Munich where he received his PhD in 1879, and his dissertation concerned the second law of thermodynamics. He was a professor of physics at the University of Berlin from 1888 to 1926. During that time, his interests in the thermodynamics of irreversible processes drew him into the problem of "blackbody radiation," which refers to the spectrum of electromagnetic radiation given off by an object that has been heated. Planck published papers in 1900 and 1901 that quantitatively accounted for blackbody spectra through a quantization assumption; however, he held to views of the *continuous* nature of matter and energy. It was Einstein's and Ehrenfest's rederivations of Planck's spectral distribution law years later that established that Planck's "resonators" (oscillators) could take on only certain energies (i.e., the quantization hypothesis). These energies were in multiples of the oscillator frequency, n, times Planck's "quantity of action," h (Planck's constant). Planck received the Nobel Prize in physics in 1918. He died in Göttingen, Germany, on October 4, 1947.

A 1913 paper written by Niels Henrik David Bohr, 2 years after his PhD dissertation, connected the discontinuous spectrum of the hydrogen atom with an energy level (quantum) hypothesis. Bohr was born in Copenhagen, Denmark, on October 7, 1885, and he died there on November 18, 1962. He spent most of his career as professor and director of the Institute for Theoretical Physics at the University of Copenhagen. Though the specific model of atomic systems he devised ultimately proved incorrect, he presented two key postulates in his 1913 papers on the model. The first was that there exist stationary energy states of atoms. The second was that a change from one stationary state to another is accomplished by the absorption or emission of one quantum of radiation, and the energy difference between the two states is equal to Planck's constant times the frequency of the radiation. With these founding postulates, Bohr not only accounted for the spectral lines that had been observed for hydrogen but also correctly determined that a certain set of lines assigned to hydrogen were in fact from helium. He was the 1922 recipient of the Nobel Prize in physics.

The next revolutionary step in quantum theory was de Broglie's suggestion that matter and radiation have dual characteristics of waves and corpuscles (particles). Louis Victor de Broglie was born in Dieppe, France, on August 15, 1892, the youngest of five children. He graduated from the Sorbonne in 1909 with degrees in mathematics and philosophy. Military service during World War I interrupted his graduate studies. About 1919, he began work in the x-ray laboratory that had been set up in the family mansion by his brother, Maurice. Their study of photoelectrons (electrons released from atoms because of incident x-rays) led him to consider whether particles exhibit wavelike properties. His 1924 doctoral dissertation at the Sorbonne postulated that wave mechanics could explain the behavior of both light and matter. This helped elucidate the Bohr model since the de Broglie wavelength of the circular-moving electrons in the Bohr model matched the path length of the electron's orbit. In 1928, he was appointed professor of physics at the University of Paris. He received the Nobel Prize in physics in 1929 for his work on wave mechanics. He died on March 19, 1987.

In 1925, Erwin Schrödinger, stimulated by de Broglie's ideas, found a way to obtain the hydrogen atom energies as eigenenergies in a differential equation. This overcame the need for unexplainable assumptions made in the Bohr model. In his second paper on the subject, published in 1926, he tied his approach to methods in mechanics developed a century earlier by W. R. Hamilton. Schrödinger, born in Vienna, Austria, on August 12, 1887, received his doctorate in physics from the University of Vienna in 1910. He served as an artillery officer during World War I and shortly thereafter was appointed to the chair formerly held by Einstein at the University of Zurich. His initial research was in statistical thermodynamics, reaction kinetics, and relativity. In 1927, he succeeded Max Planck in the theoretical physics chair at Berlin. In 1933, the year he shared the Nobel Prize in physics with P. A. M. Dirac, Schrödinger moved to Oxford because he was outraged with the new government in Germany. He returned to the University of Vienna in 1956, and died in Austria on January 4, 1961.

Paul Adrien Maurice Dirac was born in Bristol, England, on August 8, 1902. He studied electrical engineering at Bristol University from 1918 to 1921, but failing to find a job, he spent 2 more years at Bristol studying mathematics. Most of his academic career was spent at Cambridge University, where he started in 1923 as a graduate student with a strong interest in relativity and later held the position of Lucasian professor of mathematics from 1932 until his retirement in 1969. In 1971, he accepted an appointment at Florida State University. He died in Miami, Florida, on October 20, 1984. Dirac's major contributions to the quantum revolution began after Schrödinger developed an eigenequation for wavefunctions. Dirac started with a relativistic analogy and determined how explicit time dependence could be incorporated. This was the basis for a quantum formulation of absorption and emission of radiation. He also recognized that the wavefunctions for identical particles had to be either symmetric or antisymmetric, the latter condition leading to the Pauli exclusion principle (Chapter 10). Though Schrödinger had initially developed a relativistic wave equation, it was Dirac's relativistic formulation that correctly accounted for electron spin. He shared the Nobel Prize in 1933, and continued study through a long career in which he initiated the field of quantum electrodynamics, predicted the existence of antimatter and studied cosmology.

The 1932 Nobel Prize in physics was awarded to Werner Karl Heisenberg for his role in the quantum revolution. Heisenberg was born in Würzburg, Germany, on December 5, 1901. He started at the University of Munich in 1920, and in 1923 received his doctorate under the direction of Arnold Sommerfeld. He visited Bohr in 1924, and thereupon began to study the intensities of lines in the hydrogen spectrum; however, he turned instead to the problem of an anharmonic oscillator. It was then that he devised a matrix algebra approach to quantum problems called matrix mechanics. Schrödinger's wave mechanics appeared shortly thereafter, and it received greater acceptance when in March 1926, he established the equivalence of wave and matrix mechanics.

By 1927, Heisenberg, Bohr, and Max Born had developed an interpretation of the square of the wavefunction as a probability density. Schrödinger never accepted this interpretation, preferring to regard the wavefunction as a vibrational amplitude. Nonetheless, what was called the "new quantum theory" was now in place.

The quantum revolution can be characterized, in part, as a rebellion of young scientists against the established order of physics. Harvard professor Edwin C. Kemble noted in his classic 1937 text *(The Fundamental Principles of Quantum Mechanics with Elementary Applications, Dover [reprinted], New York, 1958)* a number of scientists' ages at the time of their first key contribution to the development of quantum mechanics: Bohr was 28 years old, Einstein 26, Heisenberg 24, Dirac 23, de Broglie 31, and W. Pauli 25. This was truly a remarkable, exciting period in physical science, and as later chapters will show, it was the origin of our modern, detailed knowledge of molecules.

Exercises

8.1 Evaluate the probability density of the $n = 0, 2, 4,$ and 6 (normalized) states of the harmonic oscillator at the point $x = 0$. Is there an apparent trend in these values?

8.2 At what value or values of x is the probability density function of the $n = 1$ wavefunction of the harmonic oscillator at maximum value?

8.3 Find the number of degenerate states for the lowest five energy levels of a two-dimensional oscillator of the type shown in Figure 8.2 given that $k_x = 4k_y$.

8.4 Make a sketch of the energy levels of the one-dimensional particle in a box by drawing a straight horizontal line for each level, placing the lines against a vertical energy scale. Assume the length of the box is l. To the left, sketch the same energy levels for the particle in a box if the length is $l/2$, and to the right sketch the same energy levels if the length of the box is $2l$.

8.5 Find the degeneracy of the first four energy levels of a particle in a three-dimensional box with the lengths of the sides related as $l_x = 16l_y = 4l_z$.

8.6 What are the values of the total angular momentum quantum number J for a problem with three coupled sources of angular momentum, $j_1 = 2$, $j_2 = 1$, and $j_3 = 1$?

8.7 From the definition of operator multiplication and the definition of a commutator, find a single operator that is the same as the following operators:

a. $[\widehat{B}, \widehat{A}^2]$

b. $[\widehat{B}^2, \widehat{A}]$

c. $[\widehat{B}^2, \widehat{A}^2]$

d. $[\widehat{A}^2, \widehat{B}^2]$

where $\widehat{A} = x$ and $\widehat{B} = d/dx$.

8.8 Evaluate $[e^{-3x}, d^2/dx^2]$ and $[e^{-x}, d^2/dx^2]$.

8.9 Show that the $n = 0$, 1, and 2 wavefunctions of the harmonic oscillator are orthogonal to each other.

8.10 Determine from commutation relations whether the wavefunctions of the harmonic oscillator can also be eigenfunctions of the operators $\widehat{x}$ and $\widehat{p}_x$.

8.11 What is the approximate uncertainty in the position of an election if the velocity can be measured with a standard deviation of 2.5×10^{-9} m s^{-1}?

8.12 Find the expectation value of the square of the momentum for the $n = 0$ and $n = 1$ states of the harmonic oscillator.

8.13 Find Δp for the $n = 0$ and $n = 1$ states of the harmonic oscillator.

8.14 Find the expectation value of x^2 for the $n = 0$, 1, 2, and 3 harmonic oscillator wavefunctions. What is the dependence of this quantity on the quantum number n?

8.15 Given that the expectation value $\langle x \rangle$ is zero for all harmonic oscillator states, use the result for Exercise 8.13 to develop an expression that gives Δx in terms of the n quantum number (i.e., that gives Δx as a function of n).

8.16 Find the degeneracy of the lowest four energy levels of an isotropic three-dimensional harmonic oscillator ($k_x = k_y = k_z$).

8.17 For a certain problem it has been determined that the energy levels depend on two quantum numbers, n and m, according to the expression

$$E_{n,m} = -\frac{\hbar a}{n^2} + \frac{\hbar a m^2}{4}$$

where a is a constant. If n can have only the value 1, 2, or 3, and if m can be zero or any integer from $-(n+1)$ to $n+1$, find the degeneracies of all the energy levels.

8.18 Find the expectation value, $\langle x \rangle$, and the uncertainty, Δx, for the lowest three states of the one-dimensional particle in a box.

8.19 Calculate the frequency of light emitted when an "electron in a box" of length 7.5 Å drops from the $n = 3$ to the $n = 1$ quantum state.

8.20 Express the one-dimensional particle-in-a-box wavefunctions in Equation 8.24 as a superposition of complex exponential functions. The general form was given in Equation 8.20b. With this equivalent form, find the expectation value, $\langle p \rangle$, and the uncertainty, Δp, for the lowest three states of the one-dimensional particle in a box.

8.21 An electron is trapped in a one-dimensional box. It is found that longest transition wavelength is 569.0 nm. Calculate the length of the box.

8.22 Based on the known wavefunctions for the harmonic oscillator, the wavefunctions for the particle in a box, and the continuity and smoothness requirements for all wavefunctions, construct a sketch showing the qualitative features of the three lowest energy wavefunctions for a particle trapped in a potential, $V(x)$, that is infinite at $x = 0$ and that for $x > 0$ is $kx^2/2$, where k is a constant.

8.23 If the particle of the one-dimensional particle-in-a-box problem were an electron, what would the length of the box, l, have to be in order for the transition from the lowest state to the first excited state correspond to an energy difference of 10,000 cm^{-1}? What would the value of n for H–(C≡C)$_n$–H have to be for a linear polyene to have roughly this length?

8.24 For a particle in a three-dimensional box with equal length sides in the x-, y-, and z-directions, find the degeneracy for each of the first five energy levels.

8.25 Verify the derivation of the expressions in Equations 8.48 through 8.50.

8.26 Show that $\hat{L}^2$ commutes with $\hat{L}_x$ and $\hat{L}_y$ by using the explicit spherical polar coordinate forms.

8.27 Show that for any integer m, the function $\sin^m \theta \cos \theta e^{im\varphi}$ is an eigenfunction of $\hat{L}^2$. Find the eigenvalue.

8.28 Evaluate $\left\langle Y_{20} \middle| \hat{L}_x Y_{20} \right\rangle$ and $\left\langle Y_{20} \middle| \hat{L}_x^2 Y_{20} \right\rangle$.

8.29 Verify Equation 8.60 for $l = 2$, $m = 1$ and $l' = 1$, $m' = 1$.

8.30 If it were possible to have an angular momentum source with an angular momentum quantum number of 1/2, what would be the values of the total angular momentum quantum number J for a system of two such sources coupled together? And of three and of four?

8.31 Apply the variational method to the harmonic oscillator problem using the following as the trial wavefunction:

$$\Gamma(x) = Nxe^{-\alpha x^2}$$

where
   α is the adjustable parameter
   N is the normalization constant

N has a dependence on the value of α that must be found in order to work out this problem. How much higher is this energy than that of the true ground state? Explain this result.

8.32 Apply the variational method to the harmonic oscillator problem using c_1 and c_2 as the adjustable parameters in the trial wavefunction

$$\varphi = \left[c_1 \beta x + c_2 (\beta^2 x^2 - 1/2) \right] e^{-\beta^2 x^2/2}$$

Compare with the exact wavefunctions and account for the result.

8.33 Apply the variational method to the quartic oscillator problem in Example 8.4 using the trial wavefunction in Exercise 8.28.

8.34 Find the first-order perturbation theory correction to the energies of the ground and first two excited states of the harmonic oscillator in terms of the constant g in the perturbing Hamiltonian

$$\hat{H}^{(1)} = gx^4$$

8.35 Find the first- and second-order perturbation theory corrections to the energy of the ground state of the harmonic oscillator and the first-order correction to the ground state wavefunction in terms of the constant g for the following perturbing Hamiltonians:

$$\hat{H}^{(1)} = gx^3$$

8.36 Derive a general expression for the third-order perturbation theory correction to the energy of a quantum state.

8.37 What is the second-order normalization condition in perturbation theory?

8.38 The particle-in-a-sloped-box problem is the usual particle-in-a-box problem but with the bottom of the box sloped because of a linear potential of the form $V(x) = ax$ inside the box. Treat this as a perturbation on the usual particle-in-a-box problem and find the energies and wavefunctions correct to first order for the lowest three states. Then find the second-order energy corrections.

8.39 If the lowest energy state of the particle-in-a-box problem shown in Figure 8.10 were at an energy slightly greater than V_0, in what ways would the wavefunctions of the lowest and first excited states differ from the wavefunctions of the same system but in the absence of the barrier step potential? Use a sketch to show qualitative features.

8.40 Find the first-order energy corrections for the first four states of the particle-in-a-box problem if there is a perturbing potential of the form $V'(x) = 0.1\sqrt{2/l}\,\sin(\pi x/l)$.

8.41 What is the probability that a particle in the ground state of a quantum mechanical harmonic oscillator is outside the classical limits?

8.42 Check the spherical harmonics Y_{00} and Y_{10} for normality and orthogonality.

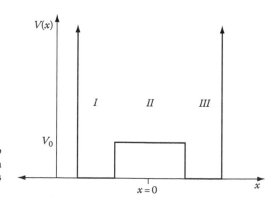

FIGURE 8.10
A particle-in-a-box problem, but with a barrier step potential. The potential is taken to be infinite, except in regions *I*, *II*, and *III*. In regions *I* and *III*, the potential is zero. In region *II*, the potential is the constant value V_0.

Bibliography

Mathematical Background
Mortimer, R. G., *Mathematics for Physical Chemistry*, 2nd edn. (Academic Press, San Diego, CA, 1999).

Introductory Level
Karplus, M. and R. N. Porter, *Atoms and Molecules* (Benjamin, Menlo Park, CA, 1971).
Levine, I. N., *Quantum Chemistry*, 6th edn. (Prentice-Hall, Upper Saddle River, NJ, 2008).
McQuarrie, D. A. and J. D. Simon, *Physical Chemistry: A Molecular Approach* (University Science Books, Sausalito, CA, 1997).
Ratner, M. A. and G. C. Schatz, *Introduction to Quantum Mechanics in Chemistry* (Prentice-Hall, Upper Saddle River, NJ, 2001).

Intermediate and Advanced Level
Atkins, P. W. and R. Friedman, *Molecular Quantum Mechanics*, 4nd edn. (Oxford University Press, New York, 2005).
Szabo, A. and N. S. Ostlund, *Modern Quantum Chemistry: Introduction to Advanced Electronic Structure Theory* (Dover Publications, New York, 1996).

9

Vibrational–Rotational Spectroscopy

Spectroscopy is a key laboratory tool for the investigation of molecular structure, dynamics, and properties. It is also useful for qualitative and quantitative analysis of many sorts of substances. This chapter begins the development of molecular spectroscopy by focusing on spectroscopic transitions among the vibrational and rotational energy levels of molecules. Using quantum mechanical fundamentals, we explore how spectroscopic techniques probe and measure atoms and molecules. We start with an idealization of a typical diatomic molecule and then continue to a more realistic treatment. The ultimate picture is the basis for interpreting infrared and microwave spectra and obtaining detailed structural and energetic information.

9.1 Molecular Spectroscopy and Transitions

Molecular spectroscopy is a subfield in which molecular behavior is studied by employing electromagnetic radiation. The radiation is used to induce transitions between molecular eigenstates. The process of absorbing or emitting radiation must take place in such a way that energy is conserved. Thus, the quantum of energy available in an absorbed or emitted photon of frequency ν,

$$E = h\nu \tag{9.1}$$

equals the energy change in the absorbing or emitting atom or molecule undergoing a transition. This means the quantity $h\nu$ equals the *difference* in the energies of two particular eigenstates of the system. A sample can absorb or emit radiation at specific frequencies that correspond to eigenenergy differences. Consequently, when we simultaneously vary the frequency of applied radiation and monitor the intensity of the radiation after passing through a sample, we observe absorption transitions as drops in intensities. The frequencies at which those intensity drops are detected are the **transition frequencies**. Monitoring the radiation in a spectroscopic experiment, directly or indirectly, is a means of investigating the energies of the states of atoms and molecules.

Whereas the transition frequencies of a molecule in some region of the electromagnetic spectrum (e.g., visible, infrared) give information on eigenenergy differences, it is quantum mechanics that relates such differences to molecular features such as the stiffness of different chemical bonds and the molecular structure. Some detective work may be involved because each measured absorption or emission frequency needs to be associated with (*assigned* to) the specific pair of states involved in the transition. The task of extracting molecular information from a spectrum usually starts with assigning frequencies.

To say that by energy conservation a transition between molecular eigenstates may occur if the photon involved in the transition has exactly the same energy does not tell *why or whether* the transition occurs. For that understanding, we need to consider the time evolution of wavefunctions and how a system undergoes a change in its quantum mechanical state.

So far in our development of quantum mechanics, attention has been limited to wavefunctions that do not evolve in time, and yet molecules may experience time-dependent interactions that cause them to change in time. An important case of this sort is the effect of electromagnetic radiation. The character of light is that of electric and magnetic fields that oscillate in space and time. When light impinges on a molecule, the oscillating fields may give rise to an interaction, that is, another element of the complete Hamiltonian, and in this way, a time-dependent interaction enters the Hamiltonian of the molecule. We shall see that the interaction with light can induce transitions among the states of a system, and the mechanisms for this are the basis for molecular spectroscopy experiments.

When time, t, is considered in a quantum mechanical problem, there is an associated operator, and like a position coordinate, the operator for time is multiplication by t. Also, like position coordinate operators that have conjugate momentum operators, the time operator has a conjugate. This means there is another operator whose commutator with t is $i\hbar$, just as in the case of the commutator of the conjugate operators of position x and momentum p_x. To find the operator that is conjugate with time, an operator equation is employed. Using f to designate an arbitrary function and G to be the operator we wish to find,

$$i\hbar = [\hat{G}, t] \Rightarrow \hat{G}tf - t\hat{G}f = i\hbar f \Rightarrow \hat{G} = i\hbar \frac{\partial}{\partial t}$$

Dimensional analysis of this operator shows that it will produce an energy quantity (namely, J-s/s = J) on applying it to a wavefunction. This operator, which will be designated as $\hat{E}$ from here on, plays an important role in the time evolution or time dependence of wavefunctions.

There is a time-dependent generalization of Postulate III (the Schrödinger equation) in which the operator $\hat{E}$ replaces E, the energy eigenvalue. This gives the **time-dependent Schrödinger equation** (TDSE):

$$\hat{H}\Psi = \hat{E}\Psi \tag{9.2a}$$

In this equation, Ψ is an explicit function of time, and as well, the Hamiltonian may have an explicit dependence on time. The generalization of Postulate III is that the solutions of the TDSE are wavefunctions that describe a system in both space and time.

The TDSE is a generalization of the time-independent Schrödinger equation (TISE), but it does not invalidate the TISE. This is because the TDSE is always a differential equation that is separable in space and time in those cases where the Hamiltonian has no explicit time dependence (which are the only kinds of cases considered until now). We demonstrate this separability in the following steps with only one spatial coordinate, x, used for simplicity. The test for separability is to assume a product form for the wavefunction and then determine if such a product form can satisfy the TDSE.

$$\Psi(x,t) = \psi(x)\phi(t) \quad \text{assumed trial product form of } \Psi$$

$$\hat{H}(x)\Psi(x,t) = i\hbar \frac{\partial}{\partial t}\Psi(x,t) \quad \text{the TDSE} \tag{9.2b}$$

$$\frac{\hat{H}(x)\psi(x)}{\psi(x)} = \frac{i\hbar}{\phi(t)}\frac{\partial\phi(t)}{\partial t} = E$$

substitution of $\psi(x)\phi(t)$ into TDSE and division by $\psi(x)\phi(t)$ left- and right-hand sides are functions of independent variables, and so each must equal a constant, which is designated E

$$\therefore \widehat{H}(x)\psi(x) = E\psi(x) \quad \text{separated equation in } x \tag{9.3}$$

$$\therefore i\hbar \frac{\partial \phi(t)}{\partial t} = E\phi(t) \quad \text{separated equation in } t \tag{9.4}$$

Therefore, with *no* explicit time dependence in the Hamiltonian, the TDSE (Equation 9.2b) is separable into, or equivalent to, a spatial differential equation, Equation 9.3, and a differential equation in time, Equation 9.4. The differential equation in the spatial coordinate(s) is seen to be the familiar TISE.

There is a general solution of the differential equation in time, Equation 9.4, that was obtained assuming a time-independent Hamiltonian:

$$\phi(t) = e^{-iEt/\hbar} \tag{9.5}$$

This function is sometimes referred to as a **phase factor** or is said to give the phase of the wavefunction. Solving the TISE, Equation 9.3, yields a spatial wavefunction $\psi(x)$, but multiplying that by the $\phi(t)$ in Equation 9.5 gives a product function that is necessarily a solution of the TDSE. Such a product wavefunction is called a **stationary-state** wavefunction because the probability density function is independent of time. Notice that $\phi^*\phi = 1$, and so, with $\Psi(x, t) = \psi(x)\phi(t)$, then $\Psi^*(x, t)\Psi(x, t) = \psi^*(x)\psi(x)$.

It is an important result that even for time-independent Hamiltonian problems, stationary-state wavefunctions are not the only possible solutions of the TDSE. In fact, any arbitrary superposition (or linear combination) of different stationary-state wavefunctions is a solution of the TDSE. To illustrate this point, let $\widehat{H}_0$ be some particular Hamiltonian with no explicit time dependence and let $\{\psi_i\}$ be the set of its spatial eigenfunctions with associated eigenvalues $\{E_i\}$.

$$\widehat{H}_0\psi_i = E_i\psi_i \tag{9.6}$$

The stationary-state wavefunctions (products of spatial wavefunctions and phase factors) for this problem are

$$\Psi_i = \psi_i e^{-iE_i t/\hbar} \tag{9.7}$$

An arbitrary superposition of these states is any linear combination of these functions with unspecified or arbitrary coefficients.

$$\Gamma = \sum_i c_i \Psi_i$$

We can show that Γ is, in fact, a solution of the TDSE by operating on it with $\widehat{H}_0$ and with the energy operator and comparing the results:

$$\widehat{H}_0 \Gamma = \sum_i c_i \widehat{H}_0 \psi_i e^{-iE_i t/\hbar} = \sum_i c_i E_i \psi_i e^{-iE_i t/\hbar}$$

$$i\hbar \frac{\partial}{\partial t} \Gamma = \sum_i c_i \psi_i (i\hbar) \frac{\partial}{\partial t} e^{-iE_i t/\hbar} = \sum_i c_i E_i \psi_i e^{-iE_i t/\hbar}$$

The two operations applied to Γ produce identical results. That is, $\widehat{H}_0 \Gamma = E\Gamma$. So, even though Γ is not an eigenfunction of the Hamiltonian, it is a solution of the TDSE. The significance of this result is that a system can be found to exist in other than a stationary state. A system can change in time even when there is no longer any time dependence in the Hamiltonian governing the system.

When the Hamiltonian has explicit time dependence, separability of the TDSE into time and spatial differential equations is usually not possible. Analytical solution of the TDSE may turn out to be a rather horrible task. We will take a narrower view of time-dependent Hamiltonian problems and treat the time dependence in the Hamiltonian as a perturbation of a system otherwise described by a time-independent Hamiltonian. In other words, we will consider any complete Hamiltonian as consisting of two pieces, a time-dependent piece, H', and a time-independent piece, H_0. We say that H' is a perturbation of the H_0 system.

As in the developments of variation theory and perturbation theory, we make use of the fact that any valid wavefunction for a system can be formed as a linear combination of the eigenfunctions of a model Hamiltonian, in this case H_0. The task of solving the time-dependent differential Schrödinger equation is then converted to the task of finding the proper linear combination. The linear combination is made from the stationary states of the TDSE involving just H_0 (Equation 9.6). Thus, for the Schrödinger equation,

$$(\widehat{H}_0 + \widehat{H}')\Phi = i\hbar \frac{\partial \Phi}{\partial t} \tag{9.8}$$

the general expansion for the wavefunction, Φ, in terms of the stationary-state wavefunctions, ψ_i, is

$$\Phi = \sum_i a_i(t)\psi_i \tag{9.9}$$

Solving Equation 9.8 means finding all the $a_i(t)$ functions. The first step is to substitute Equation 9.9 for Φ into Equation 9.8. At the same time, it is helpful to write out the stationary-state wavefunctions as the products of spatial and time functions via Equation 9.7.

$$\sum_i a_i(t) e^{-iE_i t/\hbar} \widehat{H}_0 \psi_i = \sum_i \widehat{H}' a_i(t) e^{-iE_i t/\hbar} \psi_i = i\hbar \sum_i \psi_i \frac{\partial}{\partial t} a_i(t) e^{-iE_i t/\hbar}$$

Notice that on the left-hand side of this expression, each $\widehat{H}_0 \psi_i$ term can be replaced with $E_i \psi_i$. On the right-hand side, the partial differentiation with respect to t yields exactly the same thing via differentiation of the exponential, plus an additional term. Thus, this expression of the TDSE simplifies to

$$\sum_i \hat{H}' a_i(t) \psi_i e^{-iE_i t/\hbar} = i\hbar \sum_i \dot{a}_i(t) \psi_i e^{-iE_i t/\hbar} \tag{9.10}$$

where $\dot{a}_i(t)$ means the first time derivative of that function.

Multiplying Equation 9.10 by any stationary-state wavefunction, such as Ψ_k, and integrating over all spatial coordinates simplifies the right-hand summation by the orthogonality of the wavefunctions (e.g., $\langle \Psi_i | \Psi_j \rangle = 0$).

$$\sum_i \Psi_k | \hat{H}' a_i(t) \Psi_i = i\hbar \sum_i \dot{a}_i(t) \langle \Psi_k | \Psi_i \rangle = i\hbar \dot{a}_k(t) \tag{9.11}$$

Since k can be the integer label of any state, Equation 9.11 represents a set of many coupled differential equations. If the integral quantities on the left-hand side happen to be known, it might be possible to solve the set of equations and obtain the desired $a_i(t)$ functions. Short of doing that, we restrict attention to the short-time behavior of Φ starting from a point defined as $t = 0$. At $t = 0$, the system is taken to exist in one and only one stationary state, a state that shall be designated "initial." This implies a set of values for the $a_i(t)$ functions at the instant $t = 0$:

$$a_{initial}(0) = 1 \quad \text{and} \quad a_{i \neq initial}(0) = 0 \tag{9.12}$$

Equation 9.11 is valid for any time including the specific time $t = 0$, for which we can apply the values in Equation 9.12.

$$\langle \Psi_k | \hat{H}' \Psi_{initial} \rangle = i\hbar \dot{a}_k(0) \tag{9.13}$$

We still do not know the time derivatives of the $a_k(t)$ functions, but Equation 9.13 gives a set of equations involving these time derivatives at the instant $t = 0$. This result is significant because of the different possibilities for the left-hand-side integral. For any choices of k for which the integral happens to be identically zero, the first time derivative of the $a_k(t)$ function at $t = 0$ is zero. Since $a_k(t)$ gives the weighting of the k stationary state in the Φ function, a zero value for the first derivative means that for short times at least, the weighting of the k stationary state will be unchanging. And by the conditions in Equation 9.12, this means that when k is any index other than initial, the weighting remains zero. The system is not evolving into the k stationary state (at least in the short-time limit). Therefore, whether the Φ function can evolve from some initial stationary state into another final state k because of the effect of a time-dependent perturbing Hamiltonian H' hinges (at least in the short-time limit) on whether the integral in Equation 9.13 is zero or not zero.

When H' corresponds to the interaction with electromagnetic radiation in some particular experimental arrangement, the qualitative distinction between a zero and a nonzero integral in Equation 9.13 becomes the basis for spectroscopic **selection rules**. When the integral is nonzero, the *transition* from the initial state to the k state or final state is said to be **allowed**. When the integral is identically zero, the transition is said to be **forbidden** or not allowed. In some cases, there may be different mechanisms leading to a transition, and

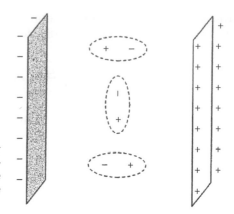

FIGURE 9.1
Dipole in a uniform electric field arising from two oppositely charged parallel plates. A dipole moment, a property of a separated positive and negative charge, interacts differently with the field at the three orientations shown since each end of the dipole is attracted toward the oppositely charged plate.

the selection rules are identified with the mechanism. For example, a transition arising via electromagnetic interaction with a molecular electric *dipole* can be said to be **dipole-allowed** or **dipole-forbidden**.

The perturbing Hamiltonian corresponding to many types of spectroscopic experiments is the interaction with the electric field of the radiation. Usually the strongest electric field interaction is with the dipole moment of a charge distribution. The energy of interaction is the negative of the dot product of the field, $\vec{E}$, and the dipole. A static uniform electric field can be generated by oppositely charging two parallel plates as shown in Figure 9.1. A simple charge distribution that possesses or gives rise to a dipole moment is an arrangement of a positive point charge and a negative charge, as in Figure 9.1. This simple illustration shows that the orientation of the charge distribution with respect to the field affects the interaction energy. At 0°, the interaction is energetically favorable, which means it takes on a negative value. At 180°, the interaction energy should be the same size but positive in sign. At 90°, the interaction is zero. This type of orientational dependence is mathematically expressed as a dot product of two vectors. One vector points from one point charge to the other (the dipole moment vector), and the other vector points from one plate to the other (the electric field vector).

With radiation oscillating at a frequency ω_r, the interaction Hamiltonian is

$$H' = -\vec{\mu} \cdot \vec{E} \cos(\omega_r t) \tag{9.14}$$

where $\vec{\mu}$ is the dipole moment vector. (Notice that the Greek letter μ is regularly used both for dipoles and for the reduced mass of a two-body system.) Let us consider the effect of this interaction for two model systems, the harmonic oscillator and the rigid rotator.

In the case of a harmonic oscillator, let us assume that a dipole moment exists because the two particles of the oscillator have partial charges, $+q$ and $-q$. With x as the coordinate for the separation between the two charges, $\mu(x) = qx$. That is, the dipole moment varies linearly with x. When the oscillator is oriented parallel to the electric field of the radiation,

$$H' = -\mu(x) E \cos(\omega_r t) = -q E \cos(\omega_r t) x \tag{9.15}$$

where E is the magnitude of the radiation field vector. According to Equation 9.13, the spatial integral of H' sandwiched between two stationary-state wavefunctions of the harmonic oscillator is zero unless the radiation can lead to a transition (in the short-time limit). Since the integration is over only the spatial coordinate(s), the essential piece of H' is simply the variable x. Thus, with ψ_n and ψ_m being harmonic oscillator spatial wavefunctions, the value of the integral $\langle \psi_n | x \psi_m \rangle$ distinguishes allowed n-state to m-state, or $n \leftrightarrow m$, transitions (nonzero integral value) from forbidden, n-state to m-state, or $n \leftrightarrow m$, transitions (zero integral value). One can show that the crucial integral in this case is zero unless $|n-m| = 1$. That is, allowed transitions for the harmonic oscillator are those for which the quantum number changes by 1. It is an allowed transition to go from the ground state ($n = 0$) to the first excited state ($m = 1$), whereas it is a forbidden transition to go from the ground state ($n = 0$) to the second excited state ($m = 2$).

Like the harmonic oscillator, the rigid rotator can be taken to possess a dipole moment if each of the two particles is charged. However, there is a difference because the separation of the particles is fixed in the rigid rotator. This means it should be modeled as having a fixed or permanent dipole moment, μ_0. Then, to treat the interaction with electromagnetic radiation we need to consider the orientation of the rotator in space. Choosing the direction of the electric field of the radiation to be the z-direction, the orientation of the rotator with respect to the direction of the electric field vector is specified by the spherical polar coordinate θ. Then, the dot product in Equation 9.14 reduces to the cosine of θ.

$$H' = -\mu_0 \cdot E \cos(\omega_r t) \cos \theta \tag{9.16}$$

The wavefunctions of the rigid rotator are the spherical harmonic functions, and thus, the crucial integrals for transitions are those of $\cos \theta$ sandwiched between spherical harmonic functions.

These integrals can be evaluated for specific spherical harmonic functions by explicit integration. It is also possible to obtain the following result that is general for any choice of an initial state (J and M quantum numbers) and a final state (J' and M' quantum numbers):

$$\langle Y_{JM} | \cos \theta Y_{J'M'} \rangle = \delta_{M,M'} \sqrt{\delta_{J,J'+1} \frac{J^2 - M^2}{4J^2 - 1} + \delta_{J,J'-1} \frac{J'^2 - M^2}{4J'^2 - 1}} \tag{9.17}$$

(Equation 9.17 can be developed conveniently by using the recursion relations of the Legendre polynomials to replace $\cos \theta Y_{J'M'}$ by a sum of $Y_{J'+1,M'}$ and $Y_{J'-1,M'}$. The orthonormality of the spherical harmonic functions then gives the result.) The immediate information from Equation 9.17 is the selection rule that the J quantum number must change by 1 between the initial and final states of a dipole allowed transition. That is, the transition is allowed only if $J = J' - 1$ or if $J = J' + 1$, and that can be stated as $|J - J'| = 1$. The M quantum number cannot change for an allowed transition according to Equation 9.17.

Selection rules for spectroscopic experiments are derived from time-dependent perturbation theory. Transitions are allowed if the integral of the perturbing Hamiltonian and initial and final stationary states is nonzero. For the harmonic oscillator, the allowed transitions are those for which the n quantum number changes by 1, provided that there exists a dipole that varies linearly with the separation distance. (N_2 has a zero dipole because of its symmetry, and there is no linear variation with distance.) For the rigid rotator, the selection rules are that the allowed transitions are a change of 1 in the J quantum number, provided that there is a nonzero permanent dipole, μ_0.

9.2 Vibration and Rotation of a Diatomic Molecule

Spectroscopy using infrared radiation probes the vibrations and rotations of simple molecules. The first such case to consider is that of a diatomic molecule because the mechanics are less complicated than those of polyatomic molecules. After separating out translational motion, the Hamiltonian is the kinetic energy operator, conveniently expressed in spherical polar coordinates, plus a potential that depends on only the separation distance of the atoms, r. The potential, $V(r)$, is something that develops because of the electrons and the chemical bonding in which they participate. In the next chapter, we will examine the quantum mechanics behind chemical bonding and the origin of the internuclear potentials, $V(r)$. For now, we will approach the quantum mechanics of a diatomic molecule by leaving the potential unspecified.

The Schrödinger equation for a vibrating and rotating diatomic molecule, or generally for a two-body problem, with μ as the reduced mass, is

$$-\frac{\hbar^2}{2\mu}\nabla^2\psi(r,\theta,\phi)+V(r)\psi(r,\theta,\phi) = E\psi(r,\theta,\phi) \tag{9.18}$$

With Equation 8.53b, the operator ∇^2 is expressed in spherical polar coordinates and in terms of the angular momentum operator, $\hat{L}^2$, to give

$$-\frac{\hbar^2}{2\mu}\left(\frac{2}{r}\frac{\partial\psi}{\partial r}+\frac{\partial^2\psi}{\partial r^2}\right)+\frac{\hat{L}^2}{2\mu r^2}\psi+V(r)\psi = E\psi \tag{9.19}$$

This differential equation is separable into radial and angular parts, where the angular wavefunctions are the spherical harmonics. That is, if we assume a product form for the eigenfunction, ψ,

$$\psi(r,\theta,\phi) = Y_{lm}(\theta,\phi)R(r) \tag{9.20}$$

applying the operator $\hat{L}^2$ (and the factor $2\mu r^2$) produces

$$\frac{\hat{L}^2}{2\mu r^2}\psi = \frac{l(l+1)\hbar^2}{2\mu r^2}Y_{lm}R \tag{9.21}$$

On substitution of the product form of ψ in Equation 9.20 into the Schrödinger equation, Equation 9.19, we have

$$-\frac{\hbar^2 Y_{lm}}{2\mu}\left(\frac{2}{r}\frac{\partial R}{\partial r}+\frac{\partial^2 R}{\partial r^2}\right)+\frac{l(l+1)\hbar^2}{2\mu r^2}Y_{lm}R+V(r)Y_{lm}R = EY_{lm}R$$

Dividing this equation by the spherical harmonic function, Y_{lm}, establishes the separability of the Schrödinger equation because the resulting differential equation is only in terms of the variable r.

$$-\frac{\hbar^2}{2\mu}\left(\frac{2}{r}\frac{\partial R}{\partial r}+\frac{\partial^2 R}{\partial r^2}-\frac{l(l+1)}{r^2}R\right)+V(r)R = ER \tag{9.22}$$

This is called the radial Schrödinger equation of the two-body problem.

Though the two-body Schrödinger equation is separable, that is, $\psi(r, \theta, \phi) = Y_{lm}(\theta, \phi)R(r)$, it is important that the radial equation, Equation 9.22, is still connected with the angular part of the problem. It incorporates the angular momentum quantum number l, and thus Equation 9.22 really represents an infinite number of differential equations corresponding to the infinite possible choices of l (namely, 0, 1, 2, …). Hence, any radial function, $R(r)$, found from solving Equation 9.22 will ultimately have to be labeled with, and distinguished by, the l value.

At this point, it becomes convenient to rewrite Equation 9.22 in terms of a new variable, s, which is the displacement from some fixed separation distance, r_e, usually chosen to be the equilibrium value of r.

$$s \equiv r - r_e \tag{9.23}$$

$s = 0$ corresponds to $r = r_e$, the potential minimum, and on a graph of V versus s, the point $s = 0$ is the bottom of the potential well. It is also convenient to introduce a new function, S, defined in terms of the still undetermined function R.

$$S(s) = S(r - r_e) \equiv rR(r) \tag{9.24}$$

Since the differential in s equals the differential in r (i.e., $ds = dr$ from Equation 9.23), then differentiating $S(s)$ with respect to s must be the same as differentiating $rR(r)$ with respect to r. Doing this twice produces the following relation between S and R:

$$\frac{\partial^2 S}{\partial s^2} = r\left(\frac{2}{r}\frac{\partial R}{\partial r} + \frac{\partial^2 R}{\partial r^2}\right) \tag{9.25}$$

The right-hand side is a part of Equation 9.22, and so with the appropriate substitutions, and then with multiplication by r, Equation 9.22 becomes

$$-\frac{\hbar^2}{2\mu}\left(\frac{\partial^2 S(s)}{\partial s^2} - \frac{l(l+1)}{(s+r_e)^2}S(s)\right) + V(s+r_e)S(s) = ES(s) \tag{9.26}$$

This is the radial Schrödinger equation expressed in terms of $S(s)$ instead of $R(r)$.

We can immediately consider the special case of Equation 9.26 when the angular momentum is zero, that is, when $l = 0$. The radial Schrödinger equation is then

$$-\frac{\hbar^2}{2\mu}\frac{\partial^2 S(s)}{\partial s^2} + V(s+r_e)S(s) = ES(s) \tag{9.27}$$

This equation is to be solved but only on specification of the potential function, V. One possibility is for the potential to be harmonic, $V(s + r_e) = ks^2/2$. Notice that Equation 9.27 then becomes identical with the standard harmonic oscillator problem, apart from the name of the displacement variable. A harmonic potential is the simplest, somewhat realistic choice for the stretching potential of a chemical bond. However, the harmonic form must be recognized as an approximation or as an idealization of the true molecular potential because, unlike a harmonic function, the potential for lengthening a chemical bond must reach a plateau or a constant potential energy once the chemical bond has completely broken and the atoms are not interacting. The harmonic function $ks^2/2$, on the other hand, goes to infinity as s gets larger; this "harmonic chemical bond" never breaks. Furthermore, even in the equilibrium region, there is no fundamental reason requiring that the potential function should have precisely the shape of a parabola. Even so, there are many cases for which a harmonic approximation is a fairly good approximation for the lower energy states. To be more realistic about the potential requires a more general functional form, such as a higher-order polynomial. This is discussed further on.

The cases for which $l > 0$ in Equation 9.26 present only a small complication of the basic treatment because the possibly troublesome $l(l + 1)$ term can be combined with the potential. That is, we can regard the potential as a combination of the $l(l + 1)$ term with the original potential.

$$V_l^{eff}(s+r_e) \equiv V(s+r_e) + \frac{l(l+1)\hbar^2}{2\mu(s+r_e)^2} \qquad (9.28)$$

Of course, instead of one potential, there is now a potential for each choice of l. Any one of these new potentials is termed an **effective potential** because the effect of the angular momentum term has been built into it.

Next, consider a power series expansion of the s dependence of the effective potential in Equation 9.28, that is, $1/(s + r_e)^2$.

$$\frac{1}{(s+r_e)^2} = \frac{1}{r_e^2} - \frac{2s}{r_e^3} + \frac{3s^2}{r_e^4} - \frac{4s^3}{r_e^5} + \cdots \qquad (9.29)$$

This is an infinite-order polynomial in s, and if the fundamental potential, V, is a polynomial as well, the pieces on the right-hand side of Equation 9.28 can be combined with it term by term. An approximation to the expansion in Equation 9.29 can be made by truncating it. The validity of the approximation depends on how large s can get relative to r_e. If s is likely to be less than 10% of r_e, for instance, the third term will be an order of magnitude smaller than the second term, the fourth term will be an order of magnitude smaller than the third term, and so on. The terms will quickly become very small, and truncation after a few should serve as a high-quality approximation. For typical chemical bonds in diatomic molecules, the classical turning point, which characterizes the extent of displacement from equilibrium, is often on the order of, or less than, 10% of the equilibrium bond length. This implies that truncating the power series at rather low order is normally a good approximation.

The most drastic truncation of the expansion in Equation 9.29 is to retain only the first term. Using this approximation in the diatomic's vibration–rotation Hamiltonian is equivalent to neglecting the coupling of the vibrational motion with the rotational motion. It is as if the system is effectively a rigid rotator, even though its ongoing vibration makes it anything but rigid. With this drastic approximation, Equation 9.26 becomes

$$-\frac{\hbar^2}{2\mu}\left(\frac{\partial^2 S(s)}{\partial s^2} - \frac{l(l+1)}{r_e^2}S(s)\right) + V(s+r_e)S(s) = ES(s)$$

It is not necessary to rewrite this with V^{eff} since the constant operator term can be brought to the right-hand side and combined with E.

$$-\frac{\hbar^2}{2\mu}\frac{\partial^2 S(s)}{\partial s^2} + V(s+r_e)S(s) = E'S(s)$$

and

$$E' \equiv E - \frac{l(l+1)\hbar^2}{2\mu r_e^2}$$

If V were harmonic, this would again be the Schrödinger equation of the harmonic oscillator, and E' would be found to be the harmonic oscillator energies.

Knowing the solutions of the harmonic oscillator problem, we will arrive at the simplest idea of the energies, E, of the rotating–vibrating diatomic system. At this point, we start using J in place of l for the angular momentum quantum number since that is the convention for molecular rotation. The possible state energies for the diatomic molecule whose potential has been approximated as harmonic and for which rotational motion is that of a rigid rotator are

$$E_{nJ} = \left(n+\frac{1}{2}\right)\hbar\omega + \frac{J(J+1)\hbar^2}{2\mu r_e^2} \tag{9.30}$$

The energy levels are given by the vibrational quantum number, n, and the rotational quantum number, J. The approximations made to arrive at this result are checked, in the end, by comparing energies predicted from Equation 9.30 with those measured in a spectroscopic experiment, and we shall consider certain specific data later.

The typical sizes of the constants in Equation 9.30 are such that a change in the vibrational quantum number, n, affects the energy more than a change in the rotational quantum number. For instance, for the molecule hydrogen fluoride, the quantity $\hbar\omega$ is around $4140\,\text{cm}^{-1}$, whereas $\hbar^2/(2\mu r_e^2)$ is $21\,\text{cm}^{-1}$ or about 200 times smaller. The energy of HF will be $4140\,\text{cm}^{-1}$ greater if n is increased from 0 to 1, but only $42\,\text{cm}^{-1}$ greater if J is increased from 0 to 1. For carbon monoxide, $\hbar\omega$ is around $2170\,\text{cm}^{-1}$ and $\hbar^2/(2\mu r_e^2)$ is about 1000 times greater. A schematic representation of the associated energy level pattern is given in Figure 9.2.

The truncation of Equation 9.29 can be made to include more terms. With the truncation after two terms, the improved approximation is

$$\frac{1}{(s+r_e)^2} \cong \frac{1}{r_e^2} - \frac{2s}{r_e^3}$$

The right-hand side is a first-order polynomial in s. So, within this approximation, the effective potential of Equation 9.28 has the form

$$V^{eff}(s+r_e) = V(s+r_e) + a + bs$$

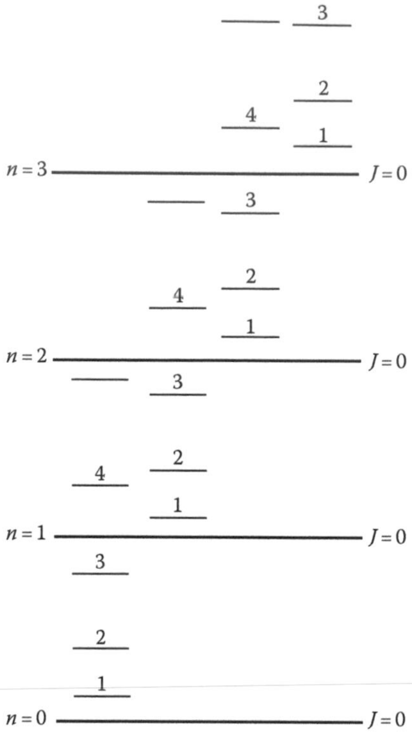

FIGURE 9.2
Energy levels of a vibrating–rotating diatomic molecule according to Equation 9.30. Each horizontal line in this figure represents an energy level with energy increasing in the vertical direction. The levels are labeled by the vibrational quantum number, n, and the rotational quantum number, J. The long, thick lines are all $J = 0$ levels, and the energy spacing between any pair of these lines is $\hbar\omega$. The diagram has been terminated at $n = 3$, but the energy levels continue on infinitely in the same pattern. Levels for which J is not zero have been drawn as short, thin lines to help organize the figure, but there is nothing fundamentally different between these and the $J = 0$ levels. These lines appear as stacks or **manifolds** originating from each $J = 0$ line. The leftmost manifold is for the states for which $n = 0$, and the rightmost manifold is for all those for which $n = 3$. The numbers above four of these lines show the sequence of the J quantum numbers. Higher J levels exist but are not shown. The degeneracy of each level is $2J + 1$. In order to show qualitative features without congestion, the levels have been drawn assuming a ratio of the vibrational frequency to the rotational constant of about 20. Usually, this ratio is greater, for example, it is about 200 for HF and 1000 for CO. Thus, the spacing between the rotational levels is very much smaller relative to the vibrational spacings depicted here.

If the original potential is harmonic, or is taken to be harmonic, the effective potential is

$$V^{eff}(s+r_e) = a + bs + \frac{1}{2}ks^2 \qquad (9.31)$$

where

$$a = \frac{\hbar^2 J(J+1)}{2\mu r_e^2}$$

$$b = -2\frac{a}{r_e}$$

This potential is harmonic since it is a quadratic polynomial in s. As shown in Figure 9.3, it is a parabola differing from the parabola of $V(s + r_e)$ in the position of its minimum and

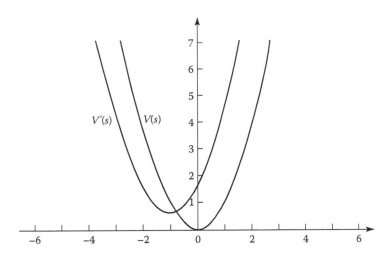

FIGURE 9.3
Plot of the function $V(s) = s^2/2$ and of the function $V'(s) = 1 + s + s^2/2$. Both functions are parabolas with the same curvature; however, the minimum of $V'(s)$ has been shifted by an amount $\delta = -1$, and the value of the potential at that point is $V'(\delta) = 1/2$ rather than zero.

not in its curvature. Instead of the minimum being $V(r_e) = 0$, it is shifted by $\delta = -b/k$. As a result, the minimum value of the potential is $V^{eff}(\delta + r_e) = a - (b^2/k) = V_0$.

The effect of shifting the minimum of the parabolic potential of a harmonic oscillator changes the energies of the eigenstates in only one way. The energy amount that the parabola vertically shifts simply adds to each eigenenergy. For instance, wavefunctions that satisfy

$$-\frac{\hbar^2}{2\mu}\frac{d^2\psi_n}{dx^2} + \frac{1}{2}kx^2\psi_n = \left(n + \frac{1}{2}\right)\hbar\omega\psi_n$$

also satisfy the Schrödinger equation,

$$-\frac{\hbar^2}{2\mu}\frac{d^2\psi_n}{dx^2} + \left(V_0 + \frac{1}{2}kx^2\right)\psi_n = \left[V_0 + \left(n + \frac{1}{2}\right)\hbar\omega\right]\psi_n$$

with V_0 being a constant. The eigenenergies here differ only by the constant V_0. Thus, for the harmonic oscillator with the effective potential of Equation 9.31, all the energy levels are shifted by whatever energy the minimum of the potential has been shifted. (When the minimum is shifted, there is no change in the energy spacings because the force constant is the same; the curvature of the parabola is unaffected by displacement of the minimum to the left or right.) Therefore, the effect of the better approximation in Equation 9.29 that we are considering can be determined from knowing the minimum of the effective potential in Equation 9.31. Substituting for a and b via Equation 9.31, δ and V_0 are

$$\delta = -\frac{\hbar^2 J(J+1)}{k\mu r_e^3} \tag{9.32}$$

$$V_0 = -\frac{\hbar^4 [J(J+1)]^2}{2k\mu^2 r_e^6} + \frac{\hbar^2 J(J+1)}{2\mu r_e^2} \tag{9.33}$$

Thus, the energy levels are now those of the harmonic oscillator but with the constant, V_0, added:

$$E_{n,J} = \left(n + \frac{1}{2}\right)\hbar\omega + \frac{\hbar^2 J(J+1)}{2\mu r_e^2} - \frac{\hbar^4 [J(J+1)]^2}{2k\mu^2 r_e^6} \tag{9.34}$$

The difference between this and Equation 9.30 is the last term, and it is said to be associated with the effect of **centrifugal distortion**.

The physical idea of centrifugal distortion is that because of centrifugal force, the rotational motion contributes to the stretching of the "spring" between the particles. The separation distance, which is taken to be fixed at r_e in the most drastic approximation of Equation 9.29, actually becomes somewhat longer as J increases. This means that the actual rotational energy is less than what Equation 9.30 dictates. The new term in Equation 9.34 provides the kind of downward correction to the energies that is expected to arise from centrifugal distortion of the bond. Notice that the J dependence of this term is quartic; it diminishes the energies of high-J states much more than low-J states.

Further improvements can be made in the vibrational–rotational energy level expression for a diatomic by carrying through the next (third) term in the truncation of Equation 9.29 via perturbation theory. This term is quadratic in the displacement coordinate, s. With second order as a chosen point of truncation of the perturbative corrections to the energy, the energy level expression is

$$E_{n,J} = \left(n + \frac{1}{2}\right)\hbar\omega + \frac{\hbar^2 J(J+1)}{2\mu r_e^2} - \frac{\hbar^4 [J(J+1)]^2}{2k\mu^2 r_e^6} + \frac{3\hbar^3}{2\omega\mu^2 r_e^4}\left(n + \frac{1}{2}\right)J(J+1) \tag{9.35}$$

The third term is the same as that in Equation 9.34 and is associated with centrifugal distortion. The last term involves both the vibrational quantum number, n, and the rotational quantum number, J, and can be regarded as a manifestation of the coupling of rotational motion with vibrational motion.

9.3 Vibrational Anharmonicity and Spectra

The most realistic vibrational potentials of molecules are not strictly harmonic. For a diatomic molecule, the stretching potential's dependence on the separation distance may turn out to be cubic, quartic, and so on. A potential is harmonic if it has only a quadratic dependence. Any higher-order dependence is an **anharmonicity** in the potential. Vibrational anharmonicity refers to the effects on energy levels and transitions of an otherwise harmonic oscillator arising from anharmonic potential terms.

An example of a potential that has the correct qualitative form for a diatomic molecule is the Morse potential, shown in Figure 9.4. It is constructed for a specific diatomic molecule by knowing (or guessing) the dissociation energy, D_e, and choosing an additional parameter, α.

$$V^{Morse}(s) = D_e(1 - e^{-\alpha s})^2 \tag{9.36}$$

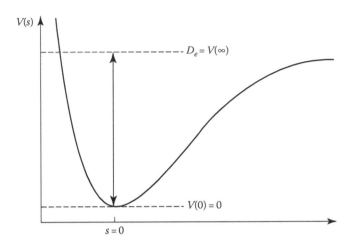

FIGURE 9.4
Functional form of the Morse potential in Equation 9.36.

where s is the coordinate corresponding to the displacement from equilibrium separation of the atoms. At $s = 0$, the potential is zero, and V^{Morse} asymptotically approaches D_e as s approaches infinity. For the region $s < 0$, the potential rises very steeply, in accord with the nature of atom–atom potentials at short range. This potential can be compared with a harmonic potential by writing its power series expansion in s.

$$V^{Morse}(s) = D_e\left(\alpha^2 s^2 - \alpha^3 s^3 + \frac{7}{12}\alpha^4 s^4 - \frac{1}{4}\alpha^5 s^5 + \cdots \right) \tag{9.37}$$

The first term is a harmonic element. For $\alpha s < 1$, the typical situation, the terms diminish in size beyond the harmonic term. Even for $\alpha s \cong 1$, the terms diminish beyond the cubic term. The cubic and quartic terms are the leading sources of anharmonicity near the equilibrium of a Morse oscillator, and it is very often the case that a cubic potential element is the most important contributor to anharmonicity effects in real molecules.

To incorporate anharmonicity effects in a description of a diatomic, a third-order polynomial function is a starting point for the potential, and we can express this generally as

$$V(s) = \frac{1}{2}ks^2 + gs^3 \tag{9.38}$$

This is a starting point because higher-order terms or other functional forms might be employed for still greater precision in representing a true potential. The cubic term (and any higher-order terms) may be well treated as a perturbation of the harmonic oscillator. The first-order corrections for all states turn out to be zero. The second-order corrections arising from the cubic term gs^3 are

$$E_n^{(2)} = -\frac{7}{16}\frac{g^2\hbar^2}{\mu k^2} - \frac{15}{4}\frac{g^2\hbar^2}{\mu k^2}\left(n + \frac{1}{2}\right)^2 \tag{9.39}$$

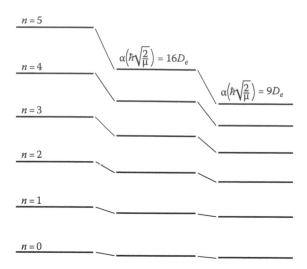

FIGURE 9.5
Representative correlation diagram of several of the low-lying energy levels of a harmonic oscillator with the energy levels of an anharmonic oscillator. The anharmonic energy levels are those given by Equation 9.39 assuming that the potential function is the power series expansion of a Morse potential truncated at the cubic term. The closer spacing of the levels with increasing vibrational quantum number is typical.

The anharmonicity corrections in Equation 9.39 are seen to be a lowering of the energy of each vibrational state relative to the harmonic picture. Furthermore, the extent of lowering increases quadratically with increasing n via the $(n + 1/2)^2$ factor. So the energy levels of an anharmonic oscillator tend to become more closely spaced at higher energies, and this is illustrated in Figure 9.5.

The states of a vibrating–rotating diatomic molecule are distinguished by three quantum numbers, n, J, and M. Under the assumption of rigid rotation and a harmonic stretching potential, the selection rules derived for a harmonic oscillator and a rigid rotator apply to a diatomic molecule. Δn, the change in n for an allowed transition, is +1 (absorption) or −1 (emission), ΔJ is +1 or −1, and $\Delta M = 0$. To see this, we need to make explicit use of the wavefunctions, which are products of radial and spherical harmonic functions:

$$\psi_{nJM}(s,\theta,\phi) = S_n(s)Y_{JM}(\theta,\phi)$$

The integral expression needed for determining the selection rules is

$$\langle \psi_{nJM} | H' \psi_{n'J'M'} \rangle = -|\vec{E}|\cos(\omega_r t)\langle S_n(s)Y_{JM}(\theta,\phi)|\mu(s)\cos\theta S_{n'}(s)Y_{J'M'}(\theta,\phi)\rangle \qquad (9.40)$$

The integral factors into integrals over the radial and angular coordinates.

$$\langle \psi_{nJM} | H' \psi_{n'J'M'} \rangle = -|\vec{E}|\cos(\omega_r t)\langle S_n(s)|\mu(s)S_{n'}(s)\rangle\langle Y_{JM}(\theta,\phi)|\cos\theta Y_{J'M'}(\theta,\phi)\rangle \qquad (9.41)$$

This equation indicates that for transitions from ψ_{nJM} to $\Psi_{n'J'M'}$ to be allowed, the integral over the coordinate s and the integral involving the spherical harmonic functions must *both* be nonzero.

The integral over the radial coordinate in Equation 9.41 can be evaluated with a power series expansion (see Appendix A) of the dipole moment function, $\mu(s)$. The expansion can be truncated at some low order thought to be appropriate for the system at hand.

$$\mu(s) = \mu_e + s \frac{d\mu}{ds}\bigg|_{s=0} + \frac{1}{2}s^2 \frac{d^2\mu}{ds^2}\bigg|_{s=0} + \cdots \tag{9.42}$$

$\mu_e = \mu(0)$ is the dipole moment when the molecule is at its equilibrium length ($s = 0$). Thus,

$$\langle S_n(s)|\mu(s)S_{n'}(s)\rangle = \mu_e \langle S_n(s)|S_{n'}(s)\rangle + \frac{d\mu}{ds}\bigg|_{s=0} \langle S_n(s)|sS_{n'}(s)\rangle + \frac{1}{2}s^2 \frac{d^2\mu}{ds^2}\bigg|_{s=0} \langle S_n(s)|s^2S_{n'}(s)\rangle + \cdots$$

If the radial wavefunctions are those of the harmonic oscillator (i.e., if the harmonic oscillator approximation of the potential is invoked), then explicit integration can be performed for any choice of n (initial state) and n' (final state). The result is the following expression, where m has been used for the reduced mass to avoid confusion with the dipole moment:

$$\langle S_n(s)|\mu(s)S_{n'}(s)\rangle = \mu_e \delta_{n,n'} + \frac{d\mu}{ds}\bigg|_{s=0} \sqrt{\frac{\hbar}{2\sqrt{mk}}} \left(\sqrt{n}\,\delta_{n,n'+1} + \sqrt{n'}\,\delta_{n+1,n'}\right)$$

$$+ \frac{1}{2}s^2 \frac{d^2\mu}{ds^2}\bigg|_{s=0} \frac{\hbar}{2\sqrt{mk}} + \left((2n+1)\delta_{n,n'} + \sqrt{n(n-1)}\,\delta_{n,n'+2} + \sqrt{n'(n'-1)}\,\delta_{n+2,n'}\right) + \cdots$$

$$\tag{9.43}$$

We interpret Equation 9.43 term by term. The first term indicates that transitions are allowed in which n and n' are the same, which means the vibrational quantum number does not have to change for a transition to take place; however, such transitions occur only if there is a nonzero equilibrium dipole moment (μ_e). The next term indicates that transitions are allowed where the difference between n and n' is 1, but only if there is a nonzero first derivative of the dipole moment function, $d\mu/ds$. Similarly, the next term after that depends on the second derivative of the dipole moment function. If it is not zero, transitions will be allowed for a change of 2 in the n quantum number. We summarize this:

A transition for which $\Delta n = 0$ is allowed if $\mu_e \neq 0$

A transition for which $\Delta n = \pm 1$ is allowed in $d\mu/ds \neq 0$

A transition for which $\Delta n = \pm 2$ is allowed if $d^2\mu/ds^2 \neq 0$

This pattern continues for greater $|\Delta n|$, and so with "curving" of the dipole moment function that gives nonzero higher-order derivative values at $s = 0$, *all* Δn transitions could become allowed.

Figure 9.6 shows the dipole moment function of the hydrogen fluoride molecule. A very typical feature of diatomic molecules is that the dipole moment function of the molecule is very nearly linear in the displacement coordinate in the vicinity of the equilibrium.

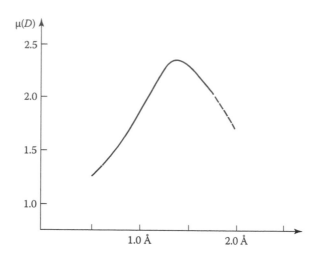

FIGURE 9.6

Dipole moment function of the hydrogen fluoride molecule as a function of the separation distance between H and F in angstroms. This curve is based on spectroscopic data. (From Sileo, R.N. and Cool, T.A., *J. Chem. Phys.*, 65, 117, 1976; Gough, T.E. et al., *Faraday Discuss. Chem. Soc.*, 71, 77, 1981.) The equilibrium separation distance is at 0.92 Å, and in the vicinity of the equilibrium, the dipole moment curve is very nearly linear.

Only at large separations, as in the case of HF, is there significant curvature. Thus, approximating the dipole moment function by truncating the expansion in Equation 9.42 after the term that is linear in s tends to be a good approximation, at least for the lower-energy vibrational states with small vibrational excursions. When this approximation is made and when the stretching potential is taken to be strictly harmonic, only the first two right-hand-side terms in Equation 9.43 remain. This is called a **doubly harmonic approximation,** and it gives us the selection rule requiring that the vibrational quantum number not be changed or change by 1; however, there must be a permanent equilibrium dipole for the first requirement and a *changing* dipole ($d\mu/ds \neq 0$) for the second. A molecule such as Cl_2 has a zero dipole moment because by symmetry it is nonpolar. The symmetry remains whether the molecule is stretched or compressed, and thus it does not have a changing dipole; the derivative of its dipole moment function, $d\mu/ds$, is zero.

From a perturbative standpoint, anharmonicity in the stretching potential mixes harmonic wavefunctions. If the anharmonicity were to produce a mixing of the zero-order $n = 1$ and $n = 2$ states, the nonzero transition integral between the harmonic $n = 0$ and $n = 1$ states would be distributed among the perturbed first and second excited states. This would make *both* the $n = 0$ to $n = 1$ and $n = 0$ to $n = 2$ transitions allowed in the anharmonically perturbed states. Anharmonicity in the potential and nonlinearity in the dipole moment function make it possible to observe vibrational transitions other than $\Delta n = 1$. However, such other transitions are usually weaker or less intense because the harmonic nature of the potential and the linearity of the dipole moment function have the major role.

Vibrational transitions are said to be **fundamental** if the vibrational quantum number changes from 0 to 1. They are said to be **overtones** if the vibrational quantum number changes from 0 to 2, 3, and so on.

The selection rules provide crucial information for relating laboratory spectral results to molecular features such as bond lengths or harmonic force constants. To do that requires that we change our thinking somewhat. So far, we have learned how to *predict* a spectrum given the molecule's potential, $V(s)$. But the stretching potentials of diatomic molecules are not known a priori. In fact, they are the very information that is to be extracted from experiment, not the

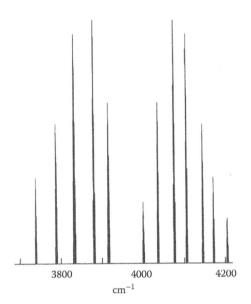

FIGURE 9.7
Idealized high-resolution spectrum of a low-density gas phase sample of hydrogen fluoride. The frequencies of the peaks are the transition frequencies, and the relative heights of the peaks correspond to the intensities. The characteristic form of an infrared spectrum of a diatomic molecule is that of two sets of lines, called branches, with diminishing peak height to the left and to the right. The branch on the left, at lower transition frequencies, is the *P*-branch, whereas the branch on the right is the *R*-branch. The separation between the two branches is roughly twice the separation between lines within the branches.

other way around. Analyzing and interpreting spectra and extracting molecular information require matching predictions, based on quantum mechanics, with laboratory measurements.

The results of a spectroscopic experiment provide a set of transition frequencies, as well as the intensities (strengths) of the spectral lines. As an example of the type of information that is available from an experiment, Figure 9.7 shows an idealization of the high-resolution infrared spectrum of the hydrogen fluoride molecule. With modern instrumental techniques, the frequencies of the lines in this spectrum can be measured with a precision of about 0.01 cm^{-1} or better. Each of the spikes or lines is at a frequency that corresponds to a particular transition between quantum states of HF. Since the frequency of radiation is proportional to the absorbed photon's energy, knowing the frequency of a transition is equivalent to knowing the energy of the transition. In fact, wavenumber units (cm^{-1}) serve as units of measure for both frequency and energy (see Appendix F). The length or height of each spike, or transition line, is related to the intensity of that transition, which is to say that the depletion of incident radiation at a given frequency depends on the likelihood that the sample will absorb at that frequency. Together, intensities and frequencies are the primary data from the spectrum that need to be matched with the results of quantum mechanical analysis.

The energy of a transition is the difference between the energies of an initial state and a final state. So, to match experimental and quantum mechanical information, we need an expression for energy differences rather than an expression for state energies. This can be accomplished by combining the information in an energy level expression with that in the selection rules. The result will be a transition energy expression that indicates where allowed transitions appear in the spectrum.

To start with the simplest level of treatment, let us use the energy level expression in Equation 9.30 while restating the doubly harmonic selection rules. There are two cases,

$\Delta n = 0$ and $\Delta n = 1$. ($\Delta n = -1$ is ignored since it corresponds to emission and not absorption of radiation.) The selection rules say that on going from some initial n,J state to another state by absorption of electromagnetic radiation,

1. $\Delta n = 1$ *and* $\Delta J = \pm 1$ (if n increases by 1, J can only increase or decrease by 1 for the transition to be allowed)
2. $\Delta n = 0$ *and* $\Delta J = 1$ (allowed transitions in which n does not change are those for which J increases by 1)

Condition 2 does not involve any vibrational excitation, and we consider it in the next section. Condition 1 presents two situations, one where J increases and one where J decreases, and thus, we obtain two transition energy expressions. Both expressions are obtained by subtracting the initial state's energy, expressed according to Equation 9.30, from the final state's energy. We shall start with J increasing by 1 from the initial to the final state. If n and J are the initial state's quantum numbers, then the allowed final state's quantum numbers are $n + 1$ and $J + 1$. With these, we find the difference in energies:

$$E_{n+1,J+1} = \left(n+\frac{3}{2}\right)\hbar\omega + \frac{(J+1)(J+2)\hbar^2}{2\mu r_e^2} - E_{n,J} = \left(n+\frac{1}{2}\right)\hbar\omega + \frac{J(J+1)\hbar^2}{2\mu r_e^2}$$

$$= \Delta E_{n,J\rightarrow n+1,J+1} = \hbar\omega + \frac{\hbar^2}{2\mu r_e^2}(2J+2) \tag{9.44}$$

This is the desired energy difference expression for the allowed transition where the initial state's J quantum number increases by 1.

The other situation is that where the J quantum number decreases by 1. This requires a separate expression for the energy difference:

$$E_{n+1,J-1} = \left(n+\frac{3}{2}\right)\hbar\omega + \frac{J(J-1)\hbar^2}{2\mu r_e^2} - E_{n,J} = \left(n+\frac{1}{2}\right)\hbar\omega + \frac{J(J+1)\hbar^2}{2\mu r_e^2} = \Delta E_{n,J\rightarrow n+1,J-1} = \hbar\omega - \frac{\hbar^2}{2\mu r_e^2}2J$$

$$\tag{9.45}$$

These two expressions for ΔE should be regarded as predictions of the positions of transition lines that may be observed in the spectrum, at least according to the particular level of treatment used for the state energies.

In both Equations 9.44 and 9.45, the same set of constants shows up in the second term. This suggests collecting them into one constant to make the expressions more concise. At the same time, it is convenient to convert the units for the energy expression to those often used by spectroscopists. This is accomplished by dividing the energy by hc (Planck's constant times the speed of light) and then multiplying by 0.01. The SI unit of E/hc is m^{-1}, and the multiplication by 0.01 leaves the energy unit as cm^{-1} (wavenumbers). Where it is not clear from the context, a tilde ($\sim$) indicates quantities expressed in wavenumbers. Thus,

$$\tilde{E} = 0.01\frac{E}{hc} \tag{9.46}$$

$$\tilde{\omega} = 0.01\frac{\hbar\omega}{hc} \tag{9.47}$$

The rotational constant, $\tilde{B}_e$, is

$$\tilde{B}_e = 0.01 \frac{1}{hc} \frac{\hbar^2}{2\mu r_e^2} = 0.01 \frac{h}{8\pi^2 \mu r_e^2 c} \tag{9.48}$$

Equations 9.44 and 9.45 are quite concise when expressed via Equations 9.46 through 9.48:

$$\Delta \tilde{E}_{n,J \to n+1, J+1} = \tilde{\omega} + 2\tilde{B}_e(J+1) \tag{9.49}$$

$$\Delta \tilde{E}_{n,J \to n+1, J-1} = \tilde{\omega} - 2\tilde{B}_e J \tag{9.50}$$

With care, one can suppress the tildes and recognize units of cm^{-1} from the form of an energy expression. For instance, Planck's constant is not found in Equation 9.49 but is found in Equation 9.44. Another way to keep track of the implied units is to write "$\Delta E/hc=$" on the left side since to make clear that the division by hc has been performed.

B_e, or for now just B, is much smaller than ω, and so the transitions in the spectrum are all clustered in the vicinity of the frequency ω. This is the first feature to identify in the spectrum. For instance, in Figure 9.7, the lines appear centered about $3960\,cm^{-1}$, and that central energy serves as a fair idea of the true value of ω.

The next task in analyzing a vibrational–rotational spectrum of a diatomic molecule is to *assign* individual lines. **Assignment** means that the transition lines have been associated with particular initial and final states. For a vibrating–rotating diatomic molecule, this means identifying the n, J quantum numbers for the initial state and the n', J' quantum numbers for the final state of each line. An ordered list of transition energies, generated with Equations 9.49 and 9.50, reveals how to make these assignments. We start with Equation 9.50 because it gives frequencies that are all below those given by Equation 9.49. We pick out a value of J and work up in frequency until reaching $J = 1$, which is where the highest frequency possible with Equation 9.50 is reached. Then, we use Equation 9.49 and work up from $J = 0$:

J (Initial) $\to$	J (Final)	ΔE	$\Delta(\Delta E)$
4	3	$\omega - 8B$	$2B$
3	2	$\omega - 6B$	$2B$
2	1	$\omega - 4B$	$2B$
1	0	$\omega - 2B$	$4B$
0	1	$\omega + 2B$	$2B$
1	2	$\omega + 4B$	$2B$
2	3	$\omega + 6B$	$2B$
3	4	$\omega + 8B$	$2B$

This table shows that there are a number of transition frequencies in the vicinity of ω but offset by some multiple of $2B$. The $\Delta(\Delta E)$ values represent the differences between transition energies. They tell how far apart adjacent spectral lines are. The pattern here is one of lines separated by $2B$, except at one crucial point where the separation is twice as much, $4B$. Examining Figure 9.7 shows that in the middle of the cluster of lines, there is a point where the separation is about twice as large as otherwise seen. Thus, the line to the left (lower frequency) of this separation is assigned to be the $J = 1$ to $J' = 0$ transition (with n changing by 1, as well). The line to the right, then, corresponds to an initial state of $J = 0$ and a final state where $J' = 1$. The remaining lines follow in the sequence in the table. The collection of lines at frequencies below the $4B$ break form the P-branch of the absorption band. The higher frequency lines form the R-branch.

A confirmation of the assignment comes from the relative intensities. Notice that the strengths of the transitions, or the heights of the lines in the spectrum, diminish to the right and to the left of the center of the band of lines. That is, the transitions with the highest initial J quantum number are the weakest. The primary reason for this has to do with the populations of the different quantum states, or the number of molecules in each of the states. The Maxwell–Boltzmann distribution law dictates that for a sample of N molecules in thermal equilibrium at a temperature T, where the molecules can exist in stationary states with energies E_i (for each of the ith states), the number of molecules in each state, N_i, is

$$N_i = N \frac{e^{-E_i/kT}}{\sum_j e^{-E_j/kT}} = \frac{N}{q_{vib}} e^{-E_i/kT} \tag{9.51}$$

where
   k is the Boltzmann constant
   q_{vib} is the partition function as in Equation 1.26

Example 9.1: Diatomic Molecule Vibrational Spectrum

Problem: Assign the features in the hypothetical infrared spectrum of a diatomic molecule shown and obtain values for ω and B.

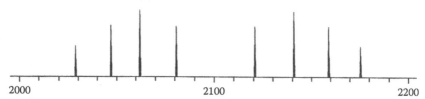

Solution: The spectrum shows a band, a set of lines, with a regular arrangement. Perhaps the first feature to identify is the gap in the lines that is seen around $2100\,\text{cm}^{-1}$. From this band center, one can assign lines: To the right are those for which J increases by 1, starting from the $J = 0 \rightarrow 1$ line. To the left are those for which J decreases by 1, starting from the $J = 1 \rightarrow 0$ line. And for the entire band, n changes from 0 to 1. Equation 9.49 applies to the lines to the right of the band center, the R-branch, and Equation 9.50 applies to the lines to the left, the P-branch. The next step is to measure each line position from the spectrum, and then using each line's assignment to apply either Equation 9.49 or 9.50 to generate expressions for ω and B.

Assignment	Line position (cm^{-1})	Equal to
$J = 4 \rightarrow 3$	2038	$\omega - 8B$
$J = 3 \rightarrow 2$	2045	$\omega - 6B$
$J = 2 \rightarrow 1$	2063	$\omega - 4B$
$J = 1 \rightarrow 0$	2082	$\omega - 2B$
$J = 0 \rightarrow 1$	2122	$\omega + 2B$
$J = 1 \rightarrow 2$	2141	$\omega + 4B$
$J = 2 \rightarrow 3$	2159	$\omega + 6B$
$J = 3 \rightarrow 4$	2176	$\omega + 8B$

This is a set of eight equations in only two unknowns. We could use least squares fitting (Appendix A), but a simpler approach is to solve the equations for only the two lines on either side of the band center (2082 and 2122 cm^{-1}). Doing so yields $B = 10$ cm^{-1} and $\omega = 2102$ cm^{-1}.

When we use the vibrational–rotational energy level expression in Equation 9.30, the population of some state with quantum numbers n, J, and M_J, relative to another population, that of the state with the same vibrational quantum number but with $J = 0$, is

$$\frac{N_{n,J,M_J}}{N_{n,0,0}} = \frac{e^{-E_{n,J}/kT}}{e^{-E_{n,0}/kT}} = e^{-\hbar^2 J(J+1)/(2\mu r_e^2/kT)}$$

As J increases, the population diminishes relative to the population of the lowest-energy J state ($J = 0$). We have already realized that transitions can be observed with any initial J quantum number, but because there are different numbers of molecules in each J state, the numbers of transitions originating from different J states differ by amounts proportional to the populations.

One further point of interest is that the transition lines do not distinguish the M_J quantum number. For any initial n,J energy level, transitions can occur from each and every one of the M_J states, and there are $2J + 1$ of these. This means that a transition strength depends on the population of an n,J energy level rather than on the population of a single state. The population of an energy level is the sum over the populations of all the states of that energy. By Equation 9.51, the populations of states of the same energy are identical. Therefore, the population of the level is simply the population of any one state times the degeneracy of the level, which is $2J + 1$. The relative intensities, I, of the lines in a high-resolution vibrational absorption band, as in Figure 9.7 and as in the spectrum of an HCl sample in Figure 9.8 and of HBr in Figure 9.9, depend on the initial state's J quantum number,

$$I_J \propto (2J+1)e^{-\chi J(J+1)} \quad \text{where} \quad \chi = \frac{\hbar^2}{\left(2\mu r_e^2 kT\right)}$$

Figure 9.10 is a plot of this function for different values of χ, treating J as a continuous variable. Notice that a curve drawn through the tops of the transition lines in Figure 9.7 would have the form of a curve in Figure 9.10. Indeed, at low resolution, where the individual lines overlap or are not resolved, the branches of an infrared absorption spectrum may resemble the curves in Figure 9.10.

Assigning lines and measuring their frequencies in the infrared spectrum of a diatomic can provide more information than values for ω and B. Since quantum mechanical analysis provides energy level expressions subject to the level of approximation chosen, in principle, a higher level of detail than ω and B is implicit in spectral data. We can approach a more detailed, more general treatment by noting that the J dependence in the energy level expression has been in terms of the quantity $J(J + 1)$, and the n dependence in terms of $(n + 1/2)$. An overview of Equations 9.30, 9.34, 9.36, and 9.39, the various energy level expressions, reveals that they all take the form of an

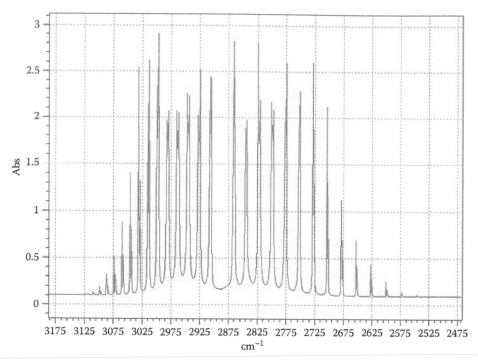

FIGURE 9.8

High-resolution infrared spectrum of the fundamental band of HCl. The horizontal axis gives the frequency in wavenumbers. The vertical axis is the absorbance in arbitrary units. The *P*-branch is on the right in this presentation of the spectrum, and the *R*-branch is on the left. Each line appears to be split in two. In fact, this is a result of the sample used for this spectrum being a mixture of $H^{35}Cl$ and $H^{37}Cl$ at natural abundance of the chlorine isotopes. The mass differences lead to slightly different rotational constants, and so the spectrum is really a superposition of the spectra of $H^{35}Cl$ and $H^{37}Cl$. (From W.M. Davis, V.G. Rivas and B.M. Ramirez, Department of Chemistry, Texas Lutheran University. With permission.)

expansion of the energy in terms of polynomials in $(n + 1/2)$ and $J(J + 1)$. That is, the energy of an n,J level has the following general form:

$$E_{n,J} = \sum_{k=0}^{\infty}\sum_{l=0}^{\infty} c_{ik}\left(n+\frac{1}{2}\right)^{k}(J(J+1))^{l} \tag{9.52}$$

(Regard $(n + 1/2)$ as a single variable, e.g., x, and $J(J + 1)$ as another independent variable, e.g., y. The polynomial expansion is then just a standard type, $E(x, y) = c_{00} + c_{10}x + c_{01}y + c_{11}xy + c_{20}x^2 + c_{02}y^2 + \cdots$) The various approximations that have been discussed lead to certain truncations of this expansion and to specific values of the constants, c_{ik}. For instance, we can say that Equation 9.30 is of the form of Equation 9.52 with $c_{10} = \hbar\omega$, $c_{01} = \hbar^2/2\mu r_e^2$, and all other c_{ik}'s equal to zero. Transition energy expressions can be developed for any desired truncation in Equation 9.52 by means of the differencing procedures in the prior sections. These give ΔE for the transition lines in terms of the c_{ik}'s. Analysis of spectra can then be carried out by first assigning lines, and then by fitting the measured transition frequencies (energies) to the ΔE expression being employed (see Appendix A). Notice how this takes the whole approach one step away from the details of the quantum mechanical analysis: The measured transition energies or frequencies must fit a simple polynomial expression, Equation 9.52, and the

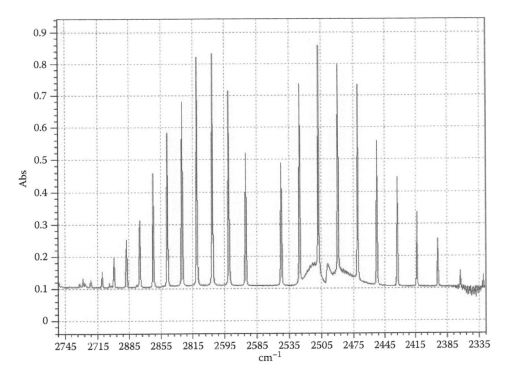

FIGURE 9.9
High-resolution infrared spectrum of the fundamental band of HBr. The vertical axis is the absorbance in arbitrary units. Notice the difference in the band center of the spectra of HBr and HCl. (From W.M. Davis, V.G. Rivas and B.M. Ramirez, Department of Chemistry, Texas Lutheran University. With permission.)

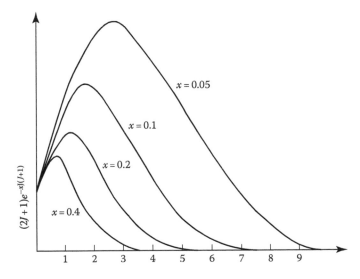

FIGURE 9.10
Plot of the relative transition intensities of the rotational lines in a diatomic molecule's vibrational absorption band. The curves are plots of the quantity $(2J + 1)^{-exJ(J+1)}$ but with J treated as a continuous variable. The numerical values of x are shown next to each curve. Notice that as the temperature increases, x becomes smaller.

main task is to find the "right" c_{ik}'s. At that stage, we can look back at the quantum mechanical analysis and relate the c_{ik}'s to various kinds of molecular information, such as the bond length and the stretching force constant.

There is well-established terminology and notation for the most important c_{ik}'s in Equation 9.52, and some of this has been introduced already. The c_{ik}'s are spectroscopic constants; that is, they are values deduced or extracted from measurement of spectra. The names and designations for the most important ones are the following. The corresponding c_{ik}'s are shown in brackets; in some cases, the conventional definition of the spectroscopic constants introduces a negative sign, and then the values correspond to $-c_{ik}$'s.

ω_e	Harmonic (equilibrium) vibrational frequency (c_{10})
$\omega_e\chi_e$	Equilibrium anharmonicity; this product is considered to be one value ($-c_{20}$)
$\omega_e y_e$	Equilibrium second anharmonicity constant; this product is considered to be one value (c_{30})
B_e	Equilibrium rotational constant (c_{01})
D_e	Equilibrium centrifugal distortion constant ($-c_{02}$)
H_e	Equilibrium second centrifugal distortion constant (c_{03})
α_e	Vibration rotation coupling constant ($-c_{11}$)
γ_e	($-c_{21}$)
β_e	($-c_{12}$)

With the named spectroscopic constants in place of the c_{ik}'s in Equation 9.52, we have

$$\tilde{E}_{n,J} = \tilde{c}_{00} + \tilde{\omega}_e\left(n+\frac{1}{2}\right) + \tilde{B}_e J(J+1) - \tilde{\omega}_e\tilde{\chi}_e\left(n+\frac{1}{2}\right)^2 - \tilde{D}_e[J(J+1)]^2 - \tilde{\alpha}_e J(J+1)\left(n+\frac{1}{2}\right)^2$$

$$+ \tilde{\omega}_e\tilde{y}_e\left(n+\frac{1}{2}\right)^3 - \tilde{\beta}_e[J(J+1)]^2\left(n+\frac{1}{2}\right) - \tilde{\gamma}_e J(J+1)\left(n+\frac{1}{2}\right)^2 + \tilde{H}_e[J(J+1)]^3 + \cdots$$

The value $\tilde{c}_{00}$ cannot be obtained by measurement of transition frequencies because it does not show up in an energy difference (ΔE) equation. It contributes, usually in a small way, to the zero-point energy of the system. Table 9.1 lists spectroscopic constants of a number of diatomics.

9.4 Rotational Spectroscopy

In the time-dependent quantum mechanical analysis of a vibrating, rotating diatomic molecule, we determined that if there is a nonzero permanent dipole moment, transition can occur for which $\Delta n = 0$ and $\Delta J = \pm 1$. These are pure rotational transitions since the vibrational quantum number does not change. $\Delta J = 1$ is a transition from a lower energy to a higher energy state, and it corresponds to absorption of radiation. $\Delta J = -1$ is a transition from a higher energy rotational state to a lower energy state, and it corresponds to an emission process. Since pure rotational transitions involve energies of the size of B rather than ω, the photon energies are much smaller than for vibrational spectra. This means

TABLE 9.1

Spectroscopic Constants (cm^{-1}) of Certain
Diatomic Molecules[a]

	$\tilde{\omega}_e$	$\tilde{B}_e$	$\tilde{\omega}_e\tilde{\chi}_e$	$\tilde{\alpha}_e$
BN	1514.6	1.666	12.3	0.025
BO	1885.4	1.7803	11.77	0.0165
BaO	669.8	0.3126	2.05	0.0014
BeO	1487.3	1.6510	11.83	0.0190
CH	2861.6	14.457	64.3	0.534
CD	2101.0	7.808	34.7	0.212
CN	2068.7	1.8996	13.14	0.0174
CO	2170.2	1.9313	13.46	0.0175
CaH	1299	4.2778	19.5	0.0963
HBr	2649.7	8.473	45.21	0.226
HCl	2989.7	10.59	52.05	0.3019
HF	4138.5	20.939	90.07	0.770
DF	2998.3	11.007	45.71	0.293
HgH	1387.1	5.549	83.01	0.312
LiH	1405.6	7.5131	23.20	0.2132
MgH	1495.7	5.818	31.5	0.1668
MgO	785.1	0.5743	5.1	0.0050
NO	1904	1.7046	13.97	0.0178
OH	3735.2	18.871	82.81	0.714
OD	2720.9	10.01	44.2	0.29
SiN	1151.7	0.7310	6.560	0.0057
SiO	1242.0	0.7263	6.047	0.0049

Source: These values have been collected in Herzberg, G., *Molecular Spectra and Molecular Structure. I. Spectra of Diatomic Molecules*, Van Nostrand Reinhold, New York, 1950.

[a] The values are for the most abundant isotopes, except for the deuterated molecules CD, DF, and OD.

the wavelengths of the radiation are much longer. It turns out that the type of radiation needed for rotational transitions is microwave and radiofrequency radiation.

Analysis of the microwave or radiofrequency absorption spectra of diatomic molecules is usually simpler than the analysis of their vibrational spectra. Instead of an infrared band of transitions arising from the populated rotational (J) levels, a single transition line corresponding to $J = 0 \rightarrow 1$ is usually observed in the microwave or radiofrequency spectrum of a diatomic molecule. Sometimes a few other transitions, such as $J = 1 \rightarrow 2$ and $J = 2 \rightarrow 3$ are also observed. The appropriate transition energy expression is obtained by taking differences using the energy level expression for the states of a diatomic molecule. Of course, since n is unchanging and since $\Delta J = +1$, the transition energy expression is rather simple: A $J = 0 \rightarrow 1$ transition is observed at a frequency of about $2B$, whereas the next transition, $J = 1 \rightarrow 2$, is at $4B$, which is twice the frequency. The next is at $6B$, and so on.

In the microwave and radiofrequency region of the electromagnetic spectrum, a doubling of the frequency from $2B$ to $4B$ may require instrumental alterations. Often, only one transition is seen for a particular equipment setup. For more massive diatomic molecules, the rotational constant, B, is small, and then it may turn out that several transitions, such as $J = 4 \rightarrow 5$, $J = 5 \rightarrow 6$, and $J = 6 \rightarrow 7$, are close enough in frequency to be observable in one scan.

It is characteristic of the technology of microwave spectroscopy that frequencies are measurable to very high precision. Until the introduction of infrared lasers, microwave spectroscopy far outran vibrational spectroscopy in the precision and accuracy of spectral measurements. The primary piece of information obtained from a microwave spectrum is the rotational constant, and given the precision available with this type of experiment, high-precision values of the rotational constant are obtained. This, in turn, implies that very precise values of the bond length of a diatomic molecule can be deduced from a microwave spectrum. In practice, measurement precision corresponding to a few parts in 10,000 is achieved.

The rotation of a rigid, linear triatomic or polyatomic molecule is mechanically equivalent to the rotation of a rigid diatomic molecule. All are the rotations of an "infinitesimally thin rod" with two or more point masses attached. The basic analysis for the rotational spectroscopy of linear polyatomic molecules follows that of diatomic molecules; however, it requires a generalization of the moment of inertia to more than two atoms. For a linear arrangement of point masses, the moment of inertia, I, about the center of mass is

$$I = \frac{\sum_{i<j} m_i m_j r_{ij}^2}{\sum_i m_i} \tag{9.53}$$

where
  r_{ij} is the distance between the i and j atoms
  m_i is the mass of the ith atom, and the sum in the numerator is over all pairs of atoms in the molecule

The denominator equals the total mass of the molecule. Notice that for a diatomic molecule, Equation 9.53 reduces to a single term, $m_1 m_2 r_{12}^2/(m_1 + m_2)$, and this is the same as μr_e^2, taking r_e to be r_{12}.

The Hamiltonian for the rotation of a rigid rod and the corresponding rotational Schrödinger equation are the same for three or more point masses as for two point masses. They are mechanically equivalent problems, and for all,

$$\frac{\hat{L}^2}{2I} \psi(\theta, \phi) = E\psi(\theta, \phi) \tag{9.54}$$

This is the generalization of the two-body rotational Schrödinger equation. It comes about merely by using I in place of the two-body-specific value of I, which is μr^2. Therefore, the wavefunctions for a rigid, rotating diatomic or linear polyatomic must be the spherical harmonic functions since they are the eigenfunctions of $\hat{L}^2$. The eigenenergies are

$$E_J = J(J+1)\frac{\hbar^2}{2I} \tag{9.55}$$

where J is the rotational quantum number.

The rotational constant is inversely related to the moment of inertia, and the moment of inertia is a function of the bond lengths. For a diatomic molecule, measurement of the rotational constant implies determination of the bond length. But for a linear triatomic molecule, there are two unknowns in the problem, the two bond lengths. One spectroscopic value, the value of B, is not sufficient to find two bond lengths. Additional information is needed.

For a linear polyatomic of $N + 1$ atoms, there are N bond lengths to be found in order to know the structure of the molecule (assuming there is a basis for already knowing that the molecule in question is linear). Finding the bond lengths is basically a problem of establishing N equations for the N unknowns. The N equations are established from measured rotational constants for *isotopically substituted* forms of the molecule. Isotopic substitution is a change in an atomic mass, and with that comes a change in the molecule's moment of inertia. We assume, though, that the bond lengths are unchanged since the chemical bonding that dictates the bond length should be unaffected by the numbers of neutrons in the nuclei. (Strictly speaking, this assumption is valid only in regard to equilibrium structures. Because of zero-point vibrational motion, the separation between atoms is an average over the vibrational excursions. The on-average length of a bond tends to be slightly different, usually greater, than the equilibrium length. The averaging depends on vibrational motion and thus depends on particle masses. As zero-point vibrational motion is taken into account in a more detailed analysis, isotopic substitution will influence the bond length, though slightly.)

For each isotopic form of a molecule with an experimentally determined B, and hence with a value for I, we can write Equation 9.53 with the bond lengths as unknowns. We need N isotopic forms to have N equations from which to find the N unknowns. A standard example is HCN. Microwave transition frequencies for the $J = 0$ to $J = 1$ transitions were found to be 88,631.6 MHz for $H^{12}C^{14}N$ and 72,414.6 MHz for $D^{12}C^{14}N$.[*] The B rotational constants are one-half of these values, and when they are converted to moments of inertia, we have $I_{HCN} = 18.937 \times 10^{-47}$ kg m^2 and $I_{DCN} = 23.178 \times 10^{-47}$ kg m^2.[†] The two equations to solve, assuming $r_{HC} = r_{DC}$ and letting m_{HCN} and m_{DCN} be the molecular masses, are

$$I_{HCN} = \frac{\left(m_H m_C r_{HC}^2 + m_C m_N r_{CN}^2 + m_H m_N (r_{HC} + r_{CN})^2\right)}{m_{HCN}}$$

$$I_{DCN} = \frac{\left(m_D m_C r_{HC}^2 + m_C m_N r_{CN}^2 + m_D m_N (r_{HC} + r_{CN})^2\right)}{m_{DCN}}$$

Using the atomic masses from Appendix E to obtain the total molecular masses (27.010899 amu for HCN and 28.017176 amu for DCN) and multiplying them by the corresponding moment of inertia yields

$$I_{HCN} m_{HCN} = 18.937 \times 10^{-47} \text{ kg m}^2 \frac{27.010899 \text{ amu}}{1.66054 \times 10^{-27} \text{ kg amu}^{-1} \times 10^{-20} \text{ m}^2 \text{ Å}^{-2}}$$

$$I_{DCN} m_{DCN} = 23.178 \times 10^{-47} \text{ kg m}^2 \frac{28.017176 \text{ amu}}{1.66054 \times 10^{-27} \text{ kg amu}^{-1} \times 10^{-20} \text{ m}^2 \text{ Å}^{-2}}$$

From Equation 9.53, we have

$$308.0356 = 12.0939 r_{HC}^2 + 168.0369 r_{CN}^2 + 14.1126 \left(r_{HC}^2 + 2 r_{HC} r_{CN} + r_{CN}^2\right)$$

$$308.0356 = 24.1692 r_{HC}^2 + 168.0369 r_{CN}^2 + 28.2036 \left(r_{HC}^2 + 2 r_{HC} r_{CN} + r_{CN}^2\right)$$

---

[*] Gordy, W., *Phys. Rev.*, 101, 599, 1956.
[†] See Ogilvie, J.F. (1998) in the bibliography for this chapter.

Solution of these two simultaneous quadratic equations may yield values of r_{HC} and r_{CN} in Å. A general approach suited to molecules with even more geometrical parameters is to evaluate the numerator of Equation 9.53 for a range of parameters (bond lengths) and for all the isotopic species. The parameter choice for which there is the smallest root mean squared difference between moment-of-inertia-total-mass products (i.e., I times m) and the numerator values becomes the starting point for a finer search through parameters. Eventually, the optimum parameters (bond lengths) are determined to some chosen precision. On a computer, this is a fast process. For the example of HCN and DCN, the bond length values obtained are 1.066 Å for r_{HC} and 1.156 Å for r_{CN}. These two values together are termed the **substitution structure** of HCN since they have been obtained on the basis of isotopic substitution.

For nonlinear molecules, three rotational constants are used in the energy level expressions. They are associated with rotations about three orthogonal axes in the nonlinear molecule called **principal axes**. One or more of the rotational constants can be measured from a microwave spectrum, and isotopic substitution studies can be employed to extract geometrical parameters (i.e., bond lengths and angles).

The analysis, so far, has assumed that the molecules are rigid, but this is only an approximation. The molecules are vibrating, and so the rotational constants represent a vibrational average of structural parameters. Also, as in the case of diatomic molecules, molecular rotation may give rise to centrifugal distortion effects.

9.5 Harmonic Picture of Polyatomic Vibrations

In low-resolution infrared spectroscopy of diatomic molecules, the rotational fine structure is lost, and some feature in a spectrum assigned to be a fundamental transition is but a single peak. The frequency of that peak is taken to be the vibrational frequency in the absence of any more precise experiments, and that is the extent of the information obtained. This low-resolution information corresponds mostly with a nonrotating picture or else a rotationally averaged picture of the molecule's dynamics; to the extent that we can analyze the data, we need only consider pure vibration. To understand the internal dynamics of polyatomic molecules, it is helpful to start with a low-resolution analysis. This means neglecting rotation, or presuming the molecules to be nonrotating.

The pure vibrations of a polyatomic molecule can be quite complicated. It is convenient to think first about the potential energy for vibrational motions in order to understand their nature. The potential energy is a function of the atomic positions. For a molecule of N atoms, there are $3N$ atomic degrees of freedom, but only $3N - 6$ ($3N - 5$ for linear molecules) are left after removing the degrees of freedom for molecular translation and rotation. Thus, there are $3N - 6$ (or $3N - 5$) coordinates that describe the structure of the molecule, not the molecule's position and orientation in space. These $3N - 6$ (or $3N - 5$) coordinates are called **internal coordinates**, and most often the bond lengths and bond angles comprise a suitable, though not unique, set. For instance, the internal structure of a water molecule can be specified by the two O–H bond lengths and the H–O–H bond angle. These constitute a set of three internal coordinates ($3N - 6 = 3$ for water).

A **force field** refers to any potential for the vibrations of a molecule expressed in terms of some chosen set of internal coordinates. In principle, we arrive at complete understanding of the vibrations of a molecule if we know the force field precisely. Thus, we think of vibrational information in relation to the force field, and this is analogous to thinking about the vibrations of a diatomic in relation to the functional form of the stretching potential, $V(x)$. Approximate force fields can be constructed (sometimes guessed) in several different ways, and they can be used for deducing or computing vibrational information. Laboratory measurement of vibrational frequencies provides the ultimate test of such computed information.

The simplest force field for a molecule is one that is harmonic. This means that the potential energy has only linear and quadratic terms involving the $3N - 6$ coordinates. Some of these terms may be cross-terms, for example, a product of two coordinates such as $r_1 r_2$. As discussed in Chapter 7, a potential that is harmonic in *all* internal coordinates can be written so that there are no cross-terms provided that the original coordinates are transformed to the normal coordinates. In this section, we will designate normal coordinates as $\{q_1, q_2, q_3, \ldots\}$, and then the classical Hamiltonian for vibration is

$$H = \frac{1}{2} \sum_{i=1}^{3N-6} \left(\dot{q}_i^2 + \omega_i^2 q_i^2 \right) \tag{9.56}$$

This Hamiltonian is that of a separable problem, and as we have already considered in Chapter 7, it is equivalent to a problem of $3N - 6$-independent harmonic oscillators. The vibrational frequencies of the oscillators are the ω_i's. The Schrödinger equation that develops from the quantum mechanical form of this Hamiltonian is also separable. The energy level expression comes from the sum of the eigenenergies of the separated harmonic oscillators or modes. For each, there is a quantum number, n_i.

$$E_{n_1, n_2, n_3 \cdots} = \sum_{i=1}^{3N-6} \left(n + \frac{1}{2} \right) \hbar \omega_i \tag{9.57}$$

The quantum numbers may take on values of 0, 1, 2,….

A vibrational state of the polyatomic is specified by a set of values for the $3N - 6$ quantum numbers in Equation 9.57. The lowest energy state is the state that has all the quantum numbers equal to zero. The energy of this state, which is $\hbar \sum \omega_i / 2$, is the zero-point vibrational energy. Excited vibrational states consist of the infinite number of states for which any or all of the quantum numbers are not zero. As illustrated for the water molecule in Figure 9.11, the energy levels for even a small polyatomic molecule become increasingly numerous at higher and higher energies above the ground state. That is, with the states arranged according to their energies, the number of states found within some small energy increment is in an overall way increasing with energy. With states arranged according to their energies, the counting of the number of states per some unit energy value is often referred to as the **density of states**. The density of vibrational states of a polyatomic molecule is a function of the energy, a function that increases with energy.

From our detailed examination of diatomic vibrational motion, we know that the harmonic picture is an approximation with notable limitations. It is an approximation that is at its best for small-amplitude displacements, and that means for low-energy states. If we restrict our attention to the low-energy states, important qualitative information can be extracted from the harmonic picture. This qualitative information is the

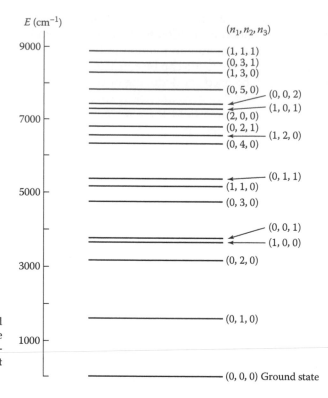

FIGURE 9.11

Representation of the low-lying vibrational state energy levels of the water molecule in a harmonic picture showing the increasing number of states per energy interval at higher energies.

nature of the normal modes of vibration. The coordinate transformation from atomic displacement coordinates to normal coordinates is a transformation that is equally valid for a classical and for a quantum mechanical picture. Therefore, we use classical notions about normal mode vibrations to understand the forms of molecular normal mode vibrations.

Let us use the carbon dioxide molecule as an example. It should have $3N - 5$ or four modes of vibration. We can represent each mode by the "direction" of the associated normal coordinate, that is, the directions in which each of the atoms move in the course of the vibration. Recall that a normal mode of vibration is the simplest motion of a system of particles, and that for a system vibrating in one mode, all the particles move in phase and at the same frequency. They reach their maximum points of displacement at the same instant, and they pass through their equilibrium positions at the same instant. If we were able to take a "freeze-frame" view of carbon dioxide vibrating purely in one of its normal modes, the directions the atoms move at the instant the particles are at their equilibrium positions would serve to describe the nature of the vibration. We could represent these directions of motion by arrows at each atom, and in fact, this is a very common way of representing molecular normal modes.

One of the normal modes of carbon dioxide is a bending motion along the O–C–O axis. The arrow representation of this mode is

$$\uparrow \downarrow \uparrow$$
$$\text{O–C–O}$$

This represents a vibration where the instantaneous direction of motion at the equilibrium point has the oxygen atoms moving up and the carbon atom moving down. There is an equivalent vibration where the atoms move the same but in a plane perpendicular to the plane of this bending motion. This makes for a degeneracy in the bending of carbon dioxide.

Another normal mode of CO_2 is that of a **breathing motion** or a **symmetric stretch**:

$$\overset{\leftarrow}{O}-C-\vec{O}$$

In this motion, the oxygen atoms move away from the carbon atom in phase. Using a classical picture for the moment, we would expect the oxygen atoms to continue to their turning points and then reverse directions and eventually pass through the equilibrium positions again. At that instant, their motions might be represented by

$$\vec{O}-C-\overset{\leftarrow}{O}$$

This arrow diagram looks different from the previous one, though it is for the same normal mode of vibration. The freeze-frame picture has been taken at a different instant when the particles are at their equilibrium. Either arrow diagram is a correct representation of the symmetric stretching vibration, and it is important not to consider the two drawings to represent different modes.

The mathematics behind the transformation to normal coordinates, which are represented by the arrow diagrams, provides certain rules that can be used to guess the qualitative form of the normal modes of molecules. First, a normal mode is a vibrational motion, and so any motion along the normal coordinate (following the arrows in the diagrams) must not lead to a rotation or translation of the molecule. For carbon dioxide, the following two diagrams are examples of pure translation and pure rotation:

$$\vec{O}\overset{\rightarrow}{-}\vec{C}-O \qquad \overset{\uparrow}{O}-C-\overset{\downarrow}{O}$$

When an arrow picture resembles these to any extent, it means that the motion is partly a rotation or translation. This implies that the motion is not purely that of a normal mode.

Normal coordinates are independent coordinates. This means that they are orthogonal just as unit vectors along the x-axis, along the y-axis, and along the z-axis of Cartesian space are orthogonal. The normal coordinates are more complicated and abstract. Even so, we can test for orthogonality in about the same way that we determine orthogonality for x, y, z unit vectors: The dot product of any two different unit vectors is zero. The dot product of a normal coordinate and another normal coordinate is found by adding up the dot products of the corresponding arrows on each atom. As an example, let us use the other stretching mode of carbon dioxide, called the **asymmetric stretch**. Its arrow diagram is

$$\vec{O}-\overset{\leftarrow}{C}-\vec{O}$$

This is orthogonal to the symmetric stretch, as seen by taking the dot product of corresponding arrows on the atoms:

$$\overset{\rightarrow 1 \leftarrow 2 \rightarrow 3}{O-C-O}$$

$$\overset{\rightarrow 4 \quad \leftarrow 5}{O-C-O}$$

Taking each of the arrows to be of unit length, the dot product of arrows 1 and 4 is +1. The dot product of the arrows on the farthest right atom (3 and 5) is −1. For the carbon atom arrows, the dot product is zero since carbon is not displaced at all in the course of the symmetric stretch. The sum of the three numbers, 1, 0, and –1, is zero. This is the mathematical statement that these two motions, or these two normal coordinates, are distinct or orthogonal.

For all molecules there is a mode that is best viewed as a breathing mode. All the atoms move outward from the equilibrium position in this mode. The symmetric stretch mode of carbon dioxide is its breathing mode. To work out arrow diagrams for all the normal modes of a molecule, we start with a diagram for the breathing mode. Then, we need to identify all other arrow diagrams that are orthogonal, making sure that the diagrams do not include any rotation or translation. If the molecule has symmetry, we can expect the arrow diagrams to reflect that in some way, and that may help systematize our search for the normal coordinate diagrams. Although this process is not necessarily unique, the qualitative information about the nature and types of vibrations will be correct.

As another example, let us find arrow diagrams for the normal modes of the water molecule. The breathing motion must be a simultaneous stretch of the two O–H bonds. If we represent this with an arrow diagram showing the hydrogens moving and the oxygen staying fixed, we will have a representation of a motion where the center of the mass of the molecule moves. Translation will be part of the motion. So, an arrow must be placed on oxygen to make sure that the motion does not include any translation. The representation of the symmetric stretching vibration is

The arrow on the oxygen is smaller than the arrows on the hydrogens because oxygen is so much more massive than the hydrogens; the center of mass remains in place for a displacement of oxygen that is relatively smaller than the hydrogen displacements. We can expect a bending motion, and again, an arrow must be used for oxygen to ensure that the motion does not include any translation or rotation.

An orthogonality test will show that this is an acceptable mode because it is orthogonal to the symmetric stretch. The number of modes is $3N - 6$ or 3. So, the remaining mode will be represented by a set of arrows that describe a motion that is orthogonal to both the symmetric stretch and the bend. After some thought, we realize that this mode is

This is an asymmetric stretch.

In general, there is no strict separation of bending and stretching vibrations. In large molecules possessing little symmetry, a normal mode may appear to involve stretching of several bonds, bending motions, and torsional motions together. Sometimes normal modes turn out to be *primarily* a stretch of one bond, or a bend about one center, and in those cases we qualitatively associate (label) a vibrational frequency with a localized

TABLE 9.2

Vibrational Frequencies (cm^{-1}) of Selected Small Molecules

	Symmetric Stretch[a]	Other Stretching Modes		Bending Modes		Other Modes	
CO_2	1333	2349		667			
H_2O	3657	3756		1595			
D_2O	2671	2788		1178			
H_2S	2615	2626		1183			
D_2S	1896	1999		855			
SO_2	1151	1362		518			
CS_2	658	1535		397			
HCN	2097	3311		712			
DCN	1925	2630		569			
NH_3	3337	3444				950	1627
ND_3	2420	2564				748	1191
HCCH	1974	3289	3374	612	730		
DCCD	1762	2439	2701	505	537		
H_2CO	1746	2783	2843	1500		1167	1249
F_2CO	1928	965	1249	584		626	774
CH_4	2917	3019				1306	1534
CD_4	2109	2259				996	1092
CF_4	909	1281				435	632
CCl_4	459	776				217	314
SiH_4	2187	2191				914	975
SiD_4	1558	1597				681	700
SiF_4	800	1032				268	389

Source: Fundamental vibrational frequencies for different vibrational modes from Lide, D.R., *Handbook of Chemistry and Physics*, 86th edn., CRC Press, Boca Raton, FL, 2005.

[a] The mode identified as the symmetric stretch mode is better described as a "breathing" mode in molecules that are not centrosymmetric.

vibration (e.g., an H–C stretch in HCN), always recognizing the vibration to be more complicated than its label. Table 9.2 lists stretching, bending, torsional, and other frequencies of a number of small polyatomic molecules.

9.6 Polyatomic Vibrational Spectroscopy

The selection rule for a diatomic molecule is that the vibrational quantum number changes by 1, at least under the harmonic approximation of the potential. Also, the dipole moment has to change in the course of the vibration or else the transition is forbidden. Carbon monoxide, for instance, has an allowed fundamental transition, whereas N_2 does not. The separation of variables that is accomplished with the normal mode analysis says that each mode can be regarded as an independent one-dimensional oscillator. Thus, we can borrow the results for the simple harmonic oscillator to conclude that a transition will be allowed if the vibrational quantum number for any single mode changes by 1 where the vibrational motion in that mode corresponds to a changing dipole moment.

The second condition for spectroscopic transitions among polyatomic vibrational levels requires that we determine if a molecule's dipole moment will change in the course of a particular normal mode vibration. Returning to the example of carbon dioxide, we can assign a partial charge to the oxygen atoms-call it δ-and a partial charge to the carbon, which then must be -2δ for the molecule to be neutral. If the atoms are displaced in the direction of the symmetric stretch mode, the contribution to the dipole moment from one oxygen will cancel the contribution from the other because one will move in the +z-direction and the other in the $-z$-direction. Thus, in the course of the symmetric stretch, the dipole moment of carbon dioxide is unchanging. This means that transitions where the symmetric stretch quantum number is changed will not be (easily) seen in an infrared absorption spectrum. The asymmetric stretch will have allowed transitions since the dipole moment changes in the course of this vibration. The two oxygens move in the same direction, and thus their contributions will not cancel. The bending motion will also give rise to a changing dipole moment.

The diatomic selection rule was found to be $|\Delta n| = 1$ when the potential is strictly harmonic. Other transitions become allowed with an anharmonic potential, but then the $|\Delta n| = 1$ transitions stand out as being stronger. For polyatomic molecules, the selection rule under the assumption of a strictly harmonic potential is $|\Delta n_i| = 1$, while $|\Delta n_j| = 0$ for all the j modes other than the i mode. This means that only one vibrational quantum number can change in any single transition event. Of course, molecules do not have *strictly* harmonic potentials, and just as for diatomics, this selection rule really predicts which transitions will be the strong ones. Other transitions become allowed because of anharmonicity. The transitions that obey this harmonic selection rule and that originate from the ground vibrational state are called **fundamental transitions**, which is the same term used with diatomic molecules. Transitions with $|\Delta n_i| > 1$ that originate in the ground state are called **overtone transitions**. If two or more vibrational quantum numbers change in a transition, it is called a **combination transition**.

The vibrational states of molecules are often designated by a list of the vibrational quantum numbers. The ordering of the list is according to the vibrational frequency. The quantum number for the highest vibrational frequency mode is first on the list. For carbon dioxide, a state is represented as (n_1, n_2, n_3), where n_1 refers to the symmetric stretch, n_2 refers to the bend, and n_3 refers to the asymmetric stretch. Only three quantum numbers are used for this type of designation because the in-plane bend and the out-of-plane bend are two modes that are degenerate; they have the same frequency, and only one quantum number is used. The ground state is $(0,0,0)$. The following is a representative list of possible transitions and their type, if appropriate:

$(0, 0, 0) \rightarrow (0, 0, 1)$	Fundamental
$(0, 0, 0) \rightarrow (1, 0, 0)$	Fundamental
$(0, 0, 0) \rightarrow (0, 2, 0)$	Overtone
$(0, 0, 0) \rightarrow (2, 0, 0)$	Overtone
$(0, 0, 0) \rightarrow (1, 0, 1)$	Combination

With the many degrees of freedom in a polyatomic molecule, the many different normal mode frequencies, and the many types of transitions that might be seen, it is evident that polyatomic vibrational spectra can be quite congested and challenging to analyze.

The analysis of a full infrared spectrum of a molecule in the gas phase usually starts by finding the fundamental transitions. Most often, these are the strongest transitions. Next, we look for **progressions**, which are a set of transitions originating from the same initial state and involving the excitation of one mode by successive quantum steps. For instance, the progression built on the fundamental transition $(0,0,0,0) \rightarrow (0,1,0,0)$ of some hypothetical molecule is the set of transitions from the ground state to $(0,2,0,0)$, and to $(0,3,0,0)$, and so on. It is useful to look for these overtone transitions because, to the extent that the harmonic picture holds, we expect them to be found in the spectrum at $2\omega_i$, $3\omega_i$, and so on, where ω_i is the fundamental transition frequency. Generally, the strongest fundamental transitions have the strongest overtone progressions. Combination bands, or the low-resolution peaks in the spectrum from combination transitions, can be identified by matching the frequencies with sums and differences of fundamental transition frequencies, and of course, allowing for the fact that the harmonic picture does not hold perfectly for real molecules.

Another type of infrared absorption band is called a **hot band**, and it arises from any transition that does not originate in the ground vibrational state. It is known that the population of excited states grows as a sample is warmed. In a room-temperature sample, it is often possible for the population of a low-lying excited vibrational state to be large enough for detectable transitions to originate from the molecules found to be in this state. Hot band transitions can occur throughout the spectrum. For instance, a hot band corresponding to the transition $(1,0,0) \rightarrow (2,0,0)$ for a linear triatomic molecule is likely to be at a frequency somewhat close to that of the fundamental transition $(0,0,0) \rightarrow (1,0,0)$. The transition moments are almost the same for the two transitions, too. However, their populations lead to important spectral differences. That fact can be exploited, for if the temperature of the sample is lowered, the hot band's intensity will diminish relative to a band that originates in the ground vibrational state. This is the consequence of diminishing the excited state's population with decreasing temperature. It is a common practice, in fact, to remove hot band congestion from a spectrum by cooling the sample.

There are vibrational transition frequencies that turn out to be characteristic of specific types of chemical bonds. For instance, the chemical bond between carbon and oxygen in a carbonyl functional group is known to be only slightly affected by what the carbonyl is attached to. The C=O bonding in formaldehyde is not sharply different from that in acetone. This means that the force constant for stretching the carbon–oxygen bond is similar in both molecules. In itself, this does not imply that there are similar vibrational frequencies for the two molecules. Even within the harmonic picture, we expect the vibration of the carbonyl to be coupled to the rest of the molecule, and that differs from one species to another. Even so, there are similar vibrational frequencies. The frequencies associated with functional groups are called **characteristic frequencies**. These are the frequencies at which we can expect vibrational transitions in molecules that contain the given functional group. Table 9.3 lists some characteristic frequencies. There is a range for each characteristic frequency because of the effect of coupling with the rest of the molecule. Characteristic

TABLE 9.3

Characteristic Vibrational Stretching Frequencies (cm^{-1}) of Certain Functional Groups[a]

C=O	1700	C≡N	2200	C–H	3000
C–O	1100	C–N	1200	O–H	3500
C=C	1650	C≡C	2200	N–H	3400

[a] The characteristic frequencies are ranges around these values, and the ranges sometimes amount to a few hundred wavenumbers.

frequencies are a useful tool in chemical analysis. A sample of an unknown that exhibits a strong absorption at $1700\,cm^{-1}$, for instance, quite probably contains a carbonyl group in its structure since that is within the range of the characteristic frequency of carbonyl stretching.

Finally, there is a fascinating vibrational motion in certain molecular systems called **inversion** or **interconversion**. An example of inversion is exhibited by the ammonia molecule. It is a pyramid-shaped molecule at its equilibrium. Let us imagine it as three protons in a plane to the left of the nitrogen so that an umbrella-opening type of vibration moves the plane of the protons to the right and that of the nitrogen to the left. If this is continued, a point is reached where the protons and the nitrogen are all in the same plane. This is called the inversion point. If the motion continues still further, the protons will be in a plane to the right of the nitrogen. But then, the molecule will have the same shape as it did originally. It will have achieved a structure that is equivalent to the original structure in terms of the internal structural parameters. The potential energy will have a unique form along this inversion pathway. It will be at a maximum at the inversion point and will have two equivalent minima corresponding to the ammonia pyramid pointing left and pointing right. This is a double-well potential and an example was shown in Figure 8.8.

The energy levels of a one-dimensional double-well problem are interesting, though working them out can be involved. As mentioned in Chapter 8, perturbation theory may be useful if the inversion barrier is low in energy relative to the ground vibrational state. And of course, a one-dimensional analysis means that motion along the inversion pathway has been separated from other vibrational motions, and that is an approximation. Figure 9.12 presents calculated energy levels for a particular double-well potential. The effect of the barrier is seen by contrasting the energy level spacing with that of a harmonic oscillator. Instead of evenly spaced levels, the perturbed system has levels brought closer together in pairs. The higher levels show the least effect. The lowest pairs of levels can be brought extremely close by a suitable barrier, and often the separation energy between such levels is termed a **splitting**. Selection rules are also affected by a potential barrier since the barrier represents a significant anharmonicity.

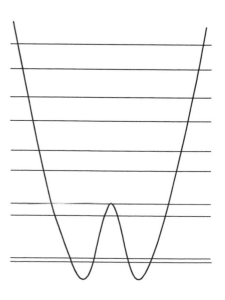

FIGURE 9.12
The 10 lowest-lying vibrational energy levels of a one-dimensional system experiencing a double-minimum potential. The potential is harmonic away from the middle, and the highest levels shown have an energy spacing characteristic of a harmonic oscillator. However, the energies of the first six states are significantly perturbed by the bump in the potential. The perturbation brings the energies of pairs of states closer together.

9.7 Conclusions

Molecular spectroscopy is the means for measuring the energy differences between molecular eigenstates via the frequencies of absorbed and/or emitted electromagnetic radiation. Incident light (electromagnetic radiation) can perturb a molecule through its time-varying electric and/or magnetic fields, and the perturbation may bring about a change in the state of the system. In the course of that change, energy is exchanged with the electromagnetic radiation in quanta whose size is $h\nu$, ν being the frequency of the radiation. Thus, measuring the frequency, ν, of absorbed or emitted radiation amounts to measuring a difference between the energies of states of a molecule.

The TDSE is a generalization of the (time-independent) Schrödinger equation that allows for a system to have a probability density that changes in time. Perturbation theory with time-dependent wavefunctions leads to an important result at first order. The likelihood of a transition from some initial state, ψ_i, to some final state, ψ_f, depends on the integral quantity $\langle \psi_i | H' \psi_f \rangle$, where H' is the part of the Hamiltonian associated with the time-dependent perturbation (e.g., radiation). If this integral is zero, the transition is forbidden to first order, and if not, it is allowed.

The vibrational-rotational states of diatomic molecules are probed in spectroscopic experiments using radiofrequency or microwave radiation (low energy, pure rotational spectroscopy) or infrared radiation (vibrational spectroscopy). The former requires a permanent dipole moment for a transition to take place and a change in the rotational quantum number of $\Delta J = 1$ ($\Delta J = -1$ in emission). The latter requires that the dipole moment change in the course of the vibrational motion and that $\Delta n = 1$ ($\Delta n = -1$ in emission) and $\Delta J = \pm 1$ (except for diatomic radicals where ΔJ can also be 0). These selection rules lead to a pattern of lines in the high-resolution vibrational spectrum, and the lines make up a vibrational band.

The transition frequencies, and hence the transition energies, of the lines in a vibrational band can be fitted to energy level expressions derived quantum mechanically. The fitting coefficients are spectroscopic parameters, such as the harmonic frequency, ω, and the rotational constant, B. These parameters characterize a molecule and can provide detailed structural information. A very approximate quantum mechanical analysis of a diatomic molecule takes the stretching potential to be strictly harmonic and the rotation to be that of a rigid set of two point masses. Under this approximation, the qualitative form of a vibrational band is found to be that of lines separated by $2B$ in two branches (P and R), with a $4B$ separation between the two branches. More general analysis includes effects of centrifugal distortion, vibrational anharmonicity, and vibration rotation coupling.

The vibrations of polyatomic molecules are complicated. Useful insight is often gained by adopting a harmonic approximation of the vibrational potential, and then the problem of molecular vibration separates into $3N - 6$ (or $3N - 5$ for linear molecules) one-dimensional oscillator problems for an N-atom molecule. Transitions from the ground state involving a change of 1 in one vibrational quantum number are fundamental transitions. Transitions from the ground state involving a change of more than 1 in one quantum number are overtones. Transitions in which two vibrational quantum numbers change are termed combinations, and transitions originating from excited states give rise to hot bands.

Point of Interest: Laser Spectroscopy

The pace of development and application of rotational, vibrational, and visible–ultraviolet spectroscopy accelerated sharply immediately after World War II. The quantum mechanical revolution was complete in physics as of the 1930s, whereas in chemistry its full effect was not manifested until a decade or so later. The quantum level understanding that then took root in chemical science generated interest, enthusiasm, and investigation in spectroscopic analysis; there was a spectroscopic revolution taking place in chemistry by the middle of the twentieth century.

In the first part of the postwar era, the highest precision for spectroscopic measurements of absorption frequencies was to six or seven significant figures in the microwave and radiofrequency regions of the spectrum, but only four or five significant figures elsewhere. In the laboratory, microwaves and radiowaves were employed with different technology than infrared and visible radiation. Usually, this long-wavelength radiation was generated with standing-wave patterns in metallic (electrically conducting) pipes with rectangular cross-sections. These were called waveguides, and their dimensions could range to 2 m in length and several cm in width and height. A terminating device was used to effect small variations in the length of the standing wave in the waveguide. Measurement of these variations to 1 mm relative to a 1 m waveguide meant a precision of 1 ppm in the measurement of the wavelength. As a result, postwar rotational spectroscopy had unrivaled precision. On top of that, the distribution of radar equipment from World War II stockpiles to research laboratories in the United States provided investigators with the instrumental "guts" to start performing rotational spectroscopy.

In 1951, Charles H. Townes had an idea for stimulated emission of microwave radiation. He and his coworkers at Columbia University used a transition of the ammonia molecule that occurred in the microwave region of its spectrum in order to create a maser (microwave amplification by stimulated emission of radiation). On producing ammonia molecules in the upper state involved in the transition, the application of microwave radiation at the transition frequency caused excited state molecules to emit a photon and drop to the ground state. Since the emitted photons were at the same frequency as that of the incident radiation, the process led to an amplification of the applied radiation.

In 1958, Townes and Arthur Schawlow of Stanford University published a report predicting that the principle behind masers could be applied in the infrared and optical regions of the electromagnetic spectrum. The first laser (light amplification by stimulated emission of radiation) was demonstrated in 1960 using a ruby crystal for the lasing material. Operation of the first gas laser was reported the following year. Today, helium–neon gas lasers are commonplace, and counting all types of lasers, the annual worldwide production is in the hundreds of millions.

Laser light is essentially light at one frequency. It is also coherent, which means all the waves are in-phase. Light from a thermal source, such as the electrically heated filament of a light bulb, is an incoherent distribution of radiation across the electromagnetic spectrum; the light source produces a continuum of frequencies. In spectroscopy, this type of light has to be frequency selected by prisms or diffraction gratings. The resolution is limited by the resolving power of the prism or grating. It is also limited by the intensity of the light source because as resolving power is increased, a narrower range of the source's output is selected. The full intensity of a laser source is at one frequency. Techniques for "tuning" laser frequencies over small but useful ranges have been developed, and in this way conventional absorption spectroscopy using a laser source becomes possible. The use of lasers in infrared and visible spectroscopy has accounted for an improvement in resolution by one, two and sometimes three orders of magnitude. Very high precision is no longer limited to rotational spectra, and new molecular knowledge continues to emerge from spectroscopic studies across the spectrum.

Exercises

9.1 Compute the frequency in Hz for electromagnetic radiation of 1, 10, 100, 1000, and 10,000 cm^{-1}. What part of the electromagnetic spectrum do these frequencies correspond to? What is the energy in J of 1 mol of photons at each of these frequencies?

9.2 The equilibrium bond length of LiH is 1.595 Å. Find the equilibrium rotational constant in cm^{-1} for LiH and for the isotopic forms, LiD and ^{6}LiH, taking the bond lengths to be identical.

9.3 Using data in Table 9.1, calculate the band center frequency (i.e., the average of the frequency of the first P-branch and first R-branch lines) for the $n = 0 \rightarrow 1, 1 \rightarrow 2, 2 \rightarrow 3$ and $3 \rightarrow 4$ vibrational bands of LiH.

9.4 Using standard bond length values as an estimate of the true bond lengths in HCCH and FCCH, obtain the rotational constants of these two molecules in MHz.

9.5 Shown here are sketches of the normal modes of acetylene. Which modes are infrared-active?

$$\overleftarrow{H} \overleftarrow{C} \overrightarrow{C} \overrightarrow{H} \qquad \overleftarrow{H} \overrightarrow{C} \overleftarrow{C} \overrightarrow{H} \qquad \overleftarrow{H} \overrightarrow{C} \overrightarrow{C} \overleftarrow{H} \qquad \overset{\uparrow}{H} \overset{\uparrow}{C} \underset{\downarrow}{C} \underset{\downarrow}{H} \qquad \overset{\uparrow}{H} \underset{\downarrow}{C} \overset{\uparrow}{C} \underset{\downarrow}{H}$$

(in-plane and out-of-plane)

9.6 Assume that at time $t = 0$, a harmonic oscillator has been prepared via some prior time-dependent perturbation such that its time-dependent wavefunction at that instant is an equally weighted superposition of the $n = 0$ and $n = 1$ time-independent wavefunctions. Write the explicit form of $\Psi(x, t)$ and derive an expression for the probability density at the point $x = 0$ as a function of time for $t > 0$.

9.7 Using first-order time-dependent perturbation theory, find selection rules for electromagnetic radiation interacting with a particle in a box, taking the particle to possess a net charge.

9.8 Verify Equation 9.17, which is needed in Equation 9.41, by writing the product, $\cos \theta \, Y_{J'M'}$ as a linear combination of other spherical harmonic functions and then using the orthogonality of the spherical harmonics to evaluate the integral.

9.9 Use the explicit $n = 0$ and $n = 2$ wavefunctions of a harmonic oscillator to confirm that a corresponding overtone transition (i.e., $\Delta n = 2$) is an allowed vibrational transition for a diatomic molecule if the dipole moment has some quadratic dependence on the displacement coordinate (s^2 dependence in Equation 9.42).

9.10 Assume that it is possible to measure a vibrational transition frequency for a diatomic for which the J quantum number remains unchanged at zero. Also assume that there is some diatomic for which the following transition frequencies (in cm^{-1}) are then obtained: 1600 for $n = 0$ to $n = 1$, 3100 for $n = 0$ to $n = 2$, and 4503 for $n = 0$ to $n = 3$. Find values for the vibrational frequency, the anharmonicity constant, and the second anharmonicity constant on the basis of these frequencies. Next, compare these values with those obtained for the vibrational frequency and the anharmonicity constant if it is assumed that the second anharmonicity constant, $\omega_e y_e$, is zero and only the first two frequencies are used.

9.11 Spectroscopic constants of CO are given in Table 9.1. Find the equilibrium bond length from the rotational constant. Then, predict the transition frequencies for the first several P- and R-branch lines in the fundamental absorption band of the isotopically substituted form, $^{13}C^{17}O$.

9.12 Calculate the relative populations of the first three rotational energy levels of CN and HF at 298 K.

9.13 The fundamental infrared band of HCl is complicated because the natural isotopic abundance of chlorine means that transitions for two isotopes can be readily observed. That is, unless isotopically purified, the spectrum of an HCl sample is actually a superposition of the spectra of two isotopes of HCl, as shown in Figure 9.8. For the first several P- and R-branch lines, calculate the separation in cm^{-1} for the corresponding lines in the spectra of the two isotopic forms.

9.14 Repeat the derivation of Equations 9.44 and 9.45 with the inclusion of an energy term, $D[J(J + 1)]^2$, associated with centrifugal distortion.

9.15 From the following general energy level expression for a diatomic,

$$\tilde{E}_{n,J} = \tilde{\omega}_e\left(n+\frac{1}{2}\right) + \tilde{B}_e J(J+1) - \tilde{D}_e[J(J+1)]^2 - \tilde{\alpha}_e J(J+1)\left(n+\frac{1}{2}\right)^2$$

find the branch separation for the fundamental vibrational band in terms of the spectroscopic constants. That is, develop an expression for the difference in transition energies between the P- and R-branch transition lines closest to the band center.

9.16 What is the bond length in CO_2 if the moment of inertia is 7.17×10^{-46} kg m^2?

9.17 Assume that the vibrational spectrum of LiH is well represented by the following energy level expression given in cm^{-1}.

$$\tilde{E}_{n,J} = 1405.65_e\left(n+\frac{1}{2}\right) + 7.513J(J+1) - 0.01J(J+1)]^2 - 23.20\left(n+\frac{1}{2}\right)^2 - 0.213\left(n+\frac{1}{2}\right)J(J+1)$$

Find the equilibrium bond length. Next, find the corresponding energy level expression for LiD, taking the potential energy function to be the same as for LiH.

9.18 Redo Example 9.1 using an energy level expression that includes centrifugal distortion.

9.19 Should there be a qualitative difference in the normal modes of vibration, both in the form of the modes and in their infrared activity, between NNO and CO_2?

9.20 Make a sketch using arrows to show a reasonable guess of the qualitative form of the normal modes of vibration for (a) formaldehyde (b) ethylene (c) hydrogen peroxide.

9.21 To give an idea of what is typical for small molecules, estimate the zero-point energies of the water molecule, of acetylene, and of formaldehyde either on the basis of actual vibrational frequencies or on the basis of characteristic infrared frequencies.

9.22 A characteristic H–C stretching frequency is $3000\,cm^{-1}$. Let us assume that this is the frequency for the normal mode of HCN that looks most like a stretch of the H–C bond. This mode might be modeled as a pseudodiatomic vibration if we were to think of the CN as one "atom" and the H as the other. Within this model, we can estimate the effects of certain changes to the molecule as if they were simply changes to the mass of the pseudodiatomic species. On that basis, what would be the frequency for this vibration if ^{13}C were substituted for ^{12}C? What would be the frequency if the group C–CN were substituted for the N? (This second case offers an idea of why, and to what extent, frequencies are "characteristic" of the bonding environment.) What would the frequency become in DCN? Compare with values in Table 9.2.

9.23 Make a sketch of the vibrational energy levels of acetylene that lie within $2500\,cm^{-1}$ of the ground vibrational state assuming harmonic behavior and using the fundamental frequencies in Table 9.2. Next to that sketch, do the same for C_2D_2.

9.24 Consider a double-well potential of the form $V(x) = x^2/2 + 3/2e^{-(\beta x)^2/2}$. If a particle of mass $m = 1$ were to experience this potential, what would be the energies of the first five states from zero-, first-, and second-order perturbation theory using the exponential part of the potential as the perturbation?

9.25 How many normal modes of vibration are there in ozone, hydrogen peroxide, and cyclohexane?

9.26 Calculate the wavelengths and frequencies for the $1 \to 2$, $3 \to 4$, $6 \to 7$, and $9 \to 10$ transitions in the pure rotational spectrum of NO.

9.27 Calculate the force constant for LiH if the fundamental vibrational absorption band occurs at $1405.6\,cm^{-1}$.

Bibliography

Bernath, P. F., *Spectra of Atoms and Molecules*, 2nd edn. (Oxford University Press, New York, 2005).

Diem, M., *Introduction to Modern Vibrational Spectroscopy* (Wiley-Interscience, New York, 1993).

Herzberg, G., *Molecular Spectra and Molecular Structure. I. Spectra of Diatomic Molecules* (Van Nostrand Reinhold, New York, 1950).

Hollas, J. M., *Modern Spectroscopy*, 2nd edn. (John Wiley, New York, 1991).

Graybeal, J. D., *Molecular Spectroscopy* (McGraw-Hill, New York, 1993).

Ogilvie, J. F., *Vibrational and Rotational Spectroscopy of Diatomic Molecules* (Academic Press, New York, 1998).

Struve, W. S., *Fundamentals of Molecular Spectroscopy* (John Wiley, New York, 1989).

Wilson, Jr. E. B., J. C. Decius, and P. C. Cross, *Molecular Vibrations. The Theory of Infrared and Raman Vibrational Spectra* (Dover, New York, 1980).

10

Electronic Structure

How electrons are distributed about nuclear centers and how they participate in chemical bonds are crucial aspects of chemistry, one dictated by the laws of quantum mechanics. This is the problem of electronic structure, using the Schrödinger equation to find wavefunctions for electrons in atoms and molecules. The atom with the fewest electrons, the hydrogen atom, is as an important model problem, and the quantum mechanical analysis of the hydrogen atom is carried out in detail in this chapter. Based on that discussion, we explore the qualitative features of the structure of more complicated atoms and molecules.

10.1 Hydrogen and One-Electron Atoms

H, He$^+$, Li^{2+}, Be^{3+}, and so on, are one-electron atoms. Each consists of only two particles, a nucleus and an electron. The quantum mechanical description of one-electron atoms is the starting point for understanding the electronic structure of atoms and of molecules. It is a problem that can be solved analytically, and it is useful to work through the details. In the case of the hydrogen atom, the nuclear mass is about 2000 times that of an electron. Thus, the proton would be expected to make small excursions about the mass center relative to any excursions of the very light electron. That is, we expect the electron to be "moving quickly" about, and in effect, orbiting the nucleus.

The Schrödinger equation for this two-body problem starts out the same as the general two-body Schrödinger equation (Equation 9.18); however, the potential function, $V(r)$, is different from that of the vibrating–rotating diatomic molecule. It is an electrostatic attraction of two point charges, and its form is

$$V(r) = -\frac{Ze^2}{r} \tag{10.1}$$

where
  Z is the integer nuclear charge or atomic number
  e is the value of the fundamental charge of the electron

This interaction between the electron and the nucleus is the product of their respective charges, in this case $+Ze$ for the nucleus and $-e$ for the electron, divided by the distance between them, r.

The potential function for the hydrogen atom is dependent only on the spherical polar coordinate r. Thus, the separation of variables carried out in going from Equations 9.18 through 9.22 remains valid for this problem. That means we immediately know that the wavefunctions for the hydrogen atom consist of some type of radial function, $R(r)$, times a spherical harmonic function, $Y_{lm}(\theta, \varphi)$. To find the radial function, we start with Equation 9.22 and use the potential function in Equation 10.1.

$$-\frac{\hbar^2}{2\mu}\left(\frac{2}{r}\frac{\partial R}{\partial r} + \frac{\partial^2 R}{\partial r^2} - \frac{l(l+1)}{r^2}R\right) - \frac{Ze^2}{r}R = ER \tag{10.2}$$

where μ is the reduced mass of the electron–nucleus two-body system. Notice that with the sizable difference in mass of the two particles, the reduced mass is very nearly equal to the mass of the lighter particle, the electron. For hydrogen,

$$\mu = \frac{m_e M_P}{m_e + M_P} = \frac{1836.15 m_e^2}{m_e + 1836.15 m_e} = \frac{1836.15 m_e}{1837.15} \cong m_e$$

where
M_P is the proton mass
m_e is the electron mass

In the vibrating–rotating two-body system, it was appropriate to approximate the $l(l + 1)$ term in the radial Schrödinger equation; however, that would not be appropriate for the hydrogen atom. The separation distance between the electron and the proton, which is given by the coordinate r, can and does vary widely in the hydrogen atom states, and so there is no basis for using the truncated power series expansion as was done for the vibrating–rotating diatomic. Fortunately, the differential equation in Equation 10.2 has known solutions, and approximation is not necessary. There are an infinite number of these solutions for each particular value of the quantum number l, and we introduce another quantum number, n, to distinguish these solutions. Both l and n must label the different eigenfunctions, for example, $R_{nl}(r)$. Conditions on the quantum number, n, that come about in solving the differential equation are that n must be a positive integer and that for some choice of l, n may take on only the values $l + 1$, $l + 2$, $l + 3$, and so on. The latter condition can be inverted so as to relate the value of l to n. Then, we have that for a given choice of n, l can be only 0, 1, 2,..., or $n - 1$.

From the separation of variables, we now have that the wavefunctions for the hydrogen atom are

$$\psi_{nlm}(r, \theta, \phi) = R_{nl}(r)Y_{lm}(\theta, \phi) \tag{10.3}$$

And the quantum numbers that distinguish the possible states must satisfy these conditions:

$$n = 1, 2, 3, \ldots \tag{10.4a}$$

$$l = 0, 1, 2, \ldots, n - 1 \tag{10.4b}$$

$$m = -1, -1+1, \ldots, 1-1, 1 \tag{10.4c}$$

The radial functions, $R_{nl}(r)$, that are the eigenfunctions of Equation 10.2 can be constructed from a set of polynomials called the **Laguerre polynomials**. The Laguerre polynomial of order k can be generated by

$$L_k(z) = e^z \frac{d^k}{dz^k}(z^k e^{-z})$$ (10.5)

Associated Laguerre polynomials can be generated by

$$L_k^j(z) = \frac{d^j}{dz^j} L_k(z)$$ (10.6)

The radial functions, expressed in terms of associated Laguerre polynomials, are

$$R_{nl}(r) = \sqrt{\frac{(n-l-1)!}{2n[(n+l)!]^3}} e^{-\rho/2} \rho^l L_{n+l}^{2l+1}(\rho)$$ (10.7)

where ρ is the variable r scaled by the constants Z, μ, e, and $\hbar$, which are in the Schrödinger equation, and by the quantum number, n.

$$\rho \equiv \frac{2Z\mu e^2}{n\hbar^2} r$$ (10.8)

The square root factor in Equation 10.7 is for normalization over the range from $r = 0$ to infinity. Recall that in spherical polar coordinates the volume element is $r^2\, dr\, \sin\theta\, d\theta\, d\varphi$, and thus, normalization of the radial equations obeys the following:

$$\int_0^\infty R_{nl}^2(r) r^2 dr = 1$$

Table 10.1 lists the explicit forms for several of these radial functions.

When the functions of Equation 10.7 are used in Equation 10.2, the energy eigenvalue associated with a particular $R_{nl}(r)$ function is found to be

$$Enl = -\frac{\mu Z^2 e^4}{2\hbar^2 n^2}$$ (10.9)

Thus, the energies of the states of the hydrogen atom depend on only the quantum number n. The lowest energy state is with $n = 1$, and for this state, the energy given by Equation 10.9 is $-109{,}678\,\mathrm{cm}^{-1}$. The next energy level occurs with $n = 2$, and this energy is one-fourth (2^{-2}) of the lowest energy or $-27{,}420\,\mathrm{cm}^{-1}$. Next is the energy level for states with $n = 3$. This energy is one-ninth (3^{-2}) of the lowest energy or $-12{,}186\,\mathrm{cm}^{-1}$. At $n = 100$, the energy of the hydrogen atom is $-11\,\mathrm{cm}^{-1}$. Clearly, as n approaches infinity, the energy approaches zero. This limiting situation corresponds to the ionization of the atom.

Since the energy of the hydrogen atom depends on only n, and since, according to Equation 10.4, there are several states with the same n, the states of the hydrogen atom are degenerate. We can use the rules in Equation 10.4 to see that the degeneracy of each level is n^2, as in Table 10.2.

TABLE 10.1

Hydrogen Atom Radial Functions

N	l	Radial Function $\left[\rho = \dfrac{2Z\mu e^2 r}{n\hbar^2} \text{ and } N = \left(\dfrac{Z\mu e^2}{\hbar^2}\right)^{3/2}\right]$
1	0	$R_{10}(r) = N2e^{-\rho/2}$
2	0	$R_{20}(r) = \dfrac{N}{2\sqrt{2}}(2-\rho)e^{-\rho/2}$
	1	$R_{21}(r) = \dfrac{N}{2\sqrt{6}}\rho e^{-\rho/2}$
3	0	$R_{30}(r) = \dfrac{N}{9\sqrt{3}}(6-6\rho+\rho^2)e^{-\rho/2}$
	1	$R_{31}(r) = \dfrac{N}{9\sqrt{6}}(4\rho-\rho^2)e^{-\rho/2}$
	2	$R_{32}(r) = \dfrac{N}{9\sqrt{30}}\rho^2 e^{-\rho/2}$
4	0	$R_{40}(r) = \dfrac{N}{96}(24-36\rho+12\rho^2-\rho^3)e^{-\rho/2}$
	1	$R_{41}(r) = \dfrac{N}{32\sqrt{15}}(20\rho-10\rho^2+\rho^3)e^{-\rho/2}$
	2	$R_{42}(r) = \dfrac{N}{96\sqrt{5}}(6\rho^2-\rho^3)e^{-\rho/2}$
	3	$R_{43}(r) = \dfrac{N}{96\sqrt{35}}\rho^3 e^{-\rho/2}$
5	0	$R_{50}(r) = \dfrac{N}{300\sqrt{5}}(120-240\rho+120\rho^2-20\rho^3+\rho^4)e^{-\rho/2}$
	1	$R_{51}(r) = \dfrac{N}{150\sqrt{30}}(120\rho-90\rho^2+18\rho^3-\rho^4)e^{-\rho/2}$
	2	$R_{52}(r) = \dfrac{N}{150\sqrt{70}}(42\rho^2-14\rho^3+\rho^4)e^{-\rho/2}$
	3	$R_{53}(r) = \dfrac{N}{300\sqrt{70}}(8\rho^3-\rho^4)e^{-\rho/2}$
	4	$R_{54}(r) = \dfrac{N}{900\sqrt{70}}\rho^4 e^{-\rho/2}$

Examination of the radial functions in Table 10.1 reveals that except for the $l = 0$ functions, all are zero-valued at $r = 0$ ($\rho = 0$). For $r > 0$, each R_{nl} function has $n-l-1$ points where the function is zero-valued. These points are roots of the polynomials in the R_{nl} functions, and they are simply the points where the radial functions change sign. They are nodes in the wavefunctions. Figure 10.1 is a plot of several of these functions. Since the quantum mechanical postulates tell us that the square of a wavefunction is the probability density, it is also interesting to notice the forms of R_{nl}^2, which are shown in Figure 10.2.

One of the features of the hydrogen atom for which observations need to be reconciled with the quantum mechanical picture is the size of the atom. The quantum mechanical description gives a probability distribution for finding the electron located about the nucleus. In analogy

TABLE 10.2

Energy Levels of the Hydrogen Atom

Level	Energy[a]	Allowed l's	Allowed m's	Degeneracy (n^2)
$n = 1$	$E_{g.s.}$	0	0	1
$n = 2$	$E_{g.s.}/4$	0	0	
		1	1, 0, −1	4
$n = 3$	$E_{g.s.}/9$	0	0	
		1	1, 0, −1	
		2	2, 1, 0, −1, −2	9
$n = 4$	$E_{g.s.}/16$	0	0	0
		1	1, 0, −1	
		2	2, 1, 0, −1, −2	
		3	3, 2, 1, 0, −1, −2, −3	16

[a] The energy is given in terms of the ground state energy, $E_{g.s.}$, which is −109,678 cm⁻¹.

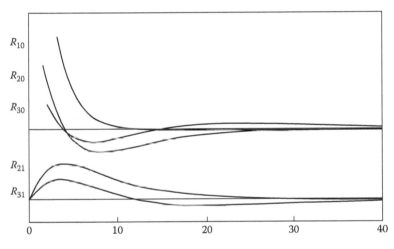

FIGURE 10.1
Radial functions, $R_{nl}(r)$, of the hydrogen (one-electron) atom. The top three functions are the $l = 0$ functions on a common vertical axis scale; the first two $l = 1$ functions are shown below on a different vertical scale. The horizontal scale is in Bohr radii (0.52918 Å).

with a classical mechanical system of two particles in an attractive potential, we regard this distribution as corresponding to the electron orbiting the nucleus (or both orbiting the center of mass). From Figure 10.2, we see that the square of the wavefunction dies away smoothly at large r; it does not end abruptly at some particular value of r. This means that there does not exist a finite sphere that entirely encompasses the hydrogen atom and defines its size. Thus, a different notion of atomic size is required, and perhaps the most reasonable one is the average separation distance between the electron and the nucleus. From the quantum mechanical postulates, such an average can be obtained from the expectation value of r. Hence, for a hydrogen atom state specified by the quantum numbers n, l, and m,

$$\langle r \rangle_{nlm} = \int \psi_{nlm}^* \, r \psi_{nlm} r^2 \sin\theta d\theta d\varphi = \int r^3 R_{nl}^2 dr \int Y_{lm}^* Y_{lm} \sin\theta d\theta d\varphi = \frac{an^2}{Z}\left(\frac{3}{2} - \frac{l(l+1)}{2n^2}\right) \quad (10.10)$$

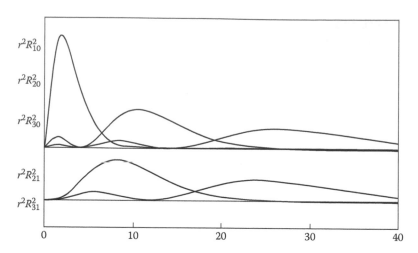

FIGURE 10.2

Radial probability functions of the hydrogen atom, $r^2 R_{nl}^2$. (The factor r^2 is included because the radial volume element is $r^2 dr$.) The number of nodes increases with both n and l.

the integration over the radial coordinate has been carried out by employing special prop-erties of the Laguerre polynomials. The integration over the angular coordinates yields 1, since the spherical harmonic functions are normalized. The constant a, called the **Bohr radius** of the hydrogen atom is obtained from the reduced mass and two fundamental constants and is equal to 0.529466 Å. Another constant called a_0 or β, is the Bohr radius for an infinitely massive nucleus. It is equal to 0.529177 Å.

$$a \equiv \frac{\hbar^2}{\mu e^2}$$

$$a_0 \equiv \frac{\hbar^2}{m_e e^2}$$

(10.11)

For the ground state of the hydrogen atom, the average value of the radial coordinate is $3a/2$ or about 0.8 Å. This result fits experience showing that atomic dimensions are on the order of angstroms. On the other hand, a hydrogen atom in a state very near the ionization limit is significantly larger. With $n = 1000$ and $l = 0$, the expectation value for r is about 0.01 cm.

The deviation from the average separation distance can be obtained from the square of the expectation value of r and from the expectation value of r^2.

$$\Delta r = \sqrt{\langle r^2 \rangle - \langle r \rangle^2}$$

The formula obtained on integration of the hydrogenic radial equations with r^2 is

$$\langle r^2 \rangle_{nlm} = \frac{a^2 n^4}{Z^2} \left(\frac{5}{2} - \frac{3l(l+1)-1}{2n^2} \right)$$

(10.12)

For the ground state, this yields $3a^2$, and thus the uncertainty in a measurement of r is $\sqrt{3}a/2$ or about 0.45 Å.

10.2 Orbital and Spin Angular Momentum

The angular parts of the hydrogen atom wavefunctions are the spherical harmonics, which are, of course, eigenfunctions of the angular momentum operators, L^2 and L_z. The associated eigenvalues, $l(l + 1)\hbar^2$ and $m\hbar^2$, respectively, give the magnitude and the z-component of the angular momentum vector arising from the orbital motion. The orbital motion of an electrically charged particle is a circulation of charge, and that must give rise to the type of magnetic field associated with a magnetic dipole source. In classical electromagnetic theory, the magnetic dipole moment,* $\vec{\mu}$, from charge flowing through a circular loop is proportional to the current and to the area of the loop, while its direction is perpendicular to the plane of the loop. We can analyze the hydrogen atom's magnetic dipole by considering it to be a charge, $-e$, flowing around a loop of some radius r, and then generalizing to the true orbital motion. The area of the hypothetical loop is πr^2, and the current is the charge times the frequency at which the charge passes through any particular point on the loop (i.e., the angular velocity, ω, divided by 2π).

$$\mu = \pi r^2 \left(\frac{-e\omega}{2\pi} \right) \frac{1}{c} = -\frac{e}{2c} r^2 \omega \tag{10.13}$$

where c is the speed of light. The angular momentum for a particle moving about a circular loop is the particle's mass times the square of the radius of the loop times ω, that is, $mr^2\omega$. By collecting $r^2\omega$ in Equation 10.13, we can introduce the angular momentum, L, and then generalize to orbital motion by allowing it to be the angular momentum vector of the hydrogen atom:

$$\vec{\mu} = -\frac{e}{2mc} \vec{L} \tag{10.14}$$

The magnetic dipole moment vector is proportional to the orbital angular momentum vector. Since angular momentum is in units of $\hbar$, it is convenient to collect it with the proportionality factor in Equation 10.14 and define a new constant, μ_B, called the **Bohr magneton**.

$$\mu_B \equiv -\frac{e\hbar}{2mc} \tag{10.15}$$

This is the basic unit or measure for electronic magnetic dipole moments in the same sense that $\hbar$ is the measuring unit for angular momentum.

If an external magnetic field is applied to an isolated hydrogen atom, the effect of the field must be incorporated into the quantum mechanical description. This means that the interaction between the magnetic dipole of the orbital motion and the external field must be added to the Hamiltonian. The classical interaction varies as the dot product of the

---

* Notice that the Greek letter μ is used for many things, reduced mass, magnetic dipole moment, and electric dipole moment. While this may be confusing, it should be clear from the context which is meant.

dipole moment and the field. The quantum mechanical operator that corresponds to this classical interaction is easy to determine because of Equation 10.14. With $\vec{H}$ as the applied magnetic field,

$$\vec{\mu} \cdot \vec{H} = \mu_x H_x + \mu_y H_y + \mu_z H_z = -\frac{e}{2mc}(L_x H_x + L_y H_y + L_z H_z)$$

Letting the orientation of the field define the z-axis in space means that the field components in the x- and y-directions are zero. Thus, the additional term in the Hamiltonian needed to account for an external field is

$$\hat{H}^{int} = -\vec{\mu} \cdot \vec{H} = \frac{\mu_B H_z \hat{L}_z}{\hbar} \tag{10.16}$$

The wavefunctions for the states of a hydrogen or one-electron atom in an external uniform magnetic field are eigenfunctions of the original Hamiltonian, which we shall now identify as H^0, with H^{int} added to it.

We should realize that since the eigenfunctions, ψ_{nlm}, of H^0 are eigenfunctions of the operator L_z, they are already eigenfunctions of $H^0 + H^{int}$.

$$\left(\hat{H}^0 + \hat{H}^{int}\right)\psi_{nlm} = (E_n + m\mu_B H_z)\psi_{nlm} \tag{10.17}$$

Thus, the eigenenergies for the atom in a magnetic field depend on the m quantum number. This means that an applied magnetic field removes the degeneracy of states with the same n and l, but with different m quantum numbers. The separation between the levels increases with the strength of the applied field, according to Equation 10.17. This also suggests why the m quantum number is often referred to as the **magnetic quantum number**.

The angular form of the orbital functions, ψ_{nlm}, or really of the spherical harmonic functions, is interesting. The $l = 0$ or s orbitals are spherically symmetric, and that means they can be represented as spheres. The forms of the higher l orbitals are more complicated. Figure 10.3 shows the form of the probability densities for certain of the $m = 0$ spherical harmonics. Notice that the number of nodal planes (planes in space where the function is zero) is equal to l. Thus, an $l = 1$, $m = 0$ or p_0 function is zero-valued everywhere in the xy-plane. The sign or phase of the function changes from one side of this plane to the other. The $l = 2$, $m = 0$ or d_0 function has two nodal planes.

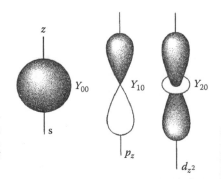

FIGURE 10.3
Representation of several of the squared spherical harmonic functions with $m = 0$. The shapes depict the surface of a contour drawn for some fixed probability density amplitude, and the alternation in shading designates regions of different phases (different signs) of the wavefunction.

Certain linear combinations of the spherical harmonic functions are easy to represent and offer a useful picture to keep in mind. The linear combinations and the letter designations for $l = 1$ and $l = 2$ orbitals are

$$Y_{11} + iY_{1-1} \rightarrow p_x$$

$$Y_{11} - iY_{1-1} \rightarrow p_y$$

$$Y_{10} \rightarrow p_z$$

$$Y_{22} + iY_{2-2} \rightarrow d_{x^2-y^2}$$

$$Y_{22} - iY_{2-2} \rightarrow d_{xy}$$

$$Y_{21} + iY_{2-1} \rightarrow d_{xz}$$

$$Y_{21} - iY_{2-1} \rightarrow d_{yz}$$

$$Y_{20} \rightarrow d_{z^2}$$

These combinations provide the real parts of the angular functions, and they are represented in Figure 10.4. Similar combinations can be made for higher l angular momentum functions.

FIGURE 10.4
Probability density representations of the real parts of the angular functions formed from linear combinations of spherical harmonic functions.

Electrons have an intrinsic angular momentum which gives rise to an intrinsic magnetic dipole moment even if $l = 0$ or even if they are not part of an atom. Very early in the history of quantum mechanics, deflection experiments that measured the magnetic moment of a moving particle by passing it through a magnetic field were established that an electron in an $l = 0$ orbital still possesses a magnetic moment even though $\vec{L} = 0$. Of considerable excitement at the time was the fact that there were two and only two possible z-components of this magnetic moment. Assuming that the magnetic moment was proportional to some angular momentum vector, the questions that arose involved the source and size of this other type of angular momentum.

The source of the angular momentum is intrinsic to the electron and is referred to as spin because the spin of a solid body about an axis is a source of angular momentum. However, the term for this intrinsic feature of an electron does not mean we should picture the electron as some mass spinning about its axis. The feature is a more subtle characteristic than that. It can be accounted for with quantum mechanics, but only if the mechanics have been adapted for relativistic effects. Spin is more a name than a description of this property.

That there are two and only two possible z-axis projections of the electron's intrinsic magnetic dipole, or equivalently, its spin angular momentum vector, is important. For orbital angular momentum, the magnetic quantum number, m, can take on values ranging from $-l$ to l. The number of such values, which is the number of different projections on the z-axis, is $2l + 1$, an odd number. For there to be an angular momentum that has an even number of projections, the associated quantum number must be a half-integer. Letting s be the quantum number for spin angular momentum, the relation $2s + 1 = 2$ requires that $s = 1/2$. (We say that the spin of an electron is 1/2.) The quantum number that gives the projection of the spin vector on the z-axis is m_s, and to avoid confusion at this point, we now designate the orbital angular momentum m quantum number m_l.

There are two spin states for an electron. For one, $m_s = 1/2$, and for the other, $m_s = -1/2$. The first state is commonly identified as being the spin-up state, and the second is the spin-down state. This is because the orientation of the spin vector with respect to a z-axis defined as the direction of an applied magnetic field is either in the +z-direction, or up, or in the $-z$-direction, or down. A shorthand designation of these electron spin states is to write α for an electron with $m_s = 1/2$, and β for $m_s = -1/2$. α and β are meant to be functions in the same way that ψ is a function of r, θ, and φ. However, α and β are abstract functions—we will not express them explicitly—of an abstract coordinate, a spin coordinate. α and β are orthonormal functions, and that means the following relations hold:

$$\langle \alpha | \alpha \rangle = 1 \qquad\qquad (10.18a)$$

$$\langle \beta | \beta \rangle = 1 \qquad\qquad (10.18b)$$

$$\langle \alpha | \beta \rangle = \langle \beta | \alpha \rangle = 0 \qquad\qquad (10.18c)$$

The integration here is over the spin coordinate, and again this is not presented as an explicit integration.

The magnetic moment associated with an electron's spin interacts with an externally applied magnetic field. This magnetic moment is proportional to the spin vector. Analogous to Equation 10.14 is

$$\vec{\mu} = -g_e \frac{e}{2mc} \vec{S} \tag{10.19}$$

This relation differs in an important way from Equation 10.14. It says that the magnetic dipole moment is proportional to the spin vector with not only a factor of $e/2mc$ but also with an additional factor, g_e. (Recall that the proportionality factor for orbital angular momentum is $e/2mc$.) The additional factor is required because the simple picture of circulation of charge that led to Equation 10.14 does not apply to the intrinsic spin of an electron. The measured value for the dimensionless constant g_e for a free electron is 2.0023.

10.3 Atomic Orbitals and Atomic States

The wavefunctions, ψ_{nlm}, for one-electron atoms are often referred to as *orbitals*. Generally, orbitals are any functions of the spatial coordinates of one electron, which for an atom means a function of the spherical polar coordinates r, θ, and φ. Atomic orbitals are labeled with

1. The value of the principal quantum number, n
2. A letter associated with the l quantum number
3. A numerical subscript which is the value of m_l

The letters* s, p, d, and f are associated with the l quantum number as s ($l = 0$), p ($l = 1$), d ($l = 2$), and f ($l = 3$). The next is g ($l = 4$), and from then on the series follows alphabetic order. Thus, the ground state orbital of the hydrogen atom is named $1s_0$ or sometimes just 1s. The $n = 2$ orbitals are 2s, $2p_1$, $2p_0$, and $2p_{-1}$.

The complete wavefunction of a hydrogen atom is a product of the spatial wavefunction and a spin function, either α or β. So with the spin of the electron incorporated, the product functions $\psi_{nlm} \alpha$ and $\psi_{nlm} \beta$ (or $1s\alpha$, $1s\beta$, etc.) are referred to as **spin-orbitals**. Spin introduces further degeneracy in the set of eigenfunctions of the original hydrogen atom Hamiltonian, H^0. The 1s orbital gives rise to the $1s\alpha$ and $1s\beta$ spin-orbitals, and so these functions are degenerate functions of H^0. Basically, spin doubles the number of states to $2n^2$.

The magnetic moments arising from electron orbital motion and from electron spin can interact. This feature of atomic structure, and of molecular electronic structure, too, is termed **spin–orbit interaction**. Since the magnetic dipoles due to spin and orbital motion are proportional to their respective angular momentum vectors, the interaction

---

* The letters s, p, and d originate in the names sharp, principal, and diffuse. These were the terms that were given to absorption and emission lines in the atomic spectra of alkali atoms on the basis of the appearance of those lines, usually on a photographic plate. Lines of similar type formed series, and it was learned that the transition frequencies measured for a series followed a simple mathematical progression. With quantum mechanics the progressions and spectral characteristics became understandable consequences of the allowed energy levels and transitions. From this association with types of lines come the orbital letters in use today.

is proportional to the dot product of the angular momentum vectors. This interaction is a small perturbation on the H^0 description of the hydrogen atom. We treat it phenomenologically at this point by introducing a proportionality constant, γ, rather than by developing a fundamental expression to be used in place of γ. The perturbing Hamiltonian for spin–orbit interaction, then, is

$$\hat{H}' = \gamma \vec{L} \cdot \vec{S} \tag{10.20}$$

We presume that the spin–orbit interaction constant, γ, is to be determined from some measurement.

Spin–orbit interaction implies a coupling of the two "motions" of spin and orbit. From the discussion in Chapter 8, we would expect that this coupling may mix the hydrogenic spin-orbital states with the resulting wavefunctions no longer assured to be eigenfunctions of the S_z and L_z operators. The interaction Hamiltonian in Equation 10.20 can be rewritten following Equation 8.65:

$$\hat{H}' = \frac{\gamma}{2}\left[\hat{J}^2 - \hat{L}^2 - \hat{S}^2\right] \tag{10.21}$$

where J refers to the total angular momentum: $\vec{J} = \vec{L} + \vec{S}$. The rules of angular momentum addition give the allowed values of the quantum number J as ranging from $l + s$ downward, in steps of 1, to $|l-s|$. For the hydrogen atom's single electron, $s = 1/2$. Thus, the spin–orbit coupled states of the hydrogen atom, or other one-electron atoms, have J quantum numbers equal to $l \pm 1/2$, except if $l = 0$, and then $J = 1/2$. These spin–orbit coupled states, which we shall designate concisely with the valid quantum numbers in a bra or ket vector, $|nJls\rangle$ or $\langle nJls|$, are eigenfunctions of the operator H' in Equation 10.21.

$$\hat{H}'|nJls\rangle = \frac{\gamma\hbar^2}{2}\left[J(J+1) - l(l+1) - s(s+1)\right]|nJls\rangle \tag{10.22}$$

Thus, the energies of the states after accounting for spin–orbit coupling follow from Equations 10.9 and 10.22.

$$E_{nJls} = -\frac{\mu Z^2 e^4}{2\hbar^2 n^2} + \frac{\gamma\hbar^2}{2}\left[J(J+1) - l(l+1) - s(s+1)\right] \tag{10.23}$$

Notice that the energies are subscripted with n, J, l, and s since these are the valid quantum numbers for states with spin and orbital motion coupled.

The **spin–orbit splitting** is the energy difference between states that are otherwise degenerate. The case with $n = 2$ and $l = 1$ is an example. In the absence of spin–orbit effects, the six hydrogen atom states, $2p_1\alpha$, $2p_1\beta$, $2p_0\alpha$, $2p_0\beta$, $2p_{-1}\alpha$, and $2p_{-1}\alpha$ are degenerate. These states can be mixed in some way because of spin–orbit interaction, and the resulting states can be distinguished according to the two possible J values, $J = 1 + 1/2 = 3/2$ and $J = 1 - 1/2 = 1/2$. (There are still six states since the $J = 3/2$ coupling is fourfold degenerate and the $J = 1/2$ coupling is doubly degenerate.) Using the formula in Equation 10.22 we can evaluate the spin–orbit energy for the two possible J values.

$$J = \frac{3}{2} : E(\text{spin}-\text{orbit}) = \frac{\gamma\hbar^2}{2}\left[\frac{3}{2}\left(\frac{3}{2}+1\right) - 1(1+1) - \frac{1}{2}\left(\frac{1}{2}+1\right)\right] = \frac{\gamma\hbar^2}{2}$$

$$J = \frac{1}{2} : E(\text{spin} - \text{orbit}) = \frac{\gamma\hbar^2}{2}\left[\frac{1}{2}\left(\frac{1}{2}+1\right) - 1(1+1) - \frac{1}{2}\left(\frac{1}{2}+1\right)\right] = -\gamma\hbar^2$$

$$\text{Energy difference: } = \frac{3\gamma\hbar^2}{2}$$

The energy difference in these two spin–orbit energies is the energy of the splitting. We can see that if the splitting were measured spectroscopically, the value of the spin–orbit interaction constant, γ, would be known.

When an external magnetic field is applied to a one-electron atom in a state for which $l > 0$, the quantum mechanical analysis of the energies becomes more complicated. The complete Hamiltonian includes the spin–orbit interaction, the interaction of the orbital magnetic dipole with the field, and the interaction of the spin dipole with the field. Instead of treating this situation generally, consider a special case: If the external field were so strong that the interaction energies with the field were much greater than the spin–orbit interaction, a good approximate description would be to neglect the spin–orbit interaction entirely. (A better approximation would be to include the spin–orbit interaction via low-order perturbation theory.) Physically, the strong field may be thought of as orienting the individual magnetic dipoles and thereby overwhelming their coupling with each other. A very strong field, then, is said to decouple the spin and orbital magnetic dipoles. On the other hand, a very weak field would not decouple the dipoles but would interact with the net magnetic dipole that results from the sum of the spin and orbital angular momentum vectors. (Again, perturbation theory could be used for a more accurate energetic analysis.) The spectra of atoms and molecules obtained with an applied magnetic field are called **Zeeman spectra**.

Many-electron atoms present a more complicated picture because of the electron–electron interaction terms that enter the Hamiltonians for the Schrödinger equations. These interaction terms are the Coulombic repulsive interaction of like charged particles. This couples the motions of the different electrons, and that precludes separation of variables. Nonetheless, important understanding of the features of the wavefunctions of many-electron atoms can be recognized on the basis of the one-electron atoms we have considered so far.

If the Schrödinger equation were separable into the coordinates of the different electrons, then the form of the resulting wavefunctions would be products of spin-orbitals. Such a product form serves as the simplest meaningful approximate description, and it is certainly of qualitative use in understanding the electronic structure of the different elements of the periodic table. Within this approximate separation, we can build up the electronic structures of atoms by assigning electrons to specific orbitals or spin-orbitals. These assignments are termed **electron occupancies** or **configurations*** in that they indicate the spatial orbitals filled by the electrons.

The **Pauli principle** for the electrons of atoms and molecules says that no two electrons can occupy the same spin-orbital, and this is obviously important in building up the orbital picture of the elements. As discussed later, this principle goes hand in hand with the indistinguishability of the electrons and with their half-integer intrinsic spin. For now, it is important to say that each distinct spatial orbital (e.g., 1s, 2s, $2p_1$, $2p_0$, $2p_{-1}$) can be occupied by at most two electrons. This is because a given spatial orbital can be combined with either an α or a β spin function to form two and only two different spin-orbitals.

---

* The term "electron configuration" can have a more specific meaning, and so the use of "electron occupancy" for the arrangement of electrons in spatial orbitals avoids possible confusion.

A set of spin-orbitals with the same n quantum number is referred to as a **shell**, and the set of spin-orbitals with the same n and l quantum numbers is a **subshell**. The Pauli principle means that an s type ($l = 0$) subshell can have an occupancy of at most two electrons. A p subshell ($l = 1$) can have an occupancy of at most 6 electrons, while a d subshell ($l = 2$) can have an occupancy of at most 10 electrons.

In a one-electron atom, the energetic ordering of the spatial orbitals is according to the principal quantum number, n. To the extent that this holds for many-electron atoms, we should expect that the orbitals are filled in order of the principal quantum number for the ground states of the elements. That is, as one goes through the periodic table, the 1s orbital is expected to be filled first, then the 2s and 2p orbitals, and then the 3s, 3p, and 3d orbitals, and so on.

The interaction between electrons affects the energetically preferred ordering and the shapes of the orbitals in several important ways. We can consider these effects even with the assumed separation of variables that underlies the orbital picture. The first important consequence of electron–electron interaction is **shielding.** We have already seen in Equation 10.10 that the average electron–nucleus separation distance for a one-electron atom increases as n^2. This simply means that an electron in the 1s orbital is closer on average to the nucleus than an electron in the 2s orbital. The 2s orbital function is closer or tighter than the 3s orbital, and so on. In a lithium atom, where we expect two electrons in the 1s spatial orbital and one in the 2s orbital, the effective potential that the electron in the 2s orbital experiences is somewhat like that of a lithium nucleus ($Z = +3e$) with a tight, negative ($-2e$) charge cloud surrounding it. In other words, we can view the electron–electron repulsion effects on the 2s electron in this system as if it were a one-electron problem with the positive nucleus being of charge less than $+3e$, and possibly as little as $+e$. The nucleus is screened or shielded by the inner two 1s electrons. A consequence is that the lithium atom's 2s orbital is more diffuse, or more spread out radially, than the 2s electron of the unshielded nucleus, that is, Li^{2+}. It is more diffuse because the effective nuclear charge is smaller.

Shielding, and electron–electron interaction generally, distinguish the energies of electrons in different subshells. For instance, the 2s orbital is usually energetically preferred relative to the 2p orbitals. (The degeneracy in the m_l quantum number remains.) Thus, the electron occupancy of the carbon atom is $1s^2 2s^2 2p^2$, which means the 1s and 2s orbitals are fully occupied and there are two electrons occupying 2p orbitals. Were the 2p orbitals preferred, the occupancy would be $1s^2 2p^4$; however, spectroscopic experiments can unambiguously demonstrate that this is not the occupancy for the ground state of carbon. The order in which orbitals are filled generally follows this pattern throughout the periodic table:

$$1s \rightarrow 2s \rightarrow 2p \rightarrow 3s \rightarrow 3p \rightarrow 4s \rightarrow 3d \rightarrow 4p \rightarrow 5s \rightarrow 4d \cdots$$

From this, we can build up the likely electron occupancies of the elements, remembering though that the orbital description implies an approximation involving the separation of variables in the Schrödinger equation.

Transitions from one electronic state of an atom to another electronic state can be induced by electromagnetic radiation. The spectra of many-electron atoms tend to be complicated because there are numerous electron occupancies and states. As well, each of the many electrons is a source of angular momentum and magnetic dipoles that interact. Angular momentum coupling rules can help sort through the manifolds of atomic electronic states.

In light elements, the coupling of magnetic dipoles is stronger among those associated with orbital motion and among those associated with spin than the coupling of dipoles due to spin and orbital momenta of individual electrons. In heavy elements, it is usually the reverse. For light elements, it is appropriate to apply angular momentum coupling rules—termed **Russell–Saunders** or *LS* coupling—to find the total orbital angular momentum, *L*, and the total electron spin, *S*. These are then coupled to form a resultant total angular momentum vector with quantum number *J*. For heavy elements, the orbital and spin vectors of the individual electrons are coupled—termed *jj* **coupling**—just as was done for the single electron of the hydrogen atom, and the resultant vectors from all the electrons are then added to form the total angular momentum vector ($\vec{J}$). We will consider the procedure for the light elements in detail.

For the purpose of working out the angular momentum coupling, electrons in the same subshell are said to be *equivalent*, while electrons in different subshells are said to be *inequivalent*. The first situation to consider is that of inequivalent electrons, and the example will be the electron occupancy

$$1s^1 2p^1 3p^1$$

While this is not an occupancy encountered for the ground states of any of the elements, it could correspond to an excited state of the lithium atom. The task is to add together the orbital angular momenta and to add together the spin angular momenta. Let us use l_1, l_2, and l_3 for the angular momenta of the three electrons. The important rule to apply is that the quantum number for a resultant angular momentum vector can take on values ranging from the sum of the quantum numbers of two sources being combined down to the absolute value of their difference, in steps of 1. Since $l_1 = 0$ and $l_2 = 1$, the quantum number for the vector sum ($\vec{L}_{12}$) of these two momenta can be only 1.

$$\vec{l}_1 + \vec{l}_2 = \vec{L}_{12} \Rightarrow L_{12} = 1$$

The angular momentum of the third electron is added to $\vec{L}_{12}$. Applying the same rule, we have

$$\vec{l}_3 + \vec{L}_{12} = \vec{L}_{total} \Rightarrow L_{total} = 1+1,...,|1-1| = 2,1,0$$

This says that there are three different possibilities for the coupling of the orbital angular momenta, and the three correspond to resultant angular momentum vectors with *L* quantum numbers of 2 or 1 or 0.

The coupling of the spins is done in the same way. The spin vectors of two electrons are coupled together to yield a resultant vector, and this vector is then coupled with the spin vector of the next electron. Clearly, the process could be continued for any number of electrons. Furthermore, since the quantum number for electron spin is always 1/2, the number of possibilities is rather limited. Applying the rule for adding angular momenta to the first two spins gives

$$\vec{s}_1 + \vec{s}_2 = \vec{S}_{12} \Rightarrow S_{12} = \frac{1}{2} + \frac{1}{2},...,\left|\frac{1}{2} - \frac{1}{2}\right| = 1,0$$

This indicates that the spins of two inequivalent electrons can be coupled in two ways. Adding the third spin gives

$$\vec{s}_3 + \vec{S}_{12} = \vec{S}_{total} \Rightarrow S_{total} = \frac{1}{2} + 1, ..., \left|\frac{1}{2} - 1\right| = 1, 0 \quad \text{and} \quad \frac{1}{2} + 0 = \frac{3}{2}, \frac{1}{2}, \frac{1}{2}$$

Notice that both possibilities for the value of S_{12} are used to find the possible values of S_{total}. Also, notice that the resulting values of S_{total} include two that are the same. This simply means that there are two distinct ways in which the spin vectors can be coupled that produce a resultant vector with an associated quantum number of 1/2.

The multiplicity associated with a given angular momentum is the number of different possible projections on the z-axis. This is always equal to 1 greater than twice the angular momentum quantum number; that is, multiplicity (J) = $2J + 1$. Spin multiplicity is equal to $2S + 1$, and the names singlet, doublet, triplet, quartet, and so on, are attached to states with spin multiplicities of 1, 2, 3, and 4, respectively. From the spin coupling just carried out, we can see that two inequivalent electrons can spin-couple to produce either a singlet state (i.e., $S = 0$) or a triplet state (i.e., $S = 1$). Three inequivalent electrons can be coupled to produce a quartet state ($S = 3/2$) or two different doublet states ($S = 1/2$).

The magnetic moment associated with the net orbital angular momentum, which we shall now call L instead of L_{total}, interacts with the magnetic moment arising from the net spin vector, which shall be S instead of S_{total}. With the resultant vector designated J, as before, we have

$$\vec{L} + \vec{S} = \vec{J} \Rightarrow J = L + S, ..., |L - S|$$

From given L and S values, a number of J values can result. These are different couplings. Also, there may be several different possibilities, not just one, for the L value and for the S value. All these are to be included in finding the J values for the resultant states. Continuing with the example of three inequivalent electrons, we may tabulate the possible J's.

Value of L: 2	Value of S: 3/2	Resultant J's: 7/2, 5/2, 3/2, 1/2
2	1/2	5/2, 3/2
2	1/2	5/2, 3/2
1	3/2	5/2, 3/2, 1/2
1	1/2	3/2, 1/2
1	1/2	3/2, 1/2
	3/2	3/2
0	1/2	1/2
0	1/2	1/2

It is clear that quite a number of distinct spin–orbit coupled states may be associated with the electron occupancy in this problem.

Term symbols in atomic spectroscopy serve to designate different electronic states and energy levels. The term symbols encode the values of J, L, and S. For the value of L, a capital letter is written: S for $L = 0$, P for $L = 1$, D for $L = 2$, F for $L = 3$, and so on, with G, H, I, etc. The spin multiplicity is written as a presuperscript, and the J value is written as a subscript. The form for these symbols, then, is

$$^{(2S+1)}L_J$$

As an example, if $L = 1$, $S = 1/2$, and $J = 3/2$, the term symbol is $^2P_{3/2}$. In the three electron example with which we have been working, the table of L, S, and J values is perfect for writing down the term symbols for the states. The first line of the preceding table translates into the term symbols $^4D_{7/2}$, $^4D_{5/2}$, $^4D_{3/2}$, and $^4D_{1/2}$. The next line gives $^2D_{5/2}$ and $^2D_{3/2}$, and so on down the table. These are the Russell–Saunders term symbols because Russell–Saunders coupling assumes that the individual orbital angular momenta of the electrons are more strongly coupled than the orbital and spin angular momenta. If the spin–orbit interaction is ignored, the terms symbols are written without the J subscript.

Equivalent electrons, those in the same n and l subshell, are more complicated in the analysis of spin and orbit coupling because of the restrictions of the Pauli principle. Essentially, the different ways in which coupling takes place are restricted. As a simple illustration of this, consider the difference between the electron occupancy $1s^2$ and the occupancy $1s^1 2s^1$. Because the two electrons in the same $1s$ orbital must have opposite spins to satisfy the Pauli principle, the net spin vector must be zero; that is, $S = 0$ only. The two inequivalent electrons in the $1s$ and $2s$ orbitals, on the other hand, can be spin-coupled two ways, as a singlet ($S = 0$) and as a triplet ($S = 1$) state.

There are a number of schemes available for finding the term symbols for equivalent electron problems. One is to make a complete list of all the possible arrangements of electrons in spin-orbitals that correspond to the given electron occupancy. For instance, if there were three electrons in the $2p$ subshell, they could be arranged among the six spin orbitals in 20 ways. To work out these 20 arrangements, ↑ and ↓ can be used under the column for a particular spatial orbital to indicate that an electron with an α and a β spin, respectively, occupies that orbital. For three electrons in a $2p$ subshell, the arrangements that are consistent with the Pauli principle are

$2p_1$	$2p_0$	$2p_{-1}$
↑↓	↑	
↑↓	↓	
↑↓		↑
↑↓		↓
↑	↑↓	
↓	↑↓	
	↑↓	↑
	↑↓	↓
↑		↑↓
↓		↑↓
	↑	↑↓
	↓	↑↓
↑	↑	↑
↓	↑	↑
↑	↓	↑
↑	↑	↓
↓	↓	↑
↓	↑	↓
↑	↓	↓
↓	↓	↓

Tables of this sort can be set up systematically for any occupancy, including those with both equivalent and inequivalent electrons. The number of arrangements in these tables is equal to the number of states of the system, although each line of the table directly corresponds to a particular state. Each row in the table is regarded as having a specific z-axis projection of the total orbital and total spin angular momentum vectors. In the first line, there are two spin-up electrons and one spin-down. The net M_S quantum number for this line is 1/2, which is the sum of the three individual m_s quantum numbers. The sum of the individual m_l quantum numbers for the three electrons, which is $1 + 1 + 0 = 2$ for the first line, is the net M_L quantum number. In other words, the z-axis projections of the individual electron momenta are added. We know that for a given angular momentum quantum number X, the largest z-axis projection is with M_X equal to X. So, if we had a list of M_X values and did not know X, the value would be found as the largest M_X value. In our table of electron arrangements, the net M_L and M_S quantum numbers determine the total L and S quantum numbers. The procedure is to go through the list to find the biggest M_L and M_S and conclude that there is a possible state with corresponding L and S quantum numbers. Then, entries on the list that go along with this (L, S) pair are eliminated, and the process is repeated to find another (L, S) pair. For each (L, S) pair, we write a Russell–Saunders term symbol. Here is the process for the $2p^3$ example.

$2p_1$	$2p_0$	$2p_{-1}$	M_L	M_S					
↑↓	↑		2	1/2	→	$(L = 2, S = 1/2)$			
↑↓	↓		2	−1/2		•			
↑↓		↑	1	1/2		•			
↑↓		↓	1	−1/2		•			
↑	↑↓		1	1/2	→		$(L = 1, S = 1/2)$		
↓	↑↓		1	−1/2			•		
	↑↓	↑	−1	1/2		•			
	↑↓	↓	−1	−1/2		•			
↑		↑↓	−1	1/2			•		
↓		↑↓	−1	−1/2			•		
	↑	↑↓	−2	1/2		•			
	↓	↑↓	−2	−1/2		•			
↑	↑	↑	0	3/2	→			$(L = 0, S = 3/2)$	
↓	↑	↑	0	1/2		•			
↑	↓	↑	0	1/2			•		
↑	↑	↓	0	1/2				•	
↓	↓	↑	0	−1/2		•			
↓	↑	↓	0	−1/2			•		
↑	↓	↓	0	−1/2				•	
↓	↓	↓	0	−3/2				•	

The first (L, S) pair is $L = 2$ and $S = 1/2$, and the associated term symbol (neglecting the J subscript) is ^2D. The next pair is $L = 1$ and $S = 1/2$, and the term symbol is ^2P. The last is $L = 0$ and $S = 3/2$, and the term symbol is ^4S.

Notice that completing a table for *any* full n and l subshell produces only one table entry from which $L = 0$ and $S = 0$. Thus, the net contribution to S and L from a full subshell in a more complicated electron occupancy does not need to be considered explicitly. Each full subshell makes a zero contribution to S and L.

The term symbols encode the values of the quantum numbers L, S, and J, and an expression for the spin–orbit interaction energy of a many-electron atom follows directly. With the Hamiltonian in Equation 10.21 and with a phenomenological constant, γ, specific to the atomic system, the energy associated with spin–orbit interaction in the absence of an external magnetic field is

$$E_{JLS}^{\text{spin–orbit}} = \frac{\gamma\hbar^2}{2}\left[J(J+1) - L(L+1) - S(S+1)\right] \tag{10.24}$$

A very useful set of rules, referred to as **Hund's rules**, predict the energetic ordering of the term symbol states that arise from a given electron occupancy. These were first developed empirically on the basis of atomic spectra. The most important of these rules is that the states are ordered energetically according to their spin multiplicity, with the greatest spin multiplicity giving the lowest energy. The second rule is that among states with the same spin multiplicity (and arising from the same electron occupancy), the energetic ordering will be according to the L quantum number, the lowest level being that with the greatest L. In the example of the $2p^3$ occupancy, Hund's rules indicate that the ^4S energy level will be lower than the ^2P and ^2D levels. Also, the ^2D will be lower than the ^2P. The term symbol for the ground state of the nitrogen atom, which has an occupancy $1s^2\,2s^2\,2p^3$, is in fact ^4S. Additional rules distinguish among the energies according to the J quantum number, but Equation 10.24 already gives a way of being quantitative about these energy differences.

The spectra of atoms arise from transitions between the possible electronic (term symbol) states. Generally, transitions from the ground state of an atom to an excited state require the energy of photons in the visible and ultraviolet regions of the electromagnetic spectrum. The selection rule is determined from the integral of the electric dipole moment operator and two atomic state wavefunctions. For one-electron atoms, this means that transitions are allowed between a state with quantum numbers $n\,l\,m_l$ and $n'l'm_l'$ if the following is nonzero:

$$\left\langle \psi_{nlm_l} \left| \vec{r} \right| \psi_{n'l'm_l'} \right\rangle$$

This represents three integral values because of the three vector components of $\vec{r}$, and a transition is allowed if any one is nonzero. It turns out that integration over the radial coordinate gives a nonzero result for any choice of n and n', but the angular coordinate integration requires that l and l' differ by 1 for the result to be nonzero. Thus, the selection rule for a one-electron atom spectrum is stated concisely as

$$|\Delta l| = 1 \tag{10.25}$$

The ground state of the hydrogen atom is ^2S, and so this selection rule means that transitions are allowed only to ^2P states. From the excited ^2P state, transitions to ^2S and ^2D states are allowed, as in Figure 10.5.

In many-electron atoms, states can have different S quantum numbers. For light elements, where the spin–orbit coupling is weak, the selection rules for atomic spectra are

$$\Delta S = 0$$
$$|\Delta L| = 1 \tag{10.26}$$

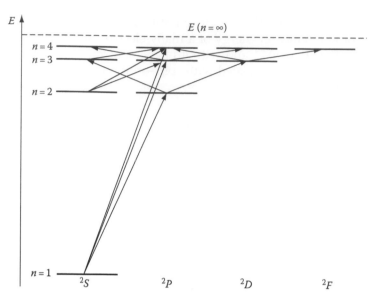

FIGURE 10.5

Lowest energy levels of the hydrogen atom (horizontal lines) and the allowed transitions. States with the same term symbol have been placed in the same column for clarity.

Furthermore, the change in L can come about only if the occupancy change corresponds to a change of 1 in the l quantum number of one and only one electron. That is, $\Delta l_i = 1$ while $\Delta l_{j\neq i} = 0$. For example, a change from the occupancy $1s^2\,2s^2\,2p^1$ to the occupancy $1s^2\,2s^1\,2p^2$ is a change in one and only one of the electrons' l values of 1. The term symbol for the first occupancy is ^2P, and so transitions are allowed to term symbol states arising from the second occupancy, but only if they are ^2S or ^2D. When we consider the fine structure of the spectra, which means the small energy differences due to spin–orbit interaction, a selection on J applies:

$$\left|\Delta J\right| = 0,1 \tag{10.27}$$

This means that J can increase or decrease by 1, or it can stay the same.

10.4 Molecules and the Born–Oppenheimer Approximation

The electronic structure of molecules is a more complicated problem than the electronic structure of atoms because instead of one central nuclear potential there are several positively charged nuclei distributed in space. Analyzing molecular electronic structure usually begins with an approximate separation of the electronic problem from the problem of nuclear motion. This separation was implicit in the treatment of vibration and rotation in Chapter 9 and in the discussion of potential energy surfaces in Chapter 6.

The general molecular Schrödinger equation, apart from electron spin interactions, is

$$(T_n + T_e + V_{nn} + V_{ee} + V_{en})\Psi = E\Psi \tag{10.28}$$

where the operators in the Hamiltonian are the kinetic energy operators of the nuclei (T_n) and the electrons (T_e), and then the potential energy operators among the pairs of nuclei (V_{nn}), among the pairs of electrons (V_{ee}), and between nuclei and electrons (V_{en}). The explicit forms of these operators are

$$T_n = -\sum_{\alpha}^{nuclei} \frac{\hbar^2}{2M_\alpha} \nabla_\alpha^2 \tag{10.29}$$

$$T_e = -\sum_{i}^{electrons} \frac{\hbar^2}{2m_e} \nabla_i^2 \tag{10.30}$$

$$V_{nn} = \sum_{\alpha > \beta} \frac{Z_\alpha Z_\beta e^2}{R_{\alpha\beta}} \tag{10.31}$$

$$V_{ee} = \sum_{i>j} \frac{e^2}{r_{ij}} \tag{10.32}$$

$$V_{en} = -\sum_{\alpha} \sum_{i} \frac{Z_\alpha e^2}{r_{i\alpha}} \tag{10.33}$$

There is a repulsive Coulombic interaction among the nuclear charges and a repulsive charge–charge interaction among the electrons. However, the interaction potential between electrons and nuclei is attractive since the particles are oppositely charged. This particular interaction couples the motions of the electrons with the motions of the nuclei.

The wavefunctions that satisfy Equation 10.28 must be functions of both the electron position coordinates and the nuclear position coordinates, and this differential equation is not separable. In principle, true solutions could be found, but the task is difficult as this is a formidable differential equation for even simple molecules. An alternative is an approximate separation of the differential equation based on the sharp difference between the mass of an electron and the masses of the nuclei. The difference suggests that the nuclei motions are sluggish relative to the electron motions. Over a brief period of time, the electrons "see" the nuclei as if fixed in space since during such a period the nuclear motions are relatively slight. The nuclei, on the other hand, "see" the electrons as something of a blur, given their relatively "faster" motions.

We can exploit the distinction between sluggish and fast particles to achieve an approximate separation of variables. First, the wavefunction is taken to be (approximately) a product of a function of nuclear coordinates only, φ, and a function of electron coordinates, ψ, *at a specific nuclear geometry*. At the specific nuclear geometry, $R = R_0$ (R collectively stands for all nuclear position coordinates), this is expressed as

$$\Psi(R_0, r_1, r_2, \ldots) \cong \varphi(R_0)\psi^{\{R_0\}}(r_1, r_2, \ldots) \tag{10.34}$$

Notice the superscript on the electronic wavefunction ψ. It designates that this function is for the choice $R = R_0$, one specific geometrical arrangement of the nuclei. If this product form of the wavefunction is to be used in the Schrödinger equation, we need to know something about the effect of the kinetic energy operators. Since T_e acts on only electron coordinates, $T_e \Psi = \varphi T_e \psi$. T_n affects both functions since ψ, the electronic wavefunction, retains a dependence on the nuclear position coordinates as parameters. However, the diffuseness of an electronic wavefunction relative to the nuclear positions, again a feature associated with the mass difference, means that φ changes more rapidly with R than does ψ. In other words, it is reasonable to approximate the effect of T_n on the electronic wavefunction as zero.

$$T_n \Psi(R, r_1, r_2, \ldots) = T_n \varphi(R) \psi^{\{R\}}(r_1, r_2, \ldots) \cong \psi^{\{R\}}(r_1, r_2, \ldots) T_n \varphi(R) \tag{10.35}$$

With this approximation, the Schrödinger equation is

$$\varphi T_e \psi + \psi T_n \varphi + \varphi \psi (V_{nn} + V_{ee} + V_{en} - E) = 0 \tag{10.36}$$

Rearrangement of the terms in this expression leads to

$$\varphi [T_e + V_{ee} + V_{en}] \psi + \psi [T_n + V_{nn} - E] = 0 \tag{10.37}$$

Of the operators in brackets acting on ψ, only one involves the nuclear position coordinates. It is a term giving the electron–nucleus attraction potential. In ψ, the nuclear coordinates are treated as parameters, and the same can be done for this operator term where it acts on ψ. Let us now assume that ψ is an eigenfunction of the operator in brackets; that is, let us write a purely electronic Schrödinger equation as

$$\left[T_e + V_{ee} + V_{en}^{\{R\}} \right] \psi^{\{R\}}(r_1, r_2, \ldots) = E^{\{R\}} \psi^{\{R\}}(r_1, r_2, \ldots) \tag{10.38}$$

The energy eigenvalue, $E^{\{R\}}$, must have a dependence on the nuclear positions, as denoted by its superscript, which follows the dependence of the operator on the left. So, at each different R, there is a different V_{en}, and consequently a different $\psi^{\{R\}}$ and a different $E^{\{R\}}$.

Substitution of Equation 10.38 into Equation 10.37 yields

$$\varphi E^{\{R\}} \psi^{\{R\}} + \psi^{\{R\}} \left[T_n + V_{nn} - E \right] \varphi = 0 \tag{10.39}$$

Rearrangement of this expression yields

$$\psi^{\{R\}} \left[T_n + (V_{nn} + E^{\{R\}}) - E \right] \varphi = 0$$

From this, we have a separated Schrödinger equation for the nuclei.

$$\left[T_n + (V_{nn} + E^{\{R\}}) \right] \varphi = E \varphi \tag{10.40}$$

The Hamiltonian in Equation 10.40 consists of a kinetic energy operator for the nuclear position coordinates, the repulsion potential between the nuclei, and an effective potential, in the form of $E^{(R)}$, that gives the energy of the electronic wavefunction as it depends on the nuclear position coordinates. The energy eigenvalue, E, on the right-hand side, is the energy of a vibrational–rotational state. This approximate separation of the molecular Schrödinger equation into a part for the electronic wavefunction and a part for the nuclear motions is the **Born–Oppenheimer approximation**.

The essential element of the Born–Oppenheimer approximation is Equations 10.35 and 10.36 in which applying the nuclear kinetic energy operator to the electronic wavefunction is approximated as zero. The physical idea is that the light, fast-moving electrons readjust to nuclear displacements instantaneously. This is the reason the approximation produces an electronic Schrödinger equation for each possible geometrical arrangement of the nuclei in the molecule. We need only know the instantaneous positions of the nuclei, not how they are moving, in order to find an electronic wavefunction, at least within this approximation.

The Born–Oppenheimer approximation is put into practice by "clamping" the nuclei of a molecule of interest. That means fixing their position coordinates to correspond to some chosen arrangement or structure. The electronic Schrödinger equation (Equation 10.38) is solved to give the electronic energy for this clamped structure. After that, perhaps, another structure is selected, and the electronic energy is found by solving Equation 10.38 once more. Eventually, enough structures might be treated so that the dependence of the electronic energy on the structural parameters is known fairly well. At such a point, the combination of $E^{(R)}$ with V_{nn} yields the effective potential for molecular vibration. It is the potential energy for the nuclei in the field created by the electrons.

The separation of the electronic and nuclear motion via the Born–Oppenheimer approximation leads to the concept introduced in Chapter 6, that of the potential energy surface. A potential energy surface gives the dependence of the electronic energy, plus nuclear repulsion, on the geometrical coordinates of the atomic centers of a molecule. It is only within the Born–Oppenheimer approximation that we can follow the dependence of the electronic energy on the atomic position coordinates, and only within the approximation does the concept of a potential energy surface exist.

Potential energy surfaces, which were presented in Chapter 6, form the conceptual framework for thinking about and analyzing many problems in chemistry. The surface, of course, is the effective potential for vibrations of the molecule. But continued to large atom–atom separations, the surface encodes the energetic information about reactions and bond breaking. It also provides a picture of the pathways for reactions.

10.5 Antisymmetrization of Electronic Wavefunctions

The electrons moving about in an atom or molecule are indistinguishable particles. The mathematical consequence is that the probability density of an electronic wavefunction of an atom or molecule must be invariant with respect to how the electron coordinates are labeled. With $\vec{r}_i$ being the designation of the position vector of the ith electron, an interchange of two electrons' position coordinates means swapping $\vec{r}_i$ and $\vec{r}_2$ in ψ. For the probability density to be unchanged, the following must hold for any choice of i and j:

$$\psi\left(\vec{r}_1, \vec{r}_2, ..., \vec{r}_i, ...\vec{r}_j, ...\right)^2 = \psi\left(\vec{r}_1, \vec{r}_2, ..., \vec{r}_j, ...\vec{r}_i, ...\right)^2 \tag{10.41}$$

For the wavefunction itself, this implies that interchange of a pair of electron coordinates can change the wavefunction only by a factor $e^{i\varphi}$ since $e^{i\varphi}(e^{i\varphi})^* = 1$ for any real number φ. This same reasoning applies to a wavefunction of any other kind of indistinguishable particle, and it turns out that the value of φ depends on the intrinsic spin of the particles. As will be verified later, for electrons and other half-integer spin particles, $\varphi = \pi$. (For other than half-integer spin particles, such as those with spin = 0, 1, 2,..., $\varphi = 0$.) Since $e^{i\pi} = -1$, all electronic wavefunctions must satisfy the following condition:

$$\psi(\vec{r}_1, \vec{r}_2, ..., \vec{r}_i, ...\vec{r}_j, ...) = -\psi(\vec{r}_1, \vec{r}_2, ..., \vec{r}_j, ...\vec{r}_i, ...) \qquad (10.42)$$

This is termed an **antisymmetrization** requirement; an electronic wavefunction is antisymmetric (changes sign) with respect to interchange of any pair of electron position coordinates.

To impose Equation 10.42 on some electronic wavefunction, a special operator, called the **antisymmetrizer**, can be used. Let us develop the form of this operator, assuming that we have at hand a normalized electronic wavefunction, Φ, for some system, and that Φ has the form of a product of independent, orthonormal functions, $u(r)$, of the electron coordinates (e.g., orbitals).

$$\Phi(\vec{r}_1, \vec{r}_2, ...) = u_1(\vec{r}_1) u_2(\vec{r}_2) \cdots$$

The interchange of a pair of position coordinates is a well-defined mathematical operation, as we have in Equations 10.41 and 10.42, and we can define an operator that performs such an interchange. Let P_{ij} be an operator that interchanges the coordinates of some electron, i, with the coordinates of some electron, j. That is,

$$P_{ij}\Phi(\vec{r}_1, \vec{r}_2, ..., \vec{r}_i, ...\vec{r}_j, ...) = \Phi(\vec{r}_1, \vec{r}_2, ..., \vec{r}_j, ...\vec{r}_i, ...)$$

From this we can construct a new function, Φ' that is assured to be antisymmetric with respect to interchange of particles 1 and 2:

$$\Phi' = \frac{1}{\sqrt{2}}(\Phi - P_{12}\Phi) \qquad (10.43)$$

Antisymmetrization is tested by applying P_{12} and finding that the negative of the function results:

$$P_{12}\Phi' = \frac{1}{\sqrt{2}}(P_{12}\Phi - P_{12}P_{12}\Phi) = -\Phi'$$

Notice that P_{12} applied twice means swapping the coordinates and then swapping them back, and that is an identity operation. The square root of 2 in Equation 10.43 is introduced so that Φ' is normalized if Φ is normalized.

$$\langle \Phi' | \Phi' \rangle = \frac{1}{2} \langle \Phi - P_{12}\Phi | \Phi - P_{12}\Phi \rangle = \frac{1}{2} \left[\langle \Phi | \Phi \rangle - \langle \Phi | P_{12}\Phi \rangle - \langle P_{12}\Phi | \Phi \rangle + \langle P_{12}\Phi | P_{12}\Phi \rangle \right]$$

$$= \frac{1}{1}(1 - 0 - 0 + 1) = 1$$

The cross-terms are zero because of the independence of the two particles' coordinates and the orthogonality of the $u(r)$ functions.

At this point it is helpful to regard Equation 10.43 as the application of an operator that antisymmetrizes the function with respect to the interchange of the first two particles.

$$\Phi' = \left\{ \frac{1}{\sqrt{2}}(1 - P_{12}) \right\} \Phi \tag{10.44}$$

The next step in imposing Equation 10.42 on Φ involves electron 3. The wavefunction must be made antisymmetric with respect to interchanging electron 1 with 3 (i.e., applying P_{13}) and to interchanging electron 2 with 3 (i.e., applying P_{23}). This can be accomplished by a generalization of the operator in Equation 10.44 to use P_{13} and P_{23} and then by applying it to Φ'.

$$\Phi'' = \left\{ \frac{1}{\sqrt{3}}(1 - P_{13} - P_{23}) \right\} \Phi' = \left\{ \frac{1}{\sqrt{3}}(1 - P_{13} - P_{23}) \right\} \left\{ \frac{1}{\sqrt{2}}(1 - P_{12}) \right\} \Phi$$

We can continue this for all the electrons. For instance, considering the fourth electron means using P_{14}, P_{24}, and P_{34}. So, for N electrons, an entirely antisymmetric wavefunction, ψ, can be constructed from an arbitrary wavefunction, Φ, by application of a sequence of operators:

$$\psi = \frac{\sqrt{N!}}{N!}(1 - P_{1N} - P_{2N} - \cdots - P_{N-1,N}) \cdots (1 - P_{13} - P_{23})(1 - P_{12})\Phi = A_N \sqrt{N!}\Phi \tag{10.45}$$

The sequence of operators and the collective normalization constant $\sqrt{N!}$ can be considered one overall antisymmetrizing operator, which we will call the **antisymmetrizer** for N electrons, A_N.

There are two important properties of the antisymmetrizer. First, if it is applied to a wavefunction that is already properly antisymmetric, it will make no change. The implication of this statement is that if the antisymmetrizer is applied twice to an arbitrary function, the same result will be achieved as if it is applied only once; the second application does not do anything. This is the condition of **idempotency**, and A_N is an idempotent operator. The operator equation that expresses this fact is

$$A_N A_N = A_N \tag{10.46}$$

(Notice that this is by no means a statement that $A_N = 1$.) The second property is that A_N commutes with the electronic Hamiltonian. The indistinguishability of the electrons is apparent in the Hamiltonian, and any interchange of electron coordinates in the Hamiltonian leaves the Hamiltonian unchanged. That is, $P_{ij}H = H$ for any choice of i and j electrons. With this, it is easy to demonstrate that the antisymmetrizer for a two-electron system commutes with the Hamiltonian:

$$\left[H, \frac{1}{\sqrt{2}}(1 - P_{12}) \right] \Phi(r_1, r_2) = \frac{1}{\sqrt{2}} \left\{ H(1 - P_{12})\Phi(r_1, r_2) - (1 - P_{12})H\Phi(r_1, r_2) \right\}$$

$$= \frac{1}{\sqrt{2}} \left\{ H\Phi(r_1, r_2) - H\Phi(r_2, r_1) - H\Phi(r_1, r_2) + H\Phi(r_2, r_1) \right\} = 0$$

In general, A_N commutes with the Hamiltonian of an N-electron system. By Theorem 8.3, the commutivity of A_N and the Hamiltonian means that wavefunctions can be found that are simultaneously eigenfunctions of the Hamiltonian and of A_N. Wavefunctions that satisfy the Schrödinger equation can be antisymmetric.

Spin-orbital functions have been defined for atoms already, and like products of spatial functions and spin functions constitute spin-orbital functions in molecules. The orbital picture of the ground state of the ammonia molecule, for example, has 10 electrons in five spatial orbitals. If these are designated ϕ_1 through ϕ_5, the spin-orbitals are

$$\phi_1\alpha, \ \phi_1\beta, \ \phi_2\alpha, \ \phi_2\beta, \ \phi_3\alpha, \ \phi_3\beta, \ \phi_4\alpha, \ \phi_4\beta, \ \phi_5\alpha, \ \phi_5\beta$$

The orbital picture implies that the wavefunction is a product,

$$\Phi(r_1, ..., r_{10}) = \phi_1(r_1)\alpha(s_1)\phi_1(r_2)\beta(s_2)...\phi_5(r_9)\alpha(s_9)\phi_5(r_{10})\beta(s_{10})$$

where the abstract spin coordinates are s_1, etc. This, of course, is not antisymmetric, but the antisymmetrizer can be applied to make it antisymmetric. It turns out that the result of applying the antisymmetrizer to a product of spin-orbitals can be expressed in the concise and useful form of a determinant (see Appendix A). In explaining this, it is convenient to adopt a special shorthand notation. We will use one symbol for a spin-orbital, and here that symbol will be u. A given spin-orbital is a function of the spin and spatial coordinates of some electron, but instead of writing $u(r_i, s_i)$, we will write $u(i)$; that is, just the electron number will be written since that is sufficient to interpret what is meant. In this shorthand notation, the product function given previously for ammonia is written as

$$\Phi(r_1, ..., r_{10}) = u_1(1)u_2(2)u_3(3)u_4(4)u_5(5)u_6(6)u_7(7)u_8(8)u_9(9)u_{10}(10)$$

In general, the orbital product form of an N-electron wavefunction is

$$\Phi(r_1, ..., r_N) = u_1(1)u_2(2)\cdots u_N(N) \tag{10.47}$$

Application of the N-electron antisymmetrizer to this product function yields a function that consists of $N!$ products combined together, as there are $N!$ products of permutation operators in Equation 10.45. These $N!$ terms can be obtained by expanding the following determinant:

$$\psi = \frac{1}{\sqrt{N!}} \begin{vmatrix} u_1(1) & u_2(1) & u_3(1) & ... & u_N(1) \\ u_1(2) & u_2(2) & u_3(2) & ... & u_N(2) \\ u_1(3) & u_2(3) & u_3(3) & ... & u_N(3) \\ ... & ... & ... & ... & ... \\ u_1(N) & u_2(N) & u_3(N) & ... & u_N(N) \end{vmatrix} = A_N\Phi \tag{10.48}$$

The diagonal of this determinant yields the product of Equation 10.47. Each column uses a different spin-orbital function, and each row uses a different electron's coordinates. ψ is called a **Slater determinant** after the inventor of this device, John C. Slater (the United States, 1900–1976).

Slater determinants are functions that are immediately seen to be properly antisymmetric with respect to exchange of the coordinates of any pair of electrons. Such an exchange corresponds to interchanging two rows of the Slater determinant, and it is a property of determinants that their value changes sign when two rows are exchanged. This means that the expanded form of the determinant, which is ψ, has an opposite sign when a pair of electron coordinates are exchanged, and that results from the corresponding exchange of rows of the determinant.

There is another important property revealed by the Slater determinant construction. If there are two identical spin-orbitals, then there will be two identical columns in the Slater determinant. It is a property of determinants that their value is zero when two columns are identical. Thus, if we consider two different electrons to be in the same spin-orbital in an original product function (Equation 10.47), anti-symmetrization (e.g., construction of the corresponding Slater determinant) will produce zero: Such a wavefunction is not permitted. Thus, antisymmetrization leads to a requirement that only one electron occupy a particular spin-orbital. This is a way of stating the Pauli exclusion principle.

10.6 Molecular Electronic Structure

The form of a many-electron wavefunction is usually complicated. The electronic Hamiltonian is not strictly separable into independent terms for different electrons because of the electron–electron repulsion operator V_{ee} in Equation 10.38. An exactly determined wavefunction cannot usually be formed as a product of one-electron functions, for example, as $u_1(\vec{r}_1)u_2(\vec{r}_2)u_3(\vec{r}_3)\ldots$, which is the orbital form. However, this orbital product form is appropriate on invoking a certain approximation to the electron–electron repulsion. The physical aspect of this approximation is to replace the electron–electron repulsion potential by an electrical potential due to the average distribution of all the electrons residing in their particular spin orbitals. In other words, the interaction between pairs of electrons is approximated by each electron interacting with a charge cloud. Each electron is treated as a separated quantum mechanical system in which the potential energy operator is that of the electron's charge interacting (1) with the positively charged nuclei and (2) with a fixed field arising from the charge cloud of the other electrons.

The approximation to the electron–electron repulsion has a mathematical effect of converting an N-electron problem of some original electronic Schrödinger equation to N one-electron problems by means of an effective one-electron potential operator, $\hat{g}$. This is combined with the operator corresponding to an electron's kinetic energy and its attraction for the nuclei, $\hat{h}$:

$$\hat{h} = -\frac{\hbar^2}{2m_e}\nabla^2 - \sum_{\alpha}^{nuclei}\frac{Z_\alpha e^2}{r_\alpha} \tag{10.49}$$

where r_α is the distance between the α nucleus and the electron. The result is a special one-electron Hamiltonian called the **Fock operator**, $\hat{F}$.

$$\hat{F} = \hat{h} + \hat{g} \tag{10.50}$$

$\hat{g}$ is associated with the effective field of all the electrons. Of course, an electron cannot be said to interact with itself, and thus we really need a field operator for the field of all the *other* electrons. It turns out that $\hat{g}$ can be constructed such that on applying the operator, the self-term, meaning the part for the interaction of the electron with its own share of the field, vanishes. As a result, the field operator $\hat{g}$ is the same for every electron. Orbital functions are obtained as eigenfunctions of the Fock operator. It is, in effect, the Hamiltonian for the system, though in the sense of a Hamiltonian for one electron at a time. The eigen-equation is

$$\hat{F}u_i = \varepsilon_i u_i \tag{10.51}$$

and the eigenfunctions, u_i, are functions of only one electron's coordinates. The eigenvalues associated with the orbitals are labeled by the same index, and they are referred to as **orbital energies**.

The field that electrons experience in this approximation can be prescribed only when the orbital wavefunctions are known. In other words, Equation 10.51 is quite different from the other eigenequations we have considered since the operator is *dependent* on the solutions. We need to know the orbitals that the electrons occupy in order to know $\hat{g}$; we need to know $\hat{g}$ in order to find the orbitals. Actually, this problem can be solved by a bootstrap procedure. From a set of guess orbitals, a corresponding $\hat{g}$ operator is formed for use in Equation 10.51. If the orbitals then obtained are not the same as the guess, they are used in constructing a new $\hat{g}$ operator. The whole process is repeated again and again until the orbitals used to construct $\hat{g}$ turn out to be the same as the eigenfunctions in Equation 10.51. This means that $\hat{g}$, which represents the effective field, is not prescribed from the outset but is determined in a *self-consistent* manner. The effective field, then, is usually termed a **self-consistent field** (SCF).

In general, separability of a Schrödinger equation comes about if there are independent, additive pieces of the Hamiltonian. The separability of an N-electron Schrödinger equation into the N one-electron equations represented by Equation 10.51 means the approximation corresponds to a many-electron Hamiltonian that is simply the sum of Fock operators for each electron, and we designate this Hamiltonian $\hat{H}_0$.

$$\hat{H}_0 = \sum_{v}^{N} \hat{F}_v \tag{10.52}$$

The Fock operators in Equation 10.52 are subscripted to indicate that each acts on independent position coordinates for the N electrons. Separability also means that the energy is a sum, and in this case, we have

$$E_0 = \sum_{i}^{N} \varepsilon_i \tag{10.53}$$

E_0 is the eigenenergy of $\hat{H}_0$. The orbital energies provide a framework for chemical energetics. For instance, since the net energy needed to form a molecule from separated nuclei and electrons is (approximately) E_0, a reaction energy can be approximately obtained as a difference in orbital energies of reactants and products.

Another energetic feature of a molecule, the **ionization potential** (IP), is the energy required to remove an electron, and according to Equation 10.53, that energy is the orbital energy, at least assuming the orbitals for the original molecule and the ionized molecule are the same. **Photoelectron spectroscopy** (PES) is a type of molecular spectroscopy in which molecules are irradiated with high-energy photons of fixed energy (i.e., monochromatic radiation). When the photon energies are greater than an ionization potential of the sample, an electron can be ejected, and its kinetic energy is the difference between the photon energy and the ionization potential. The ejected electrons are energy-analyzed in a photoelectron experiment, and the spectrum of energies represents the body of data obtained. Electrons can be ejected from the different orbitals of an atom or molecule, and so the photoelectron spectrum has sharp peaks at energies usually associated with ionization from the different orbitals. Far ultraviolet radiation is used for valence orbitals, whereas x-ray radiation is used for core orbitals. The former type of experiment is designated UPS and the latter XPS.

The self-consistent field Hamiltonian in Equation 10.52, which is an approximation of the complete Hamiltonian implicit in Equation 10.38, has certain deficiencies. Since the approximation corresponds to each electron "seeing" all the other electrons as one fixed charge cloud, it neglects the correlation of the electron distributions that should arise from their "seeing" each other as true point charges moving in space. The errors of the molecular orbital picture that result from the use of the Hamiltonian in Equation 10.52 are termed **correlation effects**. Correlation effects are known to influence the structures of molecules, the shapes of potential surfaces, electronic excitation energies, and bond formation energies. Nonetheless, a molecular orbital view (i.e., neglect of correlation) often offers immense conceptual value and qualitative insight in chemistry, as well as providing quantitative information under certain circumstances. Large-scale computer calculations are capable of evaluating correlation effects, at least in small and moderate-sized molecular systems. Wavefunctions and energies are obtained first from SCF calculations and then from treatments that go beyond SCF, with the differences being the correlation effects. A common exaggeration of electron density in bonding regions at the SCF level, for instance, is corrected as electron correlation is taken into account. As a result, there is often a net correlation effect that amounts to making a bond length somewhat longer and to making the bond less stiff, which means diminishing the force constant. For small covalent molecules, high-level treatment of electron correlation shows bond length changes that are mostly 0.01–0.03 Å and a reduction of harmonic force constants by about 5%–20%. A selection of calculated values for small molecules from the SCF level of treatment and from treatments that carefully include electron correlation effects is shown in Table 10.3. Comparison with measured values provides an assessment of the suitability of the SCF orbital picture in describing molecular systems.

An important concept rooted in the molecular orbital (SCF) approximation is that of **linear combinations of atomic orbitals** (LCAO) to form molecular orbitals. The first example is the combination of two hydrogen atom 1s orbitals to describe the electrons in the H_2 molecule. Two orthogonal linear combinations can be formed in this case. The first is a symmetric combination, $1s_l + 1s_r$, where the l and r subscripts refer to a hydrogen atom on the left (l) and on the right (r) in H–H. The other combination is antisymmetric, $1s_l - 1s_r$. This combination has a nodal plane between the two atoms, a plane where the probability density is zero. Now, we may ask, which orbital is favored energetically? There is clearly going to be greater electron density between the hydrogen nuclei in the symmetric combination than in the antisymmetric combination. An electrostatic argument, that negative charge density would be most favorable *between* two positively

TABLE 10.3

Comparison of Calculated and Measured Values of Certain Molecular Properties

		Value from SCF	Value with Correlation	Experiment
LiH	Equilibrium bond length (Å)	1.6071	1.5974	1.5957
LiH	$n = 0$ to $n = 1$ vibrational transition frequency (cm^{-1})	1389.5	1358.1	1359.8
HF	$n = 0$ to $n = 1$ vibrational transition frequency (cm^{-1})	4324	3977	3961
H_2O	Harmonic vibrational frequencies (cm^{-1})	4132	3830	3832
		1771	1677	1649
		4236	3940	3943

Source: Values from Dykstra, C.E., *Ab Initio Calculation of the Structures and Properties of Molecules*, Elsevier, Amsterdam, 1988.

charged centers rather than on the outer ends, correctly suggests that the symmetric combination is energetically favored. The symmetric combination is a bonding orbital, and the antisymmetric combination is said to have antibonding character. The electronic structure of H_2 in the vicinity of its equilibrium separation distance is essentially that of two (spin-paired, α and β) electrons in the molecular orbital that is the symmetric combination of hydrogen atom 1s orbitals.

We can consider the LCAO problem of the H_2 molecule from a perturbation theory standpoint. At infinite separation of the two atoms, the Hamiltonian is a sum of independent hydrogen atom Hamiltonians. The interaction that develops when the atoms approach each other is an attraction of each atom's electron for the other's proton plus the repulsion between the two electrons. This interaction is a perturbation, $\hat{H}_1$. There is also a repulsion between the two nuclei, and within the Born–Oppenheimer approximation, this is a constant that is added to the electronic energy at each (fixed) internuclear separation. In the absence of the perturbation, the energies of the two electrons are the hydrogen atom energies, and it is common to represent this zero-order situation by two horizontal lines alongside a vertical (orbital) energy scale. Since these two lines are placed at the true energies of the atoms at a large separation, one line is drawn on the far left and the other on the far right, as in Figure 10.6. In the presence of the perturbation the orbitals of the left atom and the right atom combine producing one with lower energy (bonding) and one with higher energy (antibonding). At the lowest order of perturbation theory that yields a nonzero effect of $\hat{H}_1$, the amount one of the new orbitals is raised in energy is equal to the amount the other is lowered. Two electrons occupying the bonding orbital corresponds to the ground electronic state of H_2. Placing one or both electrons in the antibonding orbital corresponds to an excited electronic state of the molecule. Such figures are called **orbital correlation diagrams**. They indicate how the separated atom states combine to form the molecular electronic states.

In a more complicated diatomic, the orbitals of atom (fragment) A are paired up with the orbitals of atom (fragment) B on the basis of similar energies, and the same analysis is carried out. Implicit in this picture is that electrons are interacting largely one-on-one with electrons in the other fragment. This is partly because low-order perturbation theory gives small energetic effects if the orbital energies of the two electrons from the fragments are very different in energy. (A large energy difference would lead to a sizable energy denominator in the perturbation theory expressions for the energy and wavefunction corrections.) Figure 10.7 puts this into practice for the carbon monoxide molecule. The orbital energies of carbon are represented by the series of horizontal lines on the left, and the orbital energies of oxygen are on the right. These orbitals correlate with the molecular orbitals whose approximate energies

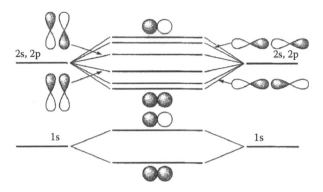

FIGURE 10.6
Orbital energy correlation diagram for the linear combination of atomic orbitals to form molecular orbitals. The vertical axis is an orbital energy scale, and the leftmost and rightmost horizontal lines represent the orbital energies of two noninteracting hydrogen atoms. In the middle of the figure are two horizontal lines representing the energies of mixed functions of the left and right hydrogen 1s orbitals. Their energies are qualitatively deduced from first-order degenerate perturbation theory which requires that the two mixed states be above and below the energies of the unmixed degenerate states by equal amounts. The qualitative form of the mixed orbitals is also shown.

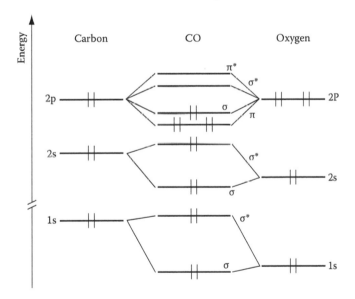

FIGURE 10.7
Orbital correlation diagram for carbon monoxide. The carbon atomic orbital energies are on the left, and the oxygen atomic orbital energies are on the right. The molecular orbitals that form from mixing of the atomic orbitals are represented by the horizontal lines in the center at their approximate orbital energies in the CO molecule. The vertical lines indicate the orbital occupancy.

are represented by the horizontal lines in the center. The correlation of atomic orbitals with molecular orbitals is indicated by dashed lines. Notice that the oxygen 1s orbital is not paired with a carbon orbital. Its orbital energy is much lower than any of the carbon atom's orbitals, and so, from low order perturbation theory, little mixing is expected with the carbon orbitals. This is in keeping with not considering an oxygen 1s orbital to be a valence orbital. It is an orbital that closely surrounds the oxygen nucleus, and we should expect it to be little affected by chemical bonding. It remains essentially an atomic orbital in character.

The correlation of atomic orbitals with molecular orbitals, as in Figure 10.7, ranks the molecular orbitals in terms of their expected energies. The lowest energy electronic state of a molecule is expected to arise when the electrons fill up the molecular orbitals from the bottom up, that is, when they occupy the most stable orbitals. A chemical bond is said to exist when two electrons occupy a bonding orbital, and again, a bonding orbital is one without a nodal plane between the two atoms. H_2 has one bond. When antibonding orbitals are occupied, the net bonding is taken to be given by a number called the bond order (BO):

$$\text{Bond order} = \frac{(\text{no. of electrons in bonding orbitals} - \text{no. of electrons in antibonding orbitals})}{2}$$

Thus, He_2 has a bond order of zero since according to Figure 10.6 two electrons would be in a bonding orbital just as in H_2, but two electrons would be in an antibonding orbital as well. Carbon monoxide has a bond order of 3 (a triple bond). The bond order gives a rough idea of the bond strength or of the relative bond strengths.

One of the key problems with the SCF approximation is in the description of breaking, or forming, a chemical bond. In the course of breaking a bond, significant changes in the orbital character of the species are not unlikely, and even the form of the wavefunction may not be the same throughout the whole process. For instance, the orbital picture of the nitrogen molecule at equilibrium is a closed shell wavefunction, and the SCF level treatment is reasonable. It dissociates, however, into two identical atoms, and their states cannot be described as closed shells. Thus, a wavefunction that is appropriate throughout must include several configurations whose importance changes with the geometrical parameters. The use of several configurations means that correlation effects are being accounted for; however, these are nondynamical correlation effects, as opposed to the dynamical correlation* effects discussed in the prior section.

One powerful approach to the problem of bond breaking is the valence bond method. To understand the basic idea of the valence bond picture, let us consider the electronic structure of the ground, singlet state of the hydrogen molecule. The form of the valence bond (or Heitler–London) wavefunction is

$$\psi(1,2) = \frac{N}{\sqrt{2}} \left(a(1)b(2) + a(2)b(1) \right) \left(\alpha(1)\beta(2) - \alpha(2)\beta(1) \right) \tag{10.54}$$

where we use the shorthand notation of electron numbers 1 and 2 in place of the spatial and spin coordinates of those electrons (e.g., $a(1)$ instead of $a(\vec{r}_1)$ and $\alpha(2)$ instead of $\alpha(\vec{s}_2)$). N is a normalization factor, the spatial functions have been designated a and b, and α and β are the usual electron spin functions. This wavefunction is not expressed in terms of Slater determinants, but it is nonetheless properly antisymmetric with respect to particle interchange. In an SCF orbital description, the spatial functions are orthogonal. In the valence bond description, that constraint is relaxed: a and b are allowed to be the same or different spatial functions. The added flexibility provides for a better description of the changing electronic structure that occurs during bond breaking and bond formation.

---

* The distinction between dynamical and nondynamical correlation is not sharp. However, as a working definition, nondynamical correlation is manifested as a few configurations entering the wavefunction with sizable importance. For dynamical correlation, there are many configurations with small expansion coefficients and a single dominant configuration.

When the two hydrogen atoms in H_2 are very far apart, perhaps $1000\,\text{Å}$, there is essentially no interaction. In this limiting case, the electronic structure is the electronic structure of two ground state hydrogen atoms. The wavefunction in Equation 10.54 can represent this limiting case, which is to say that variational optimization of ψ yields the correct 1s orbitals. We can see that the wavefunction has this flexibility by replacing a with $1s_A$, meaning a hydrogen 1s orbital on hydrogen A, and by replacing b with $1s_B$, where B is the other hydrogen.

$$\psi(1,2) = \frac{N}{\sqrt{2}}\left(1s_A(1)1s_B(2) + 1s_A(2)1s_B(1)\right)\left(\alpha(1)\beta(2) - \alpha(2)\beta(1)\right)$$

$$= \frac{N}{\sqrt{2}}\left\{\left|1s_A\alpha 1s_B\beta\right| - \left|1s_A\beta 1s_B\alpha\right|\right\} \tag{10.55}$$

Since the orbitals $1s_A$ and $1s_B$ are orthogonal, given the large A–B separation distance, rearrangement of the wavefunction into two Slater determinants in Equation 10.55 allows us to conclude that this is the form of the wavefunction for two unpaired electrons with singlet coupled spins. In other words, the wavefunction has the flexibility to take on the correct form at the separated atom limit.

The other limiting situation is the equilibrium separation of the protons in the H_2 molecule. The molecular orbital SCF picture places the two electrons in one spatial orbital, σ. The valence bond wavefunction achieves this form if both orbitals a and b in Equation 10.54 are σ.

$$\psi(1,2) = \frac{N}{\sqrt{2}}\sigma(1)\sigma(2)\left(\alpha(1)\beta(2) - \alpha(2)\beta(1)\right) \tag{10.56}$$

This is the closed-shell SCF determinant written in expanded form.

Showing that the wavefunction has the flexibility to properly describe the separated atoms and the bonded atoms means that a variational determination of the orbitals a and b at each separation distance provides an appropriate description of the system continuously from one limiting situation to the other. In contrast, the SCF orbital picture is not uniformly appropriate for the breaking of the H_2 bond. We can see this by viewing the SCF σ orbital to be essentially a linear combination of the two hydrogenic 1s orbitals:

$$\sigma = n(1s_A + 1s_B) \tag{10.57}$$

where n is a factor to ensure normalization of the σ orbital. Substituting this into the SCF wavefunction, Equation 10.56, yields

$$\psi(1,2) = N'\left(1s_A(1)1s_A(2) + 1s_A(1)1s_B(2) + 1s_B(1)1s_A(2) + 1s_B(1)1s_B(2)\right)\left(\alpha(1)\beta(2) - \alpha(2)\beta(1)\right) \tag{10.58}$$

where all the normalization factors have been collected into N'. This is the form of the SCF wavefunction everywhere, at least while restricting the σ orbital to be a linear combination of just the 1s orbitals. Even at a $1000\,\text{Å}$ separation, it would have this form. Comparing it with the proper separated limit form of Equation 10.55, shows a difference. The SCF wavefunction includes two additional terms, $1s_A(1)1s_A(2)$ and $1s_B(1)1s_B(2)$. These terms are interpreted as corresponding to an ionic arrangement of the electron density. The first places both electrons on hydrogen A, leaving bare the proton of hydrogen B. The second term is the opposite ionic form. The nonionic terms, as in Equation 10.55, correspond to a covalent sharing of the electrons. Thus, the SCF (molecular orbital) description weights the ionic and covalent terms equally for *all* A–B separation distances, whereas at the separated limit, the ionic terms must vanish. The flexibility in the valence bond description that comes about from allowing the two electrons of a bonding pair to be in different, nonorthogonal orbitals (i.e., a and b) translates into the flexibility to change the weighting between ionic and covalent parts of the electron distribution. In this way, the valence bond picture is extremely useful for describing bond breaking and formation.

With some manipulation, valence bond wavefunctions can be restated in terms of orthogonal orbitals. For the singlet ground state of H_2, it can be shown that a description equivalent to Equation 10.58 is

$$\psi = c_1 \left| \phi_1 \alpha \phi_1 \beta \right| + c_2 \left| \phi_2 \alpha \phi_2 \beta \right| \tag{10.59}$$

This wavefunction is a linear combination of two configurations, and their expansion coefficients, c_1 and c_2, are obtained as variational parameters. Each configuration places both electrons in the same spatial orbital, and ϕ_1 and ϕ_2 are orthogonal orbitals. This representation of a valence bond wavefunction proves particularly helpful in computer calculations of the wavefunctions as in the **generalized valence bond** (GVB) method.

Density functional theory (DFT) has become an extremely useful computational approach for the description of ground state properties of atoms and molecules. DFT can also be extended to bulk materials at much less computational cost than traditional, wavefunction-based methods.

DFT uses the electron density as the fundamental variable used to describe the electrons and not the electronic wavefunction. For N electrons, this means that the basic variable of the system depends only on the three spatial coordinates x, y, and z, rather than $3N$ degrees of freedom. Useful applications of DFT are based on approximations for the exchange-correlation functional. The exchange-correlation functional describes the effects of the Pauli exclusion principle and the Coulomb potential beyond a pure electrostatic interaction of the electrons. Unfortunately, the exact functional is not known for anything but the free electron gas. Still, the various approximations of DFT include electron correlation in some form, which gives this method a powerful advantage over wavefunction-based methodology.

10.7 Visible–Ultraviolet Spectra of Molecules

Sharp differences are likely to be found on exciting the electrons of a molecule to states energetically above the ground state. The molecular orbital picture has limited quantitative accuracy for predicting electronic excitation energies, yet it can be used to understand some of the qualitative features possible for excited electronic states. Consider the CO molecule and its orbital energy diagram (Figure 10.7). If an electron were promoted from one of the bonding orbitals occupied in the most stable arrangement of electrons among orbitals (i.e., the ground state) to an antibonding or nonbonding molecular orbital, the net bond order would be reduced. This means the bond would be weaker, and a manifestation of that might be a longer equilibrium C–O separation distance. The potential well would not be as deep as in the ground state, and so the curvature of the potential energy curve for C–O vibration would be different, too.

The term "spectroscopic states of molecules" refers to states that may be involved in transitions seen by spectroscopic measurement. Through such measurements, we are able to extract information about the different bonding character, about the different electronic structure, of excited states of molecules. For molecules containing atoms of other than the very heavy elements, spectroscopic states have specific net electron spins. They are frequently singlet states, meaning the spin quantum number, S, is zero, or they can be doublet states ($S = 1/2$), triplet states ($S = 1$), or states with a still greater spin quantum number. They also have certain properties that reflect geometrical symmetry of the molecule, if any (see Appendix B).

Using visible–ultraviolet spectroscopy of molecules, we examine the nature of excited electronic states since typically the absorption of a photon in this energy region electronically excites the molecule. For molecules composed of elements with low atomic number, the strongest selection rule at work in visible–ultraviolet spectroscopy is

$$\Delta S = 0 \tag{10.60}$$

That is, transitions are normally between states of the same spin. Other selection rules may relate to the geometrical symmetry of the molecule. The molecular transitions seen in the visible and ultraviolet regions of the spectrum must be transitions from one rotational–vibrational-electronic state to another. We shall consider this in detail for a hypothetical diatomic molecule for which the ground state and excited state potential curves are those shown in Figure 10.8. For both electronic state potential energy curves there are sets of vibrational states and rotational sublevels. Notice that the equilibrium distance is not the same for both curves and that the curvature (i.e., the force constant) is not the same either. Thus, there are a different vibrational frequency and a different rotational constant for each electronic state. This has to be taken into account in working out the transition frequencies.

Interpretation of molecular spectra requires a prediction of the frequencies of transition lines so as to *assign* observed lines to specific transitions. Then, the measured frequencies can be used to extract the true values for vibrational frequencies, rotational constants, and so on. The frequency of an electronic transition, as seen from Figure 10.8, depends first on the energy difference between the potential minima of the two states, and this is called the **term value**, T_e. It is the contribution to the transition energy that involves only the electronic energies. The second component of the transition energies is that due to the vibrational–rotational state energies within each of the two wells. The energies of those states depend on vibrational and rotational quantum numbers n' and J' for the initial state, and n'' and J'' for the final state. For a diatomic molecule, the simplest prediction of the pattern of transition energies that we can use to assign lines comes from using the lowest-order treatment for vibrational–rotational levels. At this level, anharmonicity, centrifugal distortion, and vibration–rotation coupling are neglected.

$$\Delta E = E_{final} - E_{initial} = E''(n'', J'') - E'(n', J') \tag{10.61a}$$

$$\Delta E = T_e + \left(n'' + \frac{1}{2}\right)\hbar\omega'' + B''J''(J''+1) - \left(n' + \frac{1}{2}\right)\hbar\omega' - BJ'(J'+1) \tag{10.61b}$$

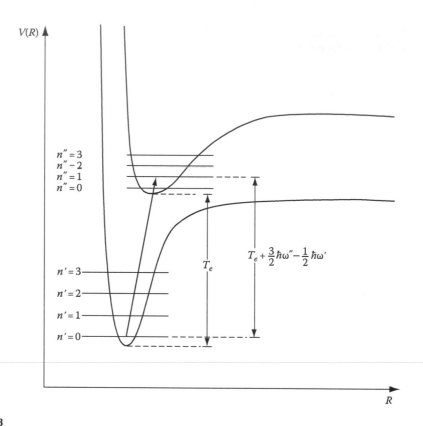

FIGURE 10.8

Potential energy curves for two electronic states, the ground state and an excited state of the same spin, of a hypothetical diatomic molecule. Each potential can be analyzed independently to yield vibrational–rotational states and energy levels. The lowest four vibrational levels are shown. In an absorption spectrum, transitions can originate from the vibrational–rotational levels of the ground electronic state and end in the vibrational–rotational levels of the excited state following the appropriate selection rules. An arrow from the $n' = 0$ to the $n'' = 1$ level depicts the initial and final states of one possible transition. The energies of the transitions are sums of the energy difference between the bottoms of the potential wells, designated T_e, and the difference between the vibrational–rotational state energies within their respective potentials.

The selection rules for J in the case of an electronic transition in a diatomic molecule is again that J can change by 1. As well, J is allowed to remain unchanged, though only when there is a change in the net electronic orbital angular momentum about the internuclear axis. This gives three cases for the part of the transition energy in Equation 10.61 associated with the rotational energy, which can be isolated in the expression as ΔE_{rot}.

$$\Delta E = T_e + \left(n'' + \frac{1}{2} \right) \hbar \omega'' - \left(n' + \frac{1}{2} \right) \hbar \omega' + \Delta E_{rot}$$

$$\Delta J = J'' - J' = 1 \Rightarrow \Delta E_{rot} = B''(J'+1)(J'+2) - B'J'(J'+1) = (B'' - B')J'^2 + (3B'' - B')J' + 2B'' \quad (10.62a)$$

$$\Delta J = J'' - J' = 0 \Rightarrow \Delta E_{rot} = (B'' - B')(J'^2 - J') \quad (10.62b)$$

$$\Delta J = J'' - J' = -1 \Rightarrow \Delta E_{rot} = (B'' - B')J'^2 - (B'' + B')J' \quad (10.62c)$$

These transition frequencies correspond to a whole series of lines for a given n' to n'' vibrational quantum number change. As in the case of infrared spectra, the transitions associated with Equation 10.62a are transitions of the P-branch, and those associated with Equation 10.62c are of the R-branch. Transitions associated with Equation 10.62b are those of a Q-branch. From the relative sizes of typical rotational constants, we should expect that ΔE_{rot} separates the lines by a small amount in relation to the size of T_e. In other words, the spectrum should show a group of lines clustered about a transition energy that is roughly

$$T_e + \left(n'' + \frac{1}{2}\right)\hbar\omega'' - \left(n' + \frac{1}{2}\right)\hbar\omega'$$

A **vibrational band** in an electronic spectrum is the clustering of lines about one such transition energy.

The rotational contributions in Equation 10.62a allow us to decipher the **rotational fine structure** of a band resolved in an experiment. In that event, some interesting features can be seen. If $B'' \approx B''$, Equation 10.62a simplifies approximately to $2B'(J' + 1)$, and this is the usual expression for the R-branch lines in the fine structure of a diatomic IR absorption spectrum. Equation 10.62b yields approximately zero for ΔE_{rot}, and so all the lines are at very nearly the same frequency. Thus, a Q-branch is a very narrow feature compared to an R-branch or a P-branch. Equation 10.62c yields approximately $-2B'J'$ for ΔE_{rot} if $B'' \approx B'$, and so these lines correspond to the P-branch. Typically, B'' and B' are somewhat different, as suggested by the different equilibrium distances for the two potentials in Figure 10.8. If $B'' > B'$, then with increasing J', the P- and R-branches will extend to higher and to lower frequencies, respectively, and the Q-branch lines will spread apart toward higher frequencies. If $B'' < B'$, the Q-branch lines will be spread apart but toward lower in frequencies with increasing J'.

The process of absorbing a photon and undergoing electronic excitation happens in a time that is extremely short, essentially instantaneous, relative to the time required to complete a vibrational cycle (i.e., the vibrational cycle if the diatomic oscillator were behaving as a classical oscillator). This means electronic transitions are essentially *vertical transitions* in that the geometrical structure of the transition is the same in the ground state as in the excited state at the instant of the transition; on a plot such as Figure 10.8, the transitions correspond to vertical arrows from the initial to the final state. From a strict quantum mechanical perspective, we should not consider an oscillator to be at any one point at one instant in time, for that is a classical mechanics view. Even so, the notion is useful. The transition is such a fast process that the quantum mechanical distribution in space is likely to persist in the excited state over the short time of the transition. One might represent this on Figure 10.8 as a wide but still vertical arrow between an initial and a final state.

According to time-dependent perturbation theory, selection rules depend on the integral of the perturbing Hamiltonian sandwiched between the initial and final state wavefunctions. The perturbing Hamiltonian involves the dipole moment operator. One way in which this operator was treated when considering infrared spectra was by a power series expansion. The first term of that expansion is the permanent dipole moment, which is a constant. Truncating the expansion to that first term leaves the integral of interest proportional to the overlap of the initial and final state wavefunctions. For the idealized situation of initial and final state vibrational potentials that are harmonic but with displaced or offset minima, as illustrated in Figure 10.9, a nonzero overlap is found for any pair of n' and n'' states. So, there is no distinguishing selection rule for the vibrational quantum number in the electronic spectra of a diatomic molecule. Transitions may be seen with $n' = n''$ and with n' different from n'' by 1, 2, 3, and so on, and these give rise to different vibrational

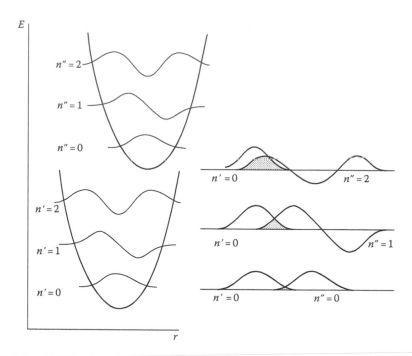

FIGURE 10.9

Vibrational wavefunctions and overlaps for two identical harmonic oscillator wavefunctions with offset minima.

bands. There are, however, differences in the relative intensities of the bands. The intensities depend on the size of the integral used to find selection rules. An approximation to that integral is the overlap of the initial and final state vibrational wavefunctions, and as Figure 10.9 shows, this overlap depends on n', n'', and the offset in the two potentials' minima. **Franck–Condon overlap** refers to this overlap of vibrational state wavefunctions and the relation to relative band intensities. It is particularly helpful in interpreting low-resolution spectra, where the rotational fine structure is not resolved, and it is helpful in determining the extent to which the potential minimum of an excited electronic state is offset from the minimum of the ground state.

Example 10.1: Diatomic Molecule Electronic Absorption Bands

Problem: For LiH, the rotational constant in the ground electronic state is 7.51 cm^{-1} and the harmonic vibrational frequency is about 1400.0 cm^{-1}. With a harmonic, rigid rotator treatment of vibrational–rotational states, predict the transition frequencies of the eight most intense rotational fine structure lines in the $n' = 0 \rightarrow n'' = 0$ vibrational band for excitation from the ground state to an electronic state of LiH with a harmonic frequency of 1100.0 cm^{-1} and (1) an excited state equilibrium bond length 8% greater than that of the ground state and then (2) 8% shorter. Take T_e to be 22,000.0 cm^{-1}.

Solution: First, the equilibrium rotational constant, B_e, is inversely proportional to the square of the equilibrium separation. So, for case 1, B_e'' is $7.51/(1.0 + 0.08)^2 = 6.44$ cm^{-1}. For case 2, B_e'' is $7.51/(1.0 - 0.08)^2 = 8.87$ cm^{-1}. With the given information, all transitions have a contribution to the transition energy of T_e plus the difference in the zero-point energies of the vibrational states in the two electronic states, that is, $(1100.0 - 1400.0)/2 = -150.0$ cm^{-1}. So, for the transitions consideration, ΔE (cm^{-1}) $= 22,000.0 - 150.0 + B_e'' J''(J'' + 1) - 7.51 J'(J' + 1)$.

The ΔE's are evaluated for the two cases and the three branches for $J' = 0, 1, 2, 3, \ldots, 8$ since these initial rotational quantum numbers should correspond to the strongest transitions in a given branch due to the Maxwell–Boltzman population of initial rotational states.

Branch	J'	J''	(1) ΔE (cm^{-1})	(2) ΔE (cm^{-1})
P	1	0	21,834.98	21,834.98
	2	1	21,817.82	21,822.68
	3	2	21,798.52	21,813.10
	4	3	21,777.08	21,806.24
	5	4	21,753.50	21,802.10
	6	5	21,727.78	21,800.68
	7	6	21,699.92	21,801.98
	8	7	21,669.92	21,806.00
Q	0	0	21,850.00	21,850.00
	1	1	21,847.86	21,852.72
	2	2	21,843.58	21,858.16
	3	3	21,837.16	21,866.32
	4	4	21,828.60	21,877.20
	5	5	21,817.90	21,890.80
	6	6	21,805.06	21,907.12
	7	7	21,790.08	21,926.16
	8	8	21,772.96	21,947.92
R	0	1	21,862.88	21,867.74
	1	2	21,873.62	21,888.20
	2	3	21,882.22	21,911.38
	3	4	21,888.68	21,937.28
	4	5	21,893.00	21,965.90
	5	6	21,895.18	21,997.24
	6	7	21,895.22	22,031.30
	7	8	21,893.12	22,068.08

These values reveal qualitative features that can be helpful in interpreting a spectrum of a diatomic molecule prior to knowing the rotational constant or bond length in the excited electronic state. Notice the sequence of transition energies in the P-branch for case 2. As the initial J quantum number increases, the lines become more closely spaced until $J' = 6$, where there is a reversal. The lines in the P-branch become clumped at and "turn around" at 21,800.68 cm^{-1}. This feature is called a **band head**. The P-branch for case 1 does not show this feature. The lines become increasingly spaced out with increasing initial J'. Next, notice the sequence of R-branch transition energies for the two cases. A band head occurs for case 1 but not case 2. So there is a qualitative difference between the two cases. If the equilibrium bond length is longer in the excited electronic state (i.e., a smaller rotational constant in the excited state), the rotational fine structure will approach (or exhibit) a band head in the R-branch, whereas if the bond length is shorter (i.e., a larger rotational constant), that feature will be in the P-branch. When enough of the vibrational fine structure can be resolved, one immediately knows qualitatively whether or not the excited state of a diatomic molecule has a longer or a shorter bond length than the ground state. Measurement of the actual transition frequencies can lead to a value for the excited state rotational constant and then to the excited state bond length. Under low resolution, a band head appears as a sharp edge on the right or on the left of an unresolved band, and that serves to distinguish the two cases.

Changes in the curvature (force constant and vibrational frequency) of one potential relative to another affect the transition energies via the ω' and ω'' values in Equation 10.61b, and they can also affect the Franck–Condon overlap. So, intensities of the bands depend on the overlap between the initial and final vibrational states, which in turn depend on the nature of the two potential curves.

Hot bands arise from transitions from excited vibrational levels of the ground electronic state, that is, $n' > 0$. They may appear at lower transition frequencies than the transition from $n' = 0$ to $n'' = 0$.

Gas phase electronic spectra of polyatomic molecules are more complicated than the spectra of diatomics. The number of vibrational modes and the possibility of combination bands usually lead to numerous vibrational bands, and these may be overlapping. Also, the rotational fine structure tends to be more complicated, as we might expect from the differences between diatomic and polyatomic infrared (IR) spectra. Conventional absorption spectra can prove to be a difficult means of measuring and assigning transitions, and so numerous experimental methods have been devised to select molecules in specific initial states and to probe the absorption or the emission spectrum with narrow frequency range lasers.

10.8 Properties and Electronic Structure

The distribution of electrons in atoms and molecules dictates numerous molecular properties. First, we should note that molecular properties are generally values that quantify certain characteristics or features of a molecule, in particular a response by the molecule to some influence. For instance, the force constant, k, of the bond of a diatomic molecule is the second-order energy response to stretching or compressing the bond. That is, the energy's quadratic dependence on the bond length varies with the value k, where

$$k = \frac{\partial^2 E}{\partial x^2}\bigg|_{x = x_{min}}$$

This property is seen to be an energy derivative, and in fact, response properties are for the most part derivatives. An energy derivative can be obtained by formal differentiation of the Schrödinger equation. This can be done for the electronic, vibrational, and rotational Schrödinger equations; in fact, it can be done for the Schrödinger equation of almost any problem. Molecular properties are not the same as bulk properties, such as heat capacities. However, bulk properties are related to the properties of individual molecules, a connection that is a focus of statistical mechanics.

Responses to applied fields are properties that are various derivatives of the molecular energy with respect to the field strengths. The electrical permanent moments of a molecule's charge distribution (dipole, quadrupole, octupole, etc.) are first derivatives of the energy, while the **polarizability** is a second derivative. The **magnetic susceptibility**, or as it is also called, the **magnetizability**, is a second derivative with respect to the strength of a magnetic field. Responses to oscillating, as opposed to static, fields are, in general, dependent on the frequency of oscillation. A polarizability, for instance, may be **frequency-dependent** because it varies with the oscillation frequency of an applied field.

Molecular response to electrical perturbations is often significant because molecules are distributions of electric charges. It is their nature to be influenced by electrical potentials or fields. The fields may arise from other molecules, external sources, or electromagnetic radiation. Thus, electrical response properties are important for intermolecular interaction, spectroscopy, aggregation phenomena, and so on.

Electrical response is generally expressed with respect to some **electrical potential** that may arise from a charge distribution or an external source. An electrical potential is a spatially varying function, $V(x,y,z)$, such that the interaction energy with a fixed point charge, q_i, is

$$E_{int} = q_i V(x_i, y_i, z_i) \tag{10.63}$$

where x_i, y_i, and z_i are the position coordinates of the point charge. The potential can be expressed as a power series expansion about a point in space. Using the coordinate system origin, $(0,0,0)$, as the point of expansion gives

$$V(x,y,z) = V(0,0,0) + x\frac{\partial V}{\partial x}\bigg|_0 + y\frac{\partial V}{\partial y}\bigg|_0 + z\frac{\partial V}{\partial z}\bigg|_0 + \frac{1}{2}\left(x^2\frac{\partial^2 V}{\partial x^2}\bigg|_0 + y^2\frac{\partial^2 V}{\partial y^2}\bigg|_0 + z^2\frac{\partial^2 V}{\partial z^2}\bigg|_0\right)$$

$$+ \frac{\partial^2 V}{\partial x \partial y}\bigg|_0 + \frac{\partial^2 V}{\partial x \partial z}\bigg|_0 + \frac{\partial^2 V}{\partial y \partial z}\bigg|_0 + \frac{1}{6}\frac{\partial^3 V}{\partial x^3}\bigg|_0 + \cdots \tag{10.64}$$

The first derivatives of V are the components of the electric field vector, and we shall designate them simply as (V_x, V_y, V_z). The second derivatives are the components of the field gradient, $(V_{xx}, V_{xy}...)$.

For a classical distribution of N fixed-point charges, the interaction energy in Equation 10.63 can be summed to give the interaction of the distribution with an external potential. If the power series expansion in Equation 10.64 is used, the following is obtained:

$$E_{int} = \sum_i^N q_i V(x_i, y_i, z_i) = V(0,0,0)\sum_i q_i + V_x\sum_i q_i x_i + V_y\sum_i q_i y_i + V_z\sum_i q_i z_i + \frac{1}{2}V_{xx}\sum_i q_i x_i^2 + \cdots$$

This suggests the definition of the **moments of a charge distribution**. The zeroth moment is the net charge

$$M_0 = \sum_i^N q_i \tag{10.65}$$

The first moment, or the dipole moment, has three elements.

$$M_x = \sum_i^N q_i x_i$$

$$M_y = \sum_i^N q_i y_i \tag{10.66}$$

$$M_z = \sum_i^N q_i z_i$$

The second moment has nine elements, six of which are unique.

$$M_{xx} = \sum_{i}^{N} q_i x_i^2$$

$$M_{xy} = M_{yx} = \frac{1}{2} \sum_{i}^{N} q_i x_i y_i$$

$$M_{yy} = \sum_{i}^{N} q_i y_i^2$$

$$M_{xz} = M_{zx} = \frac{1}{2} \sum_{i}^{N} q_i x_i z_i \tag{10.67}$$

$$M_{zz} = \sum_{i}^{N} q_i z_i^2$$

$$M_{yz} = M_{zy} = \frac{1}{2} \sum_{i}^{N} q_i y_i z_i$$

One of the simplest arrangements of separated point charges possessing a second or quadrupole moment but with a zero dipole is that of three charges, one with twice the amount of charge of the other two:

$$\bullet^+ \; O^{2-} \; \bullet^+$$

The definition of third-moment elements, fourth-moment elements, and so on, follows from the interaction energy expression. Then, the interaction energy can be conveniently expressed as a dot product:

$$E_{int} = M_0 V(0,0,0) + M_x V_x + M_y V_y + M_z V_z + M_{xx} V_{xx} + M_{xy} V_{xy} + M_{xz} V_{xz}$$

$$+ M_{yx} V_{yx} + M_{yy} V_{yy} + M_{yz} V_{yz} + M_{zx} V_{zx} + \cdots$$

$$= \left(M_0 M_x M_y M_z M_{xx} \cdots \right) \begin{pmatrix} V_0 \\ V_x \\ V_y \\ V_z \\ V_{xx} \\ \cdots \end{pmatrix} \tag{10.68}$$

(There are different conventions in the definition of the second and higher moments. Often, the factors of 1/2 in Equation 10.67 and of 1/6 in the third moment are not included. The factors are then inserted when the moment values are used in calculating an interaction.) For a distribution of fixed charges, the first derivative of the energy with respect to an element of the vector $\vec{V}$ in Equation 10.68, for example, with respect to V_x, V_y, and so on, is a particular moment element. For quantum mechanical systems, it remains true that the moments are derivatives with respect to components of the field, of the field gradient tensor, and so on. We can say that the dipole moment is a property that characterizes the first-order response of the molecular energy to a uniform applied field. The second moment is the first-order response to a field gradient.

In an atom or molecule, the electrical point charges (the electrons and the nuclei) are not fixed, and the wavefunctions can adjust in response to an applied electric potential. Consequently, there is a change in the energy that is not only first order (linear) in the elements of $\vec{V}$ but also quadratic, cubic, and so on. The quantum mechanical analysis of the effect of an applied potential requires adding an interaction to the Hamiltonian. This term uses the electric moment elements as operators. For example, the first form of Equation 10.67 serves as an expression for the operator of the x-component of the dipole moment by using $-e$ for the charges of the electrons and letting the x_i's be the x-coordinate position operators of the electrons. The interaction Hamiltonian must be the fundamental interaction of the applied potential with each of the individual particles, and so the operator follows immediately from the classical interaction energy expression.

$$\hat{H}_{int} = V_0 \hat{M}_0 + V_x \hat{M}_x + V_y \hat{M}_y + V_z \hat{M}_z + V_{xx} \hat{M}_{xx} + \cdots \tag{10.69}$$

The first term is a constant; it adds directly to the total energy and can therefore be ignored.

The Schrödinger equation for some specific choice of the elements of $\vec{V}$ is

$$\left(\hat{H}_0 + \hat{H}_{int}^{\{V\}} \right) \psi^{\{V\}} = E^{\{V\}} \psi^{\{V\}} \tag{10.70}$$

where the superscript $\{V\}$ means that a set of specific values for the elements of $\vec{V}$ were used for the interaction Hamiltonian and that as a result, the eigenenergy and the wavefunction depend parametrically on these values. Now, consider solving Equation 10.70 for three situations, first where all the elements of $\vec{V}$ are zero, second where the only nonzero element is $V_x = 1$, and third where the only nonzero element is $V_x = 2$. From these three solutions, we can obtain the energy for three values of V_x. And we use two of the energy values to find an approximate value of the first derivative.

$$\left. \frac{\partial E}{\partial V_x} \right|_{V_x=0} \approx \frac{E(V_x = 1) - E(V_x = 0)}{1 - 0}$$

This is called a **finite difference approximation**, and in this case it is an approximation for the x-component of the dipole moment. A finite difference approximation* for the second derivative, a property that is called the **polarizability**, is

$$\left. \frac{\partial^2 E}{\partial V_x^2} \right|_{V_x=0} \approx \frac{E(V_x = 2) - E(V_x = 1) / 2 - 1 - E(V_x = 1) - E(V_x = 0) / 1 - 0}{2 - 0}$$

---

* For a function $f(x)$, we can approximate the first derivative in the vicinity of two nearby points x_1 and x_2 as $[f(x_2) - f(x_1)]/(x_2 - x_1)$. This value approaches the true first derivative as x_2 approaches x_1. Likewise, higher derivatives can be so approximated.

These values can also be found analytically, without approximation, by solving equations obtained by differentiating the Schrödinger equation or by perturbation theory.

Second derivatives of the energy with respect to the elements of a uniform electric field, V_x, V_y, and V_z, make up a tensor (second rank matrix) called the **dipole polarizability**, α.

$$\alpha = \begin{pmatrix} \alpha_{xx} & \alpha_{xy} & \alpha_{xz} \\ \alpha_{yx} & \alpha_{yy} & \alpha_{yz} \\ \alpha_{zx} & \alpha_{zy} & \alpha_{zz} \end{pmatrix} \quad \text{where } \alpha_{yz} \equiv \left. \frac{\partial^2 E}{\partial V_y V_z} \right|_{\vec{V}=0} \quad \text{etc.} \tag{10.71}$$

The dipole polarizability can also be defined as a first derivative of the dipole moment because the dipole moment is found by first differentiation of the energy. This means the polarizability tells how the dipole moment vector changes to first order on application of an external uniform field. We may say that the field induces a dipole, and physically this comes about by a shift in the charge distribution. The energy of polarization from applying a field is

$$E_{pol} = \frac{1}{2}(V_x V_y V_z)\alpha \begin{pmatrix} V_x \\ V_y \\ V_z \end{pmatrix} \tag{10.72}$$

The induced dipole moment vector is

$$\begin{pmatrix} M_x^{induced} \\ M_y^{induced} \\ M_z^{induced} \end{pmatrix} = \alpha \begin{pmatrix} V_x \\ V_y \\ V_z \end{pmatrix} \tag{10.73}$$

The induced moments are combined with the permanent moments (e.g., M_x, M_y, and M_z) to yield the total dipole moment vector in the presence of the field.

As an example of the problem, consider the dipole polarizability of the hydrogen atom. If a hydrogen atom experiences a uniform field in the z-direction, V_z, the interaction Hamiltonian will be

$$\hat{H}' = -ezV_x + ez_pV_x$$

where
  z is the z-coordinate of the electron
  z_p is the z-coordinate of the proton

Taking the proton to be fixed in space makes z_p a constant. So, the interaction with the proton is a constant contribution to the energies of all the states; we can ignore its effect. The coordinate z should be transformed to spherical polar coordinates since they were used to solve for the wavefunctions of the unperturbed hydrogen atom.

$$\hat{H}' = -eV_z r \cos\theta \tag{10.74}$$

The first-order perturbation theory correction to the ground state wavefunction would involve atomic orbitals that have a nonzero matrix element with the 1s orbital via the perturbation in Equation 10.74. The $2p_0$ or $2p_z$ orbital is the lowest energy orbital to have a nonzero matrix element.

$$\left\langle \psi_{1s} \middle| \hat{H} \psi_{2p_z} \right\rangle' = -eV_z \left\langle \psi_{1s} \middle| r\cos\theta \psi_{2p_z} \right\rangle \neq 0$$

So, at first order the $2p_z$ orbital is mixed with the 1s orbital because of an external field in the z-direction.

Figure 10.10 illustrates that the mixing of the 1s and $2p_z$ hydrogen atom orbitals removes the spherical symmetry of the hydrogen atom's wavefunction and charge distribution. The negative and positive charge centers are no longer at the same point, and so a dipole moment in the z-direction has been induced. The size of this dipole moment is the polarizability of the atom times V_z. We also see that the polarizability has to do with the ability to mix the 1s and $2p_z$ orbitals. Based on this, we might compare different atoms, and for instance, we anticipate that Be is a more polarizable element than He. The He polarizability would be determined largely by the 1s–2p mixing, whereas the Be polarizability is dictated largely by the energetically more favorable 2s–2p mixing. (The 2s and 2p orbitals are much closer in energy than the 1s and 2p orbitals.)

With differentiation with respect to components of a uniform electric field, the corresponding property is the dipole moment vector. If differentiation is with respect to elements of the field gradient, it is the second or quadrupole moment. Second derivatives of the energy are polarizabilities, and if the differentiation is with respect to uniform field components, the property is the dipole polarizability. Differentiation with respect to purely field gradient components yields the quadrupole polarizability, a property that gives the size of the quadrupole moment induced by an external field gradient. There are also mixed polarizabilities. The dipole–quadrupole polarizability (derivative with respect to field and field gradient components) can give the dipole induced by a field gradient or the quadrupole induced by a uniform field. Then there are third derivatives, which are called **hyperpolarizabilities**, and fourth derivatives, which are called second hyperpolarizabilities, and so on. The list of these

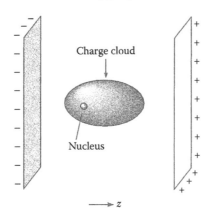

FIGURE 10.10
Influence of an external electric field in the z-direction on a hydrogen atom. We can think of the field as arising from two oppositely charged parallel plates. We can think of the electron of the hydrogen atom as being attracted toward the positively charged plate, while the proton is attracted toward the negatively charged plate. This represents a separation of the centers of negative and positive charge in the hydrogen atom, as shown schematically. The result is a nonzero dipole moment. The charge distribution of the electron is roughly that of the superposition of a 1s orbital with a 2p orbital along the z-axis. The 2p orbital adds to the wavefunction amplitude on one side of the nucleus and diminishes the amplitude on the other side where it is of opposite phase.

properties continues as far as one wishes to differentiate, but in total, they are simply values that characterize the response of an atom or molecule to an electrical potential. The electrical interactions between molecules, as well as other interactions, contribute to the attractions and repulsions that are manifested in nonideal behavior of gases.

Because of atomic and molecular response properties, there is a wealth of spectroscopic techniques beyond those discussed so far in this and the prior chapter. For instance, oppositely charged parallel plates inside a sample cell can be used to obtain rotational, vibrational, or electronic spectra with the influence of a uniform electric field, an experiment often termed **Stark effect spectroscopy** (Johannes Stark, Germany, 1874–1957). In rotational and vibrational spectra, the Stark effect can be used to find the magnitude of the dipole moment of a molecule and sometimes principal components of its dipole polarizability. An external magnetic field may be employed in addition to, or in place of, an electric field. **Zeeman spectra** are spectra obtained in the presence of a uniform magnetic field. Still another type of spectra, **Raman spectra**, involve a more complicated interaction of electromagnetic radiation than the simple dipole absorption or emission processes we have considered; it is an interaction involving the dipole polarizability. Raman spectra are generated by monitoring radiation produced or scattered after incident radiation interacts with the species of interest. The scattering process has different selection rules than simple absorption or emission.

Molecular properties may give rise to a force on atoms and molecules, individually and in bulk, because of some external influence. A force is exerted on a magnetizable substance, that is, a substance containing molecular species with appreciably sized magnetizabilities, by a suitably arranged magnetic field. Then, measurement of the force is a way of quantifying bulk magnetizability, or the magnetic susceptibility. Still another experimental approach involves letting a gas sample of molecules and an inert gas carrier freely expand into a vacuum. Those moving in other than some desired direction in the instrument are skimmed and removed. What remains is a **molecular beam**, and by application of specific, nonuniform electric fields along the path of the beam, molecules can be deflected at appropriate detection points in the instrument. The deflection will be affected by laser excitation, for instance, if it changes the molecule's electrical moments. In this way, the deflection of the beam offers another control in the process of obtaining spectroscopic information. Adding these approaches to the use of time-dependent sequences of applied radiation (pulses), the molecular spectroscopic methods available for probing rotational, vibrational, and electronic states, and for following molecular dynamics (vibrations, reactions, and so on) are extremely rich.

10.9 Conclusions

Our understanding of the electronic structure of atoms and molecules starts with knowledge of the solutions one-electron atom Schrödinger equation. These spatial eigenfunctions are atomic orbitals we designate 1s, 2s, $2p_1$, $2p_0$, $2p_{-1}$, 3s, and so on. Molecules differ from atoms by having several centers of positive charge instead of a single central charge. The mass difference between electrons and nuclei makes an approximate separation of internal nuclear motion (vibration) from electronic motion quite reliable. This is the Born–Oppenheimer approximation, and it is the basis for the concept of potential energy curves of diatomic molecules and potential energy surfaces of polyatomic molecules. Invoking the Born–Oppenheimer approximation amounts to clamping nuclei in place to find an electronic wavefunction and an electronic contribution to the molecular energy. Finding such electronic energies at numerous positions of the nuclei constitutes mapping the molecular potential energy surface.

The indistinguishability of electrons, which are spin $-1/2$ particles (fermions), leads to a requirement that electronic wavefunctions for atoms and molecules be antisymmetric with respect to the interchange of the coordinates of any pair of electrons. This is accomplished by forming a Slater determinant from individual spin-orbital functions.

An orbital product (single Slater determinant) form of a many-electron wavefunction corresponds to an approximation of the electron–electron interaction. Variational determination of such a wavefunction means invoking the self-consistent field approximation or letting each electron interact only with the averaged charge distribution of the other electrons. SCF underlies most molecular orbital arguments about molecular structure and properties. It tends to break down in describing bond breaking and formation, radical species, and excited electronic states. The valence bond picture partly relaxes the SCF approximation to provide a better description of bond breaking and of multiply bonded species.

Application of low-order perturbation theory within a molecular orbital picture leads to the concept of linear combinations of atomic orbitals to form molecular orbitals. That is, we can understand the energetics of bond formation and bond breaking in a semiquantitative manner by considering molecular orbitals to be composed of simple linear combinations of atomic orbitals of the type obtained by analytically solving the one-electron Schrödinger equation.

The charge distribution in a molecule has moments, dipole, quadrupole, and so on, and these make for interaction of the molecule with an external electric field, field gradient, and so on. An external electrical potential (field, field gradient, etc.) may affect the electron distribution, and this is characterized by polarizabilities and hyper-polarizabilities.

Atoms and molecules possess numerous excited electronic states. Electronic excitation typically requires substantially more energy than rotational or vibrational excitation. The photons for electronic spectra are usually those in the visible and ultraviolet regions of the electromagnetic spectrum. The potential energy curves of diatomic molecules and the potential energy surfaces of polyatomic molecules are likely to show differences between ground and electronically excited states. Equilibrium geometries may be different, and harmonic force constants may change with excitation. This leads to considerable complexity in electronic spectra, but it also means that such spectra serve as a probe of the potential energy surface of the excited states. For instance, in diatomic molecules, the observation of a band head in the *P*- versus the *R*-branch of a vibrational band of some electronic excitation may reveal that the bond length in the excited state is longer or shorter than in the ground state.

Because of the response of atoms and molecules to applied fields, rotational, vibrational, and electronic spectra can be influenced by external electric and/or magnetic fields. Carrying out spectroscopic experiments with versus without fields affords additional means for probing the electronic structure, properties, and dynamical behavior of molecules.

Point of Interest: John Clarke Slater

On December 22, 1900, John Clarke Slater was born in Oak Park, Illinois. At the time, his father, John Rothwell Slater, was a graduate student at the University of Chicago. In 1904, the family moved to Rochester, New York, where his father had been appointed to the English department at the University of Rochester.

Slater had an early interest in mathematics, architecture, and electrical devices. His scientific interests were strongly encouraged at home. When it came time for college, he chose to stay at home and attend the University of Rochester. He received an A.B. in science in 1920 after only 3 years of study. In his last year, he completed a senior honors thesis on the relationship between pressure and the intensities of hydrogen spectral lines, a problem that somewhat foreshadowed later endeavors. Slater completed his PhD in physics at Harvard, again in 3 years. His thesis was on the compressibility of alkali halide crystals.

With a traveling fellowship awarded by Harvard, Slater spent his first postdoctoral year at Cambridge. There, he developed a theory on radiative transitions in atoms. On discussing this idea with Neils Bohr and Hans Kramers, a joint paper on the quantum theory of radiation was published in 1924. However, Bohr and Kramers altered Slater's original idea by ascribing a virtual existence to the photons in the transitions—not the real photons that Slater believed in. In early 1925, Slater was back at Harvard and published further work of his own on radiative transitions. He presented a picture of absorption and emission of real photons coupled with energy conservation in transition processes. He also established a relationship between the width of spectral lines and the lifetimes of states.

In 1926, he began studying radiative transitions in H_2^+, and in so doing, he examined Heisenberg's ideas of symmetric and antisymmetric two-electron states in helium. When Douglas Hartree introduced the self-consistent field method for the electronic structure of atoms in 1928, Slater saw the connection with Heisenberg's two-electron states. Slater published a major paper the next year. It described a theory of complex spectra, and in it he showed that with a determinantal many-electron wavefunction (the Slater determinant) one could achieve a self-consistent field wavefunction and also have the proper symmetry for electron systems (antisymmetric with respect to particle exchange).

Slater spent the last half of 1929 in Europe, mostly with Heisenberg in Leipzig and Pauli in Zurich. He used his determinantal approach in the study of solids and metal cohesion. When he returned to the United States, he accepted an offer to head the physics department at the Massachusetts Institute of Technology (MIT). He directed the department toward solid state physics and the application of quantum mechanical methods. In 1931, he introduced "directed valence," the idea of valence bonding developed independently about the same time by Linus Pauling. Slater devised methods for electron energy bands for solids, though many of them had to wait for the early digital computers of the 1950s in order to make large-scale applications possible.

During World War II, Slater worked on microwave radar. Slater recognized that the wartime advances in technology would be the start of new applications of physics. He took to rebuilding the MIT physics department with that in mind. He set up MIT's Solid State and Molecular Theory Group and pioneered the use of digital computers for solving quantum mechanical problems. He retired from MIT in 1964 and then joined Per-Olov Löwdin's Quantum Theory Project at the University of Florida. He continued to work on problems in solid-state physics until shortly before his death on July 25, 1976. Among the honors he received for his contributions was the National Medal of Science in 1971.

Exercises

10.1 Find the energy of the $n = 1$ state of the hydrogen atom if the proton mass is infinite. This value is the infinite mass Rydberg constant, R_∞.

10.2 Find the energy of a hydrogen atom in a state with $n = 100$ and $l = 0$. Also, find the expectation value of r in this state. (Make sure to use the correct reduced mass rather than m_e.)

10.3 To get an idea of the contraction in the size of the 1s orbitals with increasing nuclear charge, Z, calculate the expectation value of r for a single electron in a 1s orbital about the nucleus of each of the rare gas elements (He^+, Ne^{9+}, etc.).

10.4 Find the atomic term symbols for the beryllium associated with the occupancy $1s^2$ $2s^1$ $2p^1$.

10.5 If the plus and minus signs in the spin and spatial parts of the wavefunction in Equation 10.58 were interchanged, what spin state would result?

10.6 Find the explicit form of the associated Laguerre polynomial, $L_3^2(z)$.

10.7 Show that $R_{30}(r)$ is normalized.

10.8 Insert $R_{41}(r)$ into Equation 10.2 and verify that the eigenenergy is that expected from Equation 10.9.

10.9 The ionization energy of a one-electron atom is the energy required to promote the electron from $n = 1$ to $n = \infty$. Find the ionization energy in cm^{-1} for the one-electron atoms He^+, Li^{2+}, C^{5+}, and Ne^{9+}. In what regions of the electromagnetic spectrum (infrared, visible, ultraviolet, etc.) are there photons of sufficient energy to ionize the electron in these species?

10.10 In terms of the parameter γ, find the energy according to Equation 10.22 associated with spin–orbit interaction for a one-electron atom in the following states: $^2P_{3/2}$, $^2P_{1/2}$, $^2D_{3/2}$, $^2F_{5/2}$, and $^2F_{7/2}$. (The subscript on each symbol is the J quantum number.)

10.11 Use the table method for electron arrangements to show that $S = 0$ and $L = 0$ for an s^2, p^6, and d^{10} occupancy. Then show that the term symbols for a $1s^2$ $2p^1$ occupancy are the same as the term symbols for a single electron in a 2p orbital.

10.12 Find the atomic term symbols associated with an atomic occupancy $1s^2$ $2s^2$ $2p^1$ $3p^1$.

10.13 Find the Russell–Saunders term symbols for the possible states of the oxygen atom with occupancy $1s^2$ $2s^2$ $2p^4$.

10.14 Find the atomic term symbol for the ground state of the nitrogen atom.

10.15 Show that the term symbols for an occupancy $3d^9$ are the same as the term symbols for the occupancy $3d^1$. Then show that the term symbols for $3d^8$ and $3d^2$ are the same. What does this suggest?

10.16 What are the term symbols for the ground state occupancies of Mn and Mn^+?

10.17 What is the expected spin multiplicity for the ground states of halogen atoms according to their expected ground state electron occupancies?

10.18 Compute the spin–orbit interaction energy in terms of the parameter γ in Equation 10.24 for the all states of an atom with occupancy $1s^2$ $2s^2$ $2p^6$ $3s^2$ $3d^2$.

10.19 For some given values for the L and S quantum numbers of an atomic state, what is the largest possible spin–orbit interaction energy? Express this in terms of the parameter γ in Equation 10.24. [Hint: Consider how J in Equation 10.24 is related to the L and S values and attempt to maximize $E_{spin-orbit}$.]

10.20 What are the allowed transitions between the ground state of the carbon atom and the excited states associated with the electron occupancy $1s^2$ $2s^1$ $2p^3$?

10.21 What should be the first allowed transition from ground state neon? (Identify the excited occupancy and the term symbols of the initial and final states.)

10.22 Find the term symbols for the states of an atom that are associated with an occupancy $1s^2$ $2s^2$ $2p^2$ $3d^1$. (Hint: Though a table procedure can be used for all three p and d electrons, it is more concise to use a table for the two equivalent p electrons and then to use rules for inequivalent electrons to complete the angular momentum coupling.)

10.23 For a collection of identical spin 3/2 particles, what is the analog of the Pauli exclusion principle? In other words, for the wavefunction to be antisymmetric and to be an orbital product form, how many particles can be found in the same orbital?

10.24 For a system of four electrons in different spin-orbitals, apply the antisymmetrizer in Equation 10.45 to a simple product of the four spin orbitals and show that this is the same function as the Slater determinant constructed from the orbital product.

10.25 Calculate the average value and uncertainty for an electron in a 3d orbital of a hydrogen atom.

10.26 Verify that a Slater determinant for the lithium atom corresponding to two $1s\alpha$ electrons and one $1s\beta$ electron vanishes.

10.27 Calculate the most probable distance from the nucleus for an electron in the ground state of Be^{3+}.

10.28 Apply the antisymmetrizing operator to the simple orbital product function $1s\alpha\ 2s\alpha\ 3s\alpha$ or else write out the terms of the Slater determinant with $1s\alpha\ 2s\alpha\ 3s\alpha$ on the diagonal. Then apply the antisymmetrizing operator to each of the terms individually and simplify the result.

10.29 Verify that the hydrogen 1s and 2s orbitals are orthogonal.

10.30 From qualitative LCAO arguments, predict the bond order of the following diatomics: LiH, HeBe, LiF, C_2^{2-}, CO^+, and NeF^+.

10.31 Repeat the analysis in Equation 10.62 with a centrifugal distortion term, $D'[J'(J' + 1)]^2$, for the initial state and $D''[J''(J'' + 1)]^2$ for the final state.

10.32 Calculate the frequency of light required to remove an electron from the ground state of a hydrogen atom. The ionization potential for atomic hydrogen is 13.5984 eV.

10.33 Explain how electronic absorption spectra can provide a bond length for N_2, whereas infrared absorption spectra cannot.

10.34 The relative intensities of rotational lines in a vibrational band of an electronic absorption spectrum are almost entirely dependent on the relative populations on the initial rotational levels (Maxwell–Boltzman distribution). Use this fact to make a sketch of the Q-branches in Example 10.1 assuming a sample temperature of 300 K. How would they appear under low resolution?

10.35 In terms of charge distributions, devise an argument to account for the fact that carbon monoxide has a dipole moment that is very nearly zero, but a sizable third (octupole) moment.

Bibliography

Herzberg, G., *Atomic Spectra and Atomic Structure* (Dover, New York, 1944).

King, G. W., *Spectroscopy and Molecular Structure* (Holt, Rinehart & Winston, New York, 1964).

Advanced Texts and Monographs

Dykstra, C. E., *Ab Initio Calculation of the Structures and Properties of Molecules* (Elsevier, Amsterdam, the Netherlands, 1988).

Hurley, A. C., *Introduction to the Electron Theory of Small Molecules* (Academic Press, New York, 1976).

Mulliken, R. S. and W. C. Ermler, *Diatomic Molecules: Results of Ab Initio Calculations* (Academic Press, New York, 1977).

Schatz, G. C. and M. A. Ratner, *Quantum Mechanics in Chemistry* (Prentice Hall, Englewood Cliffs, NJ, 1993).

Szabo, A. and N. S. Ostlund, *Modern Quantum Chemistry: Introduction to Advanced Electronic Structure Theory* (Macmillan, New York, 1982).

11

Statistical Mechanics

The connection between the quantum mechanical treatment of individual atoms and molecules and macroscopic properties and phenomena is the goal of statistical mechanical analysis. Statistical mechanics is the means for averaging contributions to properties and to energies over a large collection of atoms and molecules. The first part of the analysis is directed to the distribution of particles among available quantum states. An outcome of this analysis is the partition function, which proves to be an essential element in thermodynamics, in reaction kinetics, and in the intensity information of molecular spectra.

11.1 Probability

In earlier chapters, we considered the pressure of gas in terms of collisions with a movable wall of a container. We saw that there is a relation between a thermodynamic state property, pressure of the bulk gas, and molecular behavior. In principle, such connections can be approached in two ways. A mechanical approach is to use mechanical analysis to follow a system of many particles (atoms and molecules) in time. Macroscopic properties of a system at equilibrium correspond to the long-time average of the system's behavior, and thus, a full mechanical analysis provides everything needed to obtain pressure, internal energy, temperature, and so on. The other approach is statistical in nature, as we discussed in Chapter 1. It involves considering the probabilities of the different ways a system can exist at any instant and then averaging properties over those different ways, each weighted by its probability. A mechanical approach may seem rather clear-cut, but unfortunately the task of solving mechanical (classical or quantum) problems for huge numbers of particles over long periods of time is not usually feasible. Therefore, we need the second approach—the tools of statistical mechanics. To develop them, we need to keep in mind certain information about probabilities.

Essential to any discussion of probability is its definition. If there are N mutually exclusive and equally likely occurrences, and n_A and only n_A of them lead to some particular result, A, the mathematical probability of A is the ratio n_A/N. A way to remember this is to think about flipping a coin. There are two mutually exclusive outcomes, heads and tails, and so $N = 2$. One particular result is tails, and since it is but one of the outcomes, $n_A = 1$. The probability of tails is, therefore, 1/2. Next, we consider two theorems about relations among probabilities.

Theorem 11.1: Probabilities of Mutually Exclusive Events Are Additive

This follows from the definition of probability. If the probability of some event A is n_A/N and the probability of event B is n_B/N, the probability that A or B will occur is the sum of the occurrences leading to A or leading to B divided by N, that is, $(n_A + n_B)/N$, the sum of the probabilities.

Theorem 11.2: The Probability of Simultaneous Occurrence of Unrelated Events Is the Product of the Independent Probabilities of the Events

If the probability of some event A is n_A/N and the probability of event B is n_B/N, the probability that A and B will occur is the product of the probability of A and the probability of B, that is, $(n_A/N) \times (n_B/N)$. This also follows from the definition of probability. Consider that there are four ways to categorize all possible events: (1) A and B occur, (2) A but not B occurs, (3) B but not A occurs, and (4) neither A nor B occurs. The probability that B will not occur is 1 less the probability that B will occur $(1 - n_B/N)$, and likewise that A will not occur. With P_i designating the probability of the ith category events, several conditions exist because of the definition of the probability.

$$P_1 + P_2 + P_3 + P_4 = 1 \quad \text{total probability for everything is 1}$$

$$P_1 + P_2 = \frac{n_A}{N} \quad \text{total probability for event } A \text{ only}$$

$$P_1 + P_3 = \frac{n_B}{N} \quad \text{total probability for event } B \text{ only}$$

P_1 and P_2 must each involve n_A/N since they are probabilities related to event A. The simplest way for their sum to equal n_A/N and to satisfy the first condition is for the following to hold:

$$P_1 = \frac{n_A}{N} \frac{n_B}{N}$$

$$P_2 = \frac{n_A}{N} \left(1 - \frac{n_B}{N} \right)$$

$$P_3 = \frac{n_B}{N} \left(1 - \frac{n_A}{N} \right)$$

$$P_4 = \left(1 - \frac{n_A}{N} \right) \left(1 - \frac{n_B}{N} \right)$$

That is, the probability of a simultaneous disconnected event is a product of the probabilities of those events. For example, the probability of two coins being flipped and coming up heads is $1/2 \times 1/2 = 1/4$.

An underlying idea in the statistical mechanical analysis of chemical systems is that all quantum states with the same energy are equally probable. For any one molecule, or any one quantum mechanical system, there is no a priori reason to favor one state of a given energy over another. This is a postulate; we take it to hold so long as there are no violations in the predictions that follow from it. This idea was invoked in Chapter 1 in the statistical analysis that lead to the Maxwell–Boltzmann distribution law (Equation 1.11), and in Chapter 9 we found one direct experimental confirmation of the distribution law in the intensity pattern of infrared spectra of diatomic molecules. The law states that population diminishes exponentially with the energy that an atom or molecule possesses.

11.1.1 Classical Behavior

Sometimes classical mechanics offers a sufficient description of the dynamical behavior of molecular systems. Such a description, however, does not provide the quantum energy levels that are involved in the postulate of equal probabilities. To apply in a classical mechanical framework, the postulate of statistical mechanics requires something analogous to quantum states and their energies. We consider the analogy to show that statistical mechanics can be applied without a quantum mechanical analysis; however, the primary focus of this chapter uses quantum knowledge about molecules.

In classical mechanics, we expect to be able to measure the position and momentum of a particle at any instant of time without quantum uncertainties, and that can be the basis for the analog of a state. Let us define a box in geometrical space to be of some small volume, δv. We can envision this box as having three sides of length δx, δy, and δz. Let us make up a similar though more abstract box in the momentum coordinates with "volume" $\delta p = \delta p_x \, \delta p_y \, \delta p_z$, and let us consider the product $\delta v \delta p$ to be the volume of an abstract "position–momentum box." The postulate of equal a priori probabilities in this framework is that for some fixed choice of $\delta v \delta p$, there is equal probability of finding a particle in such a position–momentum volume anywhere. That is, the likelihood of finding the particle in some "box," $\delta v \delta p$, is the same for all boxes of that size.

The notion of an abstract volume in the space of the coordinates of a Hamiltonian (i.e., position and momentum coordinates) is formalized by the introduction of **phase space**, of which there are at least two kinds. One kind should be regarded as a six-dimensional particle space. The six dimensions are those of the three spatial coordinates and the three momentum coordinates needed in the mechanical description of a single particle. At any instant in time, a particle is at one point in this six-dimensional phase space. If there were several particles in the system, each could be associated with a distinct point in this space at every instant in time. A six-dimensional box in phase space can be referred to as having a particular volume in that space. Then, the postulate of equal probabilities is a statement that the probability that the phase space point represents a single particle is the same in any one among all like-sized boxes.

For a system of N particles, we can define a phase space of $6N$ coordinates, the position and momenta coordinates of each of the particles. In this second kind of phase space, the system as a whole is represented by a single point, whereas in the particle phase space, the system is represented by N points.

The mechanical behavior of a system of N particles can be represented by a path or trajectory through the system's phase space. At some instant, the system is at one phase space point, and at a later instant, at another point. A line connecting the points shows the evolution of the system in time. Under certain conditions, the behavior of the system can follow many different trajectories. Some volumes of phase space may be traversed by many of the possible trajectories, whereas other specific (position–momentum) volumes may be traversed only a few times. We can therefore consider that there is a density of the trajectory points in any particular volume of phase space. This density is a function of the $6N$ coordinates of the space. If we know the value of some dynamical variable at every point in phase space, we can obtain an average value for that variable by integrating the product of the dynamical variable and the density over the entire volume of the phase space. We can take as another postulate that such an average corresponds to the average behavior of the system.

11.2 Ensembles and Arrangements

The approach we are following for systems with large numbers of particles is statistical rather than mechanical averaging. Thus, we need to consider the many different ways a system can be found to exist, rather than how it evolves from one single set of initial conditions. An ensemble is a hypothetical collection of many like systems that helps us do that. Our objective is to average over all the possible ways a system exists, and these ways are included in an ensemble.

To work with ensembles, we have to distinguish between particles and systems, for we will be considering both. Let us start with a single molecule for which the electronic–vibrational–rotational Schrödinger equation is

$$H\psi_i = E_i\psi_i \tag{11.1}$$

The i subscript specifies a state among the infinite number of eigenfunctions and eigenenergies. The classical Hamiltonian from which the quantum mechanical Hamiltonian is derived is a function of position and momentum coordinates for all the particles in the molecule. Let this number of coordinates be d. A system might consist of some number, N, of these molecules. We can formally write the Schrödinger equation for the system of N molecules as

$$H\Psi_i = E_i\Psi_i \tag{11.2}$$

where the Hamiltonian, H, is based on Nd coordinates. This Hamiltonian is not necessarily a sum of the Hamiltonians of N molecules for it may include interaction potential terms among the molecules. The eigenfunctions of H will be functions of $Nd/2$ spatial coordinates. We note that solving Equation 11.1 is often quite difficult, and solving Equation 11.2 becomes extremely challenging even for $N = 2$ or 3. If N were the size of Avogadro's constant, we would have nothing currently available to write out the explicit Hamiltonian, much less solve the differential Schrödinger equation. The phase space for a single molecule in this case is a d-dimensional space, whereas the phase space for the system has Nd dimensions.

An ensemble for a system of N diatomic molecules might consist of M replicants of the system. Different types of ensembles are used to average under different sets of conditions. A **microcanonical ensemble** is one for which each replicant system in the ensemble has the same number of molecules, N, the same volume, V, and the same energy, E. That is, N, V, and E are fixed. These systems are identical from a thermodynamic perspective, but at the molecular level, they may be different. A **canonical ensemble** is one for which N, V, and the temperature, T, are the same fixed values for each replicant system in the ensemble. In a **grand canonical ensemble**, V, T, and the chemical potential are fixed. This allows N to change, as would occur in multiple-phase and reaction systems. One can construct other types of ensembles with other constraints.

Although an ensemble is a hypothetical or virtual construct, we can appreciate the physical idea of at least a canonical ensemble. Taking M as a very large number, consider there to be M different impermeable vessels of volume V. In each are placed N molecules. The vessels are in contact, and the vessel walls conduct heat. The assembly of M vessels is placed in a constant-temperature heat bath to establish some temperature T. After that is achieved, the assembly is thermally isolated from its surroundings; heat exchange among vessels can still occur, and that means the temperature of each vessel remains at T. Of course, since each vessel has N molecules in a volume V, it is N, V, and T which are constrained. Though the temperature is the same in each vessel, the molecules do not necessarily behave identically, nor is the energy of each vessel identical.

The average energy, $\langle E \rangle$, of the M vessels is the sum of their energies, each of which we take as one of the E_i's in Equation 11.2, divided by M. The same result is achieved if we know the distribution of the systems among the allowed energies. The distribution is the fraction that have a given energy. If a_i is the fraction that have energy E_i, the average energy is

$$\langle E \rangle = \sum_i a_i E_i \tag{11.3}$$

Two of the vessels (systems) in the ensemble can have the same energy. In terms of the molecule phase space, each system has N points representing it at some instant. One way for the two vessels to have the same energy is for the N points of the two vessels to be the same; however, it may be that molecule 23 in the second system is at the point of molecule 4 in the first system, and vice versa. So, the two systems are the same except that their particles are arranged differently. That there can be such different arrangements plays a role in analyzing distributions.

11.3 Distributions and the Chemical Potential

In the introduction to entropy and temperature in Chapter 1, we considered the arrangements of particles among allowed states. Dice and coins were used as examples, but these differ in an important way from the arrangements of atoms and molecules. Objects in our macroscopic world are distinguishable. As discussed in Chapter 10, electrons are indistinguishable, and so are like atoms and like molecules. There is no way to label particles at the atomic level. In the atomic world, nothing exists that is analogous to the difference between a 1996 Lincoln penny and one dated 1995. In fact, it is difficult to find anything in our macroscopic world that truly shows indistinguishability. Even three pennies, fresh from the mint with the same year and mint marks, are distinguishable by their order; we can distinguish one from another. The existence of electrons, nuclei, atoms, and molecules is different. There is neither a label nor an order that can let us keep track of, or distinguish between, two identical particles.

Distinguishability generally results in more distinct arrangements of the particles or objects. For instance, three identical pennies with heads up constitute one arrangement if the pennies are truly indistinguishable. However, think of there being tiny labels, "1," "2," and "3," on them and then there are several distinct arrangements for placing the heads-up pennies left to right on a table: 1-2-3, 1-3-2, 2-1-3, 2-3-1, 3-1-2, and 3-2-1. There are six arrangements, which is 3! more than the single arrangement of the hypothetical indistinguishable pennies. It is easy to reason that the number of arrangements of N distinguishable objects will be $N!$ more than the number of arrangements if the same objects are indistinguishable. For any given arrangement of the indistinguishable objects, there are N choices of which distinguishable particle takes the first position. Then, there are $N - 1$ choices as to which is the second, $N - 2$ for the third, and so on. The number of choices is $N(N - 1)(N - 2) \cdots = N!$.

Next consider a system with N particles distributed among some large number of available energy levels. The degeneracy of each level is designated s_i and the number of particles in each state is n_i. For distinguishable particles, there are $s_i^{n_i}$ arrangements of the particles among the different degenerate quantum states of the ith level. We multiply this factor for each level by the formula, Equation 1.5, that gives us the number of arrangements for distributing particles into groups (levels).

$$A_{distinguishable} = N! \prod_i \frac{s_i^{n_i}}{n_i!} \tag{11.4}$$

We correct this to account for indistinguishability by dividing by $N!$.

$$A_{indistinguishable} = \prod_i \frac{s_i^{n_i}}{n_i!} \tag{11.5}$$

The quantum mechanical world has a further complexity concerning distinguishability. In Chapter 10, we found that there are two kinds of indistinguishable particles, the difference related to the intrinsic spin of particles. A rigorous analysis of arrangements requires that we first consider if or under which conditions these differences are important.

The wavefunctions of particles with half-integer spins, fermions, are antisymmetric with respect to interchange of the coordinates of a pair of particles, whereas the wavefunctions of integer spin particles, bosons, are symmetric. A consequence of the antisymmetrization requirement for fermions, Equation 10.42, is that no two can exist in the same quantum mechanical state (the Pauli exclusion principle). (**Fermi–Dirac statistics** refers to statistical analysis of fermion systems, whereas **Bose–Einstein statistics** is for bosons.) It is helpful to consider a hypothetical particle, sometimes called a **boltzon** that is simply a distinguishable particle. **Boltzmann statistics** is the treatment of distinguishable particles, but corrected to be indistinguishable by incorporating a $1/N!$ factor as done in obtaining Equation 11.5 from Equation 11.4. For fermions, s_i must be greater than or equal to n_i since there can be only one particle in each of the states of a given level. This modifies the total number of arrangements, A, which we can analyze to obtain

$$A_{fermion} = \prod_i \frac{s_i(s_i-1)(s_i-2)\cdots(s_i-n_i+1)}{n_i!} \tag{11.6}$$

For bosons, the arrangements are

$$A_{boson} = \prod_i \frac{(s_i+n_i-1)(s_i+n_i-2)\cdots s_i}{n_i!} \tag{11.7}$$

If we considered the products in Equations 11.5 through 11.7 to go over only one level, we can show that

$$A_{fermion} < A_{indistinguishable} < A_{boson} \tag{11.8}$$

on the basis of the following relation among the A values for the one level:

$$s(s-1)\cdots(s-n+1) < s^n < (s+n-1)(s+n-1)\cdots s \tag{11.9}$$

Equation 11.9 uses the explicit expressions for the A numbers, but with the common $n!$ term removed. Consider the left inequality for the case $s = 8$ and $n = 2$: $8(7) < 8^2 < 9(8)$. This illustrates that the indistinguishable boltzon result may serve as an approximation to fermion or boson statistics.

The three equations for arrangements, Equations 11.5 through 11.7, can be written as one by defining a quantity, δ, such that

$$\delta = 0 \text{ for indistinguishable boltzons}$$

$$\delta = 1 \text{ for fermions}$$

$$\delta = -1 \text{ for bosons}$$

Then,

$$A = \prod_i \frac{s_i(s_i - \delta)(s_i - 2\delta)(s_i - 3\delta)\cdots(s_i - (n_i - 1)\delta)}{n_i!} \tag{11.10}$$

We can use Equations 11.5 through 11.7, individually or together as Equation 11.10, in place of Equation 1.5 and derive distribution laws (Chapter 1) allowing for the different values of δ. The distribution law derivation in Chapter 1 began by finding the differential of the logarithm of the number of arrangements, Equation 1.17. For fermions ($\delta = 1$), we express Equation 11.6 in the following equivalent form, from which the logarithm is formed and then Sterling's approximation is applied:

$$A_{fermion} = \prod_i \frac{s_i!}{n_i!(s_i - n_i)!} \tag{11.11}$$

$$\ln A_{fermion} = \sum_i \ln s_i! - \ln n_i! - \ln(s_i - n_i)! \tag{11.12}$$

$$\ln A_{fermion} = \sum_i s_i \ln s_i - s_i - (s_i - n_i)\ln(s_i - n_i) + (s_i - n_i) - n_i \ln n_i + n_i \tag{11.13}$$

The differential of this expression, recognizing that the s_i's are constants, is

$$d \ln A_{fermion} = \sum_i \left[\ln(s_i - n_i) - \ln n_i\right] dn_i \tag{11.14}$$

We incorporate the two constraints, those in Equations 1.18 and 1.19 that N and U are conserved, by adding differential quantities with undetermined multipliers α and β to

Equation 11.14. (This analysis can be invoked for different problems, and to avoid any specification at this stage, ε_i's are used to designate the quantum energies.)

$$N = \sum_i n_i \Rightarrow dN = 0 = \sum_i dn_i$$

$$U = \sum_i \varepsilon_i n_i \Rightarrow dU = 0 = \sum_i \varepsilon_i dn_i$$

To maximize $\ln A$ with constraints, αdN and βdU are subtracted from Equation 11.14 and the resulting expression for $d\ln A$ is set equal to zero.

$$0 = -\sum_i \left(\alpha + \beta \varepsilon_i - \ln(s_i - n_i) + \ln n_i\right) dn_i \tag{11.15}$$

Equation 11.15 is satisfied if each ith term is zero, which implies

$$\ln \frac{s_i - n_i}{n_i} = \alpha + \beta \varepsilon_i \tag{11.16}$$

Rearrangement and exponentiation yield

$$n_i = \frac{s_i}{1 + e^{\alpha + \beta \varepsilon_i}} \tag{11.17}$$

We can carry out the same procedure for bosons and indistinguishable boltzons and achieve a general distribution law for all three types of particles.

$$n_i = \frac{s_i}{\delta + e^{\alpha + \beta \varepsilon_i}} \tag{11.18}$$

Equation 11.14 is equivalent to the following expression, suppressing the subscript fermion knowing this development can be done generally for all particles:

$$d\ln A = \alpha dN + \beta dU \tag{11.19}$$

From this relation we have

$$\left(\frac{\partial \ln A}{\partial N}\right)_U = \alpha \tag{11.20}$$

$$\left(\frac{\partial \ln A}{\partial U}\right)_N = \beta \tag{11.21}$$

From Chapter 1, we know $\beta = 1/kT$. We also know that at constant volume and constant N, $dS = TdU$. Therefore,

$$\left(\frac{\partial S}{\partial U}\right)_{V,N} = \frac{1}{T} \Rightarrow \frac{1}{k}\left(\frac{\partial S}{\partial U}\right)_{V,N} = \frac{1}{kT} = \beta = \left(\frac{\partial \ln A}{\partial U}\right)_{N,V} \tag{11.22}$$

where the condition that V is constant has been added to the partial derivative of $\ln A$ because, in fact, the quantum states are affected by the volume, as in the particle-in-a-box problem; volume has been implicitly constrained in our analysis to this point. The conclusion drawn from Equation 11.22 is that for dS to be equal to TdU requires that entropy be related to the arrangements according to

$$S = k \ln A \tag{11.23}$$

where A is the maximum arrangement number. This is a definition of **absolute entropy**.

From the thermodynamic relation $G = U - TS + PV$, and the definition of the molecular chemical potential, g, of the system as the partial derivative of G with respect to N (here, the number of particles, not the number of moles), we can express the following differential relation:

$$g dN = dU - T dS + P dV \tag{11.24}$$

With Equation 11.23, this becomes

$$g dN = dU - kT d(\ln A) + P dV \tag{11.25}$$

From this, the partial derivative in Equation 11.20 is obtained (at constant volume)

$$\left(\frac{\partial \ln A}{\partial N} \right)_{U,V} = -\frac{g}{kT} \Rightarrow \alpha = -\frac{g}{kT} \tag{11.26}$$

Using this in Equation 11.17 yields the general distribution law.

$$n_i = \frac{s_i}{\delta + e^{(\varepsilon_i - g)/kT}} \tag{11.27}$$

In the case of indistinguishable boltzons ($d = 0$) this reduces to Equation 1.37, which when written for levels rather than states (i.e., incorporating the degeneracy factor s_i) is

$$n_i = \frac{s_i N e^{-\varepsilon_i/kT}}{q} \tag{11.28}$$

Equating the right-hand sides of Equation 11.27 with $\delta = 0$ and (11.28) yields

$$e^{-g/kT} = \frac{q}{N} \tag{11.29}$$

where q, the partition function, is

$$q = \sum_i^{levels} s_i e^{-\varepsilon_i/kT} \tag{11.30}$$

Equation 11.30 is a restatement of Equation 1.26.

11.3.1 High-Temperature Behavior

Let us now compare the general population numbers of Equation 11.27 for two levels. The first, level k, is taken to be some level with energy less than the chemical potential. The second, level m, has an energy greater than the chemical potential. From Equation 11.27, we obtain the ratio

$$\frac{n_m}{n_k} = \frac{s_m}{s_k} \frac{\delta + e^{(\varepsilon_k - g)/kT}}{\delta + e^{(\varepsilon_m - g)/kT}}$$

(11.31)

$$\varepsilon_k < g < \varepsilon_m$$

As temperature increases, the exponentials in the numerator and the denominator approach zero. Therefore,

$$\lim_{T \to \infty} \frac{n_m}{n_k} = \frac{s_m}{s_k}$$

(11.32)

This result is independent of d, and therefore, fermions and bosons are distributed in the same way as the hypothetical indistinguishable boltzons at high temperatures. We conclude that Boltzmann statistics are appropriate in that case.

11.3.2 Low-Temperature Behavior

For boltzons ($\delta = 0$), Equation 11.31 simplifies at *all* temperatures to

$$\frac{n_m}{n_k} = \frac{s_m}{s_k} e^{(\varepsilon_k - \varepsilon_m)/kT}$$

(11.33)

As T approaches zero, the exponential in Equation 11.33 approaches $e^{-\infty}$, that is, zero. All the particles go into the lowest energy state as temperature approaches zero. Let us compare that with like results for real particles, fermions and bosons, to determine if Boltzmann statistics are appropriate under low-temperature conditions.

For fermions, let us use Equation 11.27 and take all s_i values to be 1 (or all to be the same) for some hypothetical system. If we assume a large number of states such that the energy levels are essentially a continuum, we can convert Equation 11.27 to a function, f, of the energy; it is the probability that a given energy state is occupied.

$$f(\varepsilon) = \frac{1}{1 + e^{(\varepsilon - g)/kT}}$$

(11.34)

Under conditions of very low temperatures and if $\varepsilon < g$, the exponential in the denominator approaches $e^{-\infty}$, leading to $f(\varepsilon)$ approaching 1. If $\varepsilon > g$, however, the exponential approaches e^{∞} and $f(\varepsilon)$ goes to zero. As temperature is reduced, $f(\varepsilon)$ displays an increasingly sharper cutoff at the point $\varepsilon = g$; it approaches the form of a step function: $f(\varepsilon) = 1$ for $\varepsilon < g_0$ and $f(\varepsilon) = 0$ for $\varepsilon > g_0$, where g_0 is the chemical potential at $T = 0$. For valence electrons (conduction electrons) in metals, g_0 may be on the order of several hundred kJ mol^{-1}, and then even at room temperature, the distribution of electrons is very close to a step function. g_0 is called the **Fermi energy**, and g_0/k is the **Fermi temperature**.

The chemical potential is necessarily temperature-dependent, as is clear in the simple case of boltzons from Equations 11.29 and 11.30. For bosons, g approaches the lowest allowed energy (smallest ε) as T approaches 0. Since $\delta = 0$, this means that n_i in Equation 11.27 goes to zero for all the i levels above the lowest. For the lowest level, n goes to infinity, which simply means that all the particles go into their lowest quantum state. This statistical prediction is called **Bose–Einstein condensation** because on further analysis (analysis of the temperature dependence of g), it is found that this "collapse" of the distribution occurs rather suddenly at some very low, characteristic temperature. It is analogous to an ordinary phase transition. The first observation of Bose–Einstein condensation was in 1995 and was for an aggregation of rubidium atoms.

11.3.3 Dilute Behavior

A dilute system is one in which the number of particles in any level is much less than the number of particles that can be in the level. Returning to Equation 11.10, we see that with $s_i \gg n_i$, the effect of δ becomes insignificant. Consider one level with $n = 2$, first with $s = 4$ and then with $s = 100$. From Equation 11.10 we obtain

$$s = 4$$

$$A_{\delta=1} = \frac{4(3)}{2!} = 6$$

$$A_{\delta=0} = \frac{4(4)}{2!} = 8$$

$$A_{\delta=-1} = \frac{4(5)}{2!} = 10$$

$$s = 100$$

$$A_{\delta=1} = \frac{100(99)}{2!} = 4950$$

$$A_{\delta=0} = \frac{100(100)}{2!} = 5000$$

$$A_{\delta=-1} = \frac{100(101)}{2!} = 5050$$

With $s = 4$, A for $\delta = \pm 1$ differs by 25% of the value of A for $\delta = 0$, whereas with $s = 100$, the difference is only 1%. Under dilute conditions, the number of arrangements approaches the same value for all three types of particles. For sufficiently dilute systems the statistics of indistinguishable boltzons is appropriate for all particles.

11.4 Molecular Partition Functions

Equation 11.30 gives a sum over the quantum energy levels of a molecular system, the molecular partition function, when the molecular energies (Equation 11.1) are used for the ε_i's. (The parallel definition for the partition functions of systems of particles uses the energies in Equation 11.2 instead of those in Equation 11.1 for the ε_i's.) Writing the summation over states instead of levels yields

$$q = \sum_i^{states} e^{-E_i/kT} \tag{11.35}$$

The Maxwell–Boltzmann distribution law can be written in terms of q.

$$P_i = \sum_i^{states} \frac{e^{-E_i/kT}}{q} \tag{11.36}$$

P_i is the probability of being in the ith state.

Partition functions can be regarded as a normalization factor in a probability expression, for example, Equation 11.36. The same partition function—the same normalization factor—is used in the probability for all states. When we have two systems that are independent, then according to Theorem 11.2, probabilities of the simultaneous occurrence of unrelated events vary as the product of the independent probabilities of the events. This means the probability that the two systems will be in some specific set of their available states (the simultaneous unrelated events) is a product of their independent probabilities of being in those individual states. The normalization factors of the independent events are multiplied together to become the normalization factor (partition function) of the combined group of two systems. Simply, the partition functions are *multiplied* together for independent noninteracting systems.

A partition function can be written for a system of N particles using the quantum mechanical energies in Equation 11.2.

$$Q = \sum_i^{states} e^{-E_i/kT} \tag{11.37}$$

The partition function for an ensemble of M systems of N particles, taking the systems to be distinguishable for now and taking them to be independent in the sense of Theorem 11.2, is a product of partition functions.

$$Q_{distinguishable} = Q_1 Q_2 Q_3 \cdots = \prod_i Q_i \tag{11.38}$$

If the systems making up the ensemble are not distinguishable, as in the last section, we must correct with a factor of $M!$ to obtain the partition function, Q.

$$Q_{distinguishable} = \frac{1}{M!} \prod_i Q_i \tag{11.39}$$

Now consider a gas in extremely low concentration, a very dilute gas. In this limit, the gas molecules are mostly noninteracting or independent. With that, we can take each molecule as one of the M systems, or in other words, take $N = 1$. In this limit, the ensemble partition function is a product of the molecular partition functions, q_i. For other concentrations, though, this amounts to an approximation of Q.

$$Q = \frac{1}{M!} \prod_i q_i \tag{11.40}$$

In many instances, this expression for Q serves quite well.

On the basis of Equation 11.40, we recognize the importance of obtaining analytical expressions, wherever possible, for molecular partition functions. We will use certain approximations, the first being that the Schrödinger equation for a molecule is separable into parts such that there are additive contributions to the energy of the molecule in a given state.

1. $E_{nuclear}$, the contribution of the nuclear energies
2. $E_{electronic}$, the energy associated with the electrons in their distribution throughout the molecule
3. $E_{vibration}$, the energy associated with vibration of the molecule
4. $E_{rotation}$, the energy associated with rotation of the molecule
5. $E_{translation}$, the energy associated with translation of the molecule in space

That is, the energy of some molecular state represented by a collective index k is

$$E_{state(k)} = E_{nuclear(k)} + E_{electronic(k)} + E_{vibrational(k)} + E_{rotational(k)} + E_{translational(k)} \tag{11.41}$$

With this energy expression used in Equation 11.35, we find that the molecular partition function is a product of partition functions associated with the five contributions.

$$q = q_{nuclear} q_{electronic} q_{vibrational} q_{rotational} q_{translational} \tag{11.42}$$

(This results from the mathematical fact that an exponential of a sum is a product of the exponentials of each of the terms in the sum, e.g., $e^{a+b} = e^a e^b$.) We should keep in mind that Equations 11.41 and 11.42 correspond to an approximation. For instance, vibrational–rotational coupling prevents strict separation of the vibrational and rotational parts of a molecular Hamiltonian. For now, we will use Equation 11.42 to obtain q via obtaining each of the elements in the product. We shall also rely on model problems presented in Chapters 7 and 8 as idealizations of rotation and vibration of a diatomic molecule.

Example 11.1: Products of Partition Functions of Independent Systems

Problem: Show that if the quantum mechanical Hamiltonian of some system is separable into those of M identical subsystems, then the partition function of the system must be a product of the partition functions of the individual subsystems.

Solution: Let the energy eigenvalues for the j subsystem be designated E_{ji}. Equation 11.37 gives the partition function for the j subsystem

$$Q_j = \sum_i e^{-E_{ji}/kT}$$

Since the subsystems are identical in their eigenenergies, the Q_j's are identical as well.

From Chapter 8, we know that separability of the Schrödinger equation means that the complete energy eigenvalues are sums of energy eigenvalues of the separated Schrödinger equations. Thus, the energy eigenvalues for the system of M identical subsystems are given by a sum, as in

$$E_1^{sys} = E_{11} + E_{21} + E_{31} + \ldots + E_{M1}$$

$$E_2^{sys} = E_{12} + E_{21} + E_{31} + \cdots + E_{M1}$$

$$E_3^{sys} = E_{11} + E_{22} + E_{31} + \cdots + E_{M1} \text{ and so on}$$

Using these energies with Equation 11.37, we find the partition function of the system.

$$Q = \sum_J^{sys\_states} e^{-E_J^{sys}/kT}$$

$$= \left(e^{-E_{11}/kT} e^{-E_{21}/kT} \cdots \right) + \left(e^{-E_{12}/kT} e^{-E_{21}/kT} \cdots \right) + \cdots$$

Careful factoring of this open-ended expression allows Q to be put in the following form. (Notice the E-subscripts.)

$$= \left(e^{-E_{11}/kT} + e^{-E_{12}/kT} + \cdots \right) + \left(e^{-E_{21}/kT} + e^{-E_{22}/kT} + \cdots \right) + \cdots$$

$$= Q_1 Q_2 \cdots Q_M$$

a product of subsystem partition functions.

Nuclei can exist in different quantum mechanical states. A collection of protons and neutrons can be excited, the mechanics being akin to exciting vibration and rotation of the system of nucleons. Generally, excitation from a nuclear ground state to the first excited state requires an energy that is enormous compared to the energies of chemical bonds. The Maxwell–Boltzmann distribution law tells us that such excitation energies make for negligible populations of excited states at room temperature. In $q_{nuclear}$, this means that the contribution from anything but the first term is negligible for most temperatures of interest in chemistry. Then, taking the ground nuclear state energy as zero, leaves $q_{nuclear} \approx s_{nuclear}$, the degeneracy of the nuclear ground state. A similar argument holds for $q_{electronic}$ since electronic excitation energies of atoms and molecules are usually large relative to kT with T up to 400 K. However, there are some species that have low-lying excited electronic states, and for those, it may be necessary to carry several terms, as in the following expression where E_0 is the ground electronic state energy and E_1 and E_2 are the energies of the first and second excited electronic state. s_0, s_1, etc., are electronic state degeneracies:

$$q_{electronic} = s_0 e^{-E_0/kT} + s_1 e^{-E_1/kT} + s_2 e^{-E_2/kT} \tag{11.43}$$

If we define the zero of energy of the molecule to be the energy in its ground electronic state (i.e., $E_0 = 0$), we can approximate $q_{electronic}$ further, to only the first term: $q_{electronic} \approx s_0$. Most ground electronic states of organic molecules are nondegenerate, and then $s_0 = 1$.

To find $q_{vibrational}$ for a diatomic molecule, we shall use the idealization of a harmonic oscillator. The quantum mechanical energy level expression for a harmonic oscillator developed earlier in Equation 7.34 yields

$$q_{vibrational} = \sum_{n=0}^{\infty} e^{-(n+1/2)\hbar\omega/kT} \tag{11.44}$$

This has a simpler form, which we can obtain by recognizing it to be a series expansion of a simple fraction.

$$\frac{1}{1-e^{-\hbar\omega/kT}} = \sum_{n=0}^{\infty} e^{-n(\hbar\omega/kT)} \tag{11.45}$$

We also have to factor from each term in Equation 11.44, the quantity $e^{-\hbar\omega/2kT}$

$$q_{vibrational} = \frac{e^{-\hbar\omega/2kT}}{1-e^{-\hbar\omega/kT}} \tag{11.46}$$

Next is the partition function associated with rotation. Here, we will use the energy levels of the rigid rotator from Equation 8.63.

$$E_{JM} = J(J+1)\frac{\hbar^2}{2\mu R^2} = BJ(J+1) \tag{11.47}$$

where
  J is the angular momentum quantum number
  B is the rotational constant

This leads to the following partition function:

$$q_{rotational} = \sum_{J=0}^{\infty} (2J+1)e^{-BJ(J+1)/kT} \tag{11.48}$$

The $2J + 1$ factor is the degeneracy of the J-level. This factor accounts for summing over the M quantum number were the sum to be expressed as over states rather than levels (see Chapter 8). Recognizing that the energy level separations are small, at least for low values of J, we approximate the energy levels as being continuous and then replace the summation by an integral.

$$q_{rotational} = \int_0^{\infty} (2J+1)e^{-BJ(J+1)/kT}\,dJ$$

$$= \int_0^{\infty} e^{-Bx/kT}\,dx = -\frac{kT}{B}(e^{-\infty} - e^0) = \frac{kT}{B} \tag{11.49}$$

where $x = J(J + 1)$ and $dx = (2J + 1)\,dJ$ has been used in the evaluation of the integral.

A quantum mechanical treatment of translation corresponds to a particle experiencing a zero potential as it translates but being constrained to be within some volume. It is the treatment of a particle in a box. We have found that for a one-dimensional box of length l, the quantum mechanical energy expression, Equation 8.22, is quadratic in the quantum number n.

$$E_n = \frac{n^2 h^2}{8ml^2}$$

The partition function for particles behaving like the particle in a box is developed from this energy level expression.

$$q_{box} = \sum_{n=1}^{\infty} e^{-n^2 h^2/8ml^2 kT} \tag{11.50}$$

The box in which we might probe and investigate atoms and molecules is likely to have dimensions of the scale of objects in our everyday world. That means that the energy levels are extremely closely spaced and it is reasonable to approximate the summation in Equation 11.50 as an integral, just as was done for $q_{rotational}$. This is accomplished by treating the quantum number as a continuous variable.

$$q_{box} = \int_0^{\infty} e^{-n^2 h^2/8ml^2 kT} dn$$

$$= \sqrt{\frac{2\pi mkT}{h^2}} l \tag{11.51}$$

If the particle were in a three-dimensional box instead of a one-dimensional box, the partition function, q_{box}, would be a product of terms for each direction. With l_x, l_y, and l_z as the three dimensions of the box, we have

$$q_{box} = \frac{(2\pi mkT)^{3/2}}{h^3} l_x l_y l_z = \frac{(2\pi mkT)^{3/2}}{h^3} V \tag{11.52}$$

where V is the volume of the box. Since the particle-in-a-box problem is that of a freely translating particle (zero potential) inside the box, it is appropriate to take q_{box} in Equation 11.52 as the translational partition function.

11.5 Thermodynamic Functions

Thermodynamic functions can be derived from statistical mechanical analysis, and this is the ultimate bridging of molecular and macroscopic phenomena. In Chapter 1, we considered how absolute entropy is related to the number of configurations or arrangements in which a system can exist, and that the condition of equilibrium corresponds to maximizing this number. We also found an expression for the internal energy, U, a derivation we now develop somewhat more fully.

Consider the derivative of the molecular partition function, q, while recognizing that the same form of result holds for an N-particle system partition function, Q.

$$\frac{\partial q}{\partial t} = \frac{\partial}{\partial t}\sum_i e^{-E_i/kT} = \sum_i \frac{\partial}{\partial t}e^{-E_i/kT} = \sum_i \frac{E_i}{kT^2}e^{-E_i/kT} = \frac{1}{kT^2}\sum_i E_i e^{-E_i/kT} \tag{11.53}$$

The summation $\sum_i E_i e^{-E_i/kT}$ is a weighted average of the energies of the states. The mean energy is this weighted average divided by the partition function, and so the derivative of the partition function times kT^2/q is the mean or average energy. Since the fundamental expression of every partition function (versus an approximate form) is a sum of exponentials of $-E/kT$ values, then in general, we state

$$\frac{KT^2}{Q}\frac{\partial Q}{\partial t} = \frac{\sum_i E_i e^{-E_i/kT}}{Q} = \bar{E}\ (\text{or}\ \langle E\rangle) \tag{11.54}$$

We used the relation in Equation 11.53 in Chapter 1 to find the average energy of ideal gas particles. These particles are noninteracting (part of the idealization), though they can exchange kinetic energy via instantaneous collisions. This means that for an ideal gas, the ensemble partition function is a product of q's, with a factor to account for indistinguishability.

$$Q = \frac{q^N}{N!} \tag{11.55}$$

The q used in Equation 11.55 is that in Equation 11.52. This is because the monatomic ideal gas particles are point masses and there is no rotation or vibration. Now, the ensemble average energy is the thermodynamic function, U, the internal energy. Putting these items together yields an expression for U for an ideal monatomic gas.

$$U = \frac{kT^2}{Q}\frac{\partial Q}{\partial t}\ \text{by Equation 11.54}$$

$$= kT^2\frac{\partial \ln Q}{\partial t}\ \text{since}\ \frac{\partial \ln y}{\partial x} = \frac{1}{y}\frac{\partial y}{\partial x}$$

$$= kT^2\frac{\partial}{\partial t}\left(\ln\frac{q^N}{N!}\right)\ \text{by Equation 11.55}$$

$$= kT^2\frac{\partial}{\partial t}\left(N\ln q - \ln N!\right)$$

$$= NkT^2\frac{\partial}{\partial t}\ln q\ \text{since}\ N\ \text{is a constant}$$

$$= \frac{NkT^2}{q}\frac{\partial q}{\partial t}$$

$$= \frac{NkT^2 h^3}{V(2\pi mkT)^{3/2}}\frac{\partial}{\partial t}\left(\frac{V}{h^3}(2\pi mkT)^{3/2}\right)\ \text{using Equation 11.52}$$

$$= \frac{NkT^2}{T^{3/2}}\left(\frac{3}{2}T^{1/2}\right) = \frac{3}{2}NkT \tag{11.56}$$

This is the same relation given as Equation 1.37 in the development of the distribution law and the introduction of temperature, which was a classical derivation of U for an ideal gas. From an expression for U, one can use the relations among thermodynamic state functions presented in Chapters 2 and 3 to find expressions for H, G, and A.

Let us next consider a hypothetical gas of ideal, rigid diatomic particles. The difference between this and the ideal monatomic gas is that there can be rotation. We can follow the same steps that lead to Equation 11.56 except that following Equation 11.42, q must be a product.

$$q = q_{translational} q_{rotational} = \frac{V}{h^3} (2\pi mkT)^{3/2} \frac{kT}{B} \qquad (11.57)$$

$q_{rotational}$ is from Equation 11.48. From this, the following is obtained:

$$U = \frac{5}{2} NkT \qquad (11.58)$$

If we were to allow for harmonic vibration, we would obtain (see Example 11.2)

$$U = \frac{7}{2} NkT + \frac{1}{2} N\hbar\omega \qquad (11.59)$$

The temperature-independent term in Equation 11.59 is the zero-point vibrational energy contribution. Sometimes it is ignored because it is a constant that arises from taking the energy scale of a molecule to be zero at the bottom of the stretching potential. Were the zero defined to be the ground vibrational state energy, then U would be simply $7NkT/2$.

The rotation and vibration of real molecules is more complicated than harmonic vibration and rigid rotation. There is a coupling between rotation and vibration along with deviation from the quadratic form of a harmonic stretching potential, that is, vibrational anharmonicity. In principle, these features of a realistic description of a molecule can be incorporated into the partition functions. As an example of the size of the effect of anharmonicity, consider carbon monoxide. From Table 9.1, its harmonic vibrational frequency is $2170\,cm^{-1}$ and its anharmonicity constant is $13.5\,cm^{-1}$. We can find the energy of the states from these values on the basis of a harmonic energy level expression or one that includes the first anharmonicity constant. Then, evaluating the summation of state terms in the vibrational partition function state by state allows us to evaluate the difference between a harmonic and an anharmonic analysis. At $T = 1400\,K$, where kT is roughly $1000\,cm^{-1}$, we have the following evaluation of q_{vib} for CO:

	Harmonic Treatment			Anharmonic Treatment		
	Energy (cm^{-1})	Contribution to q_{vib}	Total q_{vib}	Energy (cm^{-1})	Contribution to q_{vib}	Total q_{vib}
$n = 0$	1085.0	0.33790	0.33790	1081.6	0.33905	0.33905
$n = 1$	3255.0	0.03858	0.37648	3224.6	0.03977	0.37882
$n = 2$	5425.0	0.00441	0.38089	5340.6	0.00479	0.38361
$n = 3$	7595.0	0.00050	0.38139	7429.6	0.00059	0.38420

After summing the contributions from the first four states, the effect of anharmonicity on the vibrational partition function at $T = 1400\,K$ amounts to only 0.7%. The importance of such a difference depends on the information sought, but clearly, anharmonicity tends to be a small contributor to the partition function.

Example 11.2: Internal Energy of an Ideal Diatomic Gas

Problem: Find an expression for the internal energy, U, of a system of N noninteracting (ideal) diatomic molecules that rotate rigidly and vibrate harmonically.

Solution: This problem calls for repeating the derivation leading to Equation 11.56 but using $q = q_{translational}q_{rotational}q_{vibrational}$ with the three q's being those in Equations 11.52, 11.49, and 11.46, respectively. We will let α stand for the quantity $\hbar w/2k$ in $q_{vibrational}$.

$$\frac{1}{q}\frac{\partial q}{\partial t} = \frac{1}{q_{vibrational}}\frac{1}{T^{5/2}}\frac{\partial}{\partial t}\left(T^{5/2}\frac{e^{-\alpha/T}}{1-e^{-2\alpha/T}}\right)$$

$$= \frac{5}{2}T^{-1} + \alpha T^{-2}\left(1 + \frac{2}{1-e^{-2\alpha/T}}\right)$$

U is obtained by multiplying this result by NkT^2. At the same time, we replace α by what it stands for.

$$U = \frac{NkT^2}{q}\frac{\partial q}{\partial t} = \frac{5}{2}NkT + Nk\frac{\hbar w}{2k}\left(1 + \frac{2}{1-e^{-\hbar w/kT}}\right)$$

A series expansion of the exponential term in this expression followed by truncation of the expansion after the first two terms leads to a simplification.

$$1 - e^{-\hbar w/kT} = 1 - \left(1 - \frac{\hbar w}{kT} + \cdots\right) = \frac{\hbar w}{kT}$$

Therefore,

$$U = \frac{5}{2}NkT + \frac{N\hbar w}{2} + NkT = \frac{7}{2}NkT + \frac{N\hbar w}{2}$$

Another effect on a partition function for a system arises from something we have entirely neglected, intermolecular interaction. For low-density gases, such neglect may be quite justifiable; however, in condensed phases, the interactions are clearly profound, with translational and rotational degrees of freedom becoming essentially vibrational degrees of freedom. In a pure crystal, for instance, the partition function develops solely from the atomic vibrations.

A final point about partition functions and thermodynamic variables comes from forming the differential of Equation 1.19:

$$U = \sum_i E_i n_i$$

$$(11.60)$$

$$dU = \sum_i (E_i dn_i + n_i dE_i) = \sum_i E_i dn_i + \sum_i n_i dE_i$$

From the particle-in-a-box problem, we know that expanding a box leads to a change in the energy levels, that is, $\delta\varepsilon_i$. Expansion of a box is a volume change, and a volume change against an external pressure corresponds to doing work. Since $dU = \delta q + \delta w$, the first summation on the right side of Equation 11.60 corresponds to δq, while the second summation corresponds to δw. Heat transfer is a change in the number of particles of a given energy level, dn_i, and transfer of energy by work is a change in the energies of the levels.

11.6 Heat Capacities

The constant-volume heat capacity is the derivative of the internal energy, U, with respect to temperature, Equation 3.33. Statistical mechanics affords a means of finding U in terms of molecular properties such as the mass, the rotational constant, and the vibrational frequency. Equations 11.56, 11.58, and 11.59 show that the factor multiplying NkT in U for an ideal gas increases with each internal degree of freedom (rotation and vibration), and this means the heat capacity increases as well. This analysis points to the **principle of the equipartition of energy**, which is that each position or momentum coordinate entering the Hamiltonian with a quadratic dependence contributes $NkT/2$ to U. Hence, an ideal monatomic gas (point mass with no vibration and no rotation) has $U = 3NkT/2$ because of the six coordinates of the Hamiltonian, x, y, z, p_x, p_y, and p_z, the Hamiltonian depends on only p_x^2, p_y^2, and p_z^2. An ideal gas of diatomic rigid rotators has $U = 5NkT/2$ because of the additional two degrees of freedom, rotation about two axes perpendicular to each other, and each perpendicular to the molecular axis. (Rotation about the molecular axis is not defined if the atomic masses are taken to be point masses.) There are two angular momentum coordinates in the part of the Hamiltonian associated with rotational energy adding to the three linear momentum contributions. A nonlinear polyatomic molecule has three rotational degrees of freedom since rotation of the polyatomic is a defined motion for three rather than two orthogonal axes.

Each vibration a molecule exhibits corresponds to two quadratic terms in the Hamiltonian as in Equation 7.35. There is a quadratic momentum term and a quadratic term in the associated displacement coordinate. Thus, each vibrational mode contributes $2 \times NkT/2$ to U. We can collect this information concisely by saying that for a molecular gas, U is $3NkT/2$ (translation), plus either $NkT/2$ for linear molecules or $3NkT/2$ for nonlinear molecules (rotation), plus NkT times the number of modes of vibration. The gas of an ideal diatomic molecule has $U = 7NkT/2$ as in Equation 11.59. A linear triatomic molecule has $U = 13NkT/2$ because there are three vibrational modes, one of which is twofold degenerate.

Since $C_V = (\partial U/\partial T)_V$, the equipartition principle offers a simple means of predicting C_V. For an ideal monatomic gas, $C_V = 3Nk/2$; for an ideal linear diatomic molecular gas, $C_V = 7Nk/2$; and for an ideal gas of linear triatomic molecules, $C_V = 13Nk/2$. To express this on a molar basis, we take N_A to be Avogadro's number and then $N_A k = R$, the molar gas constant. Around 300 K, C_V of neon is $1.52R$ which is very near $3R/2$. For N_2, $C_V = 2.50R$, which is the value expected for an ideal gas of diatomic molecules that can rotate but not vibrate. C_V of Cl_2 is $3.10R$, which is closer to $7R/2$. At certain higher temperatures, however, C_V of both N_2 and Cl_2 does approach $7R/2$.

Example 11.2 produced an expression for the internal energy of an ideal gas of translating, rigidly rotating, harmonic oscillators.

$$U = \frac{5}{2}NkT + Nk\frac{\hbar\omega}{2k}\left(1+\frac{2}{1-e^{-\hbar\omega/kT}}\right) \tag{11.61}$$

This can be differentiated with respect to temperature to obtain an expression for C_V which is less approximate than the value $7Nk/2$.

$$C_V = \frac{5}{2}Nk + \frac{N\hbar^2\omega^2}{kT^2}\frac{e^{-\hbar\omega/kT}}{(1-e^{-\hbar\omega/kT})^2} \tag{11.62}$$

This expression shows a complicated dependence on temperature in the second term. We can examine this dependence by introducing a variable α that is proportional to T or is a scaled temperature as in Figure 11.1: $\alpha = kT/\hbar\omega$. So, in terms of α, Equation 11.62 becomes

$$C_V = Nk\left(\frac{5}{2}+\frac{1}{\alpha^2}\frac{e^{-1/\alpha}}{(1-e^{-1/\alpha})^2}\right) \tag{11.63}$$

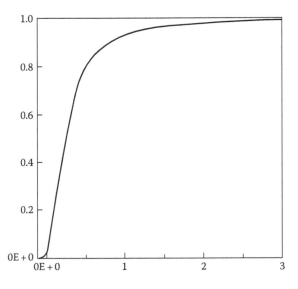

FIGURE 11.1
Plot of the function $f(\alpha) = e^{-1/\alpha}/\alpha^2(1-e^{-1/\alpha})^2$ versus α. In the expression for constant-volume heat capacity, Equation (11.63), this function is the vibrational contribution to C_V, with α being proportional to temperature.

Plotting the α-term in Equation 11.63 against a (the scaled temperature) yields a curve that rises quickly, begins to plateau at $\alpha = 1$, and approaches a value of unity in the limit of large α (high temperature). So in the limit of high temperature, the entire α-term in Equation 11.63 becomes very nearly 1.0, and then, $C_V = 7Nk/2$.

Instead of a quantum mechanical analysis, one can use classical mechanical analysis of harmonic vibration to calculate the vibrational contribution to the heat capacity. When this is performed, $q_{vibrational}$ is obtained via integration rather than the summation over quantum state energies in Equation 11.45. The result is $q_{vibrational\text{-}classical} = kT/\hbar\omega$. This is the same as the quantum mechanical result, but only in the high-temperature limit. In other words, classical mechanics and quantum mechanics reach the same result but only when the average internal energy is large (high temperature). That limit corresponds to many particles having energies that are sizable in comparison to the energy differences between quantum states. Then, the difference between discrete energies and continuous energies is not significant. The low-temperature limit, however, displays a clear difference between quantum and classical pictures. Measurement of the heat capacity at low temperatures should show no dependence on temperature if classical mechanics holds. In fact, measurements of heat capacity in the early part of the twentieth century revealed there was a dependence on temperature. The observation was made prior to the development of quantum mechanics, and it proved to be one of the observations that led to the quantum revolution.

Before concluding our discussion of the molecular and statistical view of heat capacities, there is another issue to consider. A gas of neon atoms has no rotational or vibrational degrees of freedom. Though its C_V is very nearly equal to that of a monatomic ideal gas, it is not exactly equal, and we may wonder why not. The aspect we have excluded because of the difficulty in incorporating it in an introductory analysis is interaction among gas particles, the nonideality of neon, for instance. Let us consider a hypothetical monatomic gas for which we can externally set the interaction between particles from zero (an ideal gas) to a strong, harmonically attractive potential. In the limit of the strong potential, we no longer have a gas of N atoms but a solid. The $3N$ degrees of translational freedom have been converted to $3N - 6$ vibrations, where we can ignore the -6 for large N. The equipartition principle indicates a contribution to U of $3NkT/2$ in the limit of no interaction but twice that, $3NkT$, for $3N$ vibrations, that is, the strong interaction limit. On this basis, a deviation of neon from ideality in the direction of atom–atom attractiveness would be expected to increase C_V. A slightly larger C_V (or U) than predicted by the equipartition principle may tend to be related to intermolecular attractions, and in fact, by other means we know that two neon atoms do show a very weak attraction for each other.

The effect of intermolecular interaction as in the hypothetical gas just considered has other implications. Phase transitions produce very sharp changes in heat capacity. Sublimation of solid argon to argon gas is a change from $3N$ vibrational degrees of freedom in the solid to $3N$ translational degrees of freedom in the gas. This implies a halving of C_V from $3NkT$ in the solid to $3NkT/2$ in the gas according to the equipartition principle.

Unusual types of temperature dependence of heat capacities can occur in molecular systems. A methyl group in an organic molecule often has a potential for rotating about its symmetry axis that has three peaks and three valleys over its 360° range of positions. It exhibits a vibrational motion called **hindered rotation** where it twists back and forth within one of the valleys. However, if the molecule is promoted to a higher energy level, it may be above the peaks in this torsional potential. At some point, the torsion of the methyl group is (classically) unhindered, and the group becomes a free rotor. The quantum mechanics of this type of problem are interesting and revealing, but the main point here is that a vibrational motion is in effect converted to a rotation. This happens as temperature increases and the Maxwell–Boltzmann distribution law places more and more molecules into free rotor (high-energy) states. The switch from vibration to rotation represents a loss in the contribution to U and to C_V. Hence, C_V may in fact decrease with increasing temperature in contradiction to Equation 11.63.

11.7 Conclusions

Atoms and molecules in gases and liquids are moving and colliding. At any instant, a collection of gas particles has some exchanging energy via collision, and thus the numerous species in the collection have many different energies. A single molecule stores energy through its manifold of electronic–vibrational–rotational states, and the different energies of molecules in a gas correspond to the molecules being in different states. The distribution of the molecules among the available states represents a statistical likelihood that any given molecule will be in a particular energy state. The distribution is governed by temperature and is expressed as a formula or law, for example, the Maxwell–Boltzmann distribution law.

The properties that a system exhibits are a manifestation of the statistical likelihood of each of the different ways it may exist, and statistical mechanics provides the tools for connecting those properties with the molecular level. Ensembles are the hypothetical constructions that allow us to obtain an average over the different ways a system may exist.

Partition functions are weighted sums over the states of a molecule or system, the weighting being given by a distribution law. The partition function for a collection of noninteracting entities is the product of the partition functions of the entities alone. Indistinguishability of the entities must be taken into account, and this depends on whether the entities are fermions (half-integer spin) or bosons (integer spin). Though electrons, nuclei, and atoms are either fermions or bosons, under dilute or high-temperature conditions, their behavior approaches that described by Boltzmann statistics (indistinguishable boltzons). The partition function for a collection of interacting species is not strictly a product of their individual partition functions, though that product may serve as a good approximation. Thermodynamic variables, such as the average energy, U, are derivable from the partition functions.

Through statistical mechanics, we can recognize that the temperature dependence of constant-volume heat capacities is a direct consequence of the quantization of the allowed energies of a molecule. In contrast, a classical mechanical analysis allows for a continuous range of vibrational energies, for instance, and this leads to constant, not temperature-dependent, C_V's.

We have focused on equilibrium thermodynamics. However, statistical mechanics provides the concepts and the techniques for exploring fluctuations of a system away from equilibrium, and the behavior of a system that evolves in time as in a chemical reaction. The introduction to statistical mechanics in this chapter can only set the stage for that level of analysis.

Point of Interest: Lars Onsager

Lars Onsager

Lars Onsager brought a new dimension to statistical mechanics, the study of fluctuations from equilibrium and nonequilibrium statistical mechanics. He was recognized for this work with the 1968 Nobel Prize in chemistry. He received many other awards including the American Chemical Society's Peter Debye Award (1965) and the U.S. National Medal of Science (1968).

Onsager was born in Oslo, Norway, on November 27, 1903. He was raised in Oslo, and then studied chemical engineering at Trondheim. Peter Debye invited him to be a research assistant in his lab at Zurich, and Onsager spent from 1926 to 1928 with Debye. He then came to the United States to become a teaching assistant in introductory chemistry at Johns Hopkins University. That proved to be a mismatch of Onsager's high-level approach to lecturing, and his position lasted only one term. It was at that time that he formulated the ideas that ultimately led to his receiving the Nobel Prize.

Onsager moved from Johns Hopkins to Brown University where he was an instructor charged with teaching graduate statistical mechanics. He stayed there 5 years, during which he started investigating isotope separation by thermal diffusion. In World War II, that became a crucial technological problem in making atomic bombs.

Financial conditions in 1933 kept Brown University from reappointing Onsager, and so on being offered a prestigious postdoctoral fellowship by Yale University, he accepted. On arrival at Yale, it was realized that he did not have a doctorate; his doctoral dissertation filed years before in Trondheim had not been in proper form in some way and no PhD was awarded. Yale chose to solve the problem of Onsager's appointment by awarding him a Yale PhD in chemistry on the basis of a dissertation he submitted, a dissertation he supposedly pulled from a file of his papers and writings. Onsager spent 39 years at Yale, 27 as the J. Willard Gibbs professor of theoretical chemistry. He left Yale in 1971 for the University of Miami's Center for Theoretical Studies. He had reached the mandatory retirement age, and the administration at Yale would not waive their policy that emeritus faculty could not be principal investigators on grants. At Miami, he could have grants and thereby a group of postdoctoral research associates. Onsager died in Coral Gables, Florida, on October 5, 1976.

Onsager's work covered the theory of electrolytes, superconductors, a new definition of Bose–Einstein condensation for interacting particles, electrical conductivity in ice, and his solution of a major problem in physics, the two-dimensional Ising model for the "order-disorder" transition in crystals. What he worked on during his short time at Johns Hopkins led to the Onsager reciprocal relations, which he published in 1931. Onsager established the reciprocal relations that arise wherever there are coupled flows of matter, energy, electricity, and so on. To reach that point, Onsager needed to describe irreversible processes as a relaxation process (a rate process with a rate constant) for a system to go from a non-equilibrium state to equilibrium. Onsager connected irreversible behavior with fluctuation behavior. Thirty years after his 1931 papers, the "irreversible thermodynamics" he introduced had become a widespread area of investigation.

Exercises

11.1 How many ways are there to arrange three electrons among the atomic spin-orbitals of the $n = 2$ shell corresponding to an occupancy of $2s^1 2p^2$? Verify this result by using Equation 11.6. How many more arrangements would there be if the electrons could be numbered 1, 2, and 3 and thereby distinguished from each other?

11.2 Protons and neutrons are fermions, but any two together are a boson. Thus, a nucleus with an even number of protons and neutrons is a boson. Classify the nuclei of the naturally occurring isotopes of lithium, carbon, and oxygen as bosons or fermions.

11.3 Evaluate $q_{vibrational}$ in Equation 11.46 at $T = 1, 10, 100, 300,$ and $1,000\,K$ for LiH ($\omega = 1405\,cm^{-1}$).

11.4 Find the probability that a harmonic oscillator will be in the second excited state at $T = 100$, 300, 1,000, and 10,000 K if the energy level expression is E_n (kJ mol^{-1}) $= 23.0\,(n + 1/2)$.

11.5 Find the temperatures at which $e^{-E/kT}$ is 0.001 for energies, E, of 1.0 kJ mol^{-1} (the order of a vibrational excitation energy), of 300 kJ mol^{-1} (the order of certain electronic excitation energies), and of 105 kJ mol^{-1} (a nuclear rather than chemical energy). What does this say about approximating $q_{nuclear}$ and $q_{electronic}$ as their respective ground state degeneracies?

11.6 Use the Boltzmann distribution equation to calculate the ratio of populations of energy levels separated by (a) 500 cm^{-1} and (b) 25 kJ mol^{-1}.

11.7 The volume of a typical first row atom is on the order of $10^{-8}\,m^3$. What is the value of the translational partition function at 298 and 1,500 K?

11.8 Calculate the absolute entropy of helium gas at 0°C and 1 bar.

11.9 Calculate the value of the rotational partition function for HCl gas at 273 K and at 600 K. What rotational level has the highest population under the same temperature conditions? What is the probability of this energy level?

11.10 The particle phase space for a classical one-dimensional harmonic oscillator has two coordinates, x and p. Draw a trajectory in this phase space for a single oscillatory cycle. Illustrate how the trajectory depends on the energy of the oscillator and on the mass of the particle by drawing trajectories for when the energy and then the mass are doubled.

11.11 Start with Equation 11.10 and use $\delta = 1$ to obtain Equation 11.11.

11.12 [Spreadsheet problem] Equation 11.34 can be expressed in terms of the ratio γ of the energy ε to the chemical potential g (i.e., $\gamma = \varepsilon/g$). In this form, Equation 11.34 is

$$f(\varepsilon) = f(g\gamma) = \frac{1}{1 + e^{(\gamma-1)g/kT}}$$

Evaluate $f(\varepsilon)$ for $\gamma = 0.2, 0.4, 0.8, 1.0, 1.2, 1.4, 1.6, 1.8,$ and 2.0 at a temperature such that $g/kT = 1.0$. Plot $f(\varepsilon)$ versus γ. Repeat this for $g/kT = 0.1, 2.0, 4.0$ and 10.0. On each plot, indicate the Fermi temperature and $f(\varepsilon)$ at the Fermi temperature.

11.13 Derive the series expansion $(1 - x)^{-1} = 1 + x + x^2 + x^3 + x^4 + \cdots$ (see Appendix A), and use it to verify Equation 11.45.

11.14 Evaluate the mean or average energy of a quantum mechanical harmonic oscillator system for which the energies are E_n (kJ mol^{-1}) $= 23.0 (n + 1/2)$ at $T = 1, 10, 100, 300,$ and $1,000\,$K.

11.15 Take the values for the dipole moment for the HF molecule in its lowest-lying vibrational states to be 2.91 D ($n = 0$), 3.01 ($n = 1$), 3.09 ($n = 2$), and 3.15 ($n = 3$). Take the energies of these states in cm^{-1} to be 2,000 ($n = 0$), 5,960 ($n = 1$), 9,860 ($n = 2$), and 13,700 ($n = 3$). Assuming that a collection of HF molecules can be trapped in an inert matrix so that all are parallel and are not free to reorient, calculate the average dipole for $T = 10, 100,$ and $300\,$K.

11.16 [Spreadsheet problem] (a) Evaluate $q_{vibrational}$ in Equation 11.46 for a quantum mechanical harmonic oscillator system for which the energies are E_n (kJ mol^{-1}) $= 23.0 (n + 1/2)$ at 20 temperatures from 10 to 1,000 K. (b) Evaluate $q_{vibrational}$ if the energy expression includes an anharmonicity term $-0.1(n + 1/2)^2$. Plot the harmonic and anharmonic values of $q_{vibrational}$ as a function of temperature.

11.17 [Spreadsheet problem] Compare the rotational partition function values at $T = 100\,$K and at $T = 300\,$K from Equation 11.48 with those obtained from Equation 11.49 for rotational constant values of $B = 0.2, 2.0,$ and $20.0\,$cm^{-1}. For each rotational constant and each temperature choice, evaluate and sum individual terms in Equation 11.48 until a J value is reached such that the contributions become less than 1 part in 105 of the $J = 0$ contribution. Compare the sums with the values obtained from Equation 11.49.

11.18 Redo Example 11.2 and include the next two terms before truncating the power series expansion of the exponential. Using a harmonic oscillator energy level expression of E_n (kJ mol^{-1}) $= 23.0 (n + 1/2)$, estimate the percentage error in U from truncating the expansion.

11.19 For a dilute gas on noninteracting diatomic molecules, find an expression for G in terms of the molecular partition function.

11.20 Make a plot of C_V as a function of temperature for an ideal gas composed of diatomic molecules vibrating as a harmonic oscillator according to the energy level expression of E_n (kJ mol^{-1}) = 23.0 $(n + 1/2)$.

11.21 On the basis of the equipartition principle and the assumption of ideal behavior, what is U for acetylene, for diimide, and for glyoxal?

11.22 The energies of the $n = 2$ and $n = 1$ orbitals of a certain molecule are 5.45×10^{-19} and 2.18×10^{-18} J, respectively. What are the relative populations in these levels at 0°C and 1000°C?

Bibliography

Anderson, M. H., J. R. Ensher, M. R. Matthews, C. E. Wieman, and E. A. Cornell, *Science* 289, 198 (1995); also see *Phys. Today*, August 1995, 17.

Bauman, R. P., *Modern Thermodynamics with Statistical Mechanics* (Macmillan, New York, 1992).

Chandler, D., *Introduction to Modern Statistical Mechanics* (Oxford University Press, New York, 1987).

Hecht, C. E., *Statistical Thermodynamics and Kinetic Theory* (Freeman, New York, 1990).

McQuarrie, D. A., *Statisitical Thermodynamics* (Harper & Row, New York, 1985).

12

Magnetic Resonance Spectroscopy

Electrons and certain atomic nuclei possess intrinsic magnetic moments that give rise to an interaction energy with an external magnetic field. The difference between the interaction energies of the different states can be probed by magnetic resonance spectroscopy, an immensely powerful technique now used for qualitative analysis, determination of molecular structure, and measurement of reaction rates and dynamics. Its use extends to imaging macroscopic objects, and medical applications for diagnostic work have become widespread. This chapter provides the quantum mechanical foundation for multinuclear magnetic resonance spectroscopy.

12.1 Nuclear Spin States

Atomic nuclei consist of neutrons and protons and have a structure that is as rich as the electronic structure of atoms and molecules. In chemistry, though, nuclear structure is often unimportant since it does not change in the course of a chemical reaction. It is frequently appropriate to consider nuclei to be simply positively charged point masses, as we have done to this point. However, there are some important manifestations of nuclear structure that have been exploited in developing powerful types of molecular spectroscopies.

Protons and neutrons have an intrinsic spin and an intrinsic magnetic moment. Experiments have revealed that there are just two possible orientations of their spin vectors with respect to a z-axis defined by an external field. This means that their intrinsic spin quantum number is 1/2; they are fermions. Whereas the letter S is commonly used for the electron spin quantum number, the letter I is commonly used for nuclear spin quantum numbers. Thus, $I = 1/2$ for a proton, and the allowed values for the quantum number giving the nuclear spin angular momentum projection on the z-axis, M_I, are $+ 1/2$ and $-1/2$.

The angular momentum coupling of the intrinsic proton and neutron spins in a heavy nucleus is a problem for nuclear structure theory, and it is beyond our discussion of molecules. However, we can correctly anticipate certain results on the basis of electronic structure. The first of these is that atomic nuclei can exist in different energy states. It turns out that the separation in energy between the ground state of a stable nucleus and its first excited state is usually enormous in relation to the size of chemical reaction energetics. Photons from the gamma ray region of the electromagnetic spectrum may be needed to induce transitions to excited nuclear states, and these transitions can require hundreds and thousands of times the energy needed for transition to an excited electronic state. Consequently, unless something is done to prepare a nucleus in an excited state, we can assume that all the nuclei in a molecule are in their ground state configuration in some chemical experiment.

Another result we can anticipate from electronic structure is the possible values for the total nuclear spin quantum number. For instance, the deuteron (a proton and a neutron) consists of two particles with an intrinsic spin of 1/2. Were these two particles electrons, we would know that the possible values for the total spin quantum number are 1 and 0. It turns out for the deuteron that of the two coupling possibilities, $I = 1$ and $I = 0$, the $I = 1$ spin coupling occurs for the ground state. This is a consequence of the interactions that dictate nuclear structure, which is outside the scope of this chapter. For chemical applications, the key information is simply that the deuteron is an $I = 1$ particle.

The intrinsic spins of stable nuclei have been determined experimentally, and the values have been explained with modern nuclear structure theory. Tables such as that in Appendix E are available for looking up the spin (I value) of a particular nucleus. The rules of angular momentum coupling are an aid in remembering the intrinsic spins of certain common nuclei. For instance, the helium nucleus, with its even number of protons and neutrons, has an integer spin, $I = 0$. In terms of the number of protons and neutrons, the carbon-12 nucleus is simply three helium nuclei. It should have an even-integer spin, and it also turns out to be an $I = 0$ nucleus. The carbon-13 nucleus has one more neutron and thereby has a half-integer spin, $I = 1/2$.

Nuclei with an intrinsic spin of $I > 0$ have an intrinsic magnetic moment. Just as with electron spin, the magnetic moment vector is proportional to the spin vector. A **nuclear magnetic moment** can give rise to an energetic interaction with an external magnetic field as well as with other magnetic moments in a molecule, such as those arising from electron spin.

Nuclear magnetic moments are small enough to have an almost ignorable effect on atomic and molecular electronic wavefunctions. On the other hand, the electronic structure has a measurable influence on the energies of the nuclear spin states. In this situation, it is an extremely good approximation to separate nuclear spin from the rest of a molecular wavefunction. The electronic, rotational, and vibrational wavefunctions of a molecule can be accurately determined while ignoring nuclear spin, as has been done so far. Then, the effect of the electrons and the effect of vibration and rotation can be incorporated as an external influence on the nuclear spin states.

The separated nuclear spin problem is a very special type of problem in quantum chemistry partly because the number of states is strictly limited. For example, a proton is a spin $-1/2$ particle; this means that $I = 1/2$ and that the spin multiplicity, $2I + 1$, is 2. There are only two states, one with $M_I = 1/2$ and one with $M_I = -1/2$. Then, in the hydrogen molecule where there are two protons, the number of nuclear spin states for the molecule as a whole is the product of the spin multiplicities of the two nuclei; that is, $2 \times 2 = 4$. Clearly, in large, complicated molecules, the number of nuclear spin states may be large, but finite.

Nuclear magnetic resonance (NMR) spectroscopy is concerned with the energies of the nuclear spin states and the transitions that are possible between different states. To work out the energies of the states, we need to understand the interactions that affect nuclear spin state energies and to develop an appropriate Hamiltonian.

Interactions with nuclear spins come about through the intrinsic magnetic moments of nuclei. The nuclear magnetic moment, μ, is proportional to the nuclear spin vector, I, just as the magnetic moment of an electron is proportional to its spin vector. Instead of a fundamental development of the proportionality relationship, we follow a phenomenological approach by simply using an unknown for the proportionality constant, and for now we shall call it α.

$$\vec{\mu} = \frac{\alpha \vec{I}}{\hbar}$$

(12.1)

As we have already seen with electrons, the interaction energy of a magnetic moment and a uniform external magnetic field, H, varies as the dot product of the two vectors. Thus, the interaction Hamiltonian for a bare nucleus experiencing an applied magnetic field is

$$\hat{H} = -\vec{\mu} \cdot \vec{H} \tag{12.2}$$

If there are several noninteracting nuclei experiencing the field, then the Hamiltonian is a sum of the interactions of each of the nuclei.

$$\hat{H} = -\sum_i \vec{\mu}_i \cdot \vec{H} \tag{12.3}$$

This is the basic form of the Hamiltonian for the nuclear spin state Schrödinger equation.

Nuclei embedded in molecular electronic charge distributions experience an externally applied magnetic field at a slightly altered strength. The electronic motions may tend to shield the nucleus from feeling the full strength of the field. There are also situations where the response of the electron distribution to a magnetic field amplifies the strength of the field at the nucleus. In either case, the Hamiltonian in Equation 12.3 needs to be modified to properly represent nuclei in molecules, as opposed to bare nuclei in space. Again, we can approach this phenomenologically by inserting a correction factor in Equation 12.3 without yet establishing the fundamental basis for the factor. Then, for each different nucleus there is a different correction, and this is expressed as

$$\hat{H} = -\sum_i \vec{\mu}_i \cdot (1 - \sigma_i) \cdot \vec{H} \tag{12.4}$$

The σ_i are the **nuclear magnetic shielding tensors**, these tensors being 3×3 matrices. There are up to nine elements of a shielding tensor allowing for the general ways in which components of the field and components of the nuclear magnetic dipole have their interaction affected by electron shielding. 1 is the unit or identity matrix. Multiplication of the matrix quantity in parentheses in Equation 12.4 by the field vector leads to a new vector, and it corresponds to the effective field at the nucleus.

A helpful simplification comes about from breaking up the shielding tensor into isotropic and anisotropic parts:

$$\begin{pmatrix} \sigma_{xx} & \sigma_{yy} & \sigma_{zz} \\ \sigma_{yx} & \sigma_{yy} & \sigma_{yz} \\ \sigma_{zx} & \sigma_{zy} & \sigma_{zz} \end{pmatrix} = \frac{1}{3}(\sigma_{xx} + \sigma_{yy} + \sigma_{zz}) \begin{pmatrix} 1 & 0 & 0 \\ 0 & 1 & 0 \\ 0 & 0 & 1 \end{pmatrix}$$

$$+ \begin{pmatrix} \frac{2}{3}\sigma_{xx} - \frac{1}{3}\sigma_{yy} - \frac{1}{3}\sigma_{zz} & \sigma_{xy} & \sigma_{xz} \\ \sigma_{yx} & \frac{2}{3}\sigma_{yy} - \frac{1}{3}\sigma_{xx} - \frac{1}{3}\sigma_{zz} & \sigma_{yz} \\ \sigma_{zx} & \sigma_{zy} & \frac{2}{3}\sigma_{zz} - \frac{1}{3}\sigma_{xx} - \frac{1}{3}\sigma_{yy} \end{pmatrix}$$

The isotropic part has equal diagonal elements, and so it has been given as a constant times the unit matrix, **1**. The anisotropic part is simply what remains. At this point, we will ignore the anisotropic part of the shielding tensor, which means that we will take the second matrix on the right-hand side of the preceding expression to be zero. This can be regarded as an approximation, for now, though specific experimental conditions may offer a justification. With that, the isotropic shielding reduces to a single value, a scalar quantity:

$$\sigma^{iso} = \frac{1}{3}(\sigma_{xx} + \sigma_{yy} + \sigma_{zz}) \tag{12.5}$$

Equation 12.4 simplifies on neglect of the anisotropic part of the shielding to become

$$\hat{H} = -\sum_i \vec{\mu}_i \cdot (1 - \sigma_i^{iso}) \cdot \vec{H} \tag{12.6}$$

The solutions of the Schrödinger equation with this Hamiltonian provide the basic energy level information for NMR spectroscopy.

Let us use Equation 12.6 to construct an energy level diagram for the nuclear spin states of any small molecule with its magnetic nuclei being two protons not chemically equivalent. It has four nuclear spin states, and two quantum numbers are needed to distinguish these states, the m_I numbers for the two particles. We will designate these m_{I_1} and m_{I_2}, and the states will be designated $|m_{I_1} m_{I_2}\rangle$. Using Equation 12.1 in Equation 12.6 and letting the applied field define, or be applied along, the z-axis so that $\vec{I} \cdot \vec{H} = \hat{I}_z H_z$, we find the energy eigenvalues since the spin states are eigenfunctions of the Hamiltonian.

$$H|m_{I_1} m_{I_2}\rangle = -(1-\sigma_1)\alpha_1 H_z m_{I_1}|m_{I_1} m_{I_2}\rangle - (1-\sigma_2)\alpha_2 H_z m_{I_2}|m_{I_1} m_{I_2}\rangle \tag{12.7}$$

The energies of the four states are seen to have a linear dependence on the strength of the external magnetic field, and thus, they separate in energy with increasing field strength. Since both magnetic nuclei are protons in this example, α_1 must be the same value as α_2, and we simply use α for both. A tabulation of the four states' eigenenergies from Equation 12.7 is

m_{I_1}	m_{I_2}	Energy
−1/2	−1/2	$\frac{\alpha H_z}{2}[(1-\sigma_1)+(1-\sigma_2)]$
−1/2	1/2	$\frac{\alpha H_z}{2}[(1-\sigma_1)-(1-\sigma_2)]$
1/2	−1/2	$\frac{\alpha H_z}{2}[-(1-\sigma_1)+(1-\sigma_2)]$
1/2	1/2	$\frac{\alpha H_z}{2}[-(1-\sigma_1)-(1-\sigma_2)]$

Figure 12.1 is an energy level diagram based on these energies.

In a conventional NMR experiment, transitions between the nuclear spin energy levels are induced by applying electromagnetic radiation perpendicular to the direction of the static magnetic field. The selection rules are that one m_I quantum number can increase

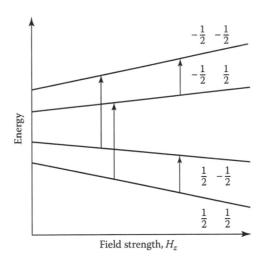

FIGURE 12.1
Nuclear spin energy levels of a molecule with two noninteracting protons in different chemical environments. The levels separate in energy with increasing strength of the external field, according to Equation 12.6. The levels are labeled by the m_I quantum numbers of the two protons. The vertical arrows indicate the allowed NMR transitions. Clearly, the energy of these transitions depends on the strength of the applied field.

or decrease by 1 while all the other m_I quantum numbers remain unchanged.* In effect, this says that absorption of energy "flips" the spin of only one nucleus at a time. From the preceding tabulation of energy levels and from Figure 12.1, we see that the selection rules correspond to allowed transitions from the lowest energy state to both of the next two energy states and that the transitions from these two states to the highest energy state are allowed. The arrows in Figure 12.1 indicate these transitions.

The transition energies are obtained by taking differences in the state energies. Using the preceding tabulation, we obtain the following transition energies:

Initial State	Final State	Transition Energy
$\lvert 1/2 \ 1/2\rangle$	$\lvert 1/2 \ -1/2\rangle$	$\alpha H_z(1 - \sigma_2)$
$\lvert 1/2 \ 1/2\rangle$	$\lvert -1/2 \ 1/2\rangle$	$\alpha H_z(1 - \sigma_1)$
$\lvert 1/2 \ -1/2\rangle$	$\lvert -1/2 \ -1/2\rangle$	$\alpha H_z(1 - \sigma_1)$
$\lvert -1/2 \ 1/2\rangle$	$\lvert -1/2 \ -1/2\rangle$	$\alpha H_z(1 - \sigma_2)$

This reveals that though there are four possible transitions, there are only two frequencies at which transitions can be detected. Furthermore, measuring the transition frequencies yields values for $\alpha(1 - \sigma_1)$ and for $\alpha(1 - \sigma_2)$. The typical field strengths used in NMR are such that the transition frequencies of the electromagnetic radiation are in the microwave-radiofrequency region of the spectrum. These are very low-energy transitions compared to vibrational or electronic excitations of a molecule. The basic experiment can be carried out in two ways. A field of fixed strength can be applied, and the frequency of the radiation varied until a transition is detected by a change in the power transmission of the radiation. Or, the frequency can be fixed and the power monitored as the field strength is varied. When this sweeping of the field strength brings a transition into resonance with the radiation frequency, a change in the power level of the radiation field is detected.

---

* The interaction that leads to a transition is between the magnetic moment of the nucleus and the magnetic field of the electromagnetic radiation. These are termed **magnetic dipole transitions**.

Thus, transition frequencies can be measured for a given field strength, or field strengths at which transitions occur at a certain frequency can be measured. The latter information can be converted to the former.

The information obtained from a low-resolution NMR scan is the chemical shielding. It is usually obtained as a shift relative to some standard or reference transition. The chemical shift is designated δ and is a dimensionless quantity. It is usually a number on the order of 10^{-6}, and so it is usually stated as being in parts per million (ppm). When the radiation frequency has been varied in the experiment, δ is given as

$$\delta \equiv \frac{\omega_{ref} - \omega}{\omega_{ref}} \tag{12.8}$$

When the field strength has been swept, δ is given as

$$\delta \equiv \frac{H_z - H_{z-ref}}{H_{z-ref}} \tag{12.9}$$

Chemical shifts are characteristic of the chemical environment. Thus, NMR spectra can serve as an analytical tool for determining functional groups that are present in a molecule, for instance.

Proton NMR spectra of organic molecules are usually referenced to the proton transitions of tetramethylsilane (TMS), taking δ of the reference, according to Equation 12.8 or 12.9, to be zero. (Another system, with values designated τ, is rarely used; the values are given as $\tau = 10 - \delta$ with $\tau(\text{TMS}) = 10\,\text{ppm}$.) The NMR signature of protons in different environments in organic molecules is a transition roughly within these ranges (δ scale):

Alkanes	0–2 ppm
Alkenes	4–8 ppm
Alkynes	2–3 ppm
Aromatic	7–8 ppm
Alcohols	10–11 ppm
Aldehydes	9–10 ppm

More extensive lists of this sort are available.

The proportionality constant introduced in Equation 12.1 varies from one nucleus to another because of the intrinsic nuclear structure. The proportionality constant, α, can be replaced by a dimensionless value, g, for a given nucleus, and a fundamental constant, μ_0, the **nuclear magneton**.

$$\vec{\mu}_i = \frac{\mu_0 g_i \vec{I}_i}{\hbar} \tag{12.10}$$

With the external magnetic field in the z-direction, we rewrite Equation 12.6

$$\hat{H} = -\mu_0 H_z \sum_i \frac{g_i(1 - \sigma_i)\hat{I}_{i_z}}{\hbar} \tag{12.11}$$

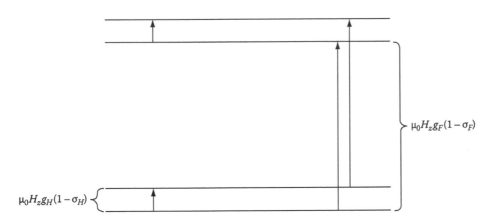

FIGURE 12.2
Energy level diagram for the four spin states of H^{19}F in an external magnetic field of fixed strength. The allowed transitions are represented by the vertical arrows. Because of the sizable difference in *g* values, the transition energies for the ^{19}F spin flip are much different than the transition energies of the proton. In practice, two different instrumental setups are required to observe the two transitions.

The variation in the nuclear *g* values has a much more profound effect on transition frequencies than does the variation in chemical shifts. This is evident from Equation 12.1. A 10% variation in σ, given that σ is on the order of 10^{-6}, has much less effect on the energy separation between states than a 10% variation in *g*. Consequently, NMR spectra are normally nuclei-specific; the instrumentation is set for a narrow range of frequencies to scan (or fields to sweep), and that is for a certain type of nucleus, such as protons. A different instrumentation setup is used for carbon-13 nuclei, or for oxygen-17, or for fluorine-19. It should be realized that even though we obtain proton NMR spectra independent of obtaining carbon-13 NMR spectra, etc., the complete Hamiltonian in Equation 12.11 is still at work and still describes the complete set of nuclear spin states. Figure 12.2 illustrates this for the case of the H^{19}F molecule. Both nuclei are spin –1/2 particles. There are four levels, but the two allowed transition frequencies are very different because of the different *g* values.

As a complicated example of nuclear spin energy levels, let us consider the diimide molecule, HNNH. At this stage, we assume no interaction between the magnetic moments of the different nuclei, and thus the Hamiltonian in Equation 12.11 is used. In either the *trans* or *cis* form of diimide, the nitrogens are in equivalent environments and the protons are in equivalent environments. Nitrogen-14 nuclei have an intrinsic spin of $I = 1$. The spin multiplicity is 3 since m_I can be –1, 0, or 1, and with the spin multiplicity of the protons being 2, the number of spin states for the molecules is $3 \times 3 \times 2 \times 2 = 36$. The energies of these states, obtained via Equation 12.11, are

$$E_{m_{I\,N-1}\,m_{I\,N-2}\,m_{I\,H-1}\,m_{I\,H-2}} = -\mu_0 H_z g_N (1-\sigma_N)(m_{I_{N-1}} + m_{I_{N-2}}) - \mu_0 H_z g_H (1-\sigma_H)(m_{I_{H-1}} + m_{I_{H-2}})$$

(12.12)

where the *N* and *H* subscripts indicate the nitrogen and hydrogen atoms, with 1 and 2 used to distinguish the like atoms. The selection rule applied to this problem is that one and only one of the *m* quantum numbers can change in a transition, and that change can be 1 or –1. This gives 84 different allowed transitions. Yet, if we systematically go through them, we can show that there are only two different transition energies. One shows up in a nitrogen-14 NMR spectrum at $\mu_0 H_z g_N (1 - \sigma_N)$. The other is in a proton NMR spectrum at $\mu_0 H_z g_H (1 - \sigma_H)$.

12.2 Nuclear Spin–Spin Coupling

The nuclear spins of different magnetic nuclei can interact and couple much the same as electron spins couple. This often leads to small energetic effects that tend to be noticeable under high resolution. The interaction that couples nuclear spins depends on their magnetic dipoles and their electronic environments. We can treat this interaction phenomenologically rather than attempt to analyze its fundamental basis. Since the dipole moments are proportional to the intrinsic spin vectors, we write the **spin–spin coupling** interaction as proportional to the dot product of two spin vectors. Therefore, for two interacting nuclei, the perturbing Hamiltonian due to spin–spin coupling is

$$\widehat{H}' = \frac{J_{12}}{\hbar^2} \vec{I}_1 \cdot \vec{I}_2 \tag{12.13}$$

where J has been introduced as the phenomenological proportionality constant, the **coupling constant**.

The spin–spin coupling interaction in Equation 12.13 is usually a small perturbation on the energies of the nuclear spin states experiencing the external field of a modern NMR instrument. It is appropriate to treat this interaction with low-order perturbation theory, particularly first-order perturbation theory. The expression for the first-order correction to an energy is the expectation value of the perturbation. For a system of two nonzero spin nuclei, the spin states are distinguished by the two quantum numbers, m_{I_1} and m_{I_2}. The corrections to the energies of these states are given by

$$
\begin{aligned}
E^{(1)}_{m_{I_1} m_{I_2}} &= \frac{J_{12}}{\hbar^2} \left\langle m_{I_1} m_{I_2} \left| \vec{I}_1 \cdot \vec{I}_2 \right| m_{I_1} m_{I_2} \right\rangle \\
&= \frac{J_{12}}{\hbar^2} \left\langle m_{I_1} m_{I_2} \left| \widehat{I}_{x_1} \widehat{I}_{x_2} + \widehat{I}_{y_1} \widehat{I}_{y_2} \right| m_{I_1} m_{I_2} \right\rangle \\
&= \frac{J_{12}}{\hbar^2} \left\langle m_{I_1} m_{I_2} \left| \widehat{I}_{z_1} \widehat{I}_{z_2} \right| m_{I_1} m_{I_2} \right\rangle
\end{aligned}
\tag{12.14}
$$

Only the last term in Equation 12.14 is nonzero, and by evaluating this integral, we obtain a simple result.

$$E^{(1)}_{m_{I_1} m_{I_2}} = J_{12} m_{I_1} m_{I_2} \tag{12.15}$$

This says that the corrections to the energy due to spin–spin coupling between two nuclei vary with the product of the m quantum numbers of the spin vectors.

As the first example of the effect of spin–spin interaction on NMR spectra consider a system with two nonequivalent protons attached to adjacent atoms. As in our previous case of two protons, the four spin states in the absence of spin–spin interaction have the following energies, these being the zero-order energies for the perturbative treatment of the spin–spin interaction:

$$E^{(0)}_{-1/2 \,-1/2} = \frac{\mu_0 H_z g_H}{2}\left[(1-\sigma_1)+(1-\sigma_2)\right]$$

$$E^{(0)}_{-1/2 \,1/2} = \frac{\mu_0 H_z g_H}{2}\left[(1-\sigma_1)-(1-\sigma_2)\right]$$

$$E^{(0)}_{1/2 \,-1/2} = \frac{\mu_0 H_z g_H}{2}\left[-(1-\sigma_1)+(1-\sigma_2)\right]$$

$$E^{(0)}_{1/2 \,1/2} = \frac{\mu_0 H_z g_H}{2}\left[-(1-\sigma_1)-(1-\sigma_2)\right]$$

The subscripts on the energies are the m_I quantum numbers of the two protons. The first-order corrections, according to Equation 12.15, are

$$E^{(1)}_{-1/2 \,-1/2} = \frac{J_{12}}{4}$$

$$E^{(1)}_{-1/2 \,1/2} = -\frac{J_{12}}{4}$$

$$E^{(1)}_{1/2 \,-1/2} = -\frac{J_{12}}{4}$$

$$E^{(1)}_{1/2 \,1/2} = \frac{J_{12}}{4}$$

Notice that the first-order corrections raise the energy of two levels and lower the energy of the other two levels. As seen in Figure 12.3, these changes in the energy levels "split" the pairs of transitions. With a, b, c, and d as designations of the four transitions, the transition energies are

$$a \text{ for } \left|-\frac{1}{2}\frac{1}{2}\right\rangle \rightarrow \left|-\frac{1}{2}-\frac{1}{2}\right\rangle \quad \Delta E_a = \mu_0 H_z g_H (1-\sigma_1)+\frac{J_{12}}{2}$$

$$b \text{ for } \left|\frac{1}{2}\frac{1}{2}\right\rangle \rightarrow \left|\frac{1}{2}-\frac{1}{2}\right\rangle \quad \Delta E_a = \mu_0 H_z g_H (1-\sigma_1)-\frac{J_{12}}{2}$$

$$c \text{ for } \left|\frac{1}{2}-\frac{1}{2}\right\rangle \rightarrow \left|-\frac{1}{2}-\frac{1}{2}\right\rangle \quad \Delta E_a = \mu_0 H_z g_H (1-\sigma_2)+\frac{J_{12}}{2}$$

$$d \text{ for } \left|\frac{1}{2}\frac{1}{2}\right\rangle \rightarrow \left|-\frac{1}{2}\frac{1}{2}\right\rangle \quad \Delta E_a = \mu_0 H_z g_H (1-\sigma_2)-\frac{J_{12}}{2}$$

Transitions a and b are moved to higher and lower frequencies, respectively, than they would be at in the absence of spin–spin interaction. The same is true for the c and d transitions.

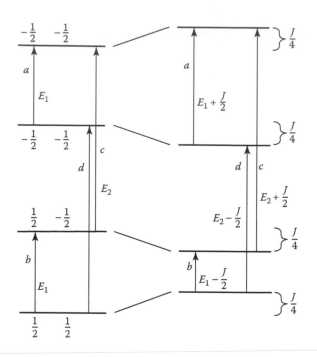

FIGURE 12.3

Energy levels for a hypothetical molecule with two protons that are nearby but in different chemical environments. On the left are the energy levels found in the absence of spin–spin interaction. On the right are the energy levels with spin–spin interaction treated via first-order perturbation theory. The vertical arrows show the allowed transitions, and the lengths of these arrows are proportional to the transition energies. Thus, the effect of the spin–spin interaction is seen to split the transitions, that is, to take each pair of transitions that would occur at the same frequency and shift one to a higher frequency and one to a lower frequency.

The separation between transition a and transition b (i.e., the difference between ΔE_a and ΔE_b) is the spin–spin coupling constant, J_{12}. Likewise, the separation between transition c and transition d is J_{12}. Figure 12.4 is a representation of the spectrum that would result.

The spin–spin interaction Hamiltonian for a system with many interacting nuclei is simply a sum of the pair interactions in Equation 12.13:

$$\hat{H}' = \sum_i \sum_{j>i} \frac{J_{ij}}{\hbar^2} \vec{I}_i \cdot \vec{I}_j \tag{12.16}$$

FIGURE 12.4

Representation of the NMR spectra for a molecule with the energy levels and transitions given in Figure 12.3. The top spectrum corresponds to the energy levels in the absence of spin–spin interaction, whereas the bottom spectrum includes the effect.

The first-order corrections to the energy are also sums of pair contributions with the form given in Equation 12.15:

$$E^{(1)} = \sum_i \sum_{j>i} \frac{J_{ij}}{\hbar^2} m_{I_i} m_{I_j} \tag{12.17}$$

The quantum number subscripts on $E^{(1)}$ have been suppressed for conciseness; they would be the m_I quantum numbers for all the nuclei. Consider a molecule whose magnetic nuclei are three protons in different chemical environments. Figure 12.5 shows how the zero-order energy levels for this system would be affected by the first-order corrections in Equation 12.17. The resulting spectrum consists of each original line split into a pair of lines which are then split again. This reveals a simple rule for interpreting proton NMR spectra: A pair of closely spaced lines may have come about because of the spin–spin interaction with a single nearby proton.

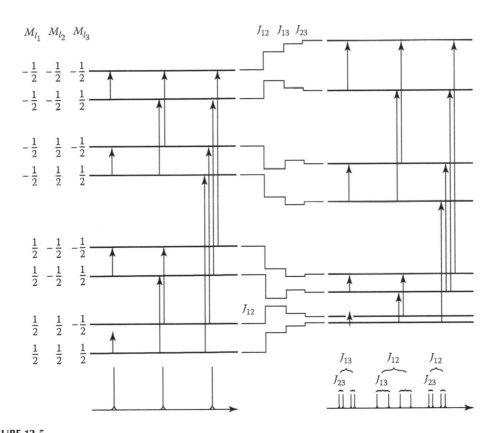

FIGURE 12.5
Energy level diagram for the nuclear spin states of a hypothetical molecule with three interacting protons in different chemical environments. The energy levels on the left neglect spin–spin interaction. The allowed transitions give rise to a spectrum of three lines, shown as a stick representation at the bottom. These three lines correspond to a spin "flip" of each of the three different nuclei, and their relative transition frequencies give the chemical shift of each. The energy levels on the right have been obtained by first including the 1–2 coupling, then the 1–3 coupling, and finally the 2–3 coupling, assuming $J_{12} > J_{13} > J_{23}$. Transition lines are drawn for the final set of levels, and the resulting stick spectrum is shown at the bottom. The values of the three coupling constants can be obtained directly from the spectrum, as shown.

We have treated spin–spin interaction phenomenologically (by employing the coupling constant, J, as a parameter instead of deriving it from fundamental interactions), and there are certain spectral features that we can anticipate. One is that the size of a coupling constant, J_{12}, varies with different pairs of nuclei. It falls off with increasing separation between the interacting nuclei to the extent that the interaction strength diminishes with increasing separation distance. In practice, proton NMR spectra generally show splitting by protons attached to adjacent atoms. Thus, splittings in spectra are intimately related to molecular structure and can serve to reveal structural features. Also, the lines of both of the interacting nuclei are split by the identical amount (i.e., J_{12}), and so finding two pairs of lines with the same splitting can identify two protons that are interacting.

Some molecules have different kinds of magnetic nuclei, a simple example being HD (deuterium-substituted H_2), and its energy levels are shown in Figure 12.6. The D nucleus has $I = 1$, and its m_I values are 1, 0, and –1. The spin states are distinguished by the m_I quantum numbers of the nuclei. From Equation 12.11 we write the zero-order energies, and from Equation 12.15, the first-order corrections. We can tabulate these energies concisely by listing the possible energy terms at the head of the column, and then with each state as a row in the table, we list the factor that should be applied to the energy term.

m_{I_D}	m_{I_H}	$\mu_0 H_z g_D(1 - \sigma_{D1})$	$\mu_0 H_z g_H(1 - \sigma_H)$	$J_{12}/2$
–1	–1/2	1	1	1
–1	1/2	1	–1	–1
0	–1/2	0	1	0
0	1/2	0	–1	0
1	–1/2	–1	1	–1
1	1/2	–1	–1	1

The energy of the state in the second row, for instance, is the sum of the first term plus the second term multiplied by –1 plus the third term multiplied by –1. Along with the energy

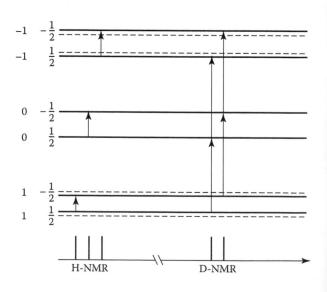

FIGURE 12.6
Nuclear spin energy levels of the HD molecule in a magnetic field. The allowed transitions, represented by vertical arrows, are those for which the m_I quantum number of the deuterium changes by 1, or the m_I quantum number of the proton changes by 1. The deuterium and proton stick spectra are shown at the bottom.

levels given in Figure 12.6 is a representation of the predicted spectra. They display the proton transition line split into three lines and the deuterium transition line split into two. This reveals a generalization of the rule about nearby protons splitting lines into pairs. It is that a nearby magnetic nuclei splits a transition line into the number of lines equal to the spin multiplicity, $2I + 1$, of the nucleus. The deuterium splits the proton transitions into three lines since $2I_D + 1 = 3$.

Equivalent nuclei are magnetic nuclei in equivalent chemical environments, such as the two protons in formaldehyde. The chemical shielding for each of these protons must be the same because of the symmetry of the molecule. The quantum mechanics for spin–spin coupling with equivalent nuclei has certain complexities that are not encountered with inequivalent nuclei, yet a rather simple rule for predicting or interpreting spectra emerges just the same. The complexity of equivalent nuclei is spin state degeneracy. The zero-order energies for the proton spin states in formaldehyde with quantum numbers (1/2 –1/2) and (–1/2 1/2) are both zero according to Equation 12.11. This is the consequence of the s for both nuclei being the same. These two states are degenerate, and degenerate first-order perturbation theory is needed to treat the spin–spin interaction.

The result from degenerate first-order perturbation theory for two interacting equivalent protons amounts to completely coupling their spins. Using the angular momentum coupling rules, two spin –1/2 protons may yield a net coupled spin with $I = 1$ or with $I = 0$. The selection rules for NMR transitions of equivalent nuclei turn out to be $\Delta I = 0$ and $\Delta M_I = \pm 1$. Figure 12.7 shows that with these selection rules, it is not possible to obtain from the NMR spectrum the J coupling constant for the equivalent protons. For formaldehyde, we expect a single proton NMR transition.

The coupling of equivalent nuclei with other nuclei can produce splittings as well as important intensity features. Consider a molecule with two equivalent protons and a nonequivalent third proton. The NMR spectrum of the third proton can be regarded as a superposition of the spectra obtained for the two coupling possibilities (namely, $I = 1$ and $I = 0$) of the equivalent protons. In the case of $I = 1$ coupling, we expect a spectrum that is qualitatively the same as the

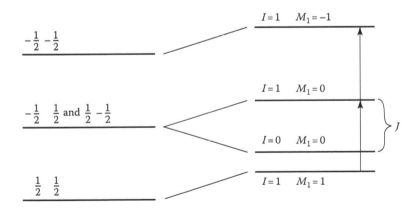

FIGURE 12.7
Energy levels and transitions for the proton spin states of formaldehyde. On the left are the levels obtained with neglect of spin–spin interaction, and they are labeled by the m_I quantum numbers of the two protons. The middle level is doubly degenerate. If spin–spin interaction is treated with first-order degenerate perturbation theory, the levels on the right result, with an assumed size of the coupling constant, J. A transition from the $I = 0$ state to an $I = 1$ state would measure J, but that is a forbidden transition. No splitting of the line in the spectrum occurs.

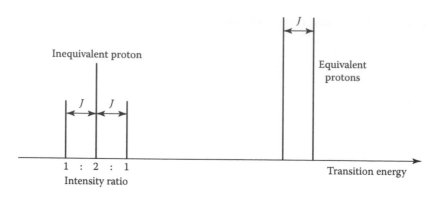

FIGURE 12.8

Representation of the NMR spectrum of a hypothetical molecule with two equivalent protons and a third, nearby proton in a different chemical environment. This spectrum may be understood as a combination of the spectrum of a system with an $I = 1$ particle interacting with the third proton and the spectrum from an $I = 0$ particle interacting with the third proton. These two pseudoparticles correspond to the possible couplings of the spins of the two equivalent nuclei.

HD molecule. In the case of $I = 0$ coupling, we expect a spectrum that looks like a spin –0 particle with a spin –1/2 particle. The single transition for this case adds to the transition intensity of the middle line of the triplet of lines from the first case. The result, shown in Figure 12.8, is that a spin flip by the third proton corresponds to three transition lines in an intensity ratio of 1:2:1. The equivalent protons' transitions are split into two lines by the third proton.

Groups of more than two equivalent nuclei can be analyzed by successive coupling of the spin angular momenta. For example, if there are three equivalent protons, the values of I_{12} (combining the spins of the first two) are 0 and 1. Adding the spin of the third means that $I_{123} = 1/2, 1/2$, and $3/2$. The spectrum of a magnetic nucleus interacting with this group is a superposition of spectra that would arise on interaction with an $I = 3/2$ nucleus and two $I = 1/2$ nuclei. It has four lines with intensity ratio 1:3:3:1. The same analysis applies to nuclei with other intrinsic spins. For instance, the I quantum number values of a group of three equivalent spin –1 nuclei are 0, 1, 1, 1, 2, 2, and 3, and the spectrum of a magnetic nucleus interacting with this group can show seven lines. The intensity pattern 1:3:6:7:6:3:1 is predicted by considering the lines to arise as a superposition.

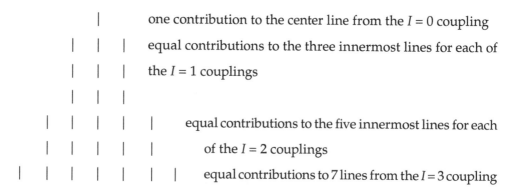

Example 12.1: NMR Energy Levels of Methane

Problem: Make a sketch of the NMR energy levels and draw arrows for the allowed NMR transitions of CH_4, CH_3D, and CH_2D_2.

Solution: In methane, the protons are equivalent. With partial deuterium substitution, the remaining protons constitute a set of equivalent nuclei and the deuterons constitute another set. Angular momentum coupling rules are used to find the total I (spin quantum number) for sets of equivalent nuclei, and then first-order perturbation theory can be used to account for the proton–deuteron coupling. In CH_4, there are four spin –1/2 particles. The first two spin couple to yield possible spin quantum numbers of 1 and 0. Coupling the next one yields 3/2, 1/2, and 1/2. Coupling the fourth yields $I = 2, 1, 1, 1, 0, 0$. The different I states have different energies because of the coupling interaction, and so they are represented as different levels; however, we do not have a basis for deciding the ordering or relative separation. The $I = 2$ state interacts with the external field, and as shown here it produces five energy levels. The $I = 1$ states separate into three levels each. Transitions among these levels are all at the same energy, one that depends not on the coupling but only on the chemical shielding. Transitions with an energy that would depend on and would reveal the coupling do not occur since $\Delta I = 0$. Here is the sketch for CH_4.

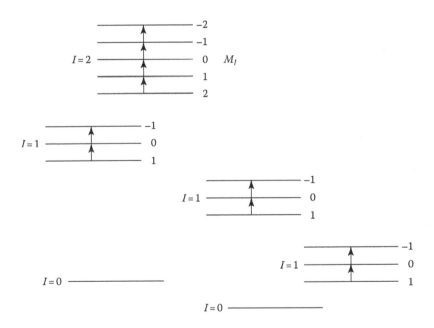

In CH_3D, the three proton spins couple together and then act as a pseudo 3/2, 1/2, and 1/2 nucleus to interact with the deuterium. We combine the energies of such pseudonuclei with the energies of the deuterium nucleus in the presence of a field and then add the coupling interaction, $JM_{H_3}M_D$. Transitions are those for which $\Delta I_{H_3} = 0$ and $\Delta M_{H_3} = \pm 1$, or $\Delta M_D = \pm 1$.

In CH_2D_2, the protons couple to have a net spin of 1 or 0, and deuterons couple to have a net spin of 2, 1, or 0. The energy level diagram shows a more complicated pattern.

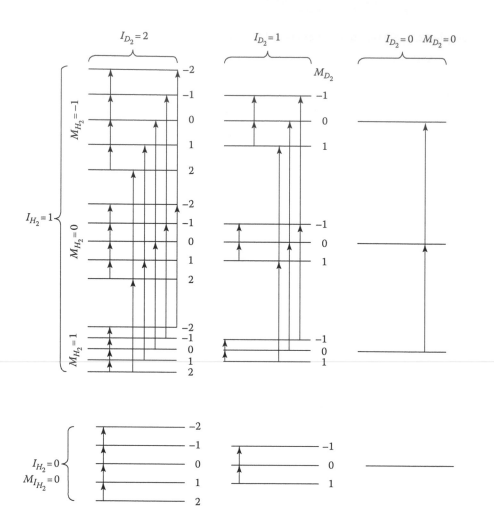

12.3 Electron Spin Resonance Spectra

If a molecule has a nonzero electronic spin (i.e., if $S > 0$), the associated magnetic moment will interact with an external magnetic field. This is the basis for electron spin resonance (ESR) spectroscopy, which is also called electron paramagnetic resonance (EPR) spectroscopy. The electron spin energy levels are separated by the magnetic field interaction, and the resulting energy differences are probed spectroscopically. The electron spin magnetic moment also interacts with nuclear magnetic moments, and it is this complication that yields certain of the useful information from ESR spectra.

Usually, electronic orbital angular momentum is zero in polyatomic molecules, and so this source of magnetic moments can often be ignored. Also, the ESR experiment is usually carried out with a liquid or solid sample, and so there is no magnetic moment arising from molecular rotation since the molecules are not freely rotating.

The general form for the interaction Hamiltonian between an isotropically sampled magnetic field $\vec{H}$ and the electron spin is

$$\hat{H} = \frac{g_e \mu_B}{\hbar} \vec{S} \cdot \vec{H} \tag{12.18}$$

where μ_B is the Bohr magneton. If magnetic nuclei are present, the Hamiltonian is a sum of this term and the nuclear Hamiltonian in Equation 12.11 along with the spin–spin interaction terms. As before, we take the direction of the magnetic field to define the z-axis, and then

$$\hat{H} = \frac{g_e \mu_B}{\hbar} \hat{S}z - \frac{\mu_0 H_z}{\hbar} \sum_i^{nuclei} g_i (1 - \sigma_i) \hat{I}_{i_z} + \sum_{i>j} \frac{J_{ij}}{\hbar^2} \vec{I}_i \cdot \vec{I}_j + \sum_i \frac{a_i}{\hbar^2} \vec{I}_i \cdot \vec{S} \tag{12.19}$$

This equation introduces certain constants, the a_i's, phenomenologically. The electron spin–nuclear spin interaction must be proportional to the dot product of the spin vectors, but the fundamental basis for the proportionality constant (a_i) is not considered.

The form of the Schrödinger equation incorporating the Hamiltonian in Equation 12.19 is the same as the form of the NMR Schrödinger equation. In both, there are z-component spin operators and spin–spin dot products. The only thing that is different in the ESR Hamiltonian is that one of the spins is electron spin and with it comes a different set of constants (e.g., g_e versus g_i). Notice that μ_B/μ_0 is the ratio of the proton mass to the electronic mass (1836.15).

In NMR, the external magnetic field effect is generally very large with respect to the size of the nuclear spin–spin interaction. This is because of the need to use a strong field in order to make the transition energies great enough to place them in the radiofrequency region of the spectrum. In ESR, the much greater size of μ_B means that the choice is available to use weaker fields; at 10,000 G, ESR transitions are typically around 10,000 MHz, whereas NMR transitions are around 10 MHz. Consequently, there are two limiting situations to consider in analyzing ESR spectra, a **high-field** case and a **low-field** case. The high-field case means that the field dominates all spin–spin interactions. The analysis, then, follows that for NMR where the spin–spin terms in the Hamiltonian are treated by first-order perturbation theory. In the low-field case, spin–spin coupling is of much greater relative importance, and first-order perturbation energies are not as appropriate.

Let us use the formyl radical HCO• as an example problem for predicting high-field and low-field ESR spectra. In this molecule, the single unpaired electron gives rise to a net electronic spin of 1/2 (i.e., $S = 1/2$). The proton spin, I, is 1/2, and so there are four spin states for the molecule. With high fields, the Hamiltonian is broken into a zero-order part and the spin–spin perturbation:

$$H^{(0)} = \frac{g_e \mu_B H_z \hat{S}_z}{\hbar} - \frac{g_H \mu_0 H_z (1 - \sigma) \hat{I}_z}{\hbar} \tag{12.20a}$$

$$H^{(1)} = \frac{a \vec{I} \cdot \vec{S}}{\hbar^2} \tag{12.20b}$$

The zero order wavefunctions, labeled by the quantum numbers m_S and m_I, are eigenfunctions of $\hat{I}_z$ and $\hat{S}_z$, (and of $\hat{I}^2$ and $\hat{S}^2$) and the associated energy values, represented in Figure 12.10, are

State	$\lvert m_S$	$m_I \rangle$	$E^{(0)}_{m_S m_I}$
	1/2	−1/2	$g_e\,\mu_B\,H_z/2 + g_H\,\mu_0\,H_z\,(1-\sigma)/2$
	1/2	1/2	$g_e\,\mu_B\,H_z/2 - g_H\,\mu_0\,H_z\,(1-\sigma)/2$
	−1/2	−1/2	$-g_e\,\mu_B\,H_z/2 + g_H\,\mu_0\,H_z\,(1-\sigma)/2$
	−1/2	1/2	$-g_e\,\mu_B\,H_z/2 - g_H\,\mu_0\,H_z\,(1-\sigma)/2$

For each of these functions, the first-order energy corrections are

$$E^{(1)}_{m_S m_I} = \left\langle m_S m_I \left| \hat{H}^{(1)} \right| m_S m_I \right\rangle$$

$$= \frac{a}{\hbar^2} \left\langle m_S m_I \left| \hat{I}_x \hat{S}_x + \hat{I}_y \hat{S}_y + \hat{I}_z \hat{S}_z \right| m_S m_I \right\rangle$$

$$= a m_S m_I \tag{12.21}$$

A sketch of the first-order energy levels is given in Figure 12.9. The high-field selection rules are $\lvert \Delta m_S \rvert = 1$ and $\lvert \Delta m_I \rvert = 0$. Thus, there are two ESR transitions, and the difference between the two transition energies is simply the coupling constant a.

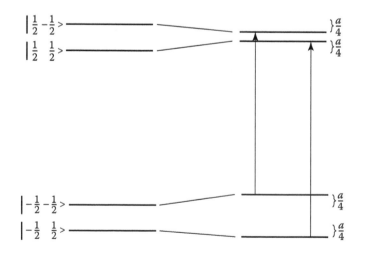

FIGURE 12.9
High-field energy levels for a system with electron spin $S = 1/2$ and with a magnetic nucleus with $I = 1/2$. The levels on the left are the zero-order energies, and the levels on the right include the first-order corrections due to electron spin–nuclear spin interaction. The two vertical arrows show the allowed transitions that would be observed in an ESR experiment. Not shown are the allowed transitions for which $\Delta M_S = 0$ since these are the NMR transitions and occur at very different energies.

If there are several magnetic nuclei in a molecule, the high-field energy level expression that follows from a first-order perturbative treatment of the general Hamiltonian in Equation 12.19 is

$$Em_Sm_{I_1}\cdots = g_e\mu_B H_z m_s - \sum_{i}^{nuclei} g_i\mu_0 H_z(1-\sigma_i)m_{I_i} + \sum_{i>j}^{nuclei} J_{ij}m_{I_i}m_{I_j} + \sum_{i}^{nuclei} a_i m_S m_{I_i} \qquad (12.22)$$

Figure 12.10 gives an example of energy levels that arise from Equation 12.22 and the resulting form of the spectrum. Notice that there is a different multiplet pattern for the ESR transitions associated with an $I = 1$ nucleus than with an $I = 1/2$ nucleus. Indeed, the patterns of lines help identify the type of magnetic nuclei much as they do in NMR spectra, and this holds in the case of equivalent magnetic nuclei, too.

ESR transition energies yield values for each of the a_i coupling constants in Equation 12.22. The relative size of these constants points to the extent of localization of the electron spin density at a given nucleus. In fact, ESR provides an ideal means for characterizing the atomic radical site in organic free radicals. If the measured coupling constants can be properly assigned to atoms in the radical, one particularly large coupling constant suggests that there is an unpaired electron (the radical electron) that is largely localized at or near that atom. The coupling constants might instead indicate that the unpaired electron is delocalized throughout the molecule, and this would give a different picture of the electronic structure of the radical.

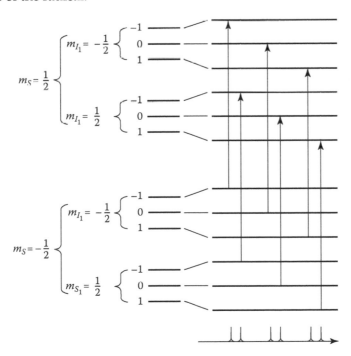

FIGURE 12.10
High-field energy levels for a system with electron spin $S = 1/2$ and magnetic nuclei with $I = 1/2$ and $I = 1$. On the left are the levels at zero order where only the interaction with the external field is included. On the right are the energy levels according to Equation 12.22, where the spin–spin interactions have been included by first-order perturbation theory. The nuclear spin–spin interaction has been exaggerated relative to typical values in order to show the effect. The vertical lines correspond to the allowed ESR transitions, and below them is a stick representation of the spectrum that corresponds to these transitions.

The low-field ESR problem is a more complicated quantum mechanical problem than the high-field case because first-order perturbation theory is not appropriate. A more suitable approach is to use linear variation theory with the Hamiltonian in Equation 12.19 and with a basis set of independent spin functions, that is, product functions that are eigenfunctions of each of the z-component operators in Equation 12.19. From the energy levels that result, the transition energies are obtained as differences in state energies. In practice, ESR spectra are sometimes analyzed by carrying out this process repeatedly for different choices of coupling constants until a satisfactory agreement between measured and calculated transition energies is achieved. In this manner, the coupling constants can be extracted from the spectra even without the simplifying aspects of the high-field case.

We can understand some of the features of low-field ESR spectra by considering the low-field energy levels to be intermediate to two limiting cases, the high-field and the zero-field limits. In the absence of an external field, spins are fully coupled. So, for a problem with an electron spin, S, and a single magnetic nucleus with spin I, the rules of angular momentum coupling dictate the energy levels. The total angular momentum is the vector sum of the two spin vectors,

$$\vec{F} = \vec{S} + \vec{I}$$

(12.23)

and the associated quantum number F must have the values

$$F = |S - I|, \ldots, S + I$$

Then, as was done for spin–orbit interaction in the hydrogen atom, we replace the spin–spin operator with

$$\vec{I} \cdot \vec{S} = \frac{1}{2}(\hat{F}^2 - \hat{I}^2 - \hat{S}^2)$$

(12.24)

In the zero-field limit, the Hamiltonian reduces to

$$\hat{H} = \frac{a}{\hbar^2}\vec{I} \cdot \vec{S}$$

and so the eigenfunctions of the Hamiltonian are the functions that are simultaneously eigenfunctions of the three angular momentum operators $\hat{F}^2$, $\hat{I}^2$, and $\hat{S}^2$. Therefore, the eigenenergies must be

$$E_{F,I,S} = \frac{a}{2}[F(F+1) - I(I+1) - S(S+1)]$$

(12.25)

Each of these energy levels has a multiplicity (degeneracy) of $2F + 1$.

For the specific situation of an electron spin $S = 1/2$ and a nuclear spin $I = 1/2$, F takes on the values of 0 and 1, and there are two energy levels. As the magnetic field is increased from zero to some high value, the energy levels change smoothly from one limiting case to the high-field case where there are four energy levels. As the field is just turned on, the degeneracy in the levels of different F values is removed. For extremely weak fields, the magnetic field can be treated with low-order perturbation theory, and then the separation of levels will depend on the value of m_F, the quantum number that gives the net z-component of the total angular momentum. Thus, the $F = 1$ level gives rise to three levels ($m_F = -1, 0, 1$) that spread apart linearly with the field strength, at least initially.

12.4 Extensions of Magnetic Resonance

As described so far, magnetic resonance is a tool for probing individual molecules. However, there is also important information that develops as a bulk property, and this has become the basis for a powerful application, magnetic resonance imaging. Here, the general idea is to measure a signal that is dependent on the spatial distribution of a species that might otherwise be studied with conventional magnetic resonance spectroscopy. The mathematical task of imaging is to map the signal into a graphical representation, an image.

To illustrate an imaging experiment, consider two long sample tubes filled with ethanol and held parallel in a larger-diameter tube that is otherwise filled with benzene. This assembly is placed in an NMR spectrometer modified so that instead of the static magnetic field being uniform across the sample the field varies in strength from one side to the other. As a consequence of the field variation the ethanol NMR transition signals vary with the orientation of two small tubes with respect to the field: If they are aligned with the applied field, one sample will experience a stronger field than the other, and so there will be signals at two distinct frequencies. If they are rotated away from this arrangement, the transition peaks will coalesce as the fields experienced by the samples in the two small tubes approach the same strength (i.e., at 90° from the original orientation). This is shown in Figure 12.11, and one can see that a spatially varying field strength encodes spatial information into the NMR spectrum.

A bulk property of a sample called **nuclear magnetization**, $\vec{M} = (M_x, M_y, M_z)$, is the sum of the individual magnetic moment vectors per unit volume. The magnetization can be measured, and its evolution in time can be followed. The characteristic times associated with the time evolution of the magnetization also encode spatial information if the field is nonuniform. There are a number of sophisticated means for measuring these times that involve a pulse sequence in the applied radiation.

Magnetic resonance imaging (MRI) translates spatially encoded NMR signals (transition frequencies and/or relaxation times) into images such as that in Figure 12.12. The technology of MRI has advanced to the point that it is perhaps the best nondestructive technique with the greatest resolution for biological systems. Images of internal structures of living systems must be carried out with x-ray or higher energy radiation or else with radiofrequency or longer wavelength radiation because there is little penetration into tissues of the radiation between these extremes. Because MRI uses radiofrequency radiation it has certain advantages over x-ray imaging. Primarily, because the photons of x-ray radiation have a substantial amount of energy relative to a chemical bond whereas radio waves have photon energies far less than a chemical bond, tissue damage is a potentially greater

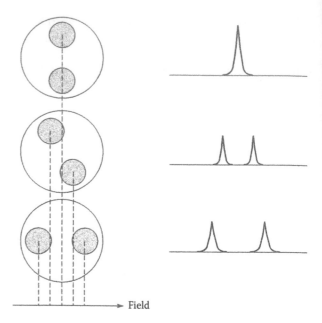

FIGURE 12.11

Two sample tubes embedded in a larger sample tube. Spectra are taken in the vicinity of the transition seen for a type of proton center in the substance in the small tubes. The field strength varies in the direction shown by the arrow. As the small tubes are rotated with respect to the applied field, the signal changes from a single peak to two peaks and back again thereby encoding spatial information in the data.

likelihood with x-rays relative to MRI. More important, conventional x-ray images come about by a scattering of x-rays from hard tissues (bone), whereas MRI is quite successful at imaging both soft and hard tissue.

The application of NMR to biomolecules would be expected to yield quite congested proton spectra, and this is the case. However, techniques involving simultaneous application of electromagnetic radiation from two sources along with suitable geometrical arrangements and sequencing of the signals and detection have produced a two-dimensional (2D) separation of the standard NMR information. In a 2D spectrum, peaks are

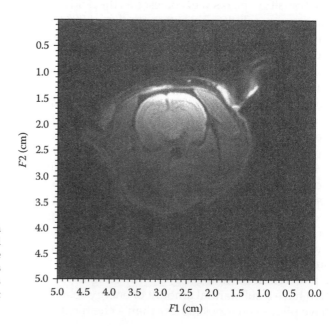

FIGURE 12.12

An MRI image of an axial section of a human brain. The bright region in the right cerebral hemisphere is an indication of a tumor. The MRI used to obtain these images operates with a 1.5 Tesla magnet, and the image is based on signals collected for a region that was 5mm thick.

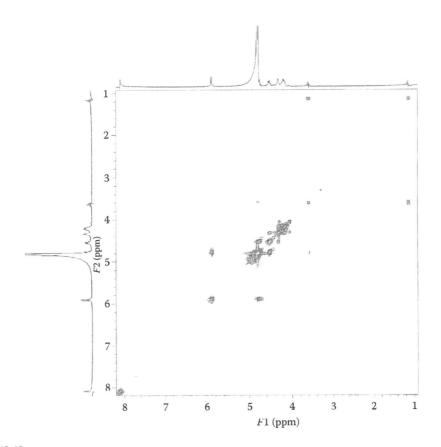

FIGURE 12.13
Two-dimensional ^{1}H NMR spectrum of guanosine triphosphate. The small circles are contours of the spectrum's "spikes," the equivalent of peaks in a 1D spectrum. (They appear as spikes when the intensities are expressed on a scale in the direction perpendicular to the plane of the paper.) Greater intensity is reflected in more contours being shown-more closely spaced contour circles. (From Davis, W.M. and Brown, D., Texas Lutheran University, Seguin, TX. With permission.)

found on an x–y grid with one axis corresponding to one frequency of radiation and the other axis to another frequency. A congested spectrum obtained in a conventional NMR experiment (one-dimensional [1D]) amounts to a projection of the 2D spectrum onto one axis. The 2D spectrum separates similar nuclei to reveal the origins of spin–spin splittings. The congestion of lines is very much diminished, and so the 2D spectrum can be used to obtain accurate shieldings and spin–spin splittings even for very large molecules. An illustration of this type of spectrum is in Figure 12.13.

12.5 Conclusions

Magnetic fields can interact with the magnetic dipoles of nuclei with intrinsic spin quantum numbers greater than zero, that is, magnetic nuclei. An external field leads to a small energy difference among the states, those distinguished by the M_I quantum number. Then, transitions among the states can be induced by electromagnetic radiation which is usually in the microwave or radiofrequency region of the spectrum. The interaction of an external magnetic field with a magnetic nucleus is an approximately separable part of the overall molecular Schrödinger equation since the magnetic moment of a nucleus has little effect on molecular electronic structure. However, even with this separation, there is a definite influence of the electron distribution on the nuclear spin state energies, an effect referred to as NMR chemical shielding. Measurement of the transition frequencies in NMR spectra can quantify chemical shieldings, and it turns out that shielding is somewhat like a fingerprint of the bonding environment of a nucleus.

Spin–spin coupling is an interaction between the magnetic moments of magnetic nuclei. With the typical field strengths of NMR spectrometers, spin–spin coupling is well described by first-order perturbation theory. The energy levels are shifted due to spin–spin coupling by an amount $J_{M_1 M_2}$, where J is the coupling constant between nucleus 1 and 2, and M_1 and M_2 are the nuclear spin M quantum numbers of the two nuclei. The coupling constant J usually appears as, and is measured as, a splitting between NMR lines. However, if nuclei are in equivalent environments, their coupling is relatively strong. In that case, the transition selection rules preclude obtaining a J value, though J-splittings are obtained for the set of equivalent nuclei interacting with still other magnetic nuclei.

Electron spin resonance spectroscopy is formally similar to NMR if one considers an unpaired electron as taking the role of a spin $-1/2$ nucleus in an NMR experiment. However, because of the much greater size of the electron magneton (the Bohr magneton) versus the nuclear magneton, much weaker external fields are employed in order to observe transitions with radiofrequency radiation. As a result, coupling interactions between the electron and magnetic nuclei may require a treatment beyond that of first-order perturbation theory.

Both ESR and NMR spectra offer information about the electron distribution in a molecule and the geometrical structure of a molecular framework. Both can provide spatial information about types of magnetic nuclei in samples, and that makes the techniques useful for sample imaging.

Point of Interest: The NMR Revolution

A 1980 review of NMR developments by J. Jonas and H. S. Gutowksy [*Annu. Rev. Phys. Chem.* 31, 1 (1980)] began with the following statement: "Within three decades, nuclear magnetic resonance spectroscopy has developed into the most important spectroscopic technique in chemistry, such that today most chemical laboratories use NMR as a routine tool." Its impact and applications have only expanded further since that review appeared. Looking back, we can see NMR as a technique that significantly changed the study of chemistry.

The history of NMR is often traced to a 1937 paper by I. I. Rabi that included the theoretical basis for magnetic resonance transitions. A professor at Columbia University, Rabi had been measuring nuclear magnetic moments by deflection of beams of atoms in a static inhomogeneous magnetic field. To better understand the transitions that were observed, Rabi worked out the transitions that would be induced by an oscillating magnetic field. In 1939 Rabi's group reported the observation of magnetic resonance of alkali molecules in a molecular beam. Rabi received the 1944 Nobel Prize in physics for his work on magnetic moments.

In 1946, observation of magnetic resonance in bulk matter was reported by Edward M. Purcell at Harvard University and by Felix Bloch at Stanford University. They shared the 1952 Nobel Prize in physics for this development.

Other major developments by physicists followed, but the next part of the story of how NMR changed chemistry is expressed succinctly in the review by Jonas and Gutowsky— "quoting from a lecture by M. E. Packard: 'Chemists got the point very quickly, thanked the physicists, and took over.'" Martin E. Packard, by the way, had worked with Bloch and gone on to lead Varian Associates of Palo Alto, California, to the first commercial production of NMR instruments.

In 1950, Herbert S. Gutowsky at the University of Illinois was able to resolve the resonances of nonequivalent fluorine-19 nuclei. With his first graduate student, Charles J. Hoffman, and an undergraduate engineering student, Robert E. McClure, shifts in proton resonances were observed in June 1950. This involved special efforts at improving the homogeneity of the magnetic fields, including hand-polishing magnet poles. Also in 1950, a paper by W. C. Dickinson and another by Warren Proctor and F. C. Yu reported that two nitrogen-14 resonances had been observed in ammonium nitrate, these being associated with the different chemical environments. The phenomenon of the chemical shift and the use of NMR for structural information about molecules were now in place.

Gutowsky and his students noticed multiplet structure in certain of their fluorine-19 spectra. They attempted to explain it as resulting from impurities or structural nonequivalences. However, by March 1951, they were convinced that the multiplets were something else. When they found a doublet in the phosphorus-31 spectrum of $POCl_2F$, Gutowksy realized that they were seeing an internuclear effect, the coupling of nuclear spins. Erwin L. Hahn and D. E. Maxwell had by then observed a beat pattern in a pulsed NMR experiment, and this was now shown to originate with the internuclear coupling. With a report by Gutowsky, David W. McCall, and Charles P. Slichter in 1953, the development of spin–spin coupling analysis for molecular structure elucidation was underway.

About 1950, Hahn's work with appropriately spaced pulses of radiofrequency radiation and a 1948 report by Nicolaas Bloembergen, Purcell, and R. V. Pound set the stage for the study of relaxation processes in NMR and for investigation of dynamic structure of liquids.

Significant advances in NMR sensitivity were made at the Eidgenössische Technische Hochschule in Zurich, Switzerland, by Richard R. Ernst and coworkers. In 1976, Ernst introduced 2D Fourier transform NMR which earned him the 1991 Nobel Prize in chemistry. Two-dimensional NMR has broadened the application of NMR by unraveling the coupling between nuclei in large, complicated molecules. It has been used for structural and conformational analysis of biological molecules and other macromolecules.

In 1973, Paul Lauterbur, then at the State University of New York at Stony Brook, proposed using NMR to generate spatial maps of spin distributions. Magnetic field gradients served to make signals spatially dependent, and thereby to produce magnetic resonance images (MRI). Coupled with Ernst's Fourier transform techniques, Lauterbur's idea and a parallel set of developments by Raymond Damadian have been the basis for the numerous commercial MRI instruments now in diagnostic use around the world.

Magnetic resonance developed rapidly in contrast to other spectroscopic techniques. Its impact on organic, inorganic, and biological chemistry has been enormous. It is notable that such impact was not the objective in the work that led to the discovery and evolution of NMR. It was a pursuit of basic knowledge, a search for answers to fundamental questions about the nature of nuclei, atoms, and molecules, that led to the development of this powerful tool.

Exercises

12.1 Assuming all nuclei are in their ground state, how many different nuclear spin states exist for the following isotopes of water: H_2O, D_2O, and $D_2{}^{17}O$?

12.2 Neglecting spin–spin coupling in the example of diimide, HNNH, of the 84 transitions, how many correspond to nitrogen spin flips and how many correspond to hydrogen spin flips?

12.3 Neglecting spin–spin coupling, develop an energy level diagram for the nuclear spin states of an isotopic form of acetylene, $DC^{13}CH$, in a magnetic field. See Appendix E for values of g_D, g_C, and g_H. Draw vertical arrows to indicate the allowed NMR transitions.

12.4 Show that $\left\langle m_{I_1} m_{I_2} \left| \hat{I}_{x1} \hat{I}_{x2} + \hat{I}_{y1} \hat{I}_{y2} \right| m_{I_1} m_{I_2} \right\rangle$ equals zero. (Use the definitions of angular momentum raising and lowering operators as in Appendix B to express the operator in terms of the raising and lowering operators. Then evaluate the integral using the properties of raising and lowering operators.)

12.5 Develop an idealized representation of the high-resolution NMR spectra predicted by first-order treatment of spin–spin interaction of the following molecules.

    a. $H_2C^{17}O$

    b. $H_2C={}^{13}CH_2$

12.6 What is the intensity pattern for a proton NMR transition split by interaction with two nearby equivalent ^{14}N nuclei?

12.7 If there are four equivalent spin $-1/2$ nuclei in a molecule and a nearby proton that has a spin–spin interaction with these nuclei, what is the intensity pattern for the NMR spectrum of the proton?

12.8 The NMR intensity-splitting pattern for a single magnetic nucleus is 1:1 with one nearby proton; it is 1:2:1 if there are two nearby equivalent protons, and 1:3:3:1 if there are three nearby equivalent protons. What would these patterns become if the protons were all replaced by deuterons?

12.9 Predict the form of the ^{13}C and the proton NMR spectra of $H-{}^{13}C\equiv{}^{13}C-{}^{13}C\equiv{}^{13}C-H$ on the basis of a first-order treatment of spin–spin coupling.

12.10 Predict the qualitative form of the high-field ESR spectra of and $H^{13}CO\bullet$ and $H^{13}C^{13}C\bullet$.

12.11 Predict the high-field ESR spectrum of the radical $H_3C-CH_2\bullet$ assuming (a) that the a_i coupling constants for CH_2 protons are much greater than for the CH_3 group protons, and (b) vice versa. Next, assume that the coupling constants are similar in size, although slightly different. Given that this is intermediate between the first two cases, what is the likely form of the spectrum?

12.12 Predict the qualitative form of the high-field ESR spectra of $HOO\bullet$, $H^{17}OO\bullet$, and $H^{17}O^{17}O\bullet$.

12.13 Consider the proton NMR spectra of the acetylenic hydrogens embedded in two identical larger molecules (e.g., the spectra of the $H-C\equiv C-$fragments) that are separated in space in by 1 cm. Make a plot of the separation between the two transition peaks for the two molecules under the condition of a magnetic field gradient along the line connecting the two molecular centers. (Assume a value for the average field strength and plot the separation in the absorption signals as a function of the size of the gradient of the field.)

Bibliography

Bovey, F. A., L. Jelinsky, and P. A. Mirau, *Nuclear Magnetic Resonance Spectroscopy* (Academic Press, New York, 1988).

Ernst, R. R., G. Bodenhausen, and A. Wokaun, *Principles of Nuclear Magnetic Resonance in One and Two Dimensions* (Oxford University Press, New York, 1987).

Keeler, J., *Understanding NMR Spectroscopy* (Wiley, West Sussex, England, U.K., 2005).

13

Introduction to Surface Chemistry

The interactions of chemical species along the interface between two phases are a very complicated subject that has wide reaching implications in chemistry, physics, materials science, and nanotechnology. The chemistry at a surface can be very different from that of the bulk phase and will always be affected by a change in either of the two interacting phases. In this chapter, we attempt to give a general introduction to this fascinating area of physical chemistry with a focus on equilibrium, kinetic, and thermodynamic processes.

13.1 Interfacial Layer and Surface Tension

In Chapter 4, we considered the thermodynamic properties of homogeneous phases with consistent properties through the entire phase. It should be obvious that this is not the case at the interface between two different phases of a system. The area of contact between two phases where the molecules of both phases are interacting is called the **interfacial layer**, which is usually considered to be a few molecules thick in a neutral species. If we are considering the bulk properties of the phase, the effect of this region on the properties of the phase can be considered to be vanishingly small.

The molecules at the surface of a liquid experience a different net attractive force than those of the bulk solution. The surface molecules tend to be attracted into the bulk liquid since the liquid's attractive force is stronger than the vapor layer above it. This phenomenon gives rise to a surface contraction (a "curving" of the surface) and a corresponding force at the surface of the liquid. This force is termed the **surface tension** of the liquid. Surface tension is responsible for the formation of liquid droplets, capillary rise, and many other physical phenomena.

Surface tension is strictly defined as the force per unit length that opposes surface area expansion. We can illustrate this definition with a simple experiment shown in Figure 13.1, where the movable bar is pulled with force F to expand a liquid film that is stretched like a soap-bubble film on a wire frame. The surface tension (γ) is

$$\gamma = \frac{F}{2L} \tag{13.1}$$

where L is the length of the bar. A factor of 2 is necessary to account for the two film surfaces. As can be seen from Equation 13.1, the units of surface tension are $N\ m^{-1}$.

To increase the area of the surface, work must be done on it. Recall the definition of work from basic physics is

$$w = F\Delta x \tag{13.2}$$

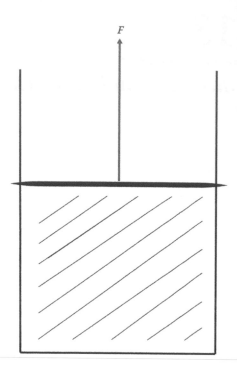

FIGURE 13.1
Diagram of an apparatus used to determine the surface tension of a liquid.

where Δx is the displacement from the initial conditions. Thus, the work required to increase the area of the film would be

$$w = F\Delta x = \gamma 2L\Delta x = 2\gamma\Delta A \qquad (13.3)$$

where ΔA is the change in the area of the film. So, as a general rule, the work done to increase the area of a surface is $\gamma\Delta A$. If work is given in joules and ΔA in square meters, we can see another valid unit for surface tension is $J\ m^{-2}$.

Measurement of surface tension can be carried out in a variety of ways. Two of the most common, simple, and relatively accurate methods of calculating surface tension are measuring the rise of liquid in a capillary (**capillary rise** method) and the pull on a thin vertical plate partially immersed in the liquid (**Wilhelmy plate** method).

In the capillary rise method, surface tension can draw liquid up the tube in a phenomenon known as capillary action. The height of the liquid column is given by

$$h = \frac{2\gamma\cos\theta}{dgr} \qquad (13.4)$$

where
   h is the height the liquid is lifted
   γ is the surface tension
   d is the density of the liquid
   g is the gravitational constant
   r is the radius of the capillary
   θ is the contact angle (the angle the tangent to the surface of the liquid makes with the
      capillary surface)

If θ is greater than 90°, as with mercury in a glass container, the liquid will be depressed rather than lifted.

The Wilhelmy plate method (Ludwig Ferdinand Wilhelmy [1812–1864]) can be used to measure surface or interfacial tension at an air–liquid or liquid–liquid interface. In this method, a thin plate is oriented perpendicular to the interface, and the force exerted on it is measured. The plate is often made from glass or platinum which may be roughened to ensure complete wetting. The plate is cleaned thoroughly and attached to a scale or balance via a thin metal wire. The force on the plate due to wetting is measured via a tensiometer or microbalance and used to calculate the surface tension (γ) using the Wilhelmy equation:

$$\gamma = \frac{F}{2L\cos\theta} \tag{13.5}$$

where
  L is the wetted length of the plate
  θ is the contact angle between the liquid phase and the plate

In normal use, the contact angle is rarely measured. Literature values can be used, or complete wetting ($\theta = 0°$) can be assumed.

Surface tension does have a dependence on temperature. As the temperature is raised, the liquid becomes more "vapor-like" until the critical temperature T_c is reached. At this point, the surface tension must be equal to 0. It should logically follow then that the surface tension of a liquid decreases as the temperature rises.

There are two useful empirical relationships that have been derived that attempt to quantify the surface tension–temperature relationship. The Eötvös relationship, named after the Hungarian physicist Loránd (Roland) Eötvös (1848–1919), has the form

$$\gamma \bar{V}^{2/3} = k_E(T_c - T) \tag{13.6}$$

where
  $\bar{V}$ is the molar volume of the liquid
  k_E is the Eötvös constant (2.1×10^{-7} J mol$^{2/3}$ K^{-1})

The density, molar mass, and the critical temperature of the liquid have to be known. At the critical point the surface tension is zero. This relationship assumes the surface tension is a linear function of the temperature, a reasonable assumption for most liquids. If the surface tension versus the temperature is plotted, a reasonably straight line will result which has a surface tension of zero at the critical temperature.

The Guggenheim-Katayama relationship has a slightly different form:

$$\gamma = \gamma^0 \left(1 - \frac{T}{T_c}\right)^n \tag{13.7}$$

where
  γ^0 is a constant for each different liquid
  n is an empirical factor

For organic liquids, n is taken to be 11/9. Both the Eötvös and Guggenheim-Katayama relationships take into account the fact that surface tension reaches 0 at the critical temperature, whereas other empirical solutions do not.

The effect of solute on the surface tension of water can be very different depending on their structure. Most nonpolar organic molecules have little or no effect. Sugar is a good example of this type of molecule. Inorganic salts tend to increase surface tension. Polar organic molecules such as alcohols, esters, ethers, and the like decrease surface tension as a function of their concentration in solution. Surfactants have a similar dependence on concentration, but they reach a minimum after which adding additional surfactant will have no effect. A complication of solutions is that the surface concentration of the solute can be very different than the concentration in the bulk solution. This difference will vary from one solute to another.

The **Gibbs isotherm** states that, in an ideal (very dilute) solution:

$$\Gamma = -\frac{1}{RT}\left(\frac{\partial \gamma}{\partial \ln C}\right)_{T,P} \tag{13.8}$$

Γ is known as surface concentration, which is the excess of solute per unit area of the surface over what would be present in the bulk solution. It has units of mol m^{-2}. C is the concentration of the substance in the bulk solution.

13.2 Adsorption and Desorption

Adsorption occurs on the surface of a solid because of the attraction of the adsorbate atoms or molecules for the solid surface. This has the effect of lowering the total potential energy of the system similar to bond formation in a molecule. The adsorbed molecules can be *mobile* where they maintain their translational degrees of freedom or *immobile* where all translational freedom is lost.

The adsorption of molecules on to a surface is a necessary prerequisite to any surface-mediated chemical process. The adsorption process is similar to a catalytic cycle and can be broken down into five basic steps:

1. Diffusion of reactants to the surface
2. Adsorption of molecules onto the surface
3. Reaction
4. Desorption of products from the surface
5. Diffusion of products from the surface

In this process, both adsorption and desorption are equally important. Both processes are considered in this section.

There are two principal modes of adsorption of molecules on surfaces: **physical adsorption (physisorption)** and **chemical adsorption (chemisorption)**. The basis of distinction between these two adsorption modes is the nature of the bonding between the molecule and the surface.

The forces causing physisorption are the same type as those that cause the condensation of a gas to form a liquid and are principally van der Waals forces. The heat evolved in the physisorption process is of the order of magnitude of the heat evolved in the process of condensing the gas, and the amount adsorbed may correspond to several monolayers. Physisorption can usually be easily reversed by lowering the pressure of the gas or the concentration of the solute, and the extent of physisorption is smaller at higher temperatures. There is normally no change in the electron density in either the adsorbate or at the substrate surface, which would correspond to a chemical bond being formed.

Chemisorption involves the formation of chemical bonds. It is therefore more specific than physisorption, but there are borderline cases in which it is not possible to make a sharp distinction between these two kinds of adsorption. Usually the enthalpy change in chemisorption is much larger than for physisorption, lying in the range of 40–200 kJ mol⁻¹. Surfaces of most adsorbents are not uniform. The energy of adsorption is different at various sites because of the arrangement of atoms and molecules in the surface. There is substantial change in the electron density of both the adsorbate and the substrate that corresponds to the chemical bond formation.

Physisorption and chemisorption may often be distinguished by the rates with which these processes occur. Equilibrium in physisorption is generally achieved rapidly and is readily reversible. Chemisorption, on the other hand, may not occur at an appreciable rate at low temperatures because the process has an activation energy. As the temperature is increased, the rate of chemisorption increases rapidly. In chemisorption, the bonding may be so tight that the original species may not be recovered. For example, when oxygen is adsorbed on graphite, heating the surface in a vacuum yields carbon monoxide.

The extent of adsorption depends greatly on the specific nature of the solid and of the molecules being adsorbed and is a function of pressure (or concentration) and temperature. If the gas being physically adsorbed is below its critical point, it is customary to plot the amount adsorbed per gram of adsorbent versus P/P_{sat}, where P_{sat} is the vapor pressure of the bulk liquid adsorbate at the temperature of the experiment. Such an adsorption isotherm is shown in Figure 13.2. The amount of gas adsorbed was determined by measuring

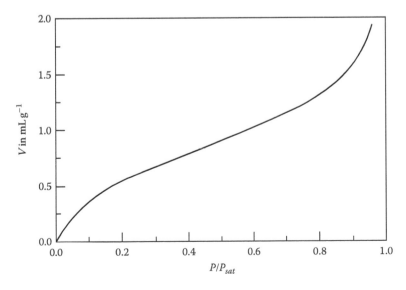

FIGURE 13.2
An idealized adsorption isotherm for a typical gas surface interaction.

the volume V of gas taken up by the adsorbent at various pressures and at a constant temperature of 89.9 K. As P/P_{sat} approaches unity, the amount adsorbed increased rapidly because at $P/P_{sat} = 1$, bulk condensation can occur. If the temperature of the experiment is above the critical temperature of the gas, P/P_{sat} cannot be calculated. When the adsorption of various gases is compared above their critical temperature, it is generally found that the smaller the amount adsorbed, the lower the critical temperature.

The rate of adsorption, R_{ads}, of a molecule onto a surface can be shown mathematically in the same form as other chemical kinetics problems we derived in Chapter 6. We can express this rate in terms of the partial pressure of the molecule in the gas phase above the surface:

$$R_{ads} = k_{ads}P^x \tag{13.9}$$

where
  x is the order of the reaction
  k_{ads} is the specific rate constant
  P is the partial pressure of the adsorbate gas

Using the Arrhenius form of the rate constant, we obtain a kinetic equation of the form:

$$R_{ads} = Ae^{-E_{ads}/RT}P^x \tag{13.10}$$

where, as before,
  E_{ads} is the activation energy for adsorption
  A is the pre-exponential (frequency) factor

The rate of adsorption will be governed by the kinetics of the diffusion of molecules to the surface and the fraction of these molecules which undergo adsorption. Therefore, we can express the rate of adsorption per unit area as a product of the flux of adsorbate molecules toward the surface (J) times the probability that the molecule will attach to the surface (A), that is,

$$R_{ads} = J \times A \tag{13.11}$$

The flux of incident molecules is given by the following equation:

$$J = \frac{P}{\sqrt{2\pi m k T}} \tag{13.12}$$

where
  m is the mass of one molecule
  k is Boltzmann's constant
  T is the temperature

The probability of attachment to the surface will always be a property of the adsorbate/ substrate system. It will depend on various factors, the most important being the amount of surface already covered by the adsorbate (θ) and the presence of any activation barrier to adsorption. Therefore, we can write

$$A = f(\varphi)e^{-E_{ads}/RT} \tag{13.13}$$

where $f(\varphi)$ is an undetermined function of the existing surface coverage of adsorbed species.

Combining the equations for A and J yields the following expression for the rate of adsorption:

$$R_{ads} = \frac{Pf(\varphi)}{\sqrt{2\pi mkT}} e^{-E_{ads}/RT} \tag{13.14}$$

This equation indicates the rate of adsorption to be first order with regard to the partial pressure of the molecule in the gas phase above the surface. It is not unrealistic to also consider that the activation energy for adsorption may itself be dependent on the surface coverage, that is, $E_{ads} = E_{ads}(\varphi)$. If we further assume that the probability of attachment is directly proportional to the concentration of vacant surface sites (a reasonable approximation for non-dissociative adsorption), then $f(\varphi)$ is proportional to the fraction of sites which are occupied.

Let us now consider both the energetics of adsorption and factors which influence the kinetics of adsorption by looking at the potential energy surface (PES) for the adsorption process. The energy will be dependent on the distance (d) of the adsorbate from a surface. This is a relatively simple model which neglects many other parameters which influence the energy of the system including the orientation of the molecule, changes in the bond angles and bond lengths of the molecule, and the position of the molecule parallel to the surface plane.

The interaction of a molecule with a given surface will also be dependent on the presence of any existing adsorbed species, whether these be surface impurities or simply pre-adsorbed molecules of the same type. As a first approximation, let us consider the interaction of an isolated molecule with a clean surface. In a limiting case of pure physisorption, the only attraction between the adsorbing species and the surface arises from van der Waals forces. These forces give rise to a shallow minimum in the PES at a relatively large distance from the surface before the repulsion from electron density overlap begins to cause a rapid increase in the total energy. There is no activation energy for this process which would prevent the atom or molecule which is approaching the surface from entering the potential well. The process is barrierless and the kinetics of physisorption are fast. Weak physisorption and associated long-range attraction will be present to in all surface-mediated systems. Where chemical bond formation between the adsorbate and substrate can occur, however, the important part of the PES is a much deeper potential well at smaller values of d. Figure 13.3 shows the PES due to physisorption and chemisorption separately. In reality, the PES for any real molecule capable of undergoing chemisorption is best described by a linear combination of the two curves, with the curves intersecting at the point at which chemisorption forces begin to dominate over those arising from physisorption alone.

The depth of the chemisorption potential well is obviously directly proportional to the strength of surface bonding and thus is a direct representation of the energy of adsorption.

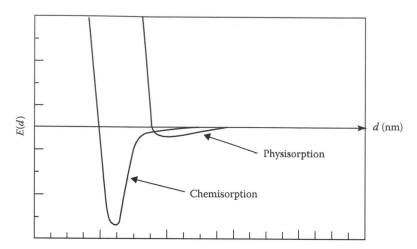

FIGURE 13.3
Comparison of the chemisorption and physisorption processes.

The location of the global minimum on the horizontal axis is the equilibrium bond distance (r_e) for the adsorbed molecule on this surface. Adsorption is considered to be an exergonic (or exothermic) process and is thus represented as ΔE_{ads} or ΔH_{ads}.

If the temperature of a surface is low enough, species could remain adsorbed indefinitely. As the temperature of the surface is increased, however, there will come a point at which the thermal energy of the adsorbed species is so high that several things could occur: The chemical species could decompose, an adsorbate could react with the substrate to yield a specific surface compound, or the species could diffuse into the bulk of the underlying solid. Alternatively, the species may desorb from the surface and return into the gas phase. The last of these options is the desorption process. In the absence of decomposition, the desorbing species will generally be the same as that originally adsorbed.

The rate of desorption, R_{des}, of an adsorbate from a surface can be expressed in the general form

$$R_{des} = k_{des}C^x \tag{13.15}$$

where
   x is kinetic order of desorption
   k_{des} is the rate constant for the desorption process
   C is the surface concentration of adsorbed species

The order of desorption can usually be predicted because we are concerned with an elementary step process. Specifically, atomic or simple molecular desorption ($X(ads) \rightarrow X(g)$) will usually be a first-order process. Recombinative molecular desorption ($2X(ads) \rightarrow X_2(g)$) will usually be a second-order process.

The rate constant for the desorption process can be expressed in an Arrhenius form,

$$k_{des} = Ae^{-E_{des}/RT} \tag{13.16}$$

where E_{des} is the activation energy for desorption. This gives the following general expression for the rate of desorption:

$$R_{des} = -\frac{dC}{dt} = Ae^{-E_{des}/RT}C^x \qquad (13.17)$$

In the particular case of simple molecular adsorption, the pre-exponential factor A can be considered the frequency of vibration of the bond between the molecule and substrate, that is, a "frequency factor." Every time this vibration occurs is essentially the molecule attempting to desorb from the surface.

13.3 Langmuir Theory of Adsorption

A gas in contact with a solid will establish an equilibrium between the molecules in the gas phase and the corresponding adsorbed species (molecules or atoms) which are bound to the surface of the solid. The position of equilibrium will depend on a number of factors such as the relative stabilities of the adsorbed and gas phase species involved, the temperature of the system, and the pressure of the gas above the surface.

The **Langmuir isotherm** (sometimes called the Langmuir Equation or Hill-Langmuir Equation) was developed by Irving Langmuir in 1916 to quantify the dependence of the surface coverage of an adsorbed gas on the pressure of the gas above the surface at a fixed temperature. The Langmuir isotherm is one of the simplest theories that provides a useful insight into the pressure dependence of the extent of surface coverage.

Langmuir considered the surface of a solid to be made up of elementary sites each of which could adsorb one gas molecule. He assumed that all the elementary sites are identical in their affinity for a gas molecule and that the presence of a gas molecule on one site does not affect the properties of neighboring sites.

If θ is the fraction of the surface occupied by gas molecules, the rate of evaporation from the surface is $r \times \theta$, where r is the rate of evaporation from the completely covered surface at a certain temperature. The rate of adsorption of molecules on the surface is proportional to the fraction of the area that is not covered $(1 - \theta)$ and to the pressure of the gas. Thus, the rate of condensation is expressed as $k_{Lan} P(1 - \theta)$, where k_{Lan} is a constant at a given temperature and includes a factor to allow for the fact that not every gas molecule that strikes an unoccupied space will stick.

At equilibrium the rate of evaporation of the adsorbed gas is equal to the rate of condensation:

$$r \times \theta = k_{Lan}P(1-\theta) \rightarrow \theta = \frac{k_{Lan}P}{r + k_{Lan}P} = \frac{K_{Lan}P}{1 + K_{Lan}P} \qquad (13.18)$$

where $K_{Lan} = k_{Lan}/r$. Since the volume V of gas adsorbed is proportional to θ, Equation 13.18 may be written as

$$V = \frac{V_{mon}}{1 + 1/K_{Lan}P} \qquad (13.19)$$

where V_{mon} is the volume of gas adsorbed in a complete monolayer. Thus, V is directly proportional to P at very low pressures where $K_{Lan}P \lll 1$. As the pressure is increased, the volume adsorbed increases and approaches the value V_{mon} asymptotically.

If we rearrange Equation 13.19, it can be seen that a linear relationship develops between $1/V$ and $1/P$.

$$\frac{1}{V} = \frac{1}{V_{mon}} + \frac{1}{V_{mon}K_{Lan}P} \qquad (13.20)$$

Thus, $1/V$ versus $1/P$ can be plotted to determine the constant K_{Lan} and V_{mon}.

Data such as that of Figure 13.2 do not show asymptotic saturation and do not give a linear plot except at low pressures. Adsorption on solids is more complicated than the Langmuir theory indicates.

The derivation of the Langmuir adsorption isotherm involves five assumptions: (1) the adsorbed gas behaves ideally in the vapor phase, (2) the adsorbed gas is confined to a monolayer, (3) the affinity of each binding site for gas molecules is the same, (4) there is no side-to-side interaction between adsorbate molecules, and (5) the adsorbed gas molecules do not move around on the surface. The first assumption is reasonable at low pressure but the second one nearly always breaks down as the pressure of the adsorbed gas is increased. The third assumption is unrealistic since most surfaces are heterogeneous; the affinity for gas molecules will be different on inequivalent crystalline faces, and different kinds of binding sites are introduced by edges, cracks, and imperfections. This heterogeneity leads to a decrease in binding energy as the surface coverage increases. The incorrectness of the fourth assumption was first shown experimentally when it was found that in certain cases the heat of adsorption may increase with the surface concentration of adsorbed molecules. This effect, which is the opposite of that expected to result from surface heterogeneity, is caused by lateral attractions of adsorbed molecules. The fifth assumption is incorrect because there are several kinds of evidence that adsorbed molecules may be mobile.

The surface area occupied by a single molecule of adsorbate on the surface may be estimated from the density of the liquefied adsorbate. For example, the area occupied by a nitrogen molecule at $-195°C$ is estimated to be $16.2\,Å^2$ on the assumption that the molecules are spherical and that they are close packed in the liquid. Thus, from the measured value of V_{mon}, the surface area of the adsorbent may be calculated. This method is widely used in the determination of the surface area of solid catalysts and adsorbents. The area values determined in this way seem, in general, to be perfectly satisfactory in spite of the rough approximations of the theory.

13.4 Temperature and Pressure Effects on Surfaces

The assumptions and simplifications of the Langmuir isotherm in the previous section lead directly to expressions for the pressure dependence of the surface coverage. In the case of a simple, reversible molecular adsorption process, the expression is, as shown in (13.18),

$$\theta = \frac{K_{Lan}P}{1 + K_{Lan}P} \qquad (13.21)$$

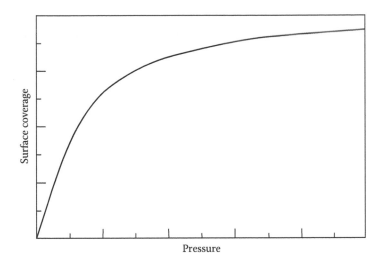

FIGURE 13.4
Typical relationship between the surface coverage and the pressure of an adsorbed gas.

where K_{Lan} is a function of temperature. This is illustrated in Figure 13.4 which shows the characteristic Langmuir variation of coverage with pressure for molecular adsorption.

At a given pressure, the extent of surface coverage is determined by the value of K_{Lan}, which in turn is dependent on both the temperature and the enthalpy of adsorption. The magnitude of the adsorption enthalpy is directly proportional to the strength of binding of the adsorbate to the surface. The value of K_{Lan} can be increased by a reduction in the system temperature and/or an increase in the strength of adsorption.

13.5 Surface Characterization Techniques

Surface spectroscopy is the most common tool used to characterize and analyze surfaces. Assuming that a technique of sufficient sensitivity can be found, another major problem that needs to be addressed in surface spectroscopy is distinguishing between signals from the surface and the bulk of the sample. It is important that the spectroscopic technique chosen be *surface specific*. The technique must be able to distinguish between signals from the bulk and the surface phase. To illustrate this, we shall look at one way in which surface specificity can be achieved that makes use of the special properties of low energy electrons. It is an approach employed in common surface spectroscopic techniques such as Auger electron spectroscopy (AES) and x-ray photoelectron spectroscopy (XPS).

Most analytical techniques used in chemistry measure all the atoms within a typical sample. By contrast, a surface-specific technique is more sensitive to those atoms which are located near the surface than it is to atoms in the bulk which are well away from the surface. Electron spectroscopic techniques are **not** completely surface specific.

Most of the signal comes from within a few atomic layers of the surface but a small part comes from much deeper into the material. AES and XPS are really surface-sensitive techniques.

The x-rays employed in these techniques penetrate on the order of micrometer into the surface of a solid. They penetrate through many layers at the surface and give no surface sensitivity. The surface sensitivity must arise from the emission and detection of the emitted electrons. Consider a hypothetical situation in which electrons of a given energy, E_0, are emitted from atoms in a solid at various depths, d, below a flat surface. Only those electrons which reach the surface and still have their initial energy (E_0) are detected in this ideal situation. When performing an XPS experiment, it is only those photoelectrons possessing characteristic emission energies and contributing to the peaks and that are considered in the analysis.

Under normal circumstances, three phenomena can prevent an emitted electron from being detected. The electron could react with another molecule before reaching the surface, be emitted in the wrong direction and never reach the surface, or lose enough energy before reaching the surface so that it reaches the detector with $E < E_0$. The process by which an electron can lose energy as it travels through the solid is known as **inelastic scattering**. Each inelastic scattering event will lead to a reduction in the energy of the electron and a change in the direction of travel (a change in momentum).

To solve this problem, consider only the electrons emitted normal to the surface. The probability of escape from a given depth, $P(d)$, is determined by the likelihood of the electron **not** being inelastically scattered before being emitted from the surface. The probability will decrease the deeper into the sample from which the electron is emitted. In order to quantify this, the inelastic scattering process must be examined in detail.

Let us now define the inelastic mean free path (IMFP) of electrons. The IMFP is defined as the average distance traveled by an electron through a solid before it is inelastically scattered. Clearly the IMFP will be dependent on the initial kinetic energy of the electron and the nature of the solid. The IMFP is mathematically defined by Equation 13.22, which gives the probability of the electron traveling a distance, d, through the solid without undergoing scattering.

$$P(d) = e^{-d/\lambda} \tag{13.22}$$

where λ is the IMFP for the electrons of energy E. Figure 13.5 illustrates this functional form of $P(d)$. The probability of escape decays very quickly and is, for all intents and purposes, zero for a distance $d > 5\lambda$.

Let us now consider many sources of electrons uniformly distributed at all distances from the surface of the solid and detect those unscattered electrons which emerge normal to the surface. What sort of distribution of the depths of these electrons can be detected? This new function, $P'(d)$, will have the same exponential form as $P(d)$ since the detection of electrons from different depths in the solid is directly proportional to the probability of electron escape from each depth. What percentage of electrons will have come from within a distance of one IMFP from the surface? Recall from our definition of probability that this is simply the integral between the limits of 0 and 1 in the exponential function divided by the integral over all space

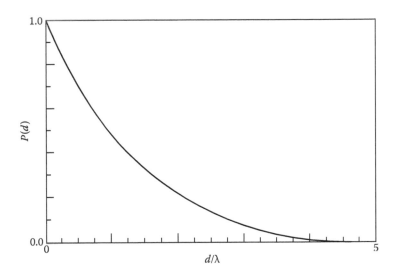

FIGURE 13.5
A plot of the probability of escape from a surface as a function of the surface depth.

$$P'(d) = \frac{\int_0^1 e^{-x}dx}{\int_0^\infty e^{-x}dx} = \frac{0.632}{1} = 63.2\%$$

where $x = d/\lambda$.

In summary, we can conclude from this answer that the majority of electrons detected come from within one IMFP of the surface. Similarly, it can be shown that >95% of the electrons detected come from within three IMFPs of the surface.

13.6 Conclusions

Atoms and molecules on the surface of a solid or on the interface between two phases experience different forces and will have different energies than those in the bulk material. This can lead to phenomena such as surface tension, which can be measured using several different methods.

Adsorption of one substance onto another can be categorized being a chemical (chemisorption) or physical (physisorption) process. Langmuir (see the following text) developed a way to describe the adsorption process that can be extended to describe the temperature and pressure effect on a surface as well.

AES is a surface-sensitive spectroscopic technique used for elemental analysis of surfaces. The technique offers high sensitivity for all elements except H and He, monitoring of the surface cleanliness of samples, and compositional analysis of the surface region of specimens, by comparison with standard samples of known composition.

XPS is a companion technique to Auger spectroscopy and can be used for quantitative measurement of the surface environment.

Point of Interest: Irving Langmuir

Irving Langmuir

Irving Langmuir was born in Brooklyn, New York, on January 31, 1881 and died on August 16, 1957. Langmuir earned his PhD degree in 1906 under Nobel laureate Walther Nernst in Göttingen, for research done using the "Nernst glower," an electric lamp invented by Nernst. His doctoral thesis was entitled "On the Partial Recombination of Dissolved Gases During Cooling." Langmuir then taught at Stevens Institute of Technology in Hoboken, New Jersey, until 1909, when he began working at the General Electric research laboratory (Schenectady, New York).

His initial contributions to science came from his study of light bulbs (a continuation of his PhD work). His first major development was the improvement of the diffusion pump, which ultimately led to the invention of the high-vacuum tube. A year later, he and Lewi Tonks discovered that the lifetime of a tungsten filament was greatly lengthened by filling the bulb with an inert gas, such as argon. He also discovered that twisting the filament into a tight coil improved its efficiency. These were important developments in the history of the incandescent light bulb. His work in surface chemistry began at this point, when he discovered that molecular hydrogen introduced into a tungsten-filament bulb dissociated into atomic hydrogen and formed a layer one atom thick on the surface of the bulb.

In 1917, he published his seminal paper on the chemistry of oil films that later became the basis for the award of the 1932 Nobel Prize in chemistry. He was the first industrial chemist to become a Nobel laureate.

Langmuir theorized that oils consisting of an aliphatic chain with a hydrophilic end group were oriented as a film one molecule thick upon the surface of water, with the hydrophilic group immersed in the water and the hydrophobic chains aggregated together on the surface. The thickness of the film could be determined from the known volume and area of the oil, which allowed investigation of the molecular configuration before spectroscopic techniques were available.

One of Langmuir's most noted publications was the famous 1919 article "The Arrangement of Electrons in Atoms and Molecules" in which, building on Gilbert N. Lewis's cubical atom theory and Walther Kossel's chemical bonding theory, he outlined his "concentric theory of atomic structure." Langmuir became embroiled in a priority dispute with Lewis over this work; Langmuir's presentation skills were largely responsible for the popularization of the theory, although the credit for the theory itself belongs mostly to Lewis.

As he continued to study filaments in vacuum and different gas environments, he began to study the emission of charged particles from hot filaments (thermionic emission). He was one of the first scientists to work with plasmas and was the first to call these ionized gases by that name, because they reminded him of blood plasma. Langmuir and Tonks discovered electron density waves in plasmas that are now known as Langmuir waves.

He introduced the concept of electron temperature and in 1924 invented the diagnostic method for measuring both temperature and density with an electrostatic probe, now called a Langmuir probe and commonly used in plasma physics. He also discovered atomic hydrogen, which he put to use by inventing the atomic hydrogen welding process, the first plasma weld ever made. Plasma welding has since been developed into gas tungsten arc welding.

He joined Katharine B. Blodgett to study thin films and surface adsorption. They introduced the concept of a monolayer (a layer of material one molecule thick) and the two-dimensional physics which describe such a surface.

The Langmuir Laboratory for Atmospheric Research near Socorro, New Mexico, was named in his honor as was the American Chemical Society journal for surface science, called *Langmuir*.

Exercises

13.1 Compare the surface area of a silver sphere that has a volume of 5.0 mL to a colloidal dispersion of silver with the same volume but each particle has a radius of 200.0 Å.

13.2 The volume of argon gas V_{mon} (measured at STP) required to form a complete monolayer on a metal surface is 134.5 cm^3 g^{-1}. Calculate the surface area per gram of the metal if each argon atom occupies 1.58×10^{-20} m^2.

13.3 Calculate the amount of work necessary to increase the surface area of water from 10.0 to 15.0 cm^2 at 25.0°C. The surface tension of water at this temperature is 72 dynes cm^{-1}.

13.4 The surface tension of methanol is 22.70 mN M^{-1} at 20°C. Estimate the surface tension at 45°C. Methanol has a critical temperature of 512.6 K.

13.5 Mercury does not wet glass and thus has a capillary depression instead of rise. Find this depression for mercury in a capillary tube with a diameter of 0.250 mm at 0°C.

13.6 Find the surface tension of an unknown organic liquid at 25°C if the liquid rises 5.0 cm in a capillary tube of diameter 0.20 mm. The density of the liquid is 0.750 g mL^{-1} and the contact angle is measured to be 15.0°.

13.7 An unknown liquid with density $0.850\,g\,mL^{-1}$ is placed in two capillary tubes of diameter 1.00 and 0.750 mm, respectively. The difference in capillary rise between the tubes is measured to be 1.25 cm. Calculate the surface tension of the liquid assuming zero contact angle.

13.8 Is there any other way to plot the Langmuir isotherm to yield a straight line?

13.9 One of Benjamin Franklin's experiments was to pour olive oil (oleic acid $CH_3(CH_2)_7CH=CH(CH_2)_7COOH$, density = $0.895\,g\,mL^{-1}$) on the surface of a pond in London and observe the oil forming a surface film. If the pond was 0.50 ac in area and 1.0 teaspoons of oil were used, calculate the area of each oleic acid molecule assuming a monolayer film was formed.

13.10 A certain material has a surface energy of $0.50\,J\,m^{-2}$. What is the total surface energy of a block of this material that measures $40.0\,cm \times 65.0\,cm \times 24.0\,cm$?

13.11 A cube of copper is divided into 27 equal smaller cubes. What is the ratio of the total surface energy of the smaller cubes to the surface energy of the original cube?

13.12 A surfactant molecule is usually represented with a hydrophilic end attached to a hydrophobic hydrocarbon "tail." Draw a sketch of a surfactant and show how it would act on a drop of oil placed in water.

13.13 The evaporation rate of water can be affected by placing a monolayer of an organic molecule such as dodecanol on the surface of the water. Explain.

13.14 Calculate the percentage of electrons that come from within a distance of three IMFP's from the surface.

Bibliography

Adamson, A. W. and A. P. Gast, *Physical Chemistry of Surfaces*, 6th edn. (Wiley Interscience, New York, 1997).
Attard, G. and C. Barnes, *Surfaces* (Oxford University Press, Oxford, U.K., 1998).
Hudson, J. B. *Surface Science: An Introduction* (Wiley Interscience, New York, 1998).

Appendix A: Mathematical Background

A.1 Power Series

A very useful mathematical tool for many problems in physical chemistry is the power series expansion of a function. For some value that depends on one or more variables, it is often important to know the linear dependence, quadratic dependence, and so on for the variables, and these are precisely the items on which a power series expansion is based.

Consider a function $f(x)$. The linear dependence of f on x at some particular point corresponds to the straight line that most closely matches the curve $f(x)$ at the particular point. That straight line is the one that is tangent to $f(x)$ at the particular point, and basic calculus tells us that the slope of that tangent line is the first derivative of f. Let $x = c$ be a particular point of interest. The first derivative of $f(x)$ is $f'(x) = df(x)/dx$, and its value at $x = c$ is $f'(c)$. We can estimate the value of f at some point near the point $x = c$ by following the tangent line. If the nearby point is $c + \delta$, then, as illustrated in Figure A.1,

$$f(c+\delta) = f(c) + \delta f'(c) \tag{A.1}$$

As δ increases, the approximation in Equation A.1 tends to be poorer. In the function shown in Figure A.1, the function's curvature causes it to deviate more and more from the tangent line as one moves further and further away from $x = c$.

An improved approximation to $f(x)$ is

$$f(x) = f(c) + \delta f'(c) + \frac{1}{2}\delta^2 f''(C) \tag{A.2}$$

Notice how this works for the following two special functions using $c = 2$ as the point of interest.

$$f(x) = x^2 \qquad\qquad g(x) = x^3$$

$$f(2) = 4 \qquad\qquad g(2) = 8$$

$$f'(x) = 2x \qquad\qquad g'(x) = 3x^2$$

$$f'(2) = 4 \qquad\qquad g'(c) = 12$$

$$f''(x) = 2 \qquad\qquad g''(x) = 6x$$

$$f''(2) = 2 \qquad\qquad g''(2) = 12$$

415

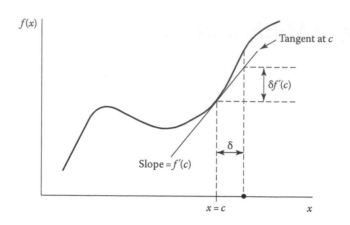

FIGURE A.1
The tangent to some function $f(x)$ at the point $x = c$ shows how f changes linearly with c. The slope of the tangent line times some incremental change in x is an estimate of the value of f at the new point. As the increment, d, increases in size, the tangent line tends to be less accurate as an estimator of the value for the function displayed.

δ is the change from the point c (i.e., 2), and it can be expressed as $\delta = x - c = x - 2$. So, applying Equation A.2 to f and then to g yields

$$f(c) + \delta f'(c) + \frac{1}{2}\delta^2 f''(C) = 4 + (x-2)(4) + \frac{1}{2}(x-2)^2(2) = x^2$$

$$g(c) + \delta g'(c) + \frac{1}{2}\delta^2 g''(C) = 8 + (x-2)(12) + \frac{1}{2}(x-2)^2(12) = 6x^2 - 12x + 8$$

Equation A.2 applied to $f(x)$ yields $f(x)$, and this is because f is quadratic in x. In contrast, Equation A.2 applied to $g(x)$ yields an expression that is quadratic in x and is only an approximation to $g(x)$. If Equation A.2 were continued to another term, a cubic term, then the expansion would yield $g(x)$ exactly.

Expressing a function as a polynomial in the variable(s) of the function constitutes a **power series** form of the function. Power series expansions are not usually applied to simple polynomials, for they already are power series. A power series expansion of $\sin(x)$, or $\exp(x)$, however, is a way of expressing those functions in terms of a polynomial in x. The order of the expansion determines the accuracy, with an infinite expansion being precisely equivalent to the expanded function. Sometimes, the function that we seek to expand is unknown, but its low-order derivatives may be known. In that situation, a truncated power series expansion provides an approximate form of the unknown function; one can find an approximation without having to know the function!

Extending Equation A.2 to all orders of dependence on x yields a very general result. This general expression for a power series expansion of some function of one variable, $f(x)$, about a particular point, $x = c$, is

$$f(x) = f(c) + (x-c)\frac{df}{dx}\bigg|_{x=c} + \frac{1}{2}(x-c)^2\frac{d^2 f}{dx^2}\bigg|_{x=c} + \frac{1}{3!}(x-c)^3\frac{d^3 f}{dx^3}\bigg|_{x=c} + \cdots \qquad \text{(A.3)}$$

For example, if $f(x) = \exp(-5x)$, evaluation of the first derivatives at some point yields the coefficients needed for the power series expansion. Here is how this works when the expansion point c is chosen to be zero.

$$f(c) = e^{-5(0)} = 1$$

$$\left.\frac{df}{dx}\right|_{x=c} = -5e^{-5(0)} = -5$$

$$\left.\frac{d^2 f}{dx^2}\right|_{x=c} = 25e^{-5(0)} = 25$$

The derivatives in Equation A.3 are simple coefficients of powers of x:

$$f(x) = 1 - 5x + \frac{25}{2}x^2 - \frac{125}{6}x^3 + \frac{625}{24}x^4 + \cdots$$

The exponential is mathematically equivalent to the infinite order polynomial on the right-hand side. An advantage of the power series expansion is that it can be truncated. If we are interested in the values of $f(x)$ for x within 0.01 of the expansion point (the chosen value of c), we can quickly see that the terms in the expansion become less important with increasing order. Go to the limit of this range, $x = c + 0.01 = 0.01$, and the term that is linear in x contributes -0.05 to the value of the function. The term that is quadratic in x will contribute $+0.00125$, and the next term is about -0.00002. Clearly, for this range of the function a good approximation is $f(x) \cong 1 - 5x$. Including the next term in the series is a better approximation, or an approximation suited to a greater range in x values, and so on. The approximate form of the function, $1 - 5x$, is certainly simpler than the exponential of the true function.

For functions of several variables the general power series expansion simply requires an expansion point (i.e., specific values for all variables) and partial derivatives.

$$f(x,y) = f(a,b) + (x-a)\left.\frac{\partial f}{\partial x}\right|_{x=a,y=b} + (y-b)\left.\frac{\partial f}{\partial y}\right|_{x=a,y=b}$$

$$+ \frac{1}{2}(x-a)^2 \left.\frac{\partial^2 f}{\partial x^2}\right|_{x=a,y=b} + \frac{1}{2}(y-b)^2 \left.\frac{\partial^2 f}{\partial y^2}\right|_{x=a,y=b}$$

$$+ (x-a)(y-b)\left.\frac{\partial^2 f}{\partial x \partial y}\right|_{x=a,y=b} + \cdots \qquad (A.4)$$

Power series expansions are so useful in physical chemistry that one may be tempted to suggest, "If in doubt, carry out a power series expansion!"

A.2 Curve Fitting

Often, a mathematical relationship between an observable and a variable does not emerge directly from an experiment. It is likely that the experiment yields a set of data points, (x,y), where x is the variable, a controllable quantity set by the experimenter, and y is the observed or measured quantity. The problem of finding a functional relation between the observable and the variable, that is, finding $y(x)$, is the problem of curve fitting.

We must realize at the outset that the data, the set of (x, y) points, are not likely to be precise. There may be measurement uncertainty that limits the precision of the values of x and y. We must realize that the number of data points is finite, and so that means a minor feature (a bump or wiggle) that occurs for the true $y(x)$ between two data points may not be revealed by the data. We won't know that it is there, and likewise we can't know that it is not there. Furthermore, the data points themselves do not identify the actual functional form. For instance, there are many different functions that could pass through a set of three data points, perhaps $\sin(x)$ or a parabola or an exponential. Thus, the task of curve fitting is nothing more than finding some analytical function that represents fairly well the set of data points whether or not there is a fundamental reason for the form of the function.

One of the most standard approaches for curve fitting is the **least squares method**. The idea behind this method is that we choose a function with a number of embedded parameters. The function tends to "hit and miss" the data points, but we can adjust the embedded parameters to make it hit more than it misses. In the least squares method, the squares of the deviations of the data points from the curve given by our chosen function are summed. This value provides a measure of how well the function represents the data. It is this value that we seek to minimize through adjustment of the embedded parameters. In the end we will have found the parameter choices that give the least squares error.

Let us assume that there are N data points: (x_1, y_1), (x_2, y_2),..., (x_N, y_N). We seek a function of x that closely represents the data. Let $f(x)$ be a guess of the function. It should be a function that has one or more adjustable parameters that constitute the set $\{c_i\}$. For any particular data point, the error in the fit is the difference between y and $f(x)$ for that point. If we were to sum the errors at each of the points, we would have a poor measure of the quality of the guess function. There could be one point where $f(x)$ is much greater than the y value, and another point where $f(x)$ is much less than the y value. The sum of these two errors might be zero, and that would not reflect the fact that $f(x)$ is not representing these two points well. For this reason, it is the square of the errors (always positively valued) at each point that should be used to measure the error. The total of the squared errors shall be designated E, and so

$$E = \sum_{i=1}^{N} \left[f(x_i) - y_i \right]^2 \tag{A.5}$$

Least squares fitting means finding the choice of parameters such that E is minimized. Clearly, E is a function of the parameters; as they are changed, E changes. When a function is at a minimum value, its first derivative (or first derivatives for multivariable functions) is zero. Recall that for a single variable function, the first derivative gives the slope of a line tangent to the curve of the function. At a minimum in the function, the curve is uphill in either direction, and so the tangent line must be flat. The slope is zero, and the first derivative is zero.

Differentiation of Equation A.5 with respect to one of the adjustable parameters yields

$$\frac{\partial E}{\partial c_j} = \sum_{i=1}^{N} 2 \left[f(x_i) - y_i \right] \frac{\partial f(x_i)}{\partial c_j} \tag{A.6}$$

At the choice of c_j parameters that gives the minimum value of E [i.e., the best $f(x)$], all the derivatives represented by Equation A.6 must be zero. Setting each equal to zero establishes a set of coupled equations that determine the optimum parameter choices. For large data sets, solution of the coupled equations is conveniently done with computer programs.

If we do not know the functional form that some set of data should obey, it is always possible to try a power series. We presume that the true function, if known, could be expanded in a power series via the various derivatives of the function at the point in the expansion. Since we don't know the function, we don't know the derivative values, but we can use curve fitting to find those values. That is, we express the function in terms of unknown parameters, c_0, c_1, and so on.

$$f(x) = c_0 + c_1 x + c_2 x^2 + c_3 x^3 + \cdots \tag{A.7}$$

This needs to be truncated at some point, and thus, we may fit the data to a quadratic function if we truncate after the c_2 term, or to a cubic function, or a quartic function, and so on. For this type of function, the adjustable parameters, c_0, c_1, c_2, and so on, enter the function linearly. This makes the first derivative expressions rather simple.

$$\frac{\partial f}{\partial c_0} = 1$$

$$\frac{\partial f}{\partial c_1} = x$$

$$\frac{\partial f}{\partial c_2} = x^2 \text{ and so on}$$

As an example, consider a quadratic fit of some set of data. Using the preceding first derivative expressions in Equation A.6 gives

$$\frac{\partial E}{\partial c_0} = \sum_{i=1}^{N} 2 \left[f(x_i) - y_i \right] = 2 \sum_{i=1}^{N} \left(c_0 + c_1 x_i + c_2 x_i^2 - y_i \right) \tag{A.8}$$

$$\frac{\partial E}{\partial c_1} = \sum_{i=1}^{N} 2 \left[f(x_i) - y_i \right] x_i = 2 \sum_{i=1}^{N} \left(c_0 x_i + c_1 x_i^2 + c_2 x_i^3 - x_i y_i \right) \tag{A.9}$$

$$\frac{\partial E}{\partial c_2} = \sum_{i=1}^{N} 2 \left[f(x_i) - y_i \right] x_i^2 = 2 \sum_{i=1}^{N} \left(c_0 x_i^2 + c_1 x_i^3 + c_2 x_i^4 - x_i^2 y_i \right) \tag{A.10}$$

Notice that the summations in these last three equations can be separated into summations over data values.

$$\sum_{i=1}^{N} \left(c_0 + c_1 x_i + c_2 x_i^2 - y_i \right) = N c_0 + c_1 \sum_{i=1}^{N} x_i + c_2 \sum_{i=1}^{N} x_i^2 - \sum_{i=1}^{N} y_i$$

It is now convenient to define total values of certain x–y products as

$$\overline{x^n y^m} \equiv \sum_{i=1}^{N} x_i^n y_i^m \tag{A.11}$$

These are simple values or numbers that can be obtained from the set of data points. Thus, Equation A.8 becomes

$$\frac{\partial E}{\partial c_0} = 2\left\{ Nc_0 + c_1\overline{x} + c_2\overline{x^2} - \overline{y} \right\}$$

Finally, setting each of the three derivatives equal to zero, we arrive at three coupled equations.

$$Nc_0 + c_1\overline{x} + c_2\overline{x^2} - \overline{y} = 0$$

$$c_0\overline{x} + c_1\overline{x^2} + c_2\overline{x^3} - \overline{xy} = 0$$

$$c_0\overline{x^2} + c_1\overline{x^3} + c_2\overline{x^4} - \overline{x^2 y} = 0$$

These are three linear equations in c_0, c_1, and c_2, and their solution gives the optimum parameter values. With this procedure, one can fit functions with more linear parameters, but then the number of coupled equations is increased.

A.3 Multivariable Partial Differentiation

For any well-behaved function of several variables, the differential of the function can be expressed in terms of the partial derivatives of the function. For instance, if $f = f(x, y, z, \ldots)$,

$$df = \frac{\partial f}{\partial x} dx + \frac{\partial f}{\partial y} dy + \frac{\partial f}{\partial z} dz + \cdots \tag{A.12}$$

The partial derivatives are obtained as if they were total derivatives of a function of one variable, the other variables being treated as if they were constant. For example,
  if

$$f(x, y) = ax^2 y + by^2 e^{-x^2}$$

then

$$\frac{\partial f}{\partial x} = 2axy - 2bxy^2 e^{-x^2}$$

and

$$\frac{\partial f}{\partial y} = 2ax^2 + 2bye^{-x^2}$$

Often, a subscript is used with a partial derivative to emphasize that one or another variable is taken as constant. So, for $f = f(x,y)$, we write

$$\frac{\partial f}{\partial x} \quad \text{or} \quad \left(\frac{\partial f}{\partial x}\right)_y$$

Both mean the same thing.

The differential, df, in Equation A.12 is the incremental change in the value of the function for an incremental change in the x variable and an incremental change in the y variable (i.e., dx and dy). The differential is said to be an **exact differential** if it is a property of the function that the second cross partial derivatives are equal. That is, if for $f = f(x,y)$,

$$\left[\frac{\partial}{\partial x}\left(\frac{\partial f}{\partial y}\right)_x\right]_y = \left[\frac{\partial}{\partial y}\left(\frac{\partial f}{\partial x}\right)_y\right]_x$$

then df is exact. The significance of a differential being exact is that the integral of the differential depends on only the end points of the integration and not on the path of integration. For example, for the function $g(x,y) = xy$, we have

$$dg = ydx + xdy$$

The definite integral of dg from the point (x_1,y_1) to (x_2,y_2) is

$$\Delta g = \int_1^2 dg = \int_{(x_1,y_1)}^{(x_2,y_2)} (ydx + xdy)$$

To find Δg, we must know how x and y are related when going from point 1 to point 2. That is, to evaluate the integral over x and over y, we must substitute a function of x for y in the term ydx, and we must substitute a function of y for x in the term xdy in order to carry out the integration. This means specifying a relation about how x and y vary with each other in going from point 1 to point 2, and that means specifying the path of integration. However, it can be shown that because dg is exact, the result of the integration is independent of the path. Δg can be found from the end points of the integration, and so it must be equal to $g(x_2,y_2) - g(x_1,y_1)$.

Often it is of interest to obtain partial derivative expressions under certain constraints. Consider the function $z = z(x,y)$. The differential relation is

$$dz = \left(\frac{\partial z}{\partial x}\right)_y dx + \left(\frac{\partial z}{\partial y}\right)_x dy \qquad (A.13)$$

When z is constrained to be constant, $dz = 0$. Using a z subscript with the differentials dx and dy to indicate this condition, Equation A.13 yields

$$0 = \left(\frac{\partial z}{\partial x}\right)_y dx_z + \left(\frac{\partial z}{\partial y}\right)_x dy_z$$

The ratio of the differentials is the partial derivative of x with respect to y under the constraint that z is constant.

$$0 = \left(\frac{\partial z}{\partial x}\right)_y \left(\frac{\partial x}{\partial y}\right)_z + \left(\frac{\partial z}{\partial y}\right)_x \Rightarrow \left(\frac{\partial x}{\partial y}\right)_z = \frac{-\left(\partial z/\partial y\right)_x}{\left(\partial z/\partial x\right)_y} \tag{A.14}$$

Likewise,

$$\left(\frac{\partial y}{\partial x}\right)_z = \frac{-\left(\partial z/\partial x\right)_y}{\left(\partial z/\partial y\right)_x} \tag{A.15}$$

In effect, we are free to treat x, y, and z as a set of related variables, and the constraint that z is constant means that only two are independent. There is no restriction as to which two variables can be considered to be independent.

Constraints can be imposed in more general ways. For instance, if there is a function $z = z(x,y)$ and another function $v = v(x,y)$, we can ask about the partial derivatives of z under the constraint that v is held constant. Saying that v is held constant amounts to establishing a relation about how x and y change together; the constraint removes a degree of freedom, and they are not independent variables. The procedure is to write Equation A.13 under the condition of constant v. This is simply

$$dz_v = \left(\frac{\partial z}{\partial x}\right)_y dx_v + \left(\frac{\partial z}{\partial y}\right)_x dy_v \tag{A.16}$$

From this, two partial derivative relations are obtained such as

$$\left(\frac{\partial z}{\partial x}\right)_v = \left(\frac{\partial z}{\partial x}\right)_y + \left(\frac{\partial z}{\partial y}\right)_x \left(\frac{\partial y}{\partial x}\right)_v \tag{A.17}$$

This can be thought of as the result of "dividing" Equation A.16 by dx_v. Another expression results from doing the same thing with dy_v instead of dx_v. According to Equation A.14, we can substitute for the last derivative in Equation A.17. Recall that Equation A.14 gives a derivative relation for constant z. We need the same relation for the function v instead of z, and on rewriting Equation A.14 with v replacing z, we obtain an expression for $(\partial y/\partial x)_v$. Substituting this into Equation A.17 yields

$$\left(\frac{\partial z}{\partial x}\right)_v = \frac{\left(\partial z/\partial x\right)_y - \left(\partial z/\partial y\right)_x \left(\partial v/\partial x\right)_y}{\left(\partial v/\partial y\right)_x} \tag{A.18}$$

As an example of this process, consider these two functions.

$$z(x,y) = x^2 + y^2$$

and

$$v(x,y) = \frac{y^2}{1+x}$$

The constraint that v is constant fixes how x and y vary. Calling that constant value v_0 means

$$v_0 + v_0 x = y^2$$

The partial derivatives needed on the right-hand side of Equation A.18 are

$$\left(\frac{\partial z}{\partial x}\right)_y = 2x$$

$$\left(\frac{\partial z}{\partial y}\right)_x = 3y^2$$

$$\left(\frac{\partial v}{\partial x}\right)_y = -\frac{y^2}{(1+x)^2}$$

$$\left(\frac{\partial v}{\partial y}\right)_x = \frac{2y}{1+x}$$

Assembling these according to Equation A.18 gives the desired derivative.

$$\left(\frac{\partial z}{\partial x}\right)_v = 2x - \frac{3y^2\left(-y^2/(1+x)^2\right)}{(2y/1+x)} = 2x + \frac{3y^3}{2+2x}$$

In this way, we can determine derivatives of functions when variables are constrained by some given relation.

A.4 Matrix Algebra

Matrices are ordered arrays of constants, variables, functions, or almost anything. A vector in a Cartesian coordinate system is a very simple ordered array; it consists of three elements which are the x-, y-, and z-components of the vector. The order of the elements tells which is which. In the vector (a,b,c), the x-component is a and not b or c. The ordering is established whether one uses a column or a row arrangement of the elements.

$$\text{Column:} \quad \begin{pmatrix} a \\ b \\ c \end{pmatrix} \qquad \text{Row:} \quad \begin{pmatrix} a & b & c \end{pmatrix}$$

Of course, there may be more than three elements in certain cases, and so it is convenient to subscript the elements with a number that gives the position in the row or column, for example, $(a_1 \ a_2 \ a_3 \ \dots \ a_n)$. A single column of elements is termed a **column matrix**. A letter with an arrow is the designation used here for a column matrix (or vector). Usually the same letter is used for the individual elements, but they are subscripted:

$$\vec{a} = \begin{pmatrix} a_1 \\ a_2 \\ a_3 \\ a_4 \\ \dots \\ a_n \end{pmatrix} \tag{A.19}$$

A row array of elements is a row matrix. Such a matrix is designated in the same way as a column matrix, except that there is a T (for transpose) as a superscript.

$$\vec{a}^T = \begin{pmatrix} a_1 & a_2 & a_3 & a_4 & \cdots & a_n \end{pmatrix} \tag{A.20}$$

Other sets of elements can be ordered by both rows and columns, and these are **square** or **rectangular matrices**. The elements of these matrices are subscripted with a row–column index, that is, with two integers that give the row and column position in the array.

$$A = \begin{pmatrix} A_{11} & A_{12} & A_{13} & \cdots & A_{1n} \\ A_{21} & A_{22} & A_{23} & \cdots & A_{2n} \\ A_{31} & A_{32} & A_{33} & \cdots & A_{3n} \\ A_{m1} & A_{m2} & A_{m3} & \cdots & A_{mn} \end{pmatrix} \tag{A.21}$$

This is an $m \times n$ matrix because there are m rows and n columns. Matrices can also be defined with three indices, or more.

Matrix addition is defined as the addition of corresponding elements of two matrices. The following two examples illustrate this definition.

$$\begin{pmatrix} 1 \\ 3 \\ 7 \\ 0 \end{pmatrix} + \begin{pmatrix} -1 \\ 0 \\ 10 \\ 2 \end{pmatrix} = \begin{pmatrix} 0 \\ 3 \\ 17 \\ 2 \end{pmatrix}$$

$$\begin{pmatrix} 0 & 3 & -i \\ 2 & -1 & 5 \end{pmatrix} + \begin{pmatrix} 0.5 & 1 & 1 \\ -2 & 0 & 5 \end{pmatrix} = \begin{pmatrix} 0.5 & 4 & 1-i \\ 0 & -1 & 10 \end{pmatrix}$$

The zero matrix is a matrix whose elements are all zero, and it is designated here as 0.

Multiplication of a matrix by a scalar (a single value) is defined as the multiplication of every element in the matrix by that scalar. That is,

$$cA = c\begin{pmatrix} A_{11} & A_{12} & A_{13} & \cdots & A_{1n} \\ A_{21} & A_{22} & A_{23} & \cdots & A_{2n} \\ A_{31} & A_{32} & A_{33} & \cdots & A_{3n} \\ A_{m1} & A_{m2} & A_{m3} & \cdots & A_{mn} \end{pmatrix} = \begin{pmatrix} cA_{11} & cA_{12} & cA_{13} & \cdots & cA_{1n} \\ cA_{21} & cA_{22} & cA_{23} & \cdots & cA_{2n} \\ cA_{31} & cA_{32} & cA_{33} & \cdots & cA_{3n} \\ cA_{m1} & cA_{m2} & cA_{m3} & \cdots & cA_{mn} \end{pmatrix}$$

Multiplication of a matrix by another matrix goes by the "row-into-column" procedure. If $C = AB$, the elements of C are given by

$$C_{ij} = \sum_{k=1}^{n} A_{ik} B_{kj} \tag{A.22}$$

where n is the number of columns of A and the number of rows of B. Recall that a vector dot product is

$$\begin{pmatrix} a_1 & a_2 & a_3 & \cdots & a_n \end{pmatrix} \begin{pmatrix} b_1 \\ b_2 \\ b_3 \\ \cdots \\ b_n \end{pmatrix} = a_1 b_1 + a_2 b_2 + a_3 b_3 + \cdots + a_n b_n$$

Thus, Equation A.6 says that the i–j element of C is obtained by a vector dot product of the ith row of A and the jth column of B:

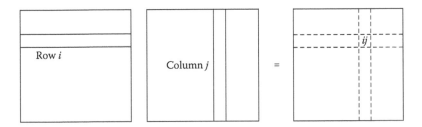

There are several features of matrix multiplication that may not be expected. First, it is not commutative, that is, in general, $AB \neq BA$. Second, it is possible for the product of two matrices to be the zero matrix even if both have nonzero elements.

The **identity matrix** for multiplication of square matrices is designated here as 1. It is a matrix whose elements are all zero except for the elements along the diagonal which are one.

$$1 = \begin{pmatrix} 1 & 0 & 0 & \cdots \\ 0 & 1 & 0 & \cdots \\ 0 & 0 & 1 & \cdots \\ \cdots & \cdots & \cdots & \cdots \end{pmatrix} \tag{A.23}$$

It is easy to show for some arbitrary matrix A, multiplication by the identity matrix returns the original matrix as the solution, that is, $A1 = 1A = A$.

The **transpose of a matrix** is formed by interchanging rows with columns. A transpose is designated here with a superscript T.

For example, if

$$A = \begin{pmatrix} 1 & 2 & 3 & 4 \\ 5 & 6 & 7 & 8 \\ 9 & 10 & 11 & 12 \\ 13 & 14 & 15 & 16 \end{pmatrix}$$

then

$$A^T = \begin{pmatrix} 1 & 5 & 9 & 13 \\ 2 & 6 & 10 & 14 \\ 3 & 7 & 11 & 15 \\ 4 & 8 & 12 & 16 \end{pmatrix}$$

The **complex conjugate of a matrix** is the matrix of complex conjugates of the elements of the matrix.

For example, if

$$B = \begin{pmatrix} b_{11} & b_{12} & b_{13} \\ b_{21} & b_{22} & b_{23} \\ b_{31} & b_{32} & b_{33} \end{pmatrix}$$

then

$$B^* = \begin{pmatrix} b_{11}^* & b_{12}^* & b_{13}^* \\ b_{21}^* & b_{22}^* & b_{23}^* \\ b_{31}^* & b_{32}^* & b_{33}^* \end{pmatrix}$$

The complex conjugate transpose of a matrix is called the **adjoint** and is designated with a superscript †. Thus, $A^\dagger = AT^*$

The **inverse** of some matrix C is C^{-1} if the following is true:

$$CC^{-1} = C^{-1}C = 1$$

C and C^{-1} are inverses of each other.

A **determinant** is a scalar quantity expressed in terms of n^2 elements, n being the order of the determinant. The elements are arranged in n rows and n columns between vertical bars. Its appearance is similar to that of a square matrix, but it is entirely different since it represents a single value. That value depends on the n^2 elements. Carrying out the evaluation involves using the **minor** of an element of a determinant. To see how this is done, let d_{ij} be an arbitrary element of the determinant Φ:

$$\Phi = \begin{vmatrix} d_{11} & d_{12} & d_{13} & \dots & d_{1n} \\ d_{21} & d_{22} & d_{23} & \dots & d_{2n} \\ d_{31} & d_{32} & d_{33} & \dots & d_{3n} \\ \dots & \dots & \dots & \dots & \dots \\ d_{n1} & d_{n2} & d_{n3} & \dots & d_{nn} \end{vmatrix}$$

The minor of the element d_{ij} is the determinant that remains after deleting the ith row and jth column from Φ. We shall designate that determinant D^{ij}. The value of the determinant Φ is the sum of the products of elements from any row, or from any column, and their minors with a particular factor of -1:

$$\Phi = \sum_{i=1}^{n} (-1)^{i+j} d_{ji} D^{ji}$$

$$\text{for } j = 1, 2, \dots, n \qquad (A.24)$$

$$\Phi = \sum_{i=1}^{n} (-1)^{i+j} d_{ij} D^{ij}$$

Of course, finding values for each of the D^{ij} determinants also requires using one of these equations. The process reaches a conclusion, though, when the minor has one only one element; the value of that determinant (the minor) is that element.

To illustrate the evaluation process, notice that a second-order determinant is evaluated from Equation A.24 in the following way:

$$\begin{vmatrix} e_{11} & e_{12} \\ e_{21} & e_{22} \end{vmatrix} = (-1)^{1+1} e_{11} E^{11} + (-1)^{1+2} e_{12} E^{12} = e_{11} e_{22} - e_{12} e_{21}$$

This is helpful because applying Equation A.24 to a third-order determinant (one with three rows and three columns) gives a sum with minors that are second-order determinants; this result can be used in working out the value for a third-order determinant. A fourth-order determinant can be evaluated with minors that are third order, and in such a stepwise fashion, a determinant of any order can be evaluated with Equation A.24. When carried out fully, the value of a determinant of order n is a sum of $n!$ products of its elements.

A.5 Matrix Eigenvalue Methods

There is a linear algebra equivalent of the differential eigenvalue equation. Instead of a differential operator acting on a function, a square matrix multiplies a column vector, and this is equal to a constant, the eigenvalue, times the vector:

$$A\vec{c} = e\vec{c} \qquad (A.25)$$

where
  A is a square matrix
  $\vec{c}$ is the eigenvector
  e is the eigenvalue

For an $N \times N$ matrix, there are N different eigenvalues and N different eigenvectors. That is, there are N different solutions to Equation A.25. Thus, the equation is better written with subscripts that distinguish the different solutions:

$$A\vec{c}_1 = e_1\vec{c}_1 \tag{A.26}$$

Because $i = 1,\ldots, N$, Equation A.26 represents the N different equations. However, from the rules for matrix multiplication, it is easy to realize that these equations can be collectively represented by one matrix equation. This is accomplished by collecting the column eigenvectors into one $N \times N$ matrix.

$$A\begin{pmatrix} \vec{c}_1 & \vec{c}_2 & \vec{c}_3 & \cdots & \vec{c}_N \end{pmatrix} = \begin{pmatrix} \vec{c}_1 & \vec{c}_2 & \vec{c}_3 & \cdots & \vec{c}_N \end{pmatrix}\begin{pmatrix} e_1 & 0 & 0 & \cdots & 0 \\ 0 & e_2 & 0 & \cdots & 0 \\ 0 & 0 & e_3 & \cdots & 0 \\ \cdots & \cdots & \cdots & \cdots & \cdots \\ 0 & 0 & 0 & \cdots & e_N \end{pmatrix} \tag{A.27}$$

With the different column vectors set next to each other, we obtain a square array, which will be designated C. The rightmost matrix in Equation A.27, which will be called E, is a **diagonal matrix** since its only nonzero elements are along the diagonal. Because of its diagonal form, matrix multiplication of E by C just leads to each column of C scaled by a corresponding diagonal element of E. From this, we have the form of the general matrix eigenvalue equation,

$$AC = CE \tag{A.28}$$

This equation is solved for some specific matrix A, and the solutions are the eigenvectors arranged as columns in C, plus the associated eigenvalues arranged on the diagonal of E.

  It is quite often the case that a matrix whose eigenvalues are sought is real and symmetric, and we consider only such matrices here. If the matrix C in Equation A.28 is known, and if its inverse can be found, then multiplication of Equation A.28 by the inverse will lead to

$$C^{-1}AC = C^{-1}CE = E \tag{A.29}$$

The process of multiplying some matrix on the right by a particular matrix, and multiplying on the left by the inverse of that matrix, is called a **transformation**. A matrix equation remains true on transformation of each of the matrices in the equation. For example, if

$$R + ST = X$$

then

$$(C^{-1}RC) + (C^{-1}SC)(C^{-1}TC) = (C^{-1}XC)$$

Notice that if we use a prime to designate a matrix transformed by C, for example,

$$R' = C^{-1}RC$$

the example of a transformed matrix equation will become simply

$$R' + S'T' = X'$$

This has exactly the same form as the original equation, and so "transformation" seems to be appropriate terminology for what has happened. The matrices have been changed by the transformation, but the matrix equation is unchanged in form.

The general matrix eigenequation, expressed in the form of Equation A.29, can now be thought of differently. We seek a matrix C that transforms the matrix A in a very particular way such that the transformed matrix is of diagonal form. The process to accomplish this is referred to simply as **diagonalization** of A. If we seek the eigenvalues and eigenvectors of some matrix, then we seek to diagonalize it.

An important category of transformation matrices is that of **unitary matrices**. These are employed for diagonalizing real, symmetric matrices, and we will use U to designate a unitary matrix in the following discussion. A unitary matrix is one whose inverse is the same as or equal to its transpose. Or, when the elements of U are complex, unitarity means the inverse is equal to the adjoint:

$$U^{-1} = U^{\dagger}$$

The property of unitarity means that a transformation can be written with either the transpose (or adjoint) or with the inverse.

$$U^{-1}AU = U^{T}AU$$

Another property of a unitary matrix is that the value of the determinant constructed from its elements is equal to 1, that is, $\det(U) = 1$.

There are a number of ways of finding unitary matrices that diagonalize real, symmetric matrices. A very useful recipe for 2×2 matrices uses sines and cosines. Starting with A as a matrix we seek to diagonalize

$$A = \begin{pmatrix} A_{11} & A_{12} \\ A_{21} & A_{22} \end{pmatrix}$$

an angle θ is determined from values in A.

$$\tan 2\theta = \frac{2A_{12}}{A_{22} - A_{11}} \tag{A.30}$$

The transformation matrix is

$$U = \begin{pmatrix} \cos\theta & \sin\theta \\ -\sin\theta & \cos\theta \end{pmatrix} \tag{A.31}$$

The eigenvalues are

$$U^T A U = \begin{pmatrix} e_1 & 0 \\ 0 & e_2 \end{pmatrix} \tag{A.32}$$

(The recipe for θ in Equation A.30 can be derived by carrying out the matrix multiplication in Equation A.32 explicitly in terms of elements of A and $\sin\theta$ and $\cos\theta$. This yields expressions for each of the four elements of the product matrix. Setting the expression for the off-diagonal element to be zero, as on the right-hand side of Equation A.32, gives the proper condition for θ.)

There is a geometrical analogy to the 2×2 diagonalization scheme. The U transformation matrix applied to a vector of x- and y-coordinates rotates the vector about the origin by the angle θ, yielding two new coordinates.

$$U \begin{pmatrix} x \\ y \end{pmatrix} = \begin{pmatrix} x' \\ y' \end{pmatrix}$$

A consequence of the unitary nature of U is that the length of the original vector is unchanged, that is,

$$\sqrt{x^2 + y^2} = \sqrt{x'^2 + y'^2}$$

Therefore, the transformation with U is a pure rotation; there is no change in length of any rotated vector. From the geometrical picture of transformations, it is common to refer to unitary transformations as **rotations**. For diagonalization problems, "rotation" is used loosely or abstractly because whatever is being rotated is whatever labels the rows and columns of the matrix being diagonalized. That is not necessarily going to be a set of directions in a geometrical space.

Diagonalization by 2×2 rotations is one means of diagonalizing larger problems. The idea is to extract a 2×2 from a large matrix and use the preceding recipe. To see what this involves, we must recall special features of matrix multiplication. Consider an 8×8 matrix, A. Let us extract the following 2×2 from it.

$$\begin{pmatrix} A_{33} & A_{36} \\ A_{63} & A_{66} \end{pmatrix}$$

And let the matrix that diagonalizes this be

$$U = \begin{pmatrix} a & b \\ c & d \end{pmatrix}$$

In order to apply the transformation given by U to the entire A matrix, it must be super-posed on the unit matrix. This means that we replace the 33, 36, 63, and 66 elements of an 8×8 unit matrix (i.e., 1) by $a, b, c,$ and d, respectively, from U. These replaced elements are in the very same positions as the elements extracted from A to make up the 2×2 for diagonalization. As illustrated here, this matrix multiplying an 8×8 square matrix alters the values of only the elements in rows 3 and 6.

The superposed unit matrix is shown on the left (only nonzero values) multiplying another matrix (shaded). The only elements that are different after the matrix multiplication are those in the rows with $a, b, c,$ and d, and these are indicated by the darkened rows on the rightmost matrix. Were the multiplication done in the reverse order, the elements of the third and sixth columns would be the ones affected:

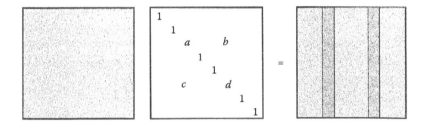

Thus, when the 2×2 U matrix is properly superposed on the unit matrix and used to transform the original A matrix, the result is a transformed A matrix:

Transformed
A matrix:

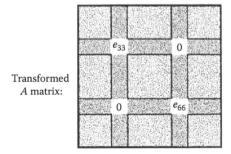

The darkened rows and columns indicate elements whose values are now different from what they were in the original A matrix. The eigenvalues of the extracted 2×2 part of the A matrix are now on the diagonal, and the corresponding off-diagonal elements are zero, as indicated. This matrix is closer to being in diagonal form.

In the next step, another 2×2 is extracted from A, and the same thing is done. A complication is that in this second 2×2 rotation, the 36 and 63 elements that became zero may change. That is, they may end up being different from the desired value of zero if the second 2×2 rotation involves the third or the sixth row and column. However, by continuing the process again and again, it is possible for all the off-diagonal elements to be as close to zero as one wishes. Typically, it is best to select the particular 2×2 for each step on the basis of the largest off-diagonal value present. This is a process that is easily coded for computers, but to illustrate how it works and to show all the details, here is an example of the diagonalization of a particular 3×3 matrix, B.

$$B = \begin{pmatrix} 5 & -3\sqrt{3} & 2 \\ -3\sqrt{3} & -1 & 2\sqrt{3} \\ 2 & 2\sqrt{3} & 20 \end{pmatrix}$$

Let us extract a 2×2 and apply Equation A.30 to find the necessary rotation. This 2×2 will be from the first two rows and columns of B:

$$\begin{pmatrix} 5 & -3\sqrt{3} \\ -3\sqrt{3} & -1 \end{pmatrix}$$

The rotation angle for diagonalizing this 2×2 is found to be $30°$ from the expression

$$\tan 2\theta = \frac{2\left(3\sqrt{3}\right)}{-1-5} = \sqrt{3}$$

The transformation matrix for the 2×2 is placed in a 3×3 matrix in the positions corresponding to the 2×2 that was extracted from B. The rest of this matrix is filled with zeros off the diagonal and ones along the diagonal. Thus,

$$U = \begin{pmatrix} \cos\theta & \sin\theta & 0 \\ -\sin\theta & \cos\theta & 0 \\ 0 & 0 & 1 \end{pmatrix} = \begin{pmatrix} \sqrt{3}/2 & 1/2 & 0 \\ -1/2 & \sqrt{3}/2 & 0 \\ 0 & 0 & 1 \end{pmatrix}$$

This is a 3×3 unitary transformatin matrix.

Transforming B with U yields a matrix that is closer to being diagonal than the original matrix B.

$$U^T B U = \begin{pmatrix} 8 & 0 & 0 \\ 0 & -4 & 4 \\ 0 & 4 & 20 \end{pmatrix}$$

The next step in the whole process is to extract another 2×2 to diagonalize. In this case, the 2×2 has to be that in the second and third rows and columns. With this second transformation matrix designated V, the rotation angle and the elements of V are

$$\tan 2\phi = \frac{2(4)}{20-(-4)} = \frac{1}{3}$$

$$\therefore \phi = 9.22°$$

$$V = \begin{pmatrix} 1 & 0 & 0 \\ 0 & \cos\phi & \sin\phi \\ 0 & -\sin\phi & \cos\phi \end{pmatrix} = \begin{pmatrix} 1 & 0 & 0 \\ 0 & 0.9817 & 0.1602 \\ 0 & -0.1602 & 0.9817 \end{pmatrix}$$

Applying this transformation to B, after transformation by U, yields

$$V^T(U^TBU)V = \begin{pmatrix} 8 & 0 & 0 \\ 0 & -4.65 & 0 \\ 0 & 0 & 20.65 \end{pmatrix}$$

Therefore, the eigenvalues of the matrix B are -4.65, 8, and 20.65. The eigenvectors are the columns of the matrix that transformed B into diagonal form, and by the previous expression, the eigenvectors are the columns of the matrix UV. The overall transformation is a product of the two rotation matrices.

B was chosen for this example in such a way that only two successive rotations were required to completely diagonalize the matrix. In general, the diagonalization of even a 3×3 matrix via 2×2 rotations may require many 2×2 rotations. This happens because a rotation that diagonalizes one 2×2 block might have the effect of changing an already-zero off-diagonal element to a nonzero value. This makes the process tedious if not done with a computer but otherwise presents no problem. The whole process can be applied to a matrix of any size; however, there are other algorithms that are computationally advantageous for many diagonalization problems encountered in quantum chemistry.

There is a special recipe, or method, for immediately obtaining just the eigenvalues of a matrix. The recipe is to subtract an unspecified parameter—call it λ—from the diagonal elements and then set the determinant formed from the elements of the matrix equal to zero. This generates a polynomial in the parameter λ. The roots of that polynomial are the eigenvalues of the matrix. Here is an example of finding the eigenvalues of a 3×3 matrix, G, by this scheme.

$$G = \begin{pmatrix} 5 & 0 & \sqrt{3} \\ 0 & 6 & 0 \\ \sqrt{3} & 0 & 7 \end{pmatrix}$$

The determinant of the elements of this matrix, with λ subtracted along the diagonal, is set to zero.

$$\begin{vmatrix} (5-\lambda) & 0 & \sqrt{3} \\ 0 & (6-\lambda) & 0 \\ \sqrt{3} & 0 & (7-\lambda) \end{vmatrix} = 0$$

Multiplying out or evaluating this determinant yields a polynomial in λ.

$$\lambda^3 - 18\lambda^2 + 104\lambda - 192 = 0$$

The eigenvalues are the roots of the polynomial. In this example, chosen for its simplicity, the polynomial is factorable:

$$(\lambda - 4)(\lambda - 6)(\lambda - 8) = 0$$

Thus, the roots are 4, 6, and 8. These are the eigenvalues of the matrix G.

When the eigenvalues are obtained in this manner, the eigenvectors or the transformation matrix that diagonalizes G remain unknown. However, the associated eigenvector of any known eigenvalue can always be obtained by going back to the basic eigenvalue equation, Equation A.26, and solving it using the known eigenvalues one at a time.

The diagonalization of a complex **Hermitian matrix**, which is one for which off-diagonal elements are related as

$$H_{ij} = H_{ji}^*$$

may require a transformation matrix, U, that is complex. Then, the transformation is not that in Equation A.32, because the adjoint of U must be used instead of the transpose of U. The eigenvalues of a Hermitian matrix are real numbers.

A.5.1 Simultaneous Diagonalization

A transformation can be found that simultaneously diagonalizes several matrices if they commute with each other. This will not necessarily be a unitary transformation. The problem of simultaneous diagonalization sometimes arises just as a need to transform several matrices to diagonal form. It also arises in a special matrix eigenvalue equation that is really a generalization of Equation A.28.

$$AC = SCE \tag{A.33}$$

Multiplying by the inverse of C yields

$$C^{-1}AC = (C^{-1}SC)E \tag{A.34}$$

If the matrix product $(C^{-1}SC)$ were equal to the unit or identity matrix, 1, the right-hand side would simplify to just E, and the problem would be the same as in Equation A.29. Notice that if this were true, the matrix S would in fact have been diagonalized, its diagonal form being simply that of the unit matrix.

The process of simultaneous diagonalization amounts to finding a transformation where for all but one matrix, the resulting diagonal form is that of the unit matrix. This can be done for any nonsingular matrix by a transformation that is nonunitary. Assuming S in Equation A.33 is a nonsingular matrix with positive eigenvalues, we can find the unitary transformation that makes it diagonal in the usual way.

$$U^{-1}SU = D$$

where
   D is diagonal
   U is unitary

Using the elements of D, a matrix designated d is constructed as

$$d = \begin{pmatrix} 1/\sqrt{D_{11}} & 0 & 0 & \cdots \\ 0 & 1/\sqrt{D_{22}} & 0 & \cdots \\ 0 & 0 & 1/\sqrt{D_{33}} & \cdots \\ \cdots & \cdots & \cdots & \cdots \end{pmatrix} \tag{A.35}$$

This matrix is a diagonal matrix, its elements being the inverse square roots of the corresponding elements of D. With this definition, it is easy to see that

$$dDd = 1 = \begin{pmatrix} 1 & 0 & 0 & \cdots \\ 0 & 1 & 0 & \cdots \\ 0 & 0 & 1 & \cdots \\ \cdots & \cdots & \cdots & \cdots \end{pmatrix}$$

This special matrix is equal to its transpose, and so it is not necessary to use a transpose symbol in the preceding expression. This matrix, though, is not a unitary transformation. When multiplied by its transpose (itself), the result is not the unit matrix. The composite or product transformation of this with U is the transformation that takes S into the unit matrix since

$$(dU^T)S(Ud) = 1 \tag{A.36}$$

If this expression is further transformed with some matrix V, the right-hand side will remain as 1 if V is unitary. Because $V^{-1}V = 1$, the transformation leaves the diagonalized S matrix as 1.

The process for solving Equation A.33 can now be seen as a three-step process:

1. Find a unitary matrix, U, that diagonalizes S.
2. Set up the nonunitary transformation of Equation A.35 from the elements of D.
3. Find another unitary matrix, V, such that the composite transformation of A takes it into diagonal form.

This process says that the matrix C that solves Equation A.33 may be constructed as a product of three matrices.

$$C = UdV$$

Substituting this into Equation A.34 and then using the result in Equation A.36,

$$(V^T dU^T)A(UdV) = (V^T dU^T)S(UdV)E = V^T \left[dU^T AUd \right] V = E$$

The matrix inside the square brackets is the matrix that V must diagonalize. Following this three-step process, the eigenvalues of A are along the diagonal of E, and the matrix S is simultaneously diagonal with its eigenvalues all 1.0.

Appendix B: Molecular Symmetry

B.1 Symmetry Operations

The geometrical arrangements of atoms in molecules can sometimes place atoms in equivalent chemical environments. For instance, the two hydrogens in a water molecule are equivalently bonded to the oxygen. In general, chemical equivalence corresponds to the existence of some type of geometrical symmetry. Analysis of symmetry does not necessarily provide any information that cannot be obtained from solving the quantum mechanical Schrödinger equation. However, it can sometimes simplify the task of solving the Schrödinger equation and more often it can provide important qualitative information in the absence of solving the equation.

Certain regular geometrical constructions possess symmetry. A square has four equal sides, whereas a rectangle may have only its pairs of sides equal in length. What can be used to distinguish the degree of symmetry in the square versus the rectangle? Consider a line passing through the center of the square and perpendicular to it. If the square is rotated 90° about that line, it will look the same. If the same were done to the rectangle, there would be a difference. The long side would be where the short side was. Hence, the operation of rotation by 90° can distinguish the two objects because they differ in symmetry, the square seeming to be a "more symmetric" object.

There are quite a number of operations (e.g., rotation by some angle about some axis) that can be performed on objects to assess their symmetry. The goal is to determine whether or not the object looks the same after the operation has been performed. For objects constructed from points in space (i.e., atoms in molecules), a fairly limited number of such operations need to be considered. The first such operations are rotations about an axis, and we identify these by the symbol C_n where n is the inverse of the fraction of a full circle to which the rotation corresponds. Thus, the operation of rotation by 90°, a quarter of a circle, is designated C_4. The operation of reflection, as if through a mirror in some chosen plane, is usually designated σ. Finally, the operation of pulling each point in some object through the object's center and continuing on in the opposite direction for the same distance is called inversion. Its symbol is i. For an arrangement of points in (x, y, z) space, with the origin as the center of inversion, the operation of inversion means that every point $\{x_j, y_j, z_j\}$ becomes $\{-x_j, -y_j, -z_j\}$.

The first step in understanding symmetry in molecules is to determine the symmetry operations that can be applied to a given molecule without changing how we see it. In other words, we seek to know what rotations, reflections, or inversions can act to interchange equivalent atoms in a molecule while leaving everything else unaffected. Consider the water molecule again. We can visualize an axis passing through the oxygen atom, in the plane of the molecule, and everywhere equidistant from both hydrogens.

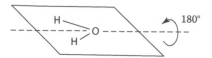

Rotation of the molecule about this axis by 180°, that is, the operation C_2 applied to the molecule, takes one hydrogen atom into the other, an interchange of equivalent centers. On the other hand, a C_4 operation changes things such that the water molecule is in a different plane. We conclude that C_2 is an operation in the collection of symmetry operations appropriate for the water molecule, whereas C_4 is not. Later, we shall refer to such a collection as a mathematical **group**.

One additional symmetry operation used for molecules is called improper rotation, and it is designated S_n. Again, the n subscript gives the inverse of the fraction of a circle of the rotation. An S_n operation is a C_n operation followed by reflection through a plane passing perpendicular to the axis of the C_n rotation. The molecule allene, H_2CCCH_2, has CH_2 groups in perpendicular planes and has an S_4 operator in its group.

Rotation of the molecule about the axis of the carbon atoms by 90° changes the planes of the two CH_2 groups. However, if a reflection operation is then performed, with the reflection plane being in the middle of the molecule, the original appearance of the molecule in space is achieved. The collection of symmetry operators appropriate for allene includes the S_4 operation.

B.2 Molecular Point Groups

The set of all distinct symmetry operations that take a molecule into an equivalent arrangement constitutes a mathematical structure called a **group**. An area of mathematics called **group theory** helps organize information we collect about molecular symmetry operations, although group theory has much broader application.

A group exists only if certain conditions are met. The first requirement is that there is an operation associated with the elements of a group. We might contemplate a mathematical group composed of all positive integers with the operation being addition. In the case of molecular symmetry, the operation is successive application of the elements, something we may term a "multiplication." We have already seen that the improper rotation operator in the allene example was the equivalent of application of the reflection operator immediately after application of the C_4 operator. In other words, there is an equality involving the elements and the multiplication process: $S_4 = C_4 \otimes \sigma$ (read "S_4 equals C_4 times σ").

Another requirement for a group is that the complete set of symmetry operations of a molecule is a closed set under the operation of multiplication $\otimes$. This means that if we perform one valid symmetry operation and then another, the result must be equivalent to having performed one single operation that is part of the set. For instance, when a C_4 rotation is applied to a square and then a C_2 rotation is applied, the result is the same as a single rotation by three-quarters of a circle. This rotation, which has the designation C_4^3, is part of the group of symmetry operations of the square, consistent with the property of group closure. Closure is helpful in identifying symmetry operations that may not have been found by inspection. If two symmetry operations have been found for some molecule, the effect of successive application of two of them can be tested to see if this multiplicative product corresponds to a third distinct symmetry operation.

A special symmetry operation that we have not yet considered has the role of an identity operation in a group. Its designation is E. If the allene molecule were rotated about the carbon atom axis by 180°, and then if this C_2 operation were performed again, there would be no interchange of equivalent atomic centers. The result would be the same as if the molecule had not been rotated at all. By closure, this "identity" operation must be in the group because it corresponds to the successive application of two symmetry operations (both C_2) in the group.

Many molecules share the same group of symmetry operations. For instance, one can readily anticipate that the following series of molecules will have the same set of symmetry operators.

$$\begin{array}{ccc} \text{H}\!\!>\!\!\text{O} & \text{H}\!\!>\!\!\text{CO} & \text{H}\!\!>\!\!\text{CCO} \\ \text{H} & \text{H} & \text{H} \end{array}$$

For these types of molecules, the operators are E, C_2, a reflection plane σ in the plane of the molecule, and a reflection plane σ' perpendicular to the plane of the molecule and including the C_2 axis. The number of distinct groups that are important for molecular problems is not very many, two dozen or so. (Of course, the absolute number of possible groups is infinite; it is the number commonly encountered that is small.) The task of finding the symmetry operations of a given molecule is then the task of finding to which **molecular point group** the molecule belongs.

Point groups are given designations, and in one convention for their designations, the names reveal key symmetry operators in the group. For instance, C_{3v} is the designation for a group that has a principal C_3 axis of rotation plus reflection planes that include this axis. Here is a short list of some commonly used molecular point groups and their symmetry operations.

C_2	E	C_2		
C_s	E	σ		
C_3	E	C_3	C_3^2	
C_{2v}	E	C_2	σ	σ'
C_{2h}	E	C_2	σ	i

A good illustration of the assignment of a symmetry group to a molecule is to compare the two isomers of glyoxal. The trans isomer has a C_2 axis that passes midway between the carbon atoms and perpendicular to the plane of the molecule. It has only one reflection plane, the molecular plane. It also has inversion symmetry, as can be seen by considering a coordinate system with its origin at the point midway between the carbons. If the coordinates of one of the hydrogens are (x_H, y_H), the coordinates of the other must be $(-x_H, -y_H)$. This means that the hydrogen atom positions are interchanged by the process of inversion, and so are the carbons, and so are the oxygens. Thus, there are four symmetry operators for the trans isomer, C_2, σ, i, and E. By referring to a list of symmetry operators associated with point groups, these four operators serve to identify trans-glyoxal as being a C_{2h} molecule.

The cis isomer of glyoxal is different from the trans. It possesses a C_2 axis in the plane of the molecule. It also has a reflection plane that includes the C_2 axis and is perpendicular to the plane of the molecule. Reflection through this plane interchanges the left and right sides of the molecule. Another reflection plane is the plane of the molecule. And, of course, there is always the identity operation, E. Thus, the list of operators for this isomer is C_2, σ, σ', and E. By checking all products of these operators (successive applications), no others appear. Therefore, cis-glyoxal belongs to the C_{2v} point group.

An important task in using symmetry arguments for molecular systems is to assign the point group appropriate for the system. This task amounts to listing all the operators that can take a molecule into an equivalent form in space. To find such a list systematically, one can do the following:

1. Check for proper rotation axes.
2. Check for reflection planes, including the plane of a molecule that is planar.
3. Check the inversion operation.
4. If the inversion operation is *not* included in the operations for the molecule, check for improper rotation axes.
5. Check for closure. Successively apply every possible pair of operations from the list deduced so far. If each pair of operations produces the same result as any single operation on the list, the closure property has been satisfied. If not, add the new operation to the list and form new pairs of operations to check for closure.
6. Compare the list of operators with the lists for different point groups and thereby assign a molecule to the group to which it belongs.

B.3 Symmetry Representations

At this point, we shall restrict our attention to a class of molecular point groups called **Abelian groups**. For a group to be Abelian means that the operations in the group commute, or that they can be applied successively without regard to order. There are many groups for which this is not the case, and the mathematical treatment of representations for non-Abelian groups is sufficiently complicated to put it beyond the scope of our development.

A corollary to Theorem 8.3 is that a set of functions can be found that are simultaneously eigenfunctions of a set of mutually commuting operators. Symmetry operators in Abelian point groups are mutually commuting. The important question is whether they commute with the molecular Hamiltonian. For the water molecule, consider the effect of a C_2 symmetry operation performed on the Hamiltonian: All the labels on kinetic and potential energy operators associated with the two hydrogen centers would be interchanged or swapped. This would change the order of additive terms in the Hamiltonian, but otherwise would not change the Hamiltonian. That it does not change the Hamiltonian is a consequence of the C_2 operator being commutative with the Hamiltonian. In general, the symmetry operators of a molecule's point group commute with the molecular Hamiltonian.

In view of Theorem 8.3 and the commutivity of the Hamiltonian and appropriate symmetry operators, we are free to seek wavefunctions that are eigenfunctions of both the symmetry operators and the Hamiltonian. In fact, we can first find functions that are eigenfunctions of the symmetry operators and then use them to construct the eigenfunctions of the Hamiltonian.

How does a function transform or change under application of a symmetry operation? And what does it mean for a function to be an eigenfunction of a symmetry operation? One of the easiest ways to answer this question is to visualize a wavefunction in three-dimensional space as if it were an object, or something almost like a real object. Imagine a 1s atomic orbital on each hydrogen in the water molecule as a fuzzy sphere or spongy ball, for instance. Applying a C_2 symmetry operator takes one sphere into the other (Figure B.1). These functions (spheres), then, are not eigenfunctions of this symmetry operator. However, if we consider a combined function of the two 1s orbitals, that is, $\varphi = 1s_A + 1s_B$ (A and B distinguishing the two hydrogens), we will have a different sort of result. φ is dumbbell-shaped. Rotate it by 180° and the same shape (function) is found. That is, $C_2\varphi = C_2(1s_A + 1s_B) = 1s_B + 1s_A = \varphi$. Thus, φ is an eigenfunction of the C_2 operator with the eigenvalue being 1. Now consider what happens if we form a different combination function, $\Gamma = 1s_A - 1s_B$. This corresponds to a dumbbell shape but with a sign change on going from one half to the other (something physical objects do not exhibit). $C_2\Gamma = C_2(1s_A - 1s_B) = 1s_B - 1s_A = -\Gamma$. This function is also an eigenfunction of C_2 but with an eigenvalue of –1.

The possible eigenvalues for symmetry operators of Abelian point groups always turn out to be either 1 or –1. That is, an eigenfunction of a symmetry operator either changes into itself or into the negative of itself on application of a symmetry operation. (A symmetry operation does not shrink the function, and so there are no eigenvalues between 1 and –1.) Useful terminology involves saying that a function is **symmetric** with respect to a particular symmetry operator if the eigenvalue is 1, and it is **antisymmetric** if the eigenvalue is –1. Arbitrary functions that one selects, such as an individual hydrogen 1s orbital in water, are

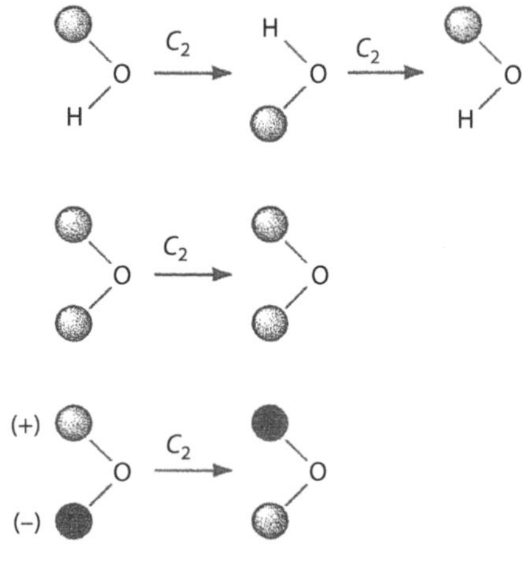

FIGURE B.1
Application of water's C_2 symmetry operator to individual hydrogen 1s orbitals rotates them into each other. Two specific linear combinations of the orbitals prove to be eigenfunctions of this symmetry operator.

not necessarily eigenfunctions of symmetry operators any more than arbitrary functions are eigenfunctions of a particular Hamiltonian. Those functions that happen to be simultaneous eigenfunctions of the symmetry operators of a molecular point group are special. Categorizing these functions according to their symmetry operator eigenvalues (i.e., according to whether they change sign or not on applying a symmetry operator) is accomplished with devices called **representations**.

A representation is merely something that behaves the same as whatever it represents. In group theory, the representations of most interest are sets of numbers that serve to represent how molecular features change on application of symmetry operations. In effect, the representations code symmetry information into numerical values. After doing that, most symmetry questions can be answered using little more than arithmetic.

Figure B.1 shows two linear combinations of hydrogen 1s atomic orbitals that are eigenfunctions of the C_2 operator but with different eigenvalues. The other operators in water's C_{2v} point group are identity (E), reflection in the plane of the molecule (σ), and reflection in a perpendicular plane (σ'). If we apply these operators to the two linear combinations, φ and Γ, the eigenvalues can be arranged in a table.

	E	C_2	σ	σ'
φ	1	1	1	1
Γ	1	−1	1	−1

Notice that Γ changes sign on reflection via the plane perpendicular to the molecular plane. Other than that case, the eigenvalues are all 1. The two lists of numbers, (1 1 1 1) for φ and (1 −1 1 −1) for Γ, are representations of the functions. The lists encode the symmetry information about the functions. These short lists of numbers tell us how each function changes on application of one of the point group's symmetry operations. Other functions may turn out to have the same representations. For instance, an oxygen 1s or 2s atomic orbital has the same symmetry representation as φ. You can show this by applying each operator to the oxygen orbital and evaluating what results.

If the number of symmetry operations in a group is finite, then it can be proven that the number of different representations for the group is also finite. Furthermore, simple rules can be developed to find all the representations. In the case of the C_{2v} point group, water's symmetry group, there happen to be two other representations. These can be illustrated by examining two linear combinations of $2p_z$ orbitals of the hydrogens, with z being the axis perpendicular to the molecular plane. The combinations and the effect of symmetry operations are shown in Figure B.2. The representations that are deduced for these functions can be added to the table for water and its C_{2v} point group.

	E	C_2	σ	σ'
φ	1	1	1	1
Γ	1	−1	1	−1
α	1	1	−1	−1
β	1	−1	−1	1

There are no other representations for this point group. All functions that belong to the C_{2v} point group can be classified as transforming one of these representations.

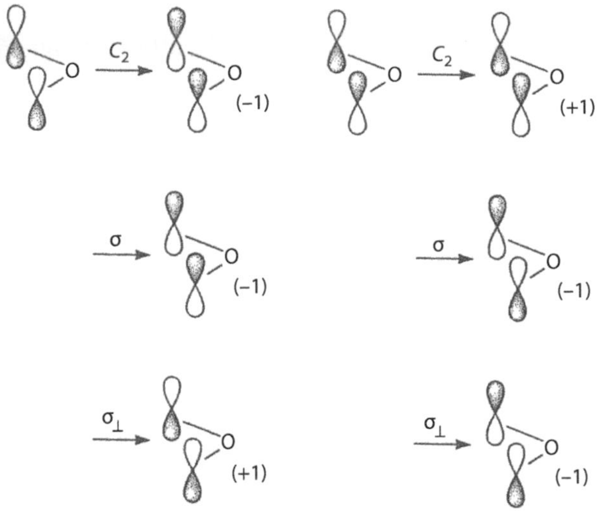

FIGURE B.2
Linear combinations formed from $2p_z$ atomic orbitals on the hydrogen atoms in the water molecule and the effect of applying symmetry operators to these functions.

A widely used convention for designating the representations of molecular point groups is the following:

1. A letter is assigned to designate the list of numbers that constitute a representation in a point group. For Abelian groups, the letter is either A or B: A if the eigenvalue for a C_2 operation is 1, and B if it is –1. If the group has no C_2 operator, all are A. Subscripts are attached to these letters.
2. If there is an inversion operator in the group, then a subscript g is attached for those representations for which the eigenvalue associated with the inversion operation is 1. A subscript u is attached for a –1 eigenvalue. (g stands for the German word gerade, meaning same, and u is for ungerade, not the same.)
3. Subscripts of 1 or 2 distinguish 1 or –1 eigenvalues, respectively, for a reflection plane operator that is perpendicular to the molecular plane if there is one.
4. If there is neither an inversion operator nor a perpendicular reflection plane operator, a prime (′) or a double-prime (″) can be added to the representation's letter designation to distinguish whether the eigenvalue with respect to some other type of reflection plane is 1 or –1, respectively.

Sometimes the use of primes versus 1 and 2 subscripts is interchanged. At the level we are following, it is not important to be able to assign designations but rather to interpret those designations we encounter. In that respect, the preceding rules are suitable. Prime versus double prime distinguishes symmetry with respect to reflection, and so does a subscript of 1 versus 2.

The water molecule's C_{2v} point group representations have already been listed. Rewriting this list with the designations gives a concise table.

	E	C_2	σ	σ'
A_1	1	1	1	1
B_2	1	–1	1	–1
A_2	1	1	–1	–1
B_1	1	–1	–1	1

The order in which the representations are presented in this table is not significant, except that it is customary to list the totally symmetric representation first. The **totally symmetric representation** is the one for which all the eigenvalues are 1. Every point group has a totally symmetric representation, including non-Abelian point groups.

Not only functions but also displacement vectors can be analyzed for symmetry properties. Displacements can be associated with movement of atoms (vibration, rotation, and translation of molecules), and these may transform as symmetry representations. Consider a vector that corresponds to simultaneously stretching both water O–H bonds by like amounts. A picture of this displacement is that of two vectors of the same magnitude on the two hydrogen centers.

If we apply the symmetry operators of water's point group to this picture, we find that it looks the same for all operations. This combination of two atomic displacement vectors transforms as the A_1 or totally symmetric representation. On the other hand, when a composite displacement consists of contracting one O–H bond by as much as the other is being stretched, then a different representation is obtained.

Applying the C_2 operator to this combination of displacement vectors reverses the directions of the two arrows, and that is a sign change in the vectors. This combination transforms as the B_2 representation. Orbital functions, displacement vectors, and other features can be analyzed for their symmetry properties in the same way.

B.4 Symmetry-Adapted Functions

For a set of orbital functions or displacement vectors that can be used in the quantum mechanical description of some molecule, it may turn out that individually they do not transform as representations of the group of that molecule. We have already seen this in the case of the water molecule in which each individual hydrogen 1s orbital was *not* an eigenfunction of all the C_{2v} symmetry operators. Rather, linear combinations made from the set of the two hydrogens' 1s functions were. A powerful feature of group theory is that it can be used to determine the necessary combinations of functions or vectors that transform as representations of the group.

The process of forming linear combinations of things such that they transform as symmetry representations is the process of symmetry adapting. Again, restricting attention to Abelian (commutative) point groups, there is a derivable formula for symmetry adapting. Let T_j designate the jth symmetry operator (such as C_2 or i or E) from the point group of some molecule of interest. Let $\Gamma_j^{\{R\}}$ be the number from the list of integers of the R representation for the jth symmetry operator. With h as the number of operators in the group (the **order of the group**), we define a special operator for each R-representation in the following way:

$$P_R \equiv \frac{1}{h}\sum_{j=1}^{h}\Gamma_j^{\{R\}}T_j \tag{B.1}$$

Application of this operator to some arbitrary function or vector either yields zero or projects out a function or vector that transforms as the R representation. Let us do this for a displacement vector corresponding to stretching one O–H bond. The symmetry operators of water's point group are applied to this vector and the results, other displacement vectors, can be presented graphically.

We next construct the four P operators according to Equation B.1:

$$P_{A_1} = \frac{1}{4}\left(E+C_2+\sigma+\sigma'\right)$$

$$P_{A_2} = \frac{1}{4}\left(E+C_2-\sigma-\sigma'\right)$$

$$P_{B_1} = \frac{1}{4}\left(E-C_2-\sigma+\sigma'\right)$$

$$P_{B_2} = \frac{1}{4}\left(E-C_2+\sigma-\sigma'\right)$$

These projection operators tell us how to combine the results of applying the individual symmetry operators to the displacement vector we are interested in. For instance, P_{B_2} and P_{A_2} yield different results when applied to the O–H displacement vector of one bond. To see this, we tabulate vertically the effect of each of the group's operators on the displacement vector. Then, we place a factor of 1/4 or –1/4 in front each arrow diagram, the factors corresponding to those that multiply the symmetry operators in the given projection operator. These diagrams are summed, and we find that we project a symmetry-adapted B_2 displacement vector from the vector for stretching one O–H bond. We obtain zero with P_{A_2}, which means there is no symmetry-adapted A_2 vector that involves the stretching of the O–H bonds.

It is not always necessary to work with all the symmetry operators and representations of a group. Certain subsets of the group of operators for a molecule may meet the requirements of a group on their own, and using the subset amounts to analyzing a problem in reduced symmetry; we then learn about a corresponding subset of the symmetry features of functions and vectors.

B.5 Character Tables

Key information about molecular symmetry is often given in tables of symmetry representations. Before considering these, we need to recognize differences in non-Abelian point groups. The mathematical analysis for non-Abelian point groups is somewhat more complicated than for Abelian groups, which is why they are not considered in detail here. Important differences are the following:

1. Some operators may commute, but there is at least one pair of operators that do not commute in a non-Abelian point group.

2. Non-Abelian point groups give rise to the possibility of degeneracy in quantum mechanical systems.

3. The symmetry-adapted functions of a non-Abelian point group do not necessarily have eigenvalues of 1 and –1 for the operators in the group. Some of the representations become matrices rather than 1 or –1 numbers. (To be general, we could think of the representations of Abelian groups as 1 × 1 matrices.) The projection operator formula requires the use of these representation matrices.

4. Less complete information than the representations of non-Abelian groups are the **characters**, which are the traces of the representation matrices. (For Abelian groups, with their 1 × 1 matrices, the characters are the same as the representations.) Characters can be used in the projection formula but only to distinguish a result of zero from a nonzero result, that is, to distinguish whether the function has any part that transforms as the representation or not. **Character tables** provide this concise information.

5. Classes of operators exist for non-Abelian groups in which all the operators of a class have the same character, though not the same representation matrix. Thus, character tables display only one column of values for all the elements in a given class. In Abelian groups, every operator is in its own class.

Table B.1 includes character tables for a number of molecular point groups. The headings for the columns are the symmetry operators grouped into classes. The horizontal rows are the characters of the representations with their designations in the leftmost column. The designation for the point group is in the upper left corner.

B.6 Application of Group Theory

The most direct information obtained from applying group theory to atomic and molecular problems is qualitative information about wavefunctions, energies, transitions, and properties. We have already considered how electronic wavefunctions can be symmetry-adapted, but there is often further insight from analysis of symmetry.

In the point groups of diatomic molecules, the classification of symmetry-adapted orbitals is according to the effect of rotations about the molecular axis. Notice that since diatomic molecules are cylindrical, rotation by any angle about the molecular axis is a symmetry operation. Orbital functions may possess nodal surfaces (surfaces where the wavefunction is zero), and these are planes that include the molecular axis as in the combination of 2p orbitals shown in Figure 10.7. Consider an end-on view of

TABLE B.1

Character Tables of Common Molecular Point Groups

C_2	E	C_2
A	1	1
B	1	−1

C_s	E	σ
A'	1	1
A''	1	−1

C_{2v}	E	C_2	σ	σ'
A_1	1	1	1	1
A_2	1	1	−1	−1
B_1	1	−1	−1	1
B_2	1	−1	1	−1

C_{2h}	E	C_2	σ	i
A_g	1	1	1	1
A_u	1	1	−1	−1
B_g	1	−1	−1	1
B_u	1	−1	1	−1

C_{3v}	E	$\{C_3, C_3^2\}$	$\{3\sigma's\}$
A_1	1	1	1
A_2	1	1	−1
E	2	−1	0

D_2	E	C_2^x	C_2^y	C_2^z
A_1	1	1	1	1
B_1	1	1	−1	−1
B_2	1	−1	1	−1
B_3	1	−1	−1	1

D_3	E	$\{C_3, C_3^2\}$	$\{3C_2's\}$
A_1	1	1	1
A_2	1	1	−1
E	2	−1	0

D_4	E	C_2	$2C_4$	C_2'	C_2''
A_1	1	1	1	1	1
A_2	1	1	1	−1	−1
B_1	1	1	−1	1	−1
B_2	1	1	−1	−1	1
E	2	−2	0	0	0

TABLE B.1 (continued)

Character Tables of Common Molecular Point Groups

D_{2d}	E	C_2	$2S_4$	$2C_2'$	$2\sigma_d$
A_1	1	1	1	1	1
A_2	1	1	1	−1	−1
B_1	1	1	−1	1	−1
B_2	1	1	−1	−1	1
E	2	−2	0	0	0

T_d	E	$8C_3$	$6S_4$	$3C_2$	$6\sigma_d$
A_1	1	1	1	1	1
A_2	1	1	−1	1	−1
E	2	−1	0	2	0
T_1	3	0	1	−1	−1
T_2	3	0	−1	−1	1

O	E	$8C_3$	$3S_2$	$6C_2$	$6C_4$
A_1	1	1	1	1	1
A_2	1	1	1	−1	−1
E	2	−1	2	0	0
T_1	3	0	−1	−1	1
T_2	3	0	−1	1	−1

D_{2h}	E	$C_2(z)$	$C_2(y)$	$C_2(x)$	i	σ_{xy}	σ_{xz}	σ_{yz}
A_g	1	1	1	1	1	1	1	1
A_u	1	1	1	1	−1	−1	−1	−1
B_{1g}	1	1	−1	−1	1	1	−1	−1
B_{2g}	1	−1	1	−1	1	−1	1	−1
B_{3g}	1	−1	−1	1	1	−1	−1	1
B_{1u}	1	1	−1	−1	−1	−1	1	1
B_{2u}	1	−1	1	−1	−1	1	−1	1
B_{3u}	1	−1	−1	1	−1	1	1	−1

the orbitals of a diatomic molecule. With zero, one, and two nodal planes, their qualitative features are

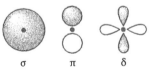

σ $\qquad$ π $\qquad$ δ

where the shaded regions are of a phase opposite that of the unshaded regions. These three correspond to different symmetry representations in the non-Abelian point group of a diatomic molecule. Notice that rotation by 180° changes the sign of the middle orbital by taking a lobe of one phase into the opposite phase, whereas rotation by 90° changes the sign of the rightmost orbital. Rotation by 180° leaves the sign of the rightmost orbital unchanged. As the number of nodal planes increases, the effects of symmetry rotations change accordingly. The designations for the different representations in the point group of a linear molecule are Σ or σ for no nodal planes, Π or π for one nodal plane, Δ or δ for two, and so on. The capital letters are used for overall symmetries of states and the small letters for orbitals.

Homonuclear diatomic molecules belong to the point group $D_{\infty h}$, which includes the inversion operator. Hence the orbitals for a molecule such as N_2 have symmetry designations with a g or u subscript (gerade or ungerade).

In the linear combinations of atomic orbitals used to describe diatomic systems, we find energetic preference for those combinations that have the fewest nodal planes between the atomic centers. A nodal plane between nuclei implies diminished electron density between the nuclei, a feature we associate with antibonding. Let us analyze several symmetry-adapted linear combinations of atomic orbitals of a homonuclear diatomic molecule starting with combinations of s orbitals.

$$s_A + s_B \longrightarrow \sigma_g \qquad s_A - s_B \longrightarrow \sigma_u$$

Atomic s orbitals can combine in two ways. The σ_g orbital is the bonding orbital, and the σ_u is antibonding. Atomic p orbitals along the molecular axis also combine to produce σ orbitals.

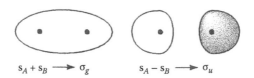

$$p_{x_A} - p_{x_B} \longrightarrow \sigma_g \qquad p_{x_A} + p_{x_B} \longrightarrow \sigma_u$$

Here, the plus combination produces the ungerade antibonding orbital. Perpendicular to the molecular axis (in either the y or z plane) are the following combinations of p orbitals.

$$p_{y_A} + p_{y_B} \longrightarrow \pi_u \qquad p_{y_A} - p_{y_B} \longrightarrow \pi_g$$

Here, the ungerade orbital is bonding and the gerade orbital is antibonding.

In diatomic molecules that do not have an inversion center, there is a qualitative similarity of the available molecular orbitals to those of a homonuclear diatomic molecule. There are always orbitals with the same number of nodal surfaces; the difference is that the nodal surfaces need not be planes and may be offset from the middle of the molecule. Though we cannot label such orbitals g or u, there are bonding and antibonding combinations qualitatively similar to those of homonuclear diatomic molecules.

Symmetry analysis can provide information about properties of the states of systems. Group theory tells us that a property will have a zero value if the operator associated with that property transforms as other than the totally symmetric representation of the group. Consider the dipole moment vector. The form of the operators of the components of the dipole is that of charge times a position coordinate (Equation 10.66). The symmetry of the component operators is deduced by applying symmetry operators to unit vectors in the directions of the position coordinate. In the case of a molecule with an inversion center, such as N_2, applying the inversion operator to x yields $-x$. Thus, the x operator and thereby the operator for the x-component of the dipole moment transform as an ungerade representation. Since the totally symmetric representation is a gerade representation, we conclude that the x-component of the dipole moment vector of N_2 is zero. The same argument applies to the y- and z-components, and thus, nitrogen has a zero-valued dipole because of its symmetry.

In contrast with nitrogen, carbon monoxide has a nonzero dipole moment for it lacks an inversion operator in its point group. Its dipole moment vector, though, has only one nonzero component, the one along the molecular axis, which we will let be the x axis. The point groups of diatomic and linear molecules have as reflecting planes all the planes that include the molecular axis. One of these planes is the xy plane. If the operation of reflection through this plane is applied to a z-position coordinate (or to a unit vector in the z direction), the result is $-z$. Hence, a z-component dipole operator is not totally symmetric. The same holds for the y-component using the xz plane as the reflection operator.

The discussion of the time-dependent perturbation theory in the beginning of Chapter 9 culminated in Equation 9.13, an equation for selection rules in a spectroscopic experiment. Transitions are allowed if the integral of the perturbing Hamiltonian between the wavefunctions of the initial and final states is nonzero. Group theory can help assess which such integrals are zero. In this case, a zero integral value occurs if the symmetry representation of the final state wavefunctions is different from the symmetry representation of the perturbing Hamiltonian acting on the initial state wavefunction.

With Abelian point groups, we find the representation of the product of two things by taking the products of their characters and determining the representation that corresponds. For example, in the C_{2v} symmetry of the water molecule, the product of something that transforms as B_1 with something that transforms as B_2 is of the A_2 representation.

C_{2v}	E	C_2	σ	σ'
A_1	1	1	1	1
A_2	1	1	-1	-1
B_1	1	-1	-1	1
B_2	1	-1	1	-1

$B_1 \times B_2 = 1\ 1\ -1\ -1$ which are the characters of the A_2 representation.

The operator of the component of the dipole moment along the C_2 axis of water transforms as A_1, the totally symmetric representation. All the A_1 characters are 1, and their products with any other representation's characters do not give any change in values. From this we conclude that transitions taking place via interaction with the dipole moment component along the C_2 axis are allowed if they are between states of the same symmetry representation (e.g., $A_1 \times A_1$, or $B_2 \times B_2$). If they are the same, the representation of the final state is the same as the representation of the perturbing operator (the dipole moment component) acting on the initial state wavefunction.

The component of the dipole moment of water perpendicular to the plane of water transforms as B_1. The value of the permanent dipole moment component in this direction is zero because this is not the totally symmetric representation. However, in transitions, there may be an interaction involving changes in this component. If the initial state were of B_1 symmetry, then the product of this dipole component operator with that state would transform as $B_1 \times B_1 = A_1$. A transition is allowed if A_1 is the symmetry of the final state. That is, transitions via this perpendicular component of the dipole moment operator are allowed if between states of B_1 symmetry and A_1 symmetry. By the same analysis, this component also gives rise to allowed transitions between A_2 and B_2 states.

The procedure for using symmetry to determine selection rules holds generally. It can be applied to vibrational transitions and to electronic transitions, and it can be applied with any interaction that a system experiences.

Appendix C: Special Quantum Mechanical Approaches

C.1 Angular Momentum Raising and Lowering Operators

There are a number of operator relations that are quite powerful in working out angular momentum problems. Certain of these involve operators called **raising operators** and **lowering operators**. These operators act on an angular momentum eigenfunction and increase the z-component of the vector by one step (raising it) or decrease it by one step (lowering it). The length of the angular momentum vector is unchanged. Raising (+) and lowering (–) operators can be generally defined for any angular momentum source, and they are

$$J_+ \equiv J_x + iJ_y \tag{C.1a}$$

$$J_- \equiv J_x - iJ_y \tag{C.1b}$$

As the context should be clear, the operator designation ^ is neglected.

The first property of the raising and lowering operators is that they commute with the operator corresponding to the square of the angular momentum. This is seen by evaluating the commutator. (Many of the raising and lowering operator expressions can be written concisely as a single equation, as is done here. These equations use the ± or ∓ symbol and are deciphered by either using the top of each symbol throughout to be one equation, or using the bottom of each symbol throughout to be the other equation.)

$$\left[J^2, J_\pm\right] = \left[J^2, \left(J_x \pm iJ_y\right)\right] = \left[J^2, J_x\right] \pm \left[J^2, J_y\right] = 0 \tag{C.2}$$

Because the J^2 operator commutes with each component operator, it must commute with the raising operator and with the lowering operator.

Next, we can work out the effect of the raising and lowering operators by applying them to an angular momentum function. Using Γ to designate any particular eigenfunction of J^2 and J_z, we have

$$J^2\Gamma = \mu\hbar^2\Gamma$$

$$J_z\Gamma = \nu\hbar\Gamma$$

μ and ν (times Planck's constant) are the eigenvalues for Γ. To understand the effect of a raising (or lowering) operator, we investigate what happens when it is applied to Γ:

$$J_+\Gamma = \Omega \tag{C.3}$$

A new function, Ω, is whatever is produced. Applying J^2 to Ω and using the commutation property in Equation C.2 shows that Ω's eigenvalue is the same as Γ's.

$$J^2\Omega = J^2\left(J_+\Gamma\right) = J_+\left(J^2\Gamma\right) = J + \left(\mu\hbar^2\Gamma\right) = \mu\hbar^2\left(J_+\Gamma\right) = \mu\hbar^2\Omega$$

This means that Ω corresponds to an angular momentum vector with the same length as the vector that corresponds to the Γ function.

Applying J_z to Ω determines the z-component of the angular momentum vector associated with this function.

$$\begin{aligned}
J_z\Omega &= J_z J_+\Gamma \quad \text{by Equation C.3}\\
&= J_z\left(J_x + iJ_y\right)\Gamma \quad \text{by Equation C.1a}\\
&= J_z J_x\Gamma + iJ_z J_y\Gamma\\
&= \left(i\hbar J_y + J_x J_z\right)\Gamma + i\left(-i\hbar J_x + J_y J_z\right)\Gamma\\
&= \hbar\left(J_x + iJ_y\right)\Gamma + \left(J_x + iJ_y\right)J_z\Gamma \quad \text{rearranged terms}\\
&= \hbar J_+\Gamma + J_+\left(\nu\hbar\right)\Gamma \quad \text{by Equation C.1a}\\
&= \hbar\left(\nu + 1\right)J_+\Gamma\\
&= \hbar\left(\nu + 1\right)\Omega
\end{aligned}$$

This shows that Ω is an eigenfunction of J_z, but its eigenvalue is bigger than the eigenvalue of Γ by ℏ. In other words, the z-component is one step up. And so the net effect of the operator J_+ is to change an angular momentum function into one with the M quantum number increased by 1. Likewise, the J_- operator produces a function whose M quantum number is 1 less than that of the original function.

If the raising operator is applied to an angular momentum function where the z-component is already at the maximum (i.e., where M = J), the result is zero. For instance,

$$L_+Y_{ll} = 0$$

$$J + |J\rangle = 0 \quad \text{where } M = J$$

Likewise, applying the lowering operator to a function with the lowest possible z-component (i.e., where M = −J) produces zero.

The application of a raising or a lowering operator changes the M quantum number of the angular momentum state, thereby changing one function into another. However, that statement does not completely describe the effect of the operator; it does not tell if the resulting function ends up multiplied by some constant. [To say that $\hat{O}$ acts on $f(x)$ to yield a function with the properties of $g(x)$ does not say that $\hat{O}f(x) = g(x)$, only that $\hat{O}f(x)$ is proportional to $g(x)$.] It happens that a raising or lowering operation does introduce a constant, but its value can be obtained using the normalization condition on angular momentum eigenfunctions. With the designation ψ_{JM} for normalized angular momentum eigenfunctions, the effect of the raising operator can be expressed in terms of an as yet unknown constant, c.

$$J_+\psi_{JM} = c\psi_{JM+1} \Rightarrow \langle J_+\psi_{JM}|J_+\psi_{JM}\rangle = \langle c\psi_{JM+1}|c\psi_{JM+1}\rangle = c^2\langle\psi_{JM+1}|\psi_{JM+1}\rangle = c^2 \tag{C.4}$$

Thus, because ψ_{JM+1} is normalized, c^2 can be found from evaluating the integral on the left-hand side of Equation C.4.

Several steps are required to find the integral that is equal to c^2. From the definition of the raising and lowering operators, each is seen to be the complex conjugate of the other. And because these are Hermitian operators, they can be rearranged in the integral we are seeking to evaluate.

$$\langle J_+\psi_{JM}|J_+\psi_{JM}\rangle = \langle\psi_{JM}|J_-J_+\psi_{JM}\rangle \tag{C.5}$$

The operator product on the right-hand side of Equation C.5 can be put in a different form by going back to the definition of the raising and lowering operators.

$$J_-J_+ = (J_x - iJ_y)(J_x + iJ_y) = J_x^2 + i(J_xJ_y - J_yJ_x) + J_y^2 = (J_x^2 + J_y^2 + J_z^2) - J_z^2 + i[J_x,J_y] = J^2 - J_z^2 - \hbar J_z \tag{C.6}$$

The last step requires evaluating the commutator. With this expression for the product of the lowering and raising operators, the integral in Equation C.5 can be evaluated because the function being acted on is an eigenfunction of the operators. This is shown in the following steps.

$$\langle\psi_{JM}|J_-J_+\psi_{JM}\rangle = \langle\psi_{JM}|(J^2 - J_z^2 - \hbar J_z)\psi_{JM}\rangle = \hbar^2[J(J+1) - M^2 - M]\langle\psi_{JM}|\psi_{JM}\rangle$$
$$= \hbar^2[J(J+1) - M^2 - M] \tag{C.7}$$

Therefore,

$$c^2 = \hbar^2[J(J+1) - M^2 - M]$$

$$c = \hbar\sqrt{J(J+1) - M^2 - M}$$

The choice of the positive root for c is by convention. (The negative root can be used so long as the choice is made consistently.) When the same analysis is carried through for the lowering operator, its c factor is obtained. Then, the complete effect of the raising and lowering operators on an arbitrary $|JM\rangle$ function is established. Written together, we have

$$J_{\pm}|J \quad M\rangle = \hbar\sqrt{J(J+1) - M^2 \mp M}\,|J \quad M \pm 1\rangle \tag{C.8}$$

The angular momentum functions designated $|j_1 m_1 j_2 m_2\rangle$ are referred to as **uncoupled states**, while the functions designated $|JM j_1 j_2\rangle$ are **coupled states**. A set of uncoupled state functions is called an **uncoupled basis**, and a set of coupled state functions is a **coupled basis**. There always exists a linear transformation between the two bases, and the functions in both are eigenfunctions of $\widehat{J_1^2}$ and $\widehat{J_2^2}$. The uncoupled functions, though, are eigenfunctions of the two operators $\widehat{J_{1z}}$ and $\widehat{J_{2z}}$, whereas the coupled functions are eigenfunctions of $\widehat{J^2}$ and $\widehat{J_z}$. This implies that we can transform from the coupled to the uncoupled basis by finding a transformation that makes the matrix representations of $\widehat{J_{1z}}$ and $\widehat{J_{2z}}$ diagonal. Or, we can transform to the coupled basis by finding a transformation that makes the representations of $\widehat{J^2}$ and $\widehat{J_z}$ diagonal.

One way of working out the transformation between coupled and uncoupled bases uses the raising and lowering operators in the last section, starting from the **parallel coupling** case, or the maximum M case. This is the one possible arrangement where $J = M = j_1 + j_2$, and for this case, the coupled state is exactly the same as the uncoupled state:

$$\left|J = (j_1 + j_2) \quad M = (j_1 + j_2) \quad j_1 \quad j_2\right\rangle = \left|j_1 \quad m_1 = j_1 \quad j_2 \quad m_2 = j_2\right\rangle$$

To this, we can apply raising or lowering operators in order to generate the relationships among the other possible states.

Since a lowering operator is formed from x- and y-component operators, a lowering operator for the total angular momentum can be obtained in terms of lowering operators of the two angular momentum sources.

$$J_x = J_{1x} + J_{2x}$$

$$J_y = J_{1y} + J_{2y}$$

$$J_+ = J_{1+} + J_{2+}$$

and

$$J_- = J_{1-} + J_{2-}$$

Applying the lowering operator to the parallel coupling case produces a new state.

$$J_-|J \quad J \quad j_1 \quad j_2\rangle_c = (J_{1-} + J_{2-})|j_1 \quad j_1 \quad j_2 \quad j_2\rangle_u \tag{C.9}$$

The left-hand side of Equation C.9 has the coupled state with a maximum z-component, which means that M has the value J. The subscript c on the ket vector is added for clarity in this discussion. The operator J_- acts on this function to lower the M value from J to $J-1$. That is, with Equation C.8, we have

$$J_- |J \quad J \quad j_1 \quad j_2 \rangle_c = \hbar \sqrt{J(J+1) - J^2 + J} |J \quad J-1 \quad j_1 \quad j_2 \rangle_c \tag{C.10}$$

The right-hand side of Equation C.9 has the uncoupled state (subscript u) where the z-components of the two sources of angular momentum are at their maximum, which means m_1 has the value j_1 and m_2 has the value j_2. One of the lowering operators being applied, the J_{1-} operator, changes the function from that with $m_1 = j_1$ to that with $m_1 = j_1 - 1$. This operator does not affect m_2.

$$J_{1-} |j_1 \quad j_1 \quad j_1 \quad j_2 \rangle_u = \hbar \sqrt{j_1(j_1+1) - j_1^2 + j_1} |j_1 \quad j_1-1 \quad j_2 \quad j_2 \rangle_u \tag{C.11}$$

A similar result is obtained from applying J_{2-}

$$J_{2-} |j_1 \quad j_1 \quad j_1 \quad j_2 \rangle_u = \hbar \sqrt{j_2(j_2+1) - j_2^2 + j_2} |j_1 \quad j_1 \quad j_2 \quad j_2-1 \rangle_u \tag{C.12}$$

The results from Equations C.10 through C.12 can be substituted into Equation C.9 to give the coupled state for $M = J - 1$ (1 down from the maximum z-component case) in terms of uncoupled functions. The process can be repeated to find the next lower M case, and so on, with the following overall result.

$$\sqrt{J(J+1) - M^2 + M} |J \quad M-1 \quad j_1 \quad j_2 \rangle_c = \sqrt{j_1(j_1+1) - m_1^2 + m_1} |j_1 \quad m_1-1 \quad j_2 \quad m_2 \rangle_u$$
$$+ \sqrt{j_2(j_2+1) - m_2^2 + m_2} |j_1 \quad m_1 \quad j_2 \quad m_2-1 \rangle_u \tag{C.13}$$

Equation C.13 is a linear transformation between the coupled and uncoupled angular momentum functions.

As a detailed example of this scheme, let us consider the case with $j_1 = 1$ and $j_2 = 1$. The possible values for the total angular momentum quantum number on coupling the two sources is $J = 2$, $J = 1$, and $J = 0$. The maximum z-component situation corresponds to $J = 2$ and $M = 2$ in the coupled states picture, and to $m_1 = 1$ and $m_2 = 1$ in the uncoupled states picture.

$$|2 \quad 2 \quad 1 \quad 1 \rangle_c = |1 \quad 1 \quad 1 \quad 1 \rangle_u \tag{C.14a}$$

Notice that since j_1 and j_2 are fixed values—they correspond to the lengths of the angular momentum vectors of the sources—it is not necessary to list them in the state designations. Thus, Equation C.14a is written more concisely with $|J\ M\rangle$ for the coupled state and $|m_1\ m_2\rangle$ for the uncoupled state:

$$|2 \quad 2 \rangle_c = |1 \quad 1 \rangle_u \tag{C.14b}$$

Applying appropriate lowering operators to both sides of Equation C.14b leads to the following steps that invoke the results in Equations C.10 through C.12.

$$J_- |2 \quad 2\rangle_c = (J_{1-} + J_{2-}) |1 \quad 1\rangle_u$$

$$\hbar\sqrt{2(2+1) - 2^2 + 2} \, |2 \quad 1\rangle_c = \hbar\sqrt{1(1+1) - 1^2 + 1} \, |0 \quad 1\rangle_u + \hbar\sqrt{1(1+1) - 1^2 + 1} \, |1 \quad 0\rangle_u \qquad \text{(C.15)}$$

$$\therefore |2 \quad 1\rangle_c = \frac{1}{\sqrt{2}} \left(|0 \quad 1\rangle_u + |1 \quad 0\rangle_u \right)$$

This says that the coupled state where $J = 2$ and $M = 1$ is a linear combination of two uncoupled states, one where $m_1 = 0$ and $m_2 = 1$, and the other where $m_1 = 1$ and $m_2 = 0$.

Applying lowering operations to Equation C.15 produces another coupled state in terms of the uncoupled states.

$$J_- |2 \quad 1\rangle_c = \frac{1}{\sqrt{2}} \left[(J_{1-} + J_{2-}) |0 \quad 1\rangle_u + (J_{1-} + J_{2-}) |0 \quad 1\rangle_u \right]$$

$$\hbar\sqrt{2(2+1) - 1^2 + 1} \, |2 \quad 0\rangle_c = \frac{\hbar}{\sqrt{2}} \left[\begin{array}{l} \sqrt{1(1+1) - 0^2 + 0} \, |-1 \quad 1\rangle_u + \sqrt{1(1+1) - 1^2 + 1} \, |0 \quad 0\rangle_u \\[4pt] + \sqrt{1(1+1) - 1^2 + 1} \, |0 \quad 0\rangle_u + \sqrt{1(1+1) - 0^2 + 0} \, |1 \quad -1\rangle_u \end{array} \right]$$

$$\therefore |2 \quad 0\rangle_c = \frac{1}{\sqrt{6}} \left[|-1 \quad 1\rangle_u + 2 |0 \quad 0\rangle_u + |1 \quad -1\rangle_u \right] \qquad \text{(C.16)}$$

This says that the coupled state where $J = 2$ and $M = 0$ is a linear combination of three uncoupled states. This process can be continued to find the coupled state $J = 2$ and $M = -1$, and then $J = 2$ and $M = -2$. Of course, this last situation is analogous to how we started, and so we arrive at

$$|2 \quad -2\rangle_c = |-1 \quad -1\rangle_u$$

An alternative to the process described is to start from this lowest M case and apply raising operators, but the results will be identical.

Application of the lowering operator to the original, maximum z-component state produced all the $J = 2$ states, but there remain $J = 1$ and $J = 0$ states. These are found, in sequence, by the condition of orthogonality. It is always true that the M quantum number of the coupled state equals the sum of the m quantum numbers of the uncoupled states that combine to form a coupled state. That is, $M = m_1 + m_2$, and this can be verified for the examples in Equations C.15 and C.16. Consequently, one of the remaining coupled states, that corresponding to $J = 1$ and $M = 1$, must be a linear combination of only the uncoupled states $|1 \quad 0\rangle_u$ and $|0 \quad 1\rangle_u$. There is only one linear combination of these that is orthogonal to the state in Equation C.15, and it is the opposite combination. Therefore, the following must be true.

$$|1 \quad 1\rangle_c = \frac{1}{\sqrt{2}} \left(|0 \quad 1\rangle_u - |1 \quad 0\rangle_u \right) \qquad \text{(C.17)}$$

From this relation, lowering operations can be performed to find the $J = 1$ and $M = 0$ coupled state function, and then the $J = 1$ and $M = -1$ function. Orthogonality is used again to find the last unknown coupled state function, that for which $J = 0$ and $M = 0$.

In this example, there are nine uncoupled states and, of course, nine coupled states. The coefficients that have been worked out are linear expansion coefficients, and so they can be arranged in a 9×9 matrix. This is the transformation (matrix) from the uncoupled basis to the coupled basis. This matrix and its inverse contain the coefficients for making up one kind of basis function from the other set. These coefficients are used in many problems in quantum physics, and they are often called **vector coupling coefficients**. Since they can be worked out once and for all for specific choices of j_1 and j_2, there are many tabulations of the values, and many alternate procedures have been devised for finding them. They are sometimes scaled by different factors for convenience in different circumstances. Names given to vector coupling coefficients under various prescriptions include Clebsch–Gordon coefficients, Racah coefficients, $3 - j$ symbols, and $6 - j$ symbols.

C.2 Matrix Methods in Variation and Perturbation Theory

A special and powerful use of variation theory is with linear variational parameters. That means that the trial wavefunction is taken to be a linear combination of functions in some chosen set. The adjustable parameters are the expansion coefficients of each of these functions. This is, of course, a specialization of the way in which variation theory can be used, but it is powerful because the resulting equations take the form of matrix expressions. Solving the Schrödinger equation becomes a problem in linear algebra, and such problems are ideally suited to computer solution.

For any problem to be treated by linear variational methods, a **basis set** must be selected at the outset. This is simply some set of functions of the coordinates for the problem under study. In the best circumstances, the functions are chosen to be part of the basis set because they are close to the anticipated form of the true wavefunction. The number of functions is not restricted. A few may be used, or many may be used. The basis set may even have an infinite number, though the discussion here will assume a finite basis set.

The linear variational wavefunction, given as a **basis set expansion**, is a linear combination of the functions in the basis. With Γ as the wavefunction and $\{\phi_i\}$ designating the set of basis functions, the expansion is

$$\Gamma = \sum_{i}^{N} c_i \phi_i \qquad (C.18)$$

N is the number of functions in the basis set. The functions in the set are distinguished by the subscript i. The coefficients, c_i, are expansion coefficients; they are the adjustable parameters. Notice that basis set expansions have already been used: The first-order perturbation theory corrections to the wavefunction were obtained as an expansion in a basis of the zero-order functions.

The expectation value of the energy of Γ is the quantity to be minimized. This is given by the following expression, subject to a normalization constraint on Γ.

$$\langle \Gamma | H \Gamma \rangle = \left\langle \sum_{i}^{N} c_i \phi_i \middle| H \sum_{j}^{N} c_j \phi_j \right\rangle = \sum_{i}^{N} \sum_{j}^{N} c_i^* \langle \phi_i | H \phi_j \rangle c_j \qquad \text{(C.19)}$$

This double-summation expression has a form that can be expressed with matrices (see Appendix A). It happens to be of the form of a row of values (coefficients) times a matrix of (integral) values times a column of coefficients, with the result being a scalar. The rank-one matrices of coefficients are column or row vectors, designated as

$$\vec{c} = \begin{pmatrix} c_1 \\ c_2 \\ c_3 \\ \dots \\ c_N \end{pmatrix}$$

and

$$\vec{c}^T = \begin{pmatrix} c_1 & c_2 & c_3 & \cdots & c_N \end{pmatrix}$$

It is also helpful to have a special symbol for the complex conjugate of the row vector: $\vec{c}^T = \vec{c}\,T^*$. The integral values are arranged into a rank-two $N \times N$ matrix designated H.

$$H = \begin{pmatrix} \langle \phi_1 | H \phi_1 \rangle & \langle \phi_1 | H \phi_2 \rangle & \langle \phi_1 | H \phi_3 \rangle & \cdots & \langle \phi_1 | H \phi_N \rangle \\ \langle \phi_2 | H \phi_1 \rangle & \langle \phi_2 | H \phi_2 \rangle & \langle \phi_2 | H \phi_3 \rangle & \cdots & \langle \phi_2 | H \phi_N \rangle \\ \langle \phi_3 | H \phi_1 \rangle & \langle \phi_3 | H \phi_2 \rangle & \langle \phi_3 | H \phi_3 \rangle & \cdots & \langle \phi_3 | H \phi_N \rangle \\ \dots & \dots & \dots & \dots & \dots \end{pmatrix}$$

H is called the **matrix representation** of the Hamiltonian operator. A matrix representation of an operator is a matrix of integral values arranged in rows and columns according to the basis functions. Clearly, the values in a matrix representation are dependent on the functions that were selected for the basis set. A different basis set implies a different matrix representation. Wherever it is important to keep track of the basis used in the representation, a superscript is added to the designation of the matrix, for example, H^ϕ, and it identifies the particular function set. Now, the quantity in Equation C.19 can be written with the coefficient vectors and the Hamiltonian matrix in a very simple form.

$$\langle \Gamma | H \Gamma \rangle = \vec{c}^t H \vec{c} = \sum_{i,j} c_i^* H_{ij} c_j$$

Again, this is the form of an N-long row of numbers times an $N \times N$ square array times an N-long column of numbers, and that produces just one value (not a vector or matrix).

The overlap of the Γ function with itself is the dot product of the coefficient vector and its complex conjugate transpose if, as we assume here, the basis functions are orthonormal.

$$\langle \Gamma | \Gamma \rangle = \vec{c}^\dagger \vec{c} = \sum_j c_j^* c_j$$

When this is divided into $\langle \Gamma | H \Gamma \rangle$, the resulting value is the energy expectation whatever the normalization of Γ. This is a way of imposing the normalization constraint.

$$W = \frac{\langle \Gamma | H \Gamma \rangle}{\langle \Gamma | \Gamma \rangle}$$

W is a function of the coefficients in $\vec{c}$. To apply variation theory, W must be minimized with respect to each of the elements of $\vec{c}$. This means taking the first derivative and setting that function to zero at the point where the coefficients lead to the minimum value of W.

$$\left. \frac{\partial W}{\partial c_i} \right|_{\vec{c} = \vec{c}_{min}} = 0 \tag{C.20}$$

There is one equation for each coefficient.

The partial differentiation of W with respect to one of the coefficients yields

$$\frac{\partial W}{\partial c_i^*} = \frac{\sum_j H_{ij} c_j}{\langle \Gamma | \Gamma \rangle} - \frac{c_i}{\langle \Gamma | \Gamma \rangle} \left\{ \frac{\langle \Gamma | H \Gamma \rangle}{\langle \Gamma | \Gamma \rangle} \right\}$$

(We can also differentiate* with respect to c_i instead of with respect to c_i^*.) Notice that the quantity in brackets is W. So, when this first-derivative function is evaluated at the coefficient values that minimize W, that is, $\vec{c}_{min}$, this bracketed quantity must equal E, the variational energy. With this replacement for the bracketed quantity, we have

$$\left. \frac{\partial W}{\partial c_i^*} \right|_{\vec{c} = \vec{c}_{min}} = \frac{1}{\langle \Gamma | \Gamma \rangle} \left(\sum_j H_{ij} c_{jmin} - c_{imin} E \right)$$

Setting this to zero following the condition for minimization of W yields

$$\sum_j H_{ij} c_{jmin} - c_{imin} E = 0$$

---

* The differentiation of W, as shown, is with respect to a complex conjugate of one of the coefficients and is carried out by treating each coefficient as independent of its complex conjugate. By carrying out the differentiation with respect to complex conjugates, equations are developed for the simple coefficients. Should the coefficients be taken to be real, which would mean the coefficient and its complex conjugate are the same, differentiation would yield the same expression, but times 2. The factor of 2, though, makes no difference in the coupled equations because of setting the derivative to zero.

Clearly, there is one such equation for each choice of the index i. The whole set of these equations can be arranged in a single matrix expression.

$$H\vec{c}_{min} = \vec{c}_{min}E \qquad\qquad (C.21)$$

This expression is an important and common one called a **matrix eigenvalue equation**. It says that the matrix H operating on (or multiplying) the vector of coefficients that minimizes W is equal to that vector times a constant, which is the energy eigenvalue. This result means that in general applying linear variation theory amounts to finding an eigenvector of the matrix representation of the Hamiltonian.

There are numerous means for solving the matrix equation given in Equation C.21. Regardless of which algorithm is applied, the process of finding the eigenvalues and the coefficient vectors is referred to as a **diagonalization** of H because it amounts to a transformation of H from the original basis to a basis where it is diagonal. Diagonalization is extremely well-suited to computers, and there are many well-crafted computer programs in use for diagonalizing matrices. Some are general routines, while others are specialized for the forms that certain matrices take on. For a matrix with N rows and N columns, it is usual for a diagonalization routine to require a number of multiplications of pairs of real numbers on the order of N^3. This means that the computational cost grows with the matrix dimension as N^3.

Matrix techniques are helpful in perturbation theory when the complication of degeneracy arises. If there is a state, ψ_j, that is degenerate with the state of interest, ψ_n, at zero order, then the first-order correction to the wavefunction requires a diagonalization. If a set of states, which may be a subset of the states of some system, happens to be degenerate with respect to some operator, then any normalized linear combination of those states will also be an eigenfunction of that operator with the same eigenvalue. For instance, consider a set of functions $\{\phi_i, i = 1, \ldots, N\}$ that are degenerate normalized eigenfunctions of a zero-order Hamiltonian. That is,

$$\hat{H}^{(0)}\phi_i = E_n^{(0)}\phi_i$$

Any linear combination of these functions is an eigenfunction of this Hamiltonian, as the following steps show:

$$\hat{H}^{(0)}\sum_{j=1}^{N}c_j\phi_j = \sum_{j=1}^{N}c_j\hat{H}^{(0)}\phi_j = \sum_{j=1}^{N}c_jE_n^{(0)}\phi_j = E_n^{(0)}\sum_{j=1}^{N}c_j\phi_j$$

Consequently, the zero-order wavefunction in the perturbation treatment for the nth state cannot be presumed to be any particular one of the ϕ_i functions. Instead, it must be regarded as an undetermined linear combination of the functions in the degenerate subset.

$$\psi_n^{(0)} = \sum_{j=1}^{N}c_j\phi_j \qquad\qquad (C.22)$$

This is substituted into the first-order perturbation theory Schrödinger equation to give

$$\hat{H}^{(1)} \sum_{j=1}^{N} c_j \phi_j + \hat{H}^{(0)} \psi_n^{(1)} = E_n^{(1)} \sum_{j=1}^{N} c_j \phi_j + E_n^{(0)} \psi_n^{(1)} \qquad (C.23)$$

Now, the solution of this equation must yield not only the first-order correction to the energy and the first-order correction to the wavefunction, but also the value of the expansion coefficients in the zero-order wavefunction, the c_j's.

In nondegenerate perturbation theory, the procedure for obtaining the first-order correction to the energy was to multiply by the zero-order wavefunction for the state of interest and integrate. The result was that the first-order correction to the energy was the expectation value of the perturbing Hamiltonian with the zero-order wavefunction. In the case of degeneracy, it is necessary to carry out this process with each of the degenerate functions. Thus, we take one function from the set and multiply Equation C.23 by it and then integrate.

$$\left\langle \phi_k \left| \hat{H}^{(1)} \sum_{j=1}^{N} c_j \phi_j \right. \right\rangle + \left\langle \phi_k \left| \hat{H}^{(0)} \right| \psi_n^{(1)} \right\rangle = E_n^{(1)} \sum_{j=1}^{N} c_j \left\langle \phi_k | \phi_j \right\rangle + E_n^{(0)} \left\langle \phi_k | \psi_n^{(1)} \right\rangle$$

Just as in the nondegenerate development, the second left-hand-side term is the same as the second right-hand-side term, and so they cancel. Because of orthogonality of the degenerate functions, the summation on the right-hand side that is multiplied by the first-order correction to the energy simplifies to c_k.

$$\sum_{j=1}^{N} \left\langle \phi_k \left| \hat{H}^{(1)} \right| \phi_j \right\rangle c_j = c_k E_n^{(1)}$$

This expression may be rewritten by using the matrix representation of the perturbing Hamiltonian in the ϕ basis.

$$\sum_{j=1}^{N} H_{kj}^{(1)\phi} c_j = c_k E_n^{(1)}$$

And when all choices of k are considered, there are N equations of this form. They can all be collected into one matrix equation by arranging the c coefficients into a column vector.

$$H^{(1)\phi} \vec{c} = \vec{c} E_n^{(1)} \qquad (C.24)$$

This happens to be the standard form of a matrix eigenvalue equation.

The interpretation of Equation C.24 is that the first-order correction to the energy of the state of interest (n) is an eigenvalue of the matrix representation of the perturbing Hamiltonian in the basis of the degenerate functions $\{\phi_i\}$. The eigenvector in Equation C.24 is the set of coefficients that tells which linear combination of degenerate functions represents the zero-order state function, that is, what values end up being used in Equation C.22. So, in the case of degenerate functions, a diagonalization procedure is used, just as in linear variation theory, except that the matrix representation is limited to just the functions that are degenerate with the state of interest. The diagonalization implies a transformation of the $\{\phi_i\}$ set of functions; they are mixed with each other. When the eigenenergies obtained via Equation C.24 are all different, the perturbation is said to have removed the degeneracy. After solving Equation C.24, the first-order correction to the wavefunction and all the higher-order corrections can be obtained as in the nondegenerate case, if the degeneracy has been removed. Or if the degeneracy has not been removed, the expressions for higher-order corrections are developed by allowing the corrections to the wavefunctions to be linear combinations of the degenerate functions, analogous to Equation C.22.

Appendix D: Table of Integrals

1. General integral relations

$$\int cf(x)dx = c\int f(x)dx$$

$$\int \left(f_1(x) + f_2(x) + \cdots\right)dx = \int f_1(x)dx + \int f_2(x)dx + \cdots$$

$$\int f(x)g(y)dxdy = \int f(x)dx \int g(y)dy$$

$$\int \left(f_1(x) + f_2(x)\right)\left(g_1(x) + g_2(x)\right)dx = \int f_1(x)g_1(x)dx + \int f_1(x)g_2(x)dx$$

$$+ \int f_2(x)g_1(x)dx + \int f_2(x)g_2(x)dx$$

$$\int \frac{dx}{x} = \ln x$$

$$\int a^x dx = \frac{a^x}{\ln a}$$

2. Integrals involving inverse polynomial functions

$$\int \frac{dx}{a+bx} = \frac{1}{b}\ln(a+bx)$$

$$\int \frac{dx}{(a+bx)^2} = \frac{1}{b(a+bx)}$$

$$\int \frac{dx}{x(a+bx)} = -\frac{1}{a}\ln\frac{a+bx}{x}$$

$$\int \frac{dx}{x(a+bx)^2} = \frac{1}{a(a+bx)} - \frac{1}{a^2}\ln\frac{a+bx}{x}$$

$$\int \frac{dx}{(a+bx)(c+dx)} = \frac{1}{ad-bc}\ln\frac{c+dx}{a+bx}$$

$$\int \frac{xdx}{a+bx^2} = \frac{1}{2b}\ln\left(x^2 + \frac{a}{b}\right)$$

3. Integrals over Gaussian functions ($c > 0$)

$$\int_0^\infty e^{-cx^2} dx = \frac{1}{2}\sqrt{\frac{\pi}{c}}$$

$$\int_0^\infty x e^{-cx^2} dx = \frac{1}{2c}$$

$$\int_0^\infty x^2 e^{-cx^2} dx = \frac{1}{4}\sqrt{\frac{\pi}{c^3}}$$

$$\int_0^\infty x^{2n+1} e^{-cx^2} dx = \frac{n!}{2c^{n+1}}$$

$$\int_0^\infty x^{2n} e^{-cx^2} dx = \frac{(2n-1)(2n-3)...(3)(1)}{2^{n+1}}\sqrt{\frac{\pi}{c^{2n+1}}}$$

$$\int_{-\infty}^0 x^n e^{-cx^2} dx = (-1)^n \int_0^\infty x^n e^{-cx^2} dx$$

4. Integrals of exp(x) ($c > 0$)

$$\int x^n e^{cx} dx = e^{cx} \sum_{j=0}^n (-1)^j \frac{n!}{(n-j)!} \frac{x^{n-j}}{c^{j+1}}$$

$$\int_0^\infty x^n e^{-cx} dx = \frac{n!}{c^{n+1}} \quad (\text{for } n = 1, 2, 3, ...)$$

$$\int \frac{e^{ax}}{x} dx = \ln x + \sum_{n=1}^\infty \frac{a^n x^n}{n(n!)}$$

$$\int \frac{dx}{a + be^{gx}} = \frac{x}{a} - \frac{1}{ag}\ln\left(a + be^{gx}\right)$$

5. Integrals over trigonometric functions

$$\int \sin(cx)dx = -\frac{1}{c}\cos(cx)$$

$$\int \cos(cx)dx = \frac{1}{c}\sin(cx)$$

$$\int \sin(a+bx)dx = -\frac{1}{b}\cos(a+bx)$$

$$\int (\sin^2 cx)dx = \frac{x}{2} - \frac{\sin(2cx)}{4c}$$

$$\int (\sin^3 cx)dx = -\frac{(\cos cx)(\sin^2 cx + 2)}{3c}$$

$$\int (\sin^n cx)dx = -\frac{(\cos cx)(\sin^{n-1} cx)}{nc} + \frac{n-1}{n}\int (\sin^{n-2} cx)dx$$

$$\int cx^n \sin(cx)dx = -x^n \cos(cx) + n\int x^{n-1}\cos(cx)dx$$

$$\int cx^n \cos(cx)dx = x^n \sin(cx) - n\int x^{n-1}\sin(cx)dx$$

$$\int x(\sin^2 cx)dx = \frac{x^2}{4} - \frac{x(\sin 2cx)}{4c} - \frac{\cos 2cx}{8c^2}$$

$$\int x^2(\sin^2 cx)dx = \frac{x^3}{6} - \left(\frac{x^2}{4c} - \frac{1}{8c^3}\right)\sin 2cx - \frac{x\cos 2cx}{4c^2}$$

$$\int \sin(ax)\sin(bx)dx = \frac{\sin[(a-b)x]}{2(a-b)} - \frac{\sin[(a+b)x]}{2(a+b)} \quad \text{for } a^2 \neq b^2$$

$$\int \cos(ax)\cos(bx)dx = \frac{\sin[(a-b)x]}{2(a-b)} + \frac{\sin[(a+b)x]}{2(a+b)} \quad \text{for } a^2 \neq b^2$$

$$\int \sin(ax)\cos(bx)dx = -\frac{\cos[(a-b)x]}{2(a-b)} + \frac{\sin[(a+b)x]}{2(a+b)} \quad \text{for } a^2 \neq b^2$$

$$\int \cos^m x \sin^n x dx = \frac{(\cos^{m-1} x)(\sin^{n+1} x)}{m+n} + \frac{m-1}{m+n}\int \cos^{m-2} x \sin^n x dx$$

Appendix E: Table of Atomic Masses and Nuclear Spins

For the nontransition elements of the periodic table, this table lists isotope masses, percent natural abundances, nuclear spins, nuclear g factors (see Equation 12.10), and electron occupancies.

Element Number	Symbol	Orbital Occupancy[a]	Isotope Mass[b]	Nuclear Spin, I_N	$g_N I_N$[b]	Natural Abundance[b]
1	H	$1s^1$	1.007825	1/2	2.79285	99.989
			2.014102	1	0.85744	0.012
			3.016049	1/2	2.97896	
2	He	$1s^2$	3.016029	1/2	−2012762	0.0001
			4.002603	0		99.999
3	Li	[He] $2s^1$	6.015123	1	0.82205	7.5
			7.016005	3/2	3.25644	92.5
4	Be	[He] $2s^2$	9.012182	3/2	−1.1776	100
5	B	[He] $2s^2\,2p^1$	10.012937	3	1.8006	19.9
			11.009305	3/2	2.6886	80.1
6	C	[He] $2s^2\,2p^2$	12.000000	0		98.90
			13.003355	1/2	0.70241	1.10
			14.003242	0		
7	N	[He] $2s^2\,2p^3$	14.003074	1	0.40376	99.63
			15.000109	1/2	−0.28319	0.37
8	O	[He] $2s^2\,2p^4$	15.994915	0		99.762
			16.999132	5/2	−1.8938	0.038
			17.999161	0		0.200
9	F	[He] $2s^2\,2p^5$	18.000938	1		
			18.998403	1/2	2.62887	100
10	Ne	[He] $2s^2\,2p^6$	19.992440	0		90.48
			20.993847	3/2	−0.66180	0.27
			21.991385	0		9.25
11	Na	[Ne] $3s^1$	21.994436	3	1.746	
			22.989768	3/2	2.217522	100
			23.990963	4	1.690	
12	Mg	[Ne] $3s^2$	23.985042	0		78.99
			24.985837	5/2	−0.85545	10.00
			25.982593	0		11.01
13	Al	[Ne] $3s^2\,3p^1$	26.981539	5/2	3.64151	100
14	Si	[Ne] $3s^2\,3p^2$	27.976927	0		92.22
			28.976495	1/2	−0.5553	4.69
			29.973770	0		3.09
15	P	[Ne] $3s^2\,3p^3$	30.973762	1/2	1.13160	100
			31.973907	1	−0.2524	

(continued)

(continued)

Element Number	Symbol	Orbital Occupancy[a]	Isotope Mass[b]	Nuclear Spin, I_N	$g_N I_N$[b]	Natural Abundance[b]
16	S	[Ne] $3s^2\ 3p^4$	31.972071	0	94.99	
			32.971459	3/2	0.64382	0.75
			33.967867	0		4.25
			34.969032	3/2	1.00	
			35.967081	0		0.01
17	Cl	[Ne] $3s^2\ 3p^5$	34.968853	3/2	0.82187	75.76
			36.965903	3/2	0.68412	24.24
18	Ar	[Ne] $3s^2\ 3p^6$	35.967546	0	0.337	
			37.962732	0	0.063	
			39.962384	0	99.600	
19	K	[Ar] $4s^1$	38.963707	3/2	0.39146	93.26
			39.963999	4	−1.29810	0.012
			40.961835	3/2	0.21487	6.73
			41.962403	2	−1.1425	
			42.960716	3/2	0.163	
20	Ca	[Ar] $4s^2$	39.962591	0		96.94
			41.958618	0		0.65
			42.958767	7/2	−1.31726	0.14
			43.955482	0		2.09
			44.956187	7/2	−1.327	
			45.953693	0		0.004
			46.954546	7/2	−1.38	
			47.952534	0		0.187
21–30	Sc to Zn	[Ca] $3d^{1-10}$				
31	Ga	[Zn] $4p^1$	66.928202	3/2	1.8507	
			67.927980	1	0.01175	
			68.925580	3/2	2.01659	60.11
			70.924701	3/2	2.56227	39.89
32	Ge	[Zn] $4p^2$	67.928094	0		
			69.924247	0		20.38
			71.922076	0		27.31
			72.923459	9/2	−0.879467	7.76
			73.921178	0		36.72
			75.921403	0		7.83
33	As	[Zn] $4p^3$	74.921597	3/2	1.43947	100
34	Se	[Zn] $4p^4$	73.922475	0		0.89
			74.922523	5/2	0.67	
			75.919212	0		9.37
			76.919913	1/2	0.53506	7.63
			77.917308	0		23.77
			78.918499	7/2	−1.018	
			79.916520	0		49.61
			81.916698	0		8.73
35	Br	[Zn] $4p^5$	78.918337	3/2	2.106400	50.69
			80.916291	3/2	2.270562	49.31

(continued)

Element Number	Symbol	Orbital Occupancy[a]	Isotope Mass[b]	Nuclear Spin, I_N	$g_N I_N$[b]	Natural Abundance[b]
36	Kr	[Zn] 4p^6	77.920365	0		0.35
			79.916379	0		2.25
			81.913484	0		11.6
			82.914136	9/2	−0.970669	11.5
			83.911507	0		57.0
			85.910611	0		17.3
37	Rb	[Kr] 5s^1	84.911794	5/2	1.353	72.17
			85.911167	2	−1.6920	
			86.909187	3/2	2.7512	27.83
38	Sr	[Kr] 5s^2	83.913430	0		0.56
			84.912933	9/2	−1.001	
			85.909267	0		9.86
			86.908884	9/2	−1.093	7.00
			87.905619	0		82.58
			88.907451	5/2	−1.149	
			89.907738	0		
39–48	Y to Cd	[Sr] 4d$^{1–10}$				
49–52	In to Te	[Cd] 5p$^{1\ 4}$				
53	I	[Cd] 5p^5	122.905589	5/2	2.82	
			124.904630	5/2	2.82	
			126.904473	5/2	2.8133	100
			128.904988	7/2	2.621	
			130.906125	7/2	2.742	
54	Xe	[Cd] 5p^6	123.905893			0.10
			125.904274	0		0.09
			127.903531	0		1.91
			128.904779	1/2	−0.7780	26.4
			129.903508	0		4.1
			130.905082	3/2	0.691862	21.2
			131.904154	0		26.9
			133.905395	0		10.4
			135.907219	0		8.9

[a] Occupancies are given relative to a core occupancy which is that of the element indicated in brackets.
[b] Values are from Lide, D.R., *Handbook of Chemistry and Physics*, 86th edn., CRC Press, Boca Raton, FL, 2005. Natural abundances are percents.

(continued)

Appendix F: Fundamental Constants and Conversion of Units

F.1 Systems of Units

A system of units of measure for purely mechanical systems can be defined by specifying the unit of measurement for just three physical quantities. Specifying a unit of measurement is simple in principle. For instance, we can make two marks on a piece of paper and declare the distance between them to be one unit of length in some system we thereby create. If we were to select or devise a unit of measurement for mass, for length, and for time, there would exist a basis for measurement of other mechanical quantities such as velocity (i.e., unit length per unit time), momentum, acceleration, energy, and so on. We can specify units of measurement for three physical quantities other than mass, length, and time in order to define a system of units, but we cannot specify more than three for mechanical systems. Electromechanical systems require another basic unit, such as charge.

In the metric system, three familiar basic units are the gram, the meter, and the second, and their symbols are g, m, and s, respectively. Their sizes are defined by international agreement to very high precision. Prefixes are used to designate units that are smaller by powers of 10 or larger by powers of 10. For each prefix, a symbol is attached to the unit symbol. A kilogram is 1000 g and its symbol is kg. The commonly encountered prefixes in chemical physics are the following.

Metric Prefix (Symbol)	Means the Prefixed Unit Is Scaled by
exa (E)	10^{18}
peta (P)	10^{15}
tera (T)	10^{12}
giga (G)	10^{9}
mega (M)	10^{6}
kilo (k)	10^{3}
deci (d)	10^{-1}
centi (c)	10^{-2}
milli (m)	10^{-3}
micro (μ)	10^{-6}
nano (n)	10^{-9}
pico (p)	10^{-12}
femto (f)	10^{-15}
atto (a)	10^{-18}

Two traditional selections of a basic unit for mass, length, and time are often encountered. The **mks** form uses meters, kilograms, and seconds. The mks energy unit, called the joule

TABLE F.1

Pressure Conversion Factors

	Pa	bar	atm	torr
1 Pa	1.0	10^{-5}	9.86923×10^{-6}	7.50062×10^{-3}
1 bar	10^5	1.0	0.986923	7.50062×10^{2}
1 atm	1.01325×10^5	1.01325	1.0	760
1 torr	1.33322×10^2	0.133322	1.31579×10^{-3}	1.0
1 psi	6.8947×10^3	6.8947×10^{-2}	6.8046×10^{-2}	51.715

(J) is the mass unit (kg), times the length unit (m) to the second power, times the inverse of the second power of time unit (s).

$$1J = 1kg \ (m/s)^2$$

The **cgs** form uses centimeters, grams, and seconds. The cgs energy unit is the erg:

$$1 \ erg = 1 \ g \ (cm/s)^2 = 10^{-3} kg \ (10^{-2} m/s)^2 = 10^{-7} J$$

The conversion between mks and cgs for any other physical quantities is obtained in a like manner.

The International System of Units (Systeme International d'Unites, abbreviated SI) is the mks system extended to other types of quantities. For instance, force is a quantity with units of energy per unit length. In the SI system, one unit of force (a newton, N) is one joule per meter. With the definition of a joule, a newton is the same as one kilogram-meter per second-squared. Thus, the SI unit of force, N, is defined or derived in terms of the three basic mks units. Likewise, the unit of power is the watt, W, and one watt is one joule per second (J s^{-1}).

Pressure is force per unit area, and the SI unit of pressure (pascal, Pa) is one Newton per meter squared (N m^{-2}). Many different measures have been in use for pressure, and Table F.1 provides the relations among certain of them.

F.2 Fundamental Constants

Among the "properties of nature," we can include the values of fundamental constants. Three are particularly important in the quantum mechanics of atoms and molecules. These are the speed of light (c), Planck's constant ($\hbar$), and the mass of an electron (m_e). The values of these constants in cgs and SI units are

$$c = 2.997925 \times 10^{10} \ cm \ s^{-1} = 2.997925 \times 10^{8} \ m \ s^{-1}$$

$$\hbar = 1.054573 \times 10^{-27} \ erg \ s^{-1} = 1.054573 \times 10^{-34} J \ s^{-1}$$

TABLE F.2

Values of Constants[a]

Constant	Symbol	Value
Avogadro's constant	N_A	6.022142×10^{23} mol^{-1}
Speed of light	c	2.997925×10^8 m s^{-1}
Electron charge	e	1.602176×10^{-19} C
		1.602176×10^{-20} emu
		1.602176×10^{-20} cm$^{1/2}$ g$^{1/2}$
		4.803204×10^{-10} cm$^{3/2}$ g$^{1/2}$ s^{-1}
Planck's constant	$\hbar$	1.054572×10^{-34} J s
		1.054572×10^{-27} erg s
	h	6.626069×10^{-34} J s
		6.626069×10^{-27} erg s
Boltzmann's constant	k	$1.3806503 \times 10^{-23}$ J K^{-1}
		$1.3806503 \times 10^{-16}$ erg K^{-1}
		0.695025 cm^{-1} K^{-1}
Gas constant	R	8.314472 J K^{-1} mol^{-1}
		8.205746×10^{-5} m^3 atm K^{-1} mol^{-1}
		8.205746×10^{-2} L-atm K^{-1} mol^{-1}
		1.986352 cal K^{-1} mol^{-1}
Faraday constant ($= N_A e$)	F	9.648534×10^4 C mol^{-1}
Electron rest mass	m_e	9.109382×10^{-31} kg
Proton rest mass	m_p	1.672622×10^{-27} kg
Neutron rest mass	m_n	1.674927×10^{-27} kg
Bohr magneton	μ_B	9.274009×10^{-24} J T^{-1}
		9.274009×10^{-21} erg G^{-1}
Nuclear magneton	μ_o	5.050783×10^{-27} J T^{-1}
		5.050783×10^{-24} erg G^{-1}
Gravitational constant	G	6.67300×10^{-11} N m^2 kg^{-2}
Acceleration of gravity	g	9.80665 m s^{-2}

[a] Where there are several values for a constant, the first one given is the SI value.

(also used: $\hbar = h/2\pi$)

$$m_e = 9.10939 \times 10^{-28} \text{g} = 9.10939 \times 10^{-31} \text{kg}$$

A very important derived constant is Avogadro's constant, N_A. It is the number of carbon atoms needed to make up 12.0 g of carbon, and its value is 6.022137×10^{23} per mole (mol^{-1}). Table F.2 lists values for a number of constants.

F.3 Atomic Units

A special system of units called atomic units (a.u.) proves particularly convenient in the quantum mechanics of the electronic structure of atoms and molecules. This is the system of units where the electromechanical measures of mass, length, time, and charge are

TABLE F.3

Energy Conversion Factors

	J (Joule)	cm^{-1}	eV	kcal mol^{-1}
1 erg	10^{-7}	5.03411×10^{15}	6.24151×10^{11}	1.4393×10^{13}
1 J	1.0	5.03411×10^{22}	6.24151×10^{18}	1.4393×10^{20}
1 cm^{-1}	1.98645×10^{-23}	1.0	1.23984×10^{-4}	2.8672×10^{-3}
1 eV	1.60218×10^{-19}	8065.54	1.0	23.061
1 h (a.u.)	4.35975×10^{-18}	2.19475×10^{5}	27.2114	627.51
1 L-atm	1.01325×10^{2}	5.10081×10^{24}	6.32421×10^{20}	1.4584×10^{15}

chosen such that three fundamental constants, m_e, $\hbar$, and e, and the Bohr radius take on values of exactly 1.0. This can be accomplished because we specify a system of units by four measures, and this choice presents no more than four constraints on those measures. The resulting atomic unit of length is called the **bohr** after Niels Bohr, and the atomic unit of energy is called the **hartree** after Douglas R. Hartree.

To illustrate the use of atomic units, let us evaluate the ground state energy of the hydrogen atom (under the assumption of an infinitely massive nucleus), which is known to be

$$E = -\frac{m_e e^4}{2\hbar^2} = -2.18 \times 10^{-11} \text{erg} = -2.1 \times 10^{-18} \text{J}$$

In atomic units, however, we have simply that $E = -1/2$ a.u. From this, we can express the conversions between atomic units of energy (hartree, h) and ergs and joules.

$$1.0 \text{ h} = 2 \ (2.18 \times 10^{-11} \text{erg}) = 4.36 \times 10^{-11} \text{erg} = 4.36 \times 10^{-18} \text{J}$$

Other energy conversions are given in Table F.3.

F.4 Electromagnetic Units and Constants

From an atomic and molecular view, one of the most basic electromagnetic quantities is the size of the charge of an electron. This fundamental quantity is usually designated e. In the International System of units e is 1.60218×10^{-19} coulombs (C). One mole of electrons has a charge of N_A times e, and this amount of charge is the SI unit known as a Faraday, F: $1.0 \text{ F} = 9.648529 \times 10^4 \text{ C mol}^{-1}$. In the cgs-emu (centimeter, gram, second-electromagnetic unit) system, the value of e is 1.60218×10^{-20} emu $= 1.60218 \times 10^{-20} \text{ cm}^{3/2} \text{ g}^{1/2}$, or 0.1 times the value in coulombs. The unit of charge is also expressed with a factor of c (speed of light) absorbed, and then e is $4.80324 \times 10^{-10} \text{ cm}^{1/2} \text{ g}^{1/2} \text{s}^{-1}$. The different choices are not complicated to recognize because a wrong interpretation yields an inconsistency in the units (i.e., dimensional analysis can be used).

The SI unit for electrical potential is the volt (V). A particle experiencing a potential has an interaction energy (in joules) that is equal to the product of the charge in SI units and the potential expressed in volts. Thus, an electron experiencing a 1 V potential has an interaction energy,

$$E = e(1.0\ V) = 1.60218 \times 10^{-19}\,C\text{-}V = 1.60218 \times 10^{-19}\,J$$

This quantity serves as a measure for a special unit of energy called the electron volt (eV). Simply, 1 eV is defined to be 1.60218×10^{-19} J.

The cgs unit of magnetic induction is the gauss (G). In the SI system, the unit is the tesla (T), and the relation is $1\ G = 10^{-4}$ T.

F.5 Frequencies

Frequencies are associated with a regular oscillation in time. They are expressed in two ways, angular frequencies and linear frequencies. Angular frequencies are in radians per unit time, whereas linear frequencies are in cycles per unit time. It is frequently the case that ω is used for an angular frequency and ν is used for a frequency expressed in cycles per unit time. The relation between the two is simple since the number of radians covered in one cycle must be 2π, the number of radians in a circle.

$$\omega = 2\pi\nu$$

An oscillation in time (t) of the form of a cosine function is written as either $\cos(\omega t)$ or $\cos(2\pi\nu t)$. That is usually the context that distinguishes angular from linear frequencies when the choice is not stated explicitly. The frequency of electromagnetic radiation is usually specified by the cycles per second (cps), a unit called the hertz (Hz), and so the values are linear frequencies.

The energy, E, of a photon of radiation is directly proportional to the frequency, and the proportionality constant is Planck's constant:

$$E = \hbar\omega = h\nu$$

Thus, the frequency scale of cycles per second may be taken to be equivalent to an energy scale in ergs (via the factor h).

In molecular spectroscopy, another frequency scale has proved to be convenient. In this case, the frequencies, ν, are divided by a fundamental constant, the speed of light, c. Then, the frequencies are given in the inverse unit of length, cm^{-1}, and 1 cm^{-1} is called a wavenumber. This amounts to using the relation between wavelength (λ) of radiation and frequency, $c = \lambda\nu$, so as to give the inverse wavelength instead of the frequency, the two being proportional:

$$\frac{1}{\lambda} = \frac{\nu}{c}$$

Wavenumbers serve as an energy scale just as well as frequencies do, and it is very common to give atomic and molecular energies in units of cm^{-1}.

TABLE F.4

The Electromagnetic Spectrum

Radiation Region	Wavelength λ (m)	Frequency Range $\nu = c/\lambda$ (s⁻¹)	Photon Energy $1/\lambda = \nu/c$ (cm⁻¹)
Radio	0.1 to 10^4	3×10^4 to 3×10^9	10^{-6} to 0.1
Microwave	10^{-2} to 0.1	3×10^9 to 3×10^{10}	0.1 to 1
Infrared	10^{-6} to 10^{-2}	3×10^{10} to 3×10^{14}	1 to 10^4
Visible	4×10^{-7} to 10^{-6}	3×10^{14} to 7.5×10^{14}	10^4 to 2.5×10^4
Ultraviolet	10^{-9} to 4×10^{-7}	7.5×10^{14} to 3×10^{17}	2.5×10^4 to 10^7
X-ray and Gamma[a]	$<10^{-9}$	$>3 \times 10^{17}$	10^7

[a] The distinction between the x-ray and gamma ray regions has changed in recent decades. Originally, the electromagnetic radiation emitted by x-ray tubes had a longer wavelength than the radiation emitted by radioactive nuclei (gamma rays). Older literature distinguished between x- and gamma radiation on the basis of wavelength, with radiation shorter than some arbitrary wavelength defined as gamma rays. However, as shorter wavelength continuous spectrum "x-ray" sources such as linear accelerators and longer wavelength "gamma ray" emitters were discovered, the wavelength bands largely overlapped. The two types of radiation are now usually defined by their origin: x-rays are emitted by electrons outside the nucleus, while gamma rays are emitted by the nucleus.

$$\tilde{E} \; (\text{cm}^{-1}) = \frac{E \; (\text{Joules})}{hc} = \frac{\nu}{c} = \frac{\omega}{2\pi c} \equiv \tilde{\omega} \; (\text{cm}^{-1})$$

The symbol $\tilde{E}$ as opposed to just E is sometimes used to indicate that the system of units for the energy is wavenumbers and not joules. A special symbol, of course, is unnecessary because an energy expression must hold in any system of units.

F.6 The Electromagnetic Spectrum

The electromagnetic spectrum has regions that are referred to by names that have arisen, in some cases, because of different instrumental techniques. Table F.4 lists the general designations and gives rough cutoff values for the ranges.

One unit of length that is still widely used for wavelengths, though not considered standard, is the angstrom, abbreviated Å. The conversion is 1.0 Å = 10^{-10} m.

Appendix G: List of Tables

2.1   Equations of State for Gases 35

2.2   Coefficients in the van der Waals Equation of State for Selected Real Gases 36

3.1   Heat Capacity Coefficients of Equation 3.48 for 1 mol of Selected Gases 73

4.1   Molar Transition Enthalpies and Transition Temperatures 93

6.1   Standard State Thermodynamic Properties 146

6.2   Reduction Potentials for Reactions in Water at 298 K Relative
        to a Hydrogen Standard 156

8.1   Associated Legendre Polynomials through $l = 4$ 220

9.1   Spectroscopic Constants (cm^{-1}) of Certain Diatomic Molecules 273

9.2   Vibrational Frequencies (cm^{-1}) of Selected Small Molecules 281

9.3   Characteristic Vibrational Stretching Frequencies (cm^{-1}) of Certain
        Functional Groups 283

10.1 Hydrogen Atom Radial Functions 294

10.2 Energy Levels of the Hydrogen Atom 295

10.3 Comparison of Calculated and Measured Values
        of Certain Molecular Properties 320

B.1   Character Tables of Common Molecular Point Groups 448

D   Table of Integrals 465

E   Table of Atomic Masses and Nuclear Spins 469

F.1   Pressure Conversion Factors 474

F.2   Values of Constants 475

F.3   Energy Conversion Factors 476

F.4   The Electromagnetic Spectrum 478

Appendix H: Points of Interest

1. James Clerk Maxwell 19
2. Intermolecular Interactions 46
3. Heat Capacities of Solids 76
4. Josiah Willard Gibbs 96
5. Gilbert Newton Lewis 119
6. Galactic Reaction Chemistry 158
7. Molecular Force Fields 188
8. The Quantum Revolution 239
9. Laser Spectroscopy 286
10. John Clarke Slater 337
11. Lars Onsager 366
12. The NMR Revolution 394
13. Irving Langmuir 412

Appendix I: Atomic Masses and Percent Natural Abundance of Light Elements

Element	Orbital Occupancy	Isotope Mass (amu)	Natural Abundance	Isotope Mass (amu)	Natural Abundance
1. H	$1s^1$	1.007825	99.985	2.014102	0.015
2. He	$1s^2$	4.002603	100		
3. Li	[He] $2s^1$	6.015121	7.5	7.016003	92.5
4. Be	[He] $2s^2$	9.012182	100		
5. B	[He] $2s^2\,2p^1$	10.012937	19.9	11.009305	80.1
6. C	[He] $2s^2\,2p^2$	12.000000	98.90	13.003355	1.10
7. N	[He] $2s^2\,2p^3$	14.003074	99.63	15.000109	0.37
8. O	[He] $2s^2\,2p^4$	15.994915	99.762	17.999160	0.200
		16.999131	0.038		
9. F	[He] $2s^2\,2p^5$	18.998403	100		
10. Ne	[He] $2s^2\,2p^6$	19.992436	90.48	21.991383	9.25
		20.993843	0.27		
11. Na	[Ne] $3s^1$	22.989768	100		
12. Mg	[Ne] $3s^2$	23.985042	78.99	25.982594	11.01
		24.985837	10.00		
13. Al	[Ne] $3s^2\,3p^1$	26.981539	100		
14. Si	[Ne] $3s^2\,3p^2$	27.976927	92.23	29.973771	3.10
		28.976495	4.67		
15. P	[Ne] $3s^2\,3p^3$	30.973762	100		
16. S	[Ne] $3s^2\,3p^4$	31.972071	95.02	33.967867	4.21
		32.971459	0.75	35.967081	0.02
17. Cl	[Ne] $3s^2\,3p^5$	34.968853	75.77	36.965903	24.23
18. Ar	[Ne] $3s^2\,3p^6$	35.967546	0.337	39.962384	99.600
19. K	[Ar] $4s^1$	38.963707	93.26	40.961835	6.88
		39.963999	0.012		
20. Ca	[Ar] $4s^2$	39.962591	96.94	42.958766	0.14
		41.958618	0.65	43.955481	2.09
31. Ga	[Zn] $4p^1$	68.925580	60.11	70.924701	39.89
32. Ge	[Zn] $4p^2$	69.924250	21.23	73.921177	35.94
		71.922079	27.66	75.921402	7.44
		72.923463	7.73		
33. As	[Zn] $4p^3$	74.921594	100		
34. Se	[Zn] $4p^4$	73.922475	0.89	77.917308	23.78
		75.919212	9.36	79.916520	49.61
		76.919913	7.63	81.916698	8.73
35. Br	[Zn] $4p^5$	78.918336	50.69	80.916289	49.31

Appendix J: Values of Constants

Avogadro's constant	N_A	$6.022142 \times 10^{23} \text{ mol}^{-1}$
Speed of light	c	$2.997925 \times 10^8 \text{ m s}^{-1}$
Electron charge	e	$1.602176 \times 10^{-19} \text{ C}$
		$1.602176 \times 10^{-20} \text{ emu}$
		$1.602176 \times 10^{-20} \text{ cm}^{1/2} \text{ g}^{1/2}$
		$4.803204 \times 10^{-10} \text{ cm}^{3/2} \text{ g}^{1/2} \text{ s}^{-1}$
Planck's constant	$\hbar$	$1.054572 \times 10^{-34} \text{ J s}$
		$1.054572 \times 10^{-27} \text{ erg s}$
	h	$6.626069 \times 10^{-34} \text{ J s}$
		$6.626069 \times 10^{-27} \text{ erg s}$
Boltzmann's constant	k	$1.3806503 \times 10^{-23} \text{ J K}^{-1}$
		$1.3806503 \times 10^{-16} \text{ erg K}^{-1}$
		$0.695025 \text{ cm}^{-1} \text{ K}^{-1}$
Gas constant	R	$8.314472 \text{ J K}^{-1} \text{ mol}^{-1}$
		$8.205746 \times 10^{-5} \text{ m}^3 \text{ atm K}^{-1} \text{ mol}^{-1}$
		$8.205746 \times 10^{-2} \text{ L-atm K}^{-1} \text{ mol}^{-1}$
		$1.986352 \text{ cal K}^{-1} \text{ mol}^{-1}$
Faraday constant $(=N_A e)$	F	$9.648534 \times 10^4 \text{ C mol}^{-1}$
Electron rest mass	m_e	$9.109382 \times 10^{-31} \text{ kg}$
Proton rest mass	m_p	$1.672622 \times 10^{-27} \text{ kg}$
Neutron rest mass	m_n	$1.674927 \times 10^{-27} \text{ kg}$
Bohr magneton	μ_B	$9.274009 \times 10^{-24} \text{ J T}^{-1}$
		$9.274009 \times 10^{-21} \text{ erg G}^{-1}$
Nuclear magneton	μ_o	$5.050783 \times 10^{-27} \text{ J T}^{-1}$
		$5.050783 \times 10^{-24} \text{ erg G}^{-1}$
Gravitational constant	G	$6.67300 \times 10^{-11} \text{ N m}^2 \text{ kg}^{-2}$
Acceleration of gravity	g	9.80665 m s^{-2}

Appendix K: The Greek Alphabet

A, α	Alpha	Z, ζ	Zeta	Λ, λ	Lambda	Π, π	Pi	Φ, φ	Phi
B, β	Beta	H, η	Eta	M, μ	Mu	P, ρ	Rho	X, χ	Chi
Γ, γ	Gamma	Θ, θ	Theta	N, ν	Nu	Σ, σ	Sigma	Ψ, ψ	Psi
Δ, δ	Delta	I, ι	Iota	Ξ, ξ	Xi	T, τ	Tau	Ω, ω	Omega
E, ε	Epsilon	K, κ	Kappa	O, o	Omicron	Y, υ	Upsilon		

Appendix K: The Greek Alphabet

Answers to Selected Exercises

1.1 $0, 7.672 \times 10^{-23}, 2.302 \times 10^{-22}$ J

1.2

[6, 6, 5, 5, 5, 5]	[5, 6, 6, 5, 5, 5]	[5, 5, 6, 5, 6, 5]
[6, 5, 6, 5, 5, 5]	[5, 6, 5, 6, 5, 5]	[5, 5, 6, 5, 5, 6]
[6, 5, 5, 6, 5, 5]	[5, 6, 5, 5, 6, 5]	[5, 5, 5, 6, 6, 5]
[6, 5, 5, 5, 6, 5]	[5, 6, 5, 5, 5, 6]	[5, 5, 5, 6, 5, 6]
[6, 5, 5, 5, 5, 6]	[5, 5, 6, 6, 5, 5]	[5, 5, 5, 5, 6, 6]

1.3 Total number of possible arrangements = 64

$$P = \frac{\left[C(4,3,1,0,0,0,0) + C(4,3,0,1,0,0,0) + C(4,3,0,0,1,0,0) + \ldots\right]}{64}$$

$$= \frac{6(5)C(4,3,1,0,0,0,0)}{64}$$

$$= \frac{20}{63} = 0.09259$$

1.5 $P_2/P_1 = \exp[(E_1 - E_2)/kT] = \exp(-6.0/1.380658) = 0.01296$

1.6 Equation 1.14: $\ln(10!) = 15.096$ Equation 1.15: $\ln(10!) = 13.026$

$$\ln(1000!) = 5912.1 \quad \ln(1000!) = 5907.8$$

$$\ln(100000!) = 1.0513 \times 10^6 \quad \ln(100000!) = 1.0513 \times 10^6$$

1.7 At $T = 300$ K, $3\,kT = 1.2426 \times 10^{-20}$ J; from Appendix E, find each atomic mass in amu and convert to kg;

300 K rms speeds (m s^{-1}) are 1367.3 (He), 611.8 (Ne), and 432.7 (Ar).

500 K rms speeds (m s^{-1}) are 1765.2 (He), 789.8 (Ne), and 558.6 (Ar).

10 K rms speeds (m s^{-1}) are 249.6 (He), 111.7 (Ne), and 79.0 (Ar).

1.8 1.014×10^{-25} kg or 61.09 amu

2.1 2.494×10^6 Pa

2.2 2.494×10^4 Pa

2.5 Particle mass = $5.3120396 \times 10^{-26}$ kg

at 300 K, $s_{max} = 394.9$ m s^{-1}, $\langle s \rangle = 445.6$ m s^{-1}, $\langle s^2 \rangle^{1/2} = 483.7$ m s^{-1}

at 400 K, $s_{max} = 456.0$ m s^{-1}, $\langle s \rangle = 514.5$ m s^{-1}, $\langle s^2 \rangle^{1/2} = 558.5$ m s^{-1}

2.6 Ideal 240.543 K; argon 240.972 K; methane 241.406 K; acetylene 242.624 K

2.7 1.3201 bar

3.1 49.033 Pa

3.3 2.305 kJ (the same for each expansion)

3.7 $b = -2.944 \times 10^{-3}\,m^3;\ 46.94\,L$

3.10 20.222

3.13 $5.763\,J\,K^{-1}$

4.2

Hydrogen	$V_c = 0.07953\,L$	$T_c = 3.297\,K$
Water	$V_c = 0.09147\,L$	$T_c = 647.2\,K$
Methane	$V_c = 0.12903\,L$	$T_c = 190.6\,K$
Benzene	$V_c = 0.3579\,L$	$T_c = 562.2\,K$

4.5 $-1.302 \times 10^{-3}\,L\,mol^{-1}$

4.11 $-5.914 \times 10^{-4}\,K$

4.13 $-1.497\,K$

5.2 For Equation 2.30, $\lim_{P \to 0}\left[aP\left(1 - b/T^2\right)\right] = 0$

5.5 $\delta T_m^\circ : H_2O - 1.027,\ H_2S - 0.123,\ CCl_4 - 1.574$

$\delta T_m^\circ : H_2O \quad 0.285,\ H_2S \quad 0.203,\ CCl_4 \quad 0.341$

5.6 0.091

6.1 At this temperature and pressure, 40.091 mol of an ideal gas has a volume of $1.0\,m^3$. That corresponds to $D = 2.4143 \times 10^{25}$ particles m^{-3}. $\lambda_{He} = 0.147\,cm;\ \lambda_{Ar} = 0.0587\,cm$

6.3 $310.4\,m\,s^{-1}$ at $250\,K;\ 345.6\,m\,s^{-1}$ at $310\,K$

6.4 a. $0.115\,kJ\,mol^{-1}$ b. $6.340\,kJ\,mol^{-1}$ c. $53.60\,kJ\,mol^{-1}$ d. $582.1\,kJ\,mol^{-1}$

6.5 a. $890.7\,kJ$

6.7 $\ln K(300) = -0.00116;\ \ln K(105) = -0.000399$

6.9 $327\,kJ\,mol^{-1}$

7.2 $100\,N\,m^{-1};$ (a) $1.048 \times 10^{16}\,s^{-1}$ (b) $2.445 \times 10^{14}\,s^{-1}$ (c) $141.4\,s^{-1}$

7.3 $2.426 \times 10^{-10}\,m;\ 2.731 \times 10^{-22}\,m\,s^{-1}$

7.5 Since $\nu = 1$, $h\nu = 6.626 \times 10^{-34}\,J$ (a very small energy separation); $n \approx 1024$

7.6 $\mu(H^{35}Cl) = 1.626653 \times 10^{-27}\,kg;\ \mu(D^{35}Cl) = 3.162355 \times 10^{-27}\,kg;$ ratio $= 0.514380$

8.1 The values are 1, 1/2, 3/8, and 5/16 times $\beta/\sqrt{\pi}$. The values diminish with increasing n.

8.2 $x = \pm 1/\beta$

8.3 Designating state quantum numbers (n_x, n_y),
Lowest energy level is nondegenerate: $(0,0)$
First excited level is nondegenerate: $(0,1)$
Next level is twofold degenerate: $(1,0), (0,2)$
Next level is twofold degenerate: $(1,1), (0,3)$
Next level is threefold degenerate: $(2,0), (1,2), (0,4)$

8.6 0, 1, 1, 2, 2, 2, 3, 3, 4

9.1 2.9979×10^{10} Hz (microwave); 2.9979×10^{11} Hz (infrared); 2.9979×10^{12} Hz (infrared); 2.9979×10^{13} Hz (infrared); 2.9979×10^{14} Hz (visible red)

9.2 LiH: $7.519\,\text{cm}^{-1}$; 6LiH: $7.677\,\text{cm}^{-1}$; LiD $4.234\,\text{cm}^{-1}$

9.3 For $n = 0 \to 1$, the band center is at $1359.0\text{ cm}^{-1} = \tilde{\omega}_e - 2\tilde{\omega}_e\tilde{\chi}_e - \tilde{\alpha}_e$

9.5 In order: inactive, inactive, active, active, inactive

10.1 $2.179873 \times 10^{-18}\,\text{J} = 109.737\,\text{cm}^{-1}$

10.2 $E = 2.178686 \times 10^{-22}\,\text{J} = 10.9677\,\text{cm}^{-1}$; $\langle r \rangle = 7.93766 \times 10^{-7}\,\text{m}$

10.3 $\langle r \rangle_{\text{He}} = 3.968828 \times 10^{-11}\,\text{m}$; $\langle r \rangle_{\text{Ne}} = 1.587531 \times 10^{-11}\,\text{m}$; $\langle r \rangle_{\text{Ar}} = 0.881962 \times 10^{-11}\,\text{m}$

10.4 ^3P and ^1P

10.5 $S = 1$, $M_S = 0$ (triplet)

11.1 30; 180

11.2 Fermions: ^{7}Li, ^{13}C, 17O; bosons: ^{6}Li, ^{12}C, ^{16}O, ^{18}O

11.4 10^{-26} at $100\,\text{K}$; 10^{-8} at $300\,\text{K}$; 0.003707 at $1000\,\text{K}$; 0.1390 at $10{,}000\,\text{K}$

11.5 $17.41\,\text{K}$; $5223\,\text{K}$; $1.741 \times 10^6\,\text{K}$

12.1 4 for H_2O; 9 for D_2O; 54 for $D_2{}^{17}O$

12.2 36 for H; 24 for N

13.10 $0.52\,\text{m}^2$

13.11 3:1

Index

A

Abelian group, 440
Absolute entropy, 9, 39, 65, 150, 351
Activation
    barrier, 131, 212, 405
    energy, 142, 403
Activity, 101, 115, 119
    coefficient, 109, 115, 121
    quotient, 112, 154
Adiabatic, 57
    expansion, 58, 60, 72
Adjoint, 426
Adsorption, 402–403
Allowed transitions, 253, 374, 452
Angular momentum, 216, 225, 371
    coupling, 226, 304, 371
    operators, 224, 297, 390
    quantum numbers, 224, 255,
        306, 357
    vectors, 224, 301
Anharmonicity, 260, 272, 325, 360
Anode, 152
Antisymmetrization, 313, 348
Antisymmetrizer, 314, 340
Approximation
    Born–Oppenheimer, 310, 336
    doubly-harmonic, 264
    power series, 415
    self-consistent field, 318, 338
    steady state, 141
    Stirling's, 13, 21
Arrhenius, S., 142
Atmospheric pressure, 69
Attractive forces, 47, 104
Auger electron spectroscopy, 409
Azeotropes, 110

B

Band center, 268
Band head, 329
Barrier, 127, 131, 189, 212, 284, 405
Basis set expansion, 459
Berthelot gas, 35
Bimolecular reaction, 136
Bloch, F., 395
Bloembergen, N., 395

Boiling point elevation, 116
Bohr magneton, 297
Bohr, N., 216, 476
Bohr radius, 296
Boltzmann, L., 8, 47, 77, 97
Boltzmann's constant, 9, 143, 404
Boltzon, 348
Bond lengths, 2, 188, 264, 405
Bond order, 169, 322
Born, M., 241
Born–Oppenheimer approximation, 310
Bose–Einstein condensation, 353, 367
Bose–Einstein statistics, 348
Bosons, 348
Bound states, 6, 209
Boyle's law, 25
Boyle, R., 25
Bragg, W.H., 77
Bragg, W.L., 77
Breathing motion, 279
Brewer, L., 120
Brillouin, L., 239
Buckingham, A., 48

C

Canonical ensemble, 346
Capillary rise, 400
Carnot cycle, 61
Carnot, S., 59
Catalyst, 408
Cathode, 152
Centrifugal distortion, 260, 325
Characteristic frequency, 282
Characters/character tables, 446
Charge distribution, 252, 330, 373
Charles, J., 24
Charles' law, 24
Chemical
    activity, 101
    potential, 86, 101, 346
    shift, 376
Chemisorption, 402–403, 406
Clapeyron, B., 90
Clapeyron equation, 89
Classical
    mechanics, 3, 165, 201, 327, 345
    turning points, 184, 212

Clausius–Clapeyron equation, 92
Clausius inequaility, 67
Clausius, R., 67
Coefficients
    activity, 115
    diffusion, 135
    fugacity, 103
    Joule–Thomson, 74
    thermal conductivity, 134
    thermal expansion, 37
    vector coupling, 459
    viscosity, 20, 133
Collision(s)
    elastic, 30
    inelastic, 30
    probability, 132
Column matrix, 424
Combination transition, 282
Commutator, 194, 248, 453
Complex reaction, 136, 158, 162
Composition variables, 86
Compressibility, 37, 338
    factor, 49, 122
Configurations, 7, 41, 65, 358
Conservation of energy, 37, 65, 145
Continuum, 212, 286, 352
Coordinate(s)
    generalized, 166
    internal, 276
    normal, 186, 277
    spherical polar, 166, 217, 254, 293
    transformation, 171, 217, 278
Correlation diagram, orbital, 320
Correlation effects, electron, 319
Coupled basis, 456
Coupling constant, 272, 378
Critical
    phenomena, 46, 85
    point, 84, 401
    temperature, 85, 401
Cross section, 51, 129
Curve fitting, 34, 417

D

Dalton's atomic theory, 1, 78
Dalton's law, 106
Damadian, R., 395
de Broglie, L., 177
de Broglie, M., 239
de Broglie relation, 177
de Broglie wavelength, 177, 240
Debye, P., 76

Debye, Peter, Award, 366
Degeneracy, 5, 206, 258, 293, 348, 383
Degrees of freedom, 16, 81, 128, 276, 324, 361, 402
Del squared, 211
Desorption, 402–403
Determinant, 316, 426
Density of states, 277
Diagonalization, 429, 462
Diagonal matrix, 428
Diatomic molecule, 5, 125, 247, 291, 345, 447
Dielectric constant, 77
Dieterici equation, 49
Diffraction, 77, 176, 286
Diffusion, 20, 133, 366, 402
    coefficient, 134
Dipole
    allowed transitions, 252
    forbidden transitions, 252
    magnetic, 297, 373
    moment, 77, 252, 291, 378
    polarizability, 334
Dirac, P., 194
Distillation, 110
    fractional, 111
Doubly harmonic approximation, 264
Dulong, P., 76
Dulong–Petit rule, 77

E

Effective potential, 256, 304
Effective pressure, 103
Ehrenfest, P., 239
Eigenfunction, 178, 195, 247, 292, 346, 374, 440, 453
Eigenvector, 428, 462
Einstein, A., 1, 77, 120, 177, 237
Elastic collision, 30
Electrochemical
    cell, 152
    potentials, 120
Electrolyte, 151, 367
Electron
    configurations, 303
    correlation, 319
    occupancies, 303
    paramagnetic resonance, 386
    spin, 241, 300, 371
    spin resonance, 386
Elementary reaction, 135
Emission lines, 301
Energy
    free, 40, 68, 112, 151
    Gibbs, 39, 68, 86, 105, 148

Helmholtz, 39
internal, 17, 30, 54, 86, 144, 343
kinetic, 3, 27, 54, 126, 211, 254, 311, 359, 410
levels, 1, 72, 184, 198, 247, 345, 374
orbital, 319
rotational, 5, 260, 326, 362
Engine, 59
Enthalpy
formation, 148
molar, 86, 148
reaction, 114, 145
standard, 145
vaporization, 92, 117
Entropy, 7, 39, 51, 82, 105, 150, 341
Enzymes, 161
Eötvös relationship, 401
Equations of motion, 165, 204
Hamilton's, 173
Equilibrium, 51, 81, 101, 125, 343, 399
constant, 111, 148
phase, 82, 115
Equipartition of energy, 362
Ernst, R., 395
Eutectic point, 117
Exact differential, 52, 421
Excited states, 4, 68, 264, 325, 356
Expectation value, 200, 295, 378, 459
External field, 297, 371

F

Faraday, M., 153
Faraday's constant, 153
Fermi–Dirac statistics, 348
Fermi energy, 352
Fermion, 337, 348, 371
Fermi temperature, 352
Fine structure, 276, 310
Finite difference, 333
First law, 37, 65, 145
First order reaction, 135
Flux, 133, 404
Fock operator, 317
Forbidden transition, 253
Force constant, 168, 202, 259, 319
Force field, 188, 276
Fractional distillation, 111
Franck–Condon overlap, 328
Freezing point, 82, 101
Frequency
characteristic, 283
factor, 404
transition, 283, 320

Fuel cell, 155
Fugacity, 103
Fugacity coefficient, 103
Fundamental transition, 276
Fusion, 92

G

Galaxy, 158
Galvanic cell, 152
Gas constant, 10
molar, 25, 363
Gauss, C., 181
Gaussian function, 181, 228, 466
Generalized valence bond, 324
Gibbs–Duhem equation, 87
Gibbs energy, 39, 68, 86, 105, 148
Gibbs–Helmholtz equation, 113
Gibbs, J., 39, 96
Gibbs isotherm, 402
Grand canonical ensemble, 346
Gravitational force, 23, 51
Ground state, 4, 68, 183, 198, 253, 296, 356, 371
Group
Abelian, 440
representations, 443
symmetry, 439
theory, 438
Guggenheim–Katayama relationship, 401
Gutowsky, H., 395

H

Hahn, E., 395
Half-life, 138
Half-reactions, 152
Hamiltonian, 165, 196, 248, 297, 345, 372
Hamilton, W., 165
Hard-sphere collisions, 129
Harmonic oscillator, 3, 167, 197, 252, 328, 357
Hartree, D., 338
Hartree energy unit, 476
Heat capacity, 69, 94, 162, 362
constant pressure, 71
constant volume, 71, 362
Heat pump, 62
Heisenberg's uncertainty relationship, 201
Heisenberg, W., 178
Heitler–London wavefunction, 322
Helmholtz energy, 39
Helmholtz, H., 39
Henry's law, 109
Henry's law constant, 109

Henry, W., 109
Hermite, C., 181
Hermite polynomials, 181
    recursion relation, 182
Hermitian
    matrix, 434
    operator, 195, 455
Hess, G., 145
Hess's law, 145
Hindered rotation, 364
Hot band, 283, 330
Hund's rules, 309
Hyperpolarizability, 335

I

Ideal gas, 10, 23, 54, 81, 102, 129, 359
Ideal gas law, 23, 64
Ideal solution, 108
Idempotency, 315
Impact parameter, 132
Improper rotation, 438
Indistinguishability, 303, 347
Inelastic mean free path, 410
Inelastic scattering, 410
Inertia, moment of, 5, 216, 274
Inexact differential, 52
Infrared
    radiation, 254
    selection rules, 264
    spectroscopy, 276
Intensity pattern, NMR, 384
Intensive, 36, 87, 189
Interconversion, 284
Interfacial layer, 399
Intermolecular
    forces, 33, 81
    interaction, 46, 82, 115, 143, 331, 361
Internal coordinates, 276
Internal energy, 17, 30, 54, 86, 144, 343
Internal pressure, 30, 70
Interstellar medium, 158
Intrinsic volume, 38
Inversion temperature, 75
Ionization potential, 319
Irreversible process, 55, 239, 367
Isenthalpic, 75
Isotherms, 25, 56, 82, 102, 148
Isothermal
    compressibility, 37
    expansion, 56
    process, 63, 102, 148
Isotropic oscillator, 242

J

jj coupling, 305
Jonas, J., 394
Joule, J., 73
Joule–Thomson coefficient, 74
Joule–Thomson expansion, 73

K

Kemble, E.C., 241
Kinetic energy, 3, 27, 54, 94, 126, 166, 211, 254,
    311, 359, 410
Kinetics
    gas, 19, 27
    reaction, 158, 240, 343
Klemperer, W., 48
Kramers, H., 338

L

Lagrange, J., 165
Lagrange multiplier, 14
Laguerre polynomials, 293
Langmuir, I., 412
Langmuir isotherm, 407
Laplacian operator, 218
Laser, 274, 330
Lauterbur, P., 395
Least squares method, 269, 418
Le Chatelier, H., 112
Le Chatelier's principle, 112
Legendre polynomials, 219, 253
Lewis, G., 119
Linear combination of atomic orbitals, 321
Liquefaction, 47, 75
London, F., 48
LS coupling, 305

M

Mach, E., 47
Magnetic
    dipole, 297, 373
        transitions, 375
    field, 248, 297, 371
    moment, 300, 371
        nuclear, 372
    quantum number, 298
    resonance imaging, 391
    susceptibility, 330
Magnetizability, 330, 336
Manifold, 132, 258, 304, 365
Mass, reduced, 130, 173, 221, 252, 292

Matrix representation, 456
Maxwell–Boltzmann law, 12, 142, 268, 344
Maxwell, J., 19
Maxwell relations, 41
McCall, D., 395
Mean free path, 131, 410
Mechanical equivalence, 174
Mechanically separable, 81, 204
Melting point, 81, 121
Microcanonical ensemble, 346
Microstates, 7
Microwave radiation, 285
Milky Way, 159
Minimum energy path, 127
Molar
    enthalpy, 86, 148
    entropy, 90, 106, 157
    volume, 89, 102, 144, 401
Molecular beam, 336, 394
Molecular dynamics, 188, 336
Molecularity, 136
Molecular mechanics, 188
Molecular point group, 438
Mole fraction, 87, 107
Moment of inertia, 5, 216, 274
Momentum
    angular, 5, 216, 297, 371
        operator, 224, 254, 297, 390
Morse potential, 260
Multiplicity, 225, 306, 372

N

Nernst equation, 154
Nernst, W., 154, 412
New quantum theory, 241
Newton, I., 165
Newton's laws, 3, 133, 165
Nonideal behavior, 33, 95, 336
Normal coordinates, 186, 277
Normalization, 199, 297
Normalization factor, 200, 322, 354
Normal modes, 187, 206, 278
Normal phase transition temperature, 82
Noyes, A., 120
Nuclear
    magnetic
        moment, 372
        resonance, 372
        shielding, 373
    magnetization, 391
    magneton, 376
    spin, 371

O

Occupancies, electron, 303
One-electron atoms, 291
Onsager, L., 366
Operator(s)
    addition, 193
    angular momentum, 224,
        297, 390
    antisymmetrizer, 314
    dipole moment, 309
    Hermitian, 195
    idempotent, 315
    Laplacian, 218
    lowering, 396, 453
    multiplication, 193
    raising, 396, 453
Orbitals
    antibonding, 320
    bonding, 320
    linear combination, 319
Orbital energies, 319
Orthogonal functions, 195
Osmosis, 118
Osmotic pressure, 117
Overtone, 264

P

Packard, M., 395
Partial differentiation, 37, 69, 180, 250,
    420, 461
Partial molar values, 86, 106
Partial pressure, 106, 404
Particle diffraction, 176
Particle-in-a-box, 206, 351
Partition function, 15, 27, 268, 343
Pauling, L., 338
Pauli principle, 241, 303, 348
Pauli, W., 241
Perfect gas, 72
Perturbation theory, 228, 232, 250, 303, 378,
    451, 459
Petit, A., 76
Phase(s)
    boundary, 82, 111
    change, 81, 101, 158, 209
    condensed, 84, 101, 143, 361
    diagram, 81
    equilibrium curves, 82, 115
    factor, 221, 249
    rule, 87, 152
    space, 345

transition, 90, 353
  first-order, 93
  second-order, 93
pH meter, 155
Photoelectric effect, 2, 177
Photoelectron spectroscopy, 319, 409
Photon, 247, 309, 371, 471
Physisorption, 402–403, 406
Piston, 23, 51
Pitzer, K., 120
Planck, M., 177, 239
Planck's constant, 4, 177, 201, 266, 454, 474
Polarizability, 330
  dipole, 334
  quadrupole, 335
Population, 12, 68, 268, 329, 345
Postulate(s), 11
  quantum mechanical, 179
Potential
  boundary, 209
  discontinuous, 208
  double-well, 233, 284
  effective, 256, 304
  electrical, 125, 317
  electrochemical, 120
  energy surface, 125, 189, 310, 405
  harmonic, 168, 255
  ionization, 319
  Morse, 260
  particle-in-a-box, 207
  step, 206
Pound, R., 395
Power series, 34, 72, 134, 233, 256, 292, 415
Pressure
  atmospheric, 69
  internal, 30, 70
  osmotic, 117
Pressure–volume work, 51
Principal axes, 276
Probability, 8, 27, 343, 404
  collision, 132
  density, 27, 198, 249, 294
Progressions, 282
Property
  optical, 20, 77
  transport, 133
Purcell, E., 395

Q

Quanta, vibrational, 4
Quantized, 3, 77, 179, 216

Quantum
  numbers, 4, 205, 253, 292
    magnetic, 298
  postulates, 179
  revolution, 78, 120, 239, 364
  states, 4, 71, 132, 224, 265, 343
    excited, 4, 68, 264, 325, 356
    ground, 4, 68, 183, 198, 253, 296, 356, 371
  tunneling, 131, 212
Quartic oscillator, 230

R

Rabi, I., 394
Radial functions, hydrogen, 293
Radiofrequency, 273, 375
Radiotelescopes, 159
Raman spectra, 336
Randall, M., 120
Raoult's law, 108
Raoult, F., 108
Rate
  collision, 129
  constant, 133, 367, 404
  equation, 135
  laws, 138
  reaction, 131, 371
  solvent effects, 142
Rate-determining step, 141
Reaction
  bimolecular, 136
  complex, 136
  coordinate, 128
  diffusion-controlled, 144
  elementary, 135
  exoergic, 128
  first order, 135
  half-life, 138
  path, 128
  quotient, 112
  rate, 131, 371
  redox, 152
  second order, 135
  termolecular, 136
  unimolecular, 136
  zero order, 135
Rectangular matrix, 424
Reduced mass, 130, 173, 221, 252, 292
Reflection, 214, 437
Refrigerator, 62
Representation, group, 443
  totally symmetric, 444

Reverse osmosis, 118
Reversible process, 56
Reversibility, 55, 141
Rigid rotator, 5, 216, 257, 328, 357
Rotational constant, 258, 325, 357
Rotational fine structure, 276, 327
Rotational transitions, 272
Row matrix, 424
Russell–Saunders coupling, 305
Rydberg constant, 6, 338

S

Saddle point, 126
Salt bridge, 155
Saturn, 19
Scattering, 132, 336, 392, 410
Schawlow, A., 286
Scherer, P., 78
Schrödinger, E., 178
Schrödinger equation, 180
    radial, 254, 292
    time-dependent, 248
    time-independent, 248
Second law, 65, 239
Second order reaction, 135
Selection rules, 251, 309, 374, 451
Self-consistent field, 318
Separability, 171, 203, 248, 318, 356
Shell, 304, 367
Shielding
    isotropic, 274
Simulation, computer, 132, 189
SI units, 474
Slater determinant, 316
Slater, J., 337
Slichter, C., 395
Snyder, L., 159
Solvent cage, 143
Speed
    mean, 29, 94, 133, 189
    most probable, 29
    relative, 130
    rms, 18, 29
Spherical harmonic functions, 219, 253, 292
Spherical polar coordinates, 166, 217, 254, 293
Spin, 371
    electron, 241, 371
    multiplicity, 306, 372
    nuclear, 371
    orbitals, 301, 367
    –orbit interaction, 301, 390
    –orbit splitting, 302

Spin–spin coupling interaction, 378
Splitting, 284, 303, 382
Spontaneous
    phase change, 89
    process, 8, 65
    unimolecular reaction, 136
Spring constant, 168, 205
Square matrix, 46
Standard
    cell voltage, 154
    deviation, 201
    state, 101, 144
Stark effect, 336
Stark, J., 336
State
    bound, 6, 209
    coupled, 227, 302, 456
    functions, 35, 51, 81, 114, 147, 360
    ground, 4, 68, 183, 198, 253, 296, 356, 371
    nuclear spin, 371
    product-function, 205
    standard, 101, 144
    stationary, 240, 249
    steady, 141
    transition, 126
    variables, 36, 53, 81, 125
    unbound, 212
    uncoupled, 456
Steady-state approximation, 141
Step potential, 206
Stirling's approximation, 13
Sublimation, 91, 364
Subshell, 304
Substitution structure, 276
Surface
    Cartesian, 125
    contours, 125
    potential energy, 125, 189, 310, 405
        slice, 125, 223
    tension, 399, 401
        temperature dependence, 401
    thermodynamic, 97
Symmetric stretch, 187, 279
Symmetry, 40, 206, 253, 325, 364, 383, 437
    group, 439

T

Temperature
    boiling point, 90, 117
    critical, 85, 401
    freezing point, 101
    inversion, 75

melting point, 91
    normal phase transition, 82
Termolecular reaction, 136
Term symbols, 306
Term value, 325
Thermal conductivity, 133
    coefficient, 134
Thermodynamic
    compass, 41, 70, 87, 150
    state functions, 35, 51, 81, 114, 147, 360
Third law, 68
Townes, C., 159, 286
Trajectory, 133, 345
Transformation
    coordinate, 171, 217, 278
    matrix, 430
Transition
    allowed, 253, 275, 310
    dipole, 375
    enthalpy, 90
    first-order, 94
    forbidden, 253
    frequency, 283, 320
    fundamental, 276
    magnetic dipole, 375
    overtone, 264
    rotational, 272
    second-order, 93
    spectroscopic, 247
    state, 126
        structure, 126
        theory, 128
    temperatures, 82, 93
    vertical, 327
    volume, 90
Transmission, 214, 375
Transport properties, 133
Transpose, matrix, 424, 461
Triple point, 82
Tunneling, 131, 212
Turning points, 169, 198, 279

U

Ultraviolet, 286, 309, 478
Unbound state, 212
Uncertainty, 201, 296, 418
Uncoupled basis, 456
Unimolecular reaction, 136

V

Valence bond, 47, 322
van der Waals
    clusters, 48
    coefficients, 50
    equation of state, 35, 84
    gas, 35, 70, 84
van der Waals, J., 46
van't Hoff equation, 114
Vapor pressure, 92, 107, 403
Variational method, 228, 459
Variation principle, 228
Variation theory, 228, 250, 390
Vector coupling coefficients, 459
Velocity, 3, 27, 59, 130, 165, 198, 297
Vibrational
    band, 285, 327
    frequencies, 188, 205, 277, 320
    quanta, 4
    transitions, 264
        fundamental, 264
        overtone, 264
Vibrational–rotational coupling, 355
Virial coefficients, 34
Viscosity, 20, 133
    coefficient, 135
Voltage
    standard cell, 154
    zero-current, 154
Volume
    molar, 89, 102, 144, 401
    transition, 90

W

Wave character, 176
Wave constant, 210
Wavefunction,
    harmonic oscillator, 205, 328
    Heitler–London, 322
    hydrogen atom, radial, 263
    many-electron, 317
    overlap, 327
    particle-in-a-box, 207
    relativistic, 241, 300
    time-dependent, 248
    time-independent, 248
    valence bond, 322

Wavelength
    de Broglie, 177, 240
Wave vector, 211
Wilhelmy plate, 400
Work, 51

X

X-ray, 77, 240, 319, 391–392
X-ray photoelectron spectroscopy, 409

Z

Zeeman spectra, 303
Zero-current voltage, 154
Zero order reaction, 135
Zero point energy, 184, 272

Milton Keynes UK
Ingram Content Group UK Ltd.
UKHW051929141024
449569UK00027B/1415

9 781138 113992